AA001089

2008 Symposium on VLSI Technology

Honolulu, HI
17-19 June 2008

IEEE Catalog Number: CFP08VTS-PRT
ISBN 13:   978-1-4244-1802-2

Copyright © 2008 by The Institute of Electrical and Electronics Engineers, Inc.
All Rights Reserved

Copyright and Reprint Permissions: Abstracting is permitted with credit to the source. Libraries are permitted to photocopy beyond the limit of U.S. copyright law for private use of patrons those articles in this volume that carry a code at the bottom of the first page, provided the per-copy fee indicated in the code is paid through Copyright Clearance Center, 222 Rosewood Drive, Danvers, MA 01923.

For other copying, reprint or republications permission, write to IEEE Copyrights Manager, IEEE Operations Center, 445 Hoes Lane, Piscataway, New Jersey USA 08854. All rights reserved.

IEEE Catalog Number: CFP08VTS-PRT

ISBN 13:   978-1-4244-1802-2

ISSN: 90-655131

Additional Copies of This Publication Are Available from:

IEEE Service Center
445 Hoes Lane
Piscataway, NJ 08854
Phone: (800) 678-IEEE
      (732) 981-1393
Fax: (732) 981-9667
E-mail: customer-service@ieee.org

2008 Symposium on VLSI Technology

Honolulu, Hawaii
17-19 June 2008

IEEE Catalog Number: CFP08VTS-POD
ISBN: 978-1-42441-802-2

Table of Contents

Has the Sun Finally Risen on Photovoltaics? 1
Mark R. Pinto

Silicon Smart Microchips for Intelligent Sensing 5
M. Ishida, T. Kawano, H. Takao, K. Sawada

FinFET Performance Advantage at 22nm: An AC perspective 9
M. Guillorn, J. Chang, A. Bryant, N. Fuller, O. Dokumaci, X. Wang, J. Newbury, K. Babich, J. Ott, B. Haran,
R. Yu, C. Lavoie, D. Klaus, Y. Zhang, E. Sikorski, W. Graham, B. To, M. Lofaro, J. Tornello, D. Koli, B. Yang,
A. Pyzyna, D. Neumeyer, M. Khater, A. Yagishita, H. Kawasaki, W. Haensch

**Flexible and Robust Capping-Metal Gate Integration Technology Enabling Multiple-VT CMOS in
MuGFETs** 11
A. Veloso, L. Witters, M. Demand, I. Ferain, N. J. Son, B. Kaczer, P. J. Roussel, E. Simoen, T. Kauerauf,
C. Adelmann, S. Brus, O. Richard, H. Bender, T. Conard, R. Vos, R. Rooyackers, S. Van Elshocht, N. Collaert,
K. De Meyer, S. Biesemans, M. Jurczak

**Novel Integration Process and Performances Analysis of Low STandby Power (LSTP) 3D Multi-Channel
CMOSFET (MCFET) on SOI with Metal / High-K Gate Stack** 13
E. Bernard, T. Ernst, B. Guillaumot, N. Vulliet, V. Barral, V. Maffini-Alvaro, F. Andrieu, C. Vizioz,
Y. Campidelli, P. Gautier, J. M. Hartmann, R. Kies, V. Delaye, F. Aussenac, T. Poiroux, P. Coronel, A. Souifi,
T. Skotnicki, S. Deleonibus

**Three-Dimensional Stress Engineering in FinFETs for Mobility/On-Current Enhancement and Gate
Current Reduction** 15
M. Saitoh, A. Kaneko, K. Okano, T. Kinoshita, S. Inaba, Y. Toyoshima, K. Uchida

**Experimental Study of Single Source-Heterojunction MOS Transistors (SHOTs) under Quasi-Ballistic
Transport** 17
T. Mizuno, Y. Moriyama, T. Tezuka, N. Sugiyama and S. Takagi

Advanced DSS MOSFET Technology for Ultrahigh Performance Applications 19
M. Awano, H. Onoda, K. Miyashita, K. Adachi, Y. Kawase, K. Miyano, H. Yoshimura, T. Nakayama

**A New Source/Drain Germanium-Enrichment Process Comprising Ge Deposition and Laser-Induced
Local Melting and Recrystallization for P-FET Performance Enhancement** 21
F. Liu, H.-S. Wong, K.-W. Ang, M. Zhu, X. Wang, D. M.-Y. Lai, P.-C. Lim, B. L. H. Tan, S. Tripathy, S.-A. Oh,
G. S. Samudra, N. Balasubramanian, Y.-C. Yeo

**Novel and Cost-Efficient Single Metallic Silicide Integration Solution with Dual Schottky-Barrier
Achieved by Aluminum Inter-Diffusion for FinFET CMOS Technology with Enhanced Performance** 23
R. T.-P. Lee, A. T.-Y. Koh, W.-W. Fang, K.-M. Tan, A. E.-J. Lim, T.-Y. Liow, S.-Y. Chow, A. M. Yong,
H. S. Wong, G.-Q. Lo, G. S. Samudra, D.-Z. Chi, Y.-C. Yeo

**Experimental Study of Mobility in [110]- and [100]-Directed Multiple Silicon Nanowire GAA MOSFETs
on (100) SOI** 25
J. Chen, T. Saraya, K. Miyaji, K. Shimizu, T. Hiramoto

**Performance Breakthrough in 8 nm Gate Length Gate-All-Around Nanowire Transistors Using Metallic
Nanowire Contacts** 27
Y. Jiang, T. Y. Liow, N. Singh, L. H. Tan, G. Q. Lo, D. S. H. Chan, D. L. Kwong

**5 nm Gate Length Nanowire-FETs and Planar UTB-FETs with Pure Germanium Source/Drain Stressors
and Laser-Free Melt-Enhanced Dopant (MeltED) Diffusion and Activation Technique** 29
T.-Y. Liow, K.-M. Tan, R. T. P. Lee, M. Zhu, B. L.-H. Tan, G. S. Samudra, N. Balasubramanian, Y.-C. Yeo

TSNWFET for SRAM Cell Application: Performance Variation and Process Dependency 31
S. D. Suk, Y. Y. Yeoh, M. Li, K. H. Yeo, S.-H. Kim, D.-W. Kim, D. Park, W.-S. Lee

Novel Vth Tuning Process for HfO2 CMOS with Oxygen-Doped TaCx 33
W. Mizubayashi, K. Akiyama, W. Wang, M. Ikeda, K. Iwamoto, Y. Kamimuta, A. Hirano, H. Ota, T. Nabatame,
A. Toriumi

Table of Contents

Novel Process to Pattern Selectively Dual Dielectric Capping Layers Using Soft-Mask Only 35
T. Schram, S. Kubicek, E. Rohr, S. Brus, C. Vrancken, S.-Z. Chang, V.S. Chang, R. Mitsuhashi, Y. Okuno,
A. Akheyar, H.-J. Cho, J.C. Hooker, V. Paraschiv, R. Vos, F. Sebai, M. Ercken, P. Kelkar, A. Delabie,
C. Adelmann, T. Witters, L-A. Ragnarsson, C. Kerner, T. Chiarella, M. Aoulaiche, Moon-Ju Cho, T. Kauerau,
K.De Meyer, A. Lauwers, T. Hoffmann, P.P. Absil, S. Biesemans

**Single Metal/Single Dielectric Gate Stack Realizing Triple Effective Workfunction for Embedded Memory
Application** 37
K. Manabe, K. Masuzaki, T. Ogura, T. Nakagawa, M. Saitoh, H. Sunamura, T. Tatsumi, H. Watanabe

**Improved FET Characteristics by Laminate Design Optimization of Metal Gates - Guidelines for
Optimizing Metal Gate Stack Structure** 39
M. Kadoshima, T. Matsuki, N. Mise, M. Sato, M. Hayashi, T. Aminaka, E. Kurosawa, M. Kitajima, S. Miyazaki,
K. Shiraishi, T. Chikyo, K. Yamada, T. Aoyama, Y. Nara, Y. Ohji

Fundamentals and Extraction of Velocity Saturation in Sub-100nm (110)-Si and (100)-Ge 41
L. Pantisano, L. Trojman, J. Mitard, B. DeJaeger, S. Severi, G. Eneman, G. Crupi, T. Hoffmann, I. Ferain,
M. Meuris, M. Heyns

**Fermi-Level Depinning in Metal/Ge Schottky Junction and Its Application to Metal Source/Drain Ge
NMOSFET** 43
M. Kobayashi, A. Kinoshita, K. Saraswat, H.-S. P. Wong, Y. Nishi

**The Effects of Ge Composition and Si Cap Thickness on Hot Carrier Reliability of Si/Si1-xGex/Si p-
MOSFETs with High-K/Metal Gate** 45
W.-Y. Loh, P. Majhi, S.-H. Lee, J.-W. Oh, B. Sassman, C. Young, G. Bersuker, B.-J. Cho, C.-S. Park, C.-Y. Kang,
P. Kirsch, B.-H. Lee, H. R. Harris, H.-H. Tseng, R. Jammy

**Impact of Source-to-Channel Carrier Injection Properties on Device Performance of Sub-100nm Metal
Source/Drain Ge-pMOSFETs** 47
H. Takeda, T. Yamamoto, T. Ikezawa, M. Kawada, S. Takagi, M. Hane

**Low VT Metal-Gate/High-k nMOSFETs - PBTI Dependence and VT Tune-Ability on La/Dy-Capping
Layer Locations and Laser Annealing Conditions** 49
S. Z. Chang, T. Y. Hoffmann, H. Y. Yu, M. Aoulaiche, E. Rohr, C. Adelmann, B. Kaczer, A. Delabie, P. Favia,
S. Van Elshocht, S. Kubicek, T. Scharm, T. Witters, L.-A. Ragnarsson, X. P. Wang, H.-J. Cho, M. Mueller,
T. Chiarella, P. Absil, S. Biesemans

**Impact of the Different Nature of Interface Defect States on the NBTI and 1/f Noise of High-k / Metal
Gate pMOSFETs between (100) and (110) Crystal Orientations** 51
M. Sato, Y. Sugita, T. Aoyama, Y. Nara, Y. Ohji

**Physical Understanding of the Reliability Improvement of Dual High-k CMOSFETs with the Fifth
Element Incorporation into HfSiON Gate Dielectrics** 53
M. Sato, N. Umezawa, N. Mise, S. Kamiyama, M. Kadoshima, T. Morooka, T. Adachi, T. Chikyow, K. Yamabe,
K. Shiraishi, S. Miyazaki, A. Uedono, K. Yamada, T. Aoyama, T. Eimori, Y. Nara, Y. Ohji

**Guidelines to Improve Mobility Performances and BTI Reliability of Advanced High-K/Metal Gate
Stacks** 55
X. Garros, M. Cassé, G. Reimbold, F. Martin, C. Leroux, A. Fanton, O. Renault, V. Cosnier, F. Boulanger

Electron Trapping: An Unexpected Mechanism of NBTI and Its Implications 57
J. P. Campbell, K. P. Cheung, J. S. Suehle, A. Oates

Id Fluctuations by Stochastic Single-Hole Trappings in High-k Dielectric p-MOSFETs 59
S. Kobayashi, M. Saitoh, K. Uchida

**Roles of Oxygen Vacancy in HfO2/Ultra-Thin SiO2 Gate Stacks - Comprehensive Understanding of VFB
Roll-Off** 61
K. Akiyama, W. Wang, W. Mizubayashi, M. Ikeda, H. Ota, T. Nabatame, A. Toriumi

Table of Contents

Mechanisms Limiting EOT Scaling and Gate Leakage Currents of High-k/Metal Gate Stacks Directly on SiGe and a Method to Enable sub-1nm EOT 63
J. Huang, P. D. Kirsch, J. Oh, S.H. Lee, J. Price, P. Majhi, H.R. Harris, D. C. Gilmer, D. Q. Kelly, P. Sivasubramani, G. Bersuker, D. Heh, C. Young, C.S. Park, Y. N. Tan, N. Goel, C. Park, P.Y. Hung, P. Lysaght, K. J. Choi, B. J. Cho, H.-H. Tseng, B. H. Lee, R. Jammy

Fully Integrated and Functioned 44nm DRAM Technology for 1GB DRAM 65
Hyunjin Lee, Dae-Young Kim, Bong-Ho Choi, Gyu-Seong Cho, Sung-Woong Chung, Wan-Soo Kim, Myoung-Sik Chang, Young-Sik Kim, Junki Kim, Tae-Kyun Kim, Hyung-Hwan Kim, Hae-Jung Lee, Han-Sang Song, Sung-Kye Park, Jin-Woong Kim, Sung-Joo Hong, Sung-Wook Park

A Cost Effective 32nm High-K/Metal Gate CMOS Technology for Low Power Applications with Single-Metal/Gate-First Process 67
X. Chen, S. Samavedam, V. Narayanan, K. Stein, C. Hobbs, C. Baiocco, W. Li, D. Jaeger, M. Zaleski, H. S. Yang, N. Kim, Y. Lee, D. Zhang, L. Kang, J. Chen H. Zhuang, A. Sheikh, J. Wallner, M. Aquilino, J. Han, Z. Jin, J. Li, G. Massey, S. Kalpat, R. Jha, N. Moumen, R. Mo, S. Kirshnan, X. Wang, M. Chudzik, M. Chowdhury, D. Nair, C. Reddy, Y. W. Teh, C. Kothandaraman, D. Coolbaugh, S. Pandey, D. Tekleab, A. Thean, M. Sherony, C. Lage, J. Sudijono, R. Lindsay, J. H. Ku, M. Khare, A. Steegen

Variability Aware Modeling and Characterization in Standard Cell in 45 nm CMOS with Stress Enhancement Technique 69
H. Aikawa, E. Morifuji, T. Sanuki, T. Sawada, S. Kyoh, A. Sakata, M. Ohta, H. Yoshimura, T. Nakayama, M. Iwai, F. Matsuoka

A Scaled Floating Body Cell (FBC) Memory with High-k+Metal Gate on Thin-Silicon and Thin-BOX for 16-nm Technology Node and Beyond 71
I. Ban, U. E. Avci, D. L. Kencke, P. L. D. Chang

On the Dynamic Resistance and Reliability of Phase Change Memory 73
B. Rajendran, M.-H. Lee, M. Breitwisch, G. W. Burr, Y.-H. Shih, R. Cheek, A. Schrott, C.-F. Chen, M. Lamorey, E. Joseph, Y. Zhu, R. Dasaka, P. L. Flaitz, F. H. Baumann, H.-L. Lung, C. Lam

10.2 Two-Bit Cell Operation in Diode-Switch Phase Change Memory Cells with 90nm Technology 75
D.-H. Kang, J.-H. Lee, J. H. Kong, D. Ha, J. Yu, C. Y. Um, J. H. Park, F. Yeung, J. H. Kim, W. I. Park, Y. J. Jeon, M. K. Lee, J. H. Park, Y. J. Song, J. H. Oh, G. T. Jeong and H. S. Jeong

10.3 A Unified Physical Model of Switching Behavior in Oxide-Based RRAM 77
N. Xu, B. Gao, L. F. Liu, B. Sun, X. Y. Liu, R. Q. Han, J. F. Kang and B. Yu

10.4 An Endurance-Free Ferroelectric Random Access Memory as a Non-Volatile RAM 79
D. J. Jung, W. S. Ahn, Y. K. Hong, H. H. Kim, Y. M. Kang, J. Y. Kang, E. S. Lee, H. K. Ko, S. Y. Kim, W. W. Jung, J. H. Kim, S. K. Kang, J. Y. Jung, H. S. Kim, D. Y. Choi, S. Y. Lee, K. H. A, C. Wei and H. S. Jeong

11.1 A New Direct Low-k/Cu Dual Damascene (DD) Contact Lines for Low-loss (LL) CMOS Device Platforms 81
J. Kawahara, M. Ueki, M. Tagami, K. Yako, H. Yamamoto, F. Ito, H. Nagase, S. Saito, N. Furutake, T. Onodera, T. Takeuchi, H. Nakamura, K. Arita, K. Motoyama, E. Nakazawa, K. Fujii, M. Sekine, N. Okada and Y. Hayashi

11.2 A Novel CVD-SiBCN Low-K Spacer Technology for High-Speed Applications 83
C. H. Ko, T. M. Kuan, K. Zhang, G. Tsai, S. M. Seutter, C. H. Wu, T. J. Wang, C. N. Ye, H. W. Chen, C. H. Ge, K. H. Wu and W. C. Lee

11.3 A Proposal of New Concept Milli-Second Annealing: Flexibly-Shaped-Pulse Flash Lamp Annealing (FSP-FLA) for Fabrication of Ultra Shallow Junction with Improvement of Metal Gate High-k CMOS Performance 85
T. Onizawa, S. Kato, T. Aoyama, Y. Nara and Y. Ohji

11.4 Steep Channel & Halo Profiles Utilizing Boron-Diffusion-Barrier Layers (Si:C) for 32 nm Node and Beyond 87
A. Hokazono, H. Itokawa, N. Kusunoki, I. Mizushima, S. Inaba, S. Kawanaka and Y. Toyoshima

Table of Contents

12.1 Scaling Evaluation of BE-SONOS NAND Flash Beyond 20 nm 89
H.-T. Lue, T.-H. Hsu, S. C. Lai, Y. H. Hsiao, W. C. Peng, C. W. Liao, Y. F. Huang, S. P. Hong, M. T. Wu, F. H. Hsu, N. Z. Lien, S. Y. Wang, L. W. Yang, T. Yang, K. C. Chen, K. Y. Hsieh, R. Liu and C.-Y. Lu

12.2 Highly Scalable NAND Flash Memory with Robust Immunity to Program Disturbance Using Symmetric Inversion-Type Source and Drain Structure 91
C.-H. Lee, J. Choi, Y. Park, C. Kang, B.-I. Choi, H. Kim, H. Oh and W.-S. Lee

12.3 Vertical Structure NAND Flash Array Integration with Paired FinFET Multi-Bit Scheme for High-Density NAND Flash Memory Application 93
J.-M. Koo, T.-E. Yoon, T. Lee, S. Byun, Y.-G. Jin, W. Kim, S. Kim, J. Park, J. Cho, J.-D. Choe, C.-H. Lee, J. J. Lee, J.-W. Han, Y. Kang, S. Park, B. Kwon, Y.-J. Jung, I. Yoo and Y. Park

12.4 Novel 3-D Structure for Ultra High Density Flash Memory with VRAT (Vertical-Recess-Array-Transistor) and PIPE (Planarized Integration on the same PlanE) 95
J. Kim, A. J. Hong, M. Ogawa, S. Ma, E. B. Song, Y.-S. Lin, J. Han, U.-I. Chung and K. L. Wang

13.1 Channel-Stress Study on Gate-Size Effects for Damascene-Gate pMOSFETs with Top-Cut Compressive Stress Liner and eSiGe 97
S. Mayuzumi, S. Yamakawa, D. Kosemura, M. Takei, J. Wang, T. Ando, Y. Tateshita, M. Tsukamoto, H. Wakabayashi, T. Ohno, A. Ogura and N. Nagashima

13.2 45nm High-k + Metal Gate Strain-Enhanced Transistors 99
C. Auth, A. Cappellani, J.-S. Chun, A. Dalis, A. Davis, T. Ghani, G. Glass, T. Glassman, M. Harper, M. Hattendorf, P. Hentges, S. Jaloviar, S. Joshi, J. Klaus, K. Kuhn, D. Lavric, M. Lu, H. Mariappan, K. Mistry, B. Norris, N. Rahhal-Orabi, P. Ranade, J. S

13.3 Strain Enhanced Low-VT CMOS featuring La/Al-Doped HfSiO/TaC and 10ps Invertor Delay 101
S. Kubicek, T. Schram, E. Rohr, V. Paraschiv, R. Vos, M. Demand, C. Adelmann, T. Witters, L. Nyns, A. Delabie, L.-Å. Ragnarsson, T. Chiarella, C. Kerner, A. Mercha, B. Parvais, M. Aoulaiche, C. Ortolland, H. Yu, A. Veloso, L. Witters, R. Singanamalla, T. Kauerauf, S.Brus, C.Vrancken, V.S.Chang, S-Z.Chang, R.Mitsuhashi, Y.Okuno, A.Akheyar, H.-J.Cho, J.Hooker, B. J. O'Sullivan, S.Van Elshocht, K.De Meyer, M.Jurczak, P.Absil, S.Biesemans and T.Hoffmann

13.4 Impact of Tantalum Composition in TaC/HfSiON Gate Stack on Device Performance of Aggressively Scaled CMOS Devices with SMT and Strained CESL 103
M. Goto, K. Tatsumura, S. Kawanaka, K. Nakajima, R. Ichihara, Y. Yoshimizu, H. Onoda, K. Nagatomo, T. Sasaki, T. Fukushima, A. Nomachi, S. Inumiya, H. Oguma, K. Miyashita, H. Harakawa, S. Inaba, T. Ishida, A. Azuma, T. Aoyama, M. Koyama, K. Eguchi and Y. Toyoshima

14.1 Embedded Split-Gate Flash Memory with Silicon Nanocrystals for 90nm and Beyond 105
G. Chindalore, J. Yater, H. Gasquet, M. Suhail, S.-T. Kang, C. M. Hong, N. Ellis, G. Rinkenberger, J. Shen, M. Herrick, W. Malloch, R. Syzdek, K. Baker and K.-M. Chang

14.2 Gate-All-Around Single Silicon Nanowire MOSFET with 7 nm Width for SONOS NAND Flash Memory 107
K. H. Yeo, K. H. Cho, M. Li, S. D. Suk, Y.-Y. Yeoh, M.-S. Kim, H. Bae, J.-M. Lee, S.-K. Sung, J. Seo, B. Park, D.-W. Kim, D. Park and W.-S. Lee

14.3 A Novel Junction-Free BE-SONOS NAND Flash 109
H.-T. Lue, E.-K. Lai, Y. H. Hsiao, S. P. Hong, M. T. Wu, F. H. Hsu, N. Z. Lien, S. Y. Wang, L. W. Yang, T. Yang, K. C. Chen, K. Y. Hsieh, R. Liu and C.-Y. Lu

14.4 Enhanced Endurance of Dual-Bit SONOS NVM Cells Using the GIDL Read Method 111
A. Padilla, S. Lee, D. Carlton and T.-J. K. Liu

15.1 Planar Bulk+ Technology Using TiN/Hf-Based Gate Stack for Low Power Applications 113
G. Bidal, F. Boeuf, S. Denorme, N. Loubet, C. Laviron, F. Leverd, S. Barnola, T. Salvetat, V. Cosnier, F. Martin, M. Grosjean, P. Perreau, D. Chanemougame, S. Haendler, M. Marin, M. Rafik, D. Fleury, C. Leyris, L. Clement, M. Sellier, S. Monfray, J. Bougu

vi

Table of Contents

15.2 High Performance Sub-35 nm Bulk CMOS with Hybrid Gate Structures of NMOS ; Dopant Confinement Layer (DCL) / PMOS ; Ni-FUSI by Using Flash Lamp Anneal (FLA) in Ni-Silicidation 115
H. Ohta, K. Kawamura, H. Fukutome, M. Tajima, K. Okabe, K. Ikeda, K. Hosaka, Y. Momiyama, S. Satoh and T. Sugii

15.3 Cost-Effective Ni-Melt-FUSI Boosting 32-nm Node LSTP Transistors 117
H. Fukutome, K. Kawamura, H. Ohta, K. Hosaka, T. Sakoda, Y. Morisaki and Y. Momiyama

15.4 Design and Demonstration of Very High-k (k~50) HfO2 for Ultra-Scaled Si CMOS 119
S. Migita, Y. Watanabe, H. Ota, H. Ito, Y. Kamimuta, T. Nabatame and A. Toriumi

16.1 Analyses of 5sigma Vth Fluctuation in 65nm-MOSFETs Using Takeuchi Plot 121
T. Tsunomura, A. Nishida, F. Yano, A. T. Putra, K. Takeuchi, S. Inaba, S. Kamohara, K. Terada, T. Hiramoto and T. Mogami

16.2 Reduction of Vth Variation by Work Function Optimization for 45-nm Node SRAM Cell 123
G. Tsutsui, K. Tsunoda, N. Kariya, Y. Akiyama, T. Abe, S. Maruyama, T. Fukase, M. Suzuki, Y. Yamagata and K. Imai

16.3 45nm Low-Power CMOS SoC Technology with Aggressive Reduction of Random Variation for SRAM and Analog Transistors 125
S. Ekbote, K. Benaissa, B. Obradovic, S. Liu, H. Shichijo, F. Hou, T. Blythe, T. W. Houston, S. Martin, R. Taylor, A. Singh, H. Yang and G. Baldwin

16.4 Understanding and Prediction of EWF Modulation Induced by Various Dopants in the Gate Stack for a Gate-First Integration Scheme 127
X. P. Wang, H. Y. Yu, Y.-C. Yeo, M.-F. Li, S.-Z. Chang, H.-J. Cho, S. Kubicek, D. Wouters, G. Groeseneken and S. Biesemans

17.1 Smallest Vth Variability Achieved by Intrinsic Silicon on Thin BOX (SOTB) CMOS with Single Metal Gate 129
Y. Morita, R. Tsuchiya, T. Ishigaki, N. Sugii, T. Iwamatsu, T. Ipposhi, H. Oda, Y. Inoue, K. Torii and S. Kimura

17.2 Selenium Co-Implantation and Segregation as a New Contact Technology for Nanoscale SOI N-FETs Featuring NiSi:C Formed on Silicon-Carbon (Si:C) Source/Drain Stressors 131
H.-S. Wong, F.-Y. Liu, K.-W. Ang, S.-M. Koh, A. T.-Y. Koh, T.-Y. Liow, R. T.-P. Lee, A. E.-J. Lim, W.-W. Fang, M. Zhu, L. Chan, N. Balasubramaniam, G. Samudra and Y.-C. Yeo

17.3 Mobility of Strained and Unstrained Short Channel FD-SOI MOSFETs: New Insight by Magnetoresistance 133
M. Cassé, F. Rochette, N. Bhouri, F. Andrieu, D. K. Maude, M. Mouis, G. Reimbold and F. Boulanger

17.4 On Implementation of Embedded Phosphorus-Doped SiC Stressors in SOI nMOSFETs 135
Z. Ren, G. Pei, J. Li, B. F. Yang, R. Takalkar, K. Chan, G. Xia, Z. Zhu, A. Madan, T. Pinto, T. Adam, J. Miller, A. Dube, L. Black, J. W. Weijtmans, B. Yang, E. Harley, A. Chakravarti, T. Kanarsky, R. Pal, I. Lauer, D.-G. Park and D. Sadana

18.1 A Designer Friendly 45nm High Performance Technology with In-Situ C-Doped e-SiGe & Dual Stress Liner in SRAM 137
R. Khamankar, C. Bowen, H. Bu, D. Corum, I. Fujii, Y. Gu, B. Hornung, T. Kim, B. Kirkpatrick, K. Kirmse, A. Krishnan, C. Lin, L. Liu, T. Lowry, C. Montgomery, O. Olubuyide, S. Prins, D. Riley, S. Yu, J. Blatchford, C. Machala, C. O'Brien, G. Shinn and T. Grider

18.2 Higher Hole Mobility Induced by Twisted Direct Silicon Bonding (DSB) 139
M. Hamaguchi, H. Yin, K. L. Saenger, C. Y. Sung, R. Hasumi, R. Iijima, K. Ohuchi, Y. Takasu, J. A. Ott, H. Kang, M. Biscardi, J. Li, A. G. Domenicucci, Z. Zhu, P. Ronsheim, R. Zhang, N. Rovedo, H. Utomo, K. Fogel, J. P. de Souza, D. K. Sadana, M. Takayana

18.3 32nm Device Architecture Optimization for Critical Path Speed Improvement 141
R. Gwoziecki, S. Kohler and F. Arnaud

Table of Contents

18.4 Strain Additivity in III-V Channels for CMOSFETs Beyond 22nm Technology Node 143
S. Suthram, Y. Sun, P. Majhi, I. Ok, H. Kim, H. R. Harris, N. Goel, S. Parthasarathy, A. Koehler, T. Acosta, T. Nishida, H.-H. Tseng, W. Tsai, J. Lee, R. Jammy and S. E. Thompson

19.1 Laser-Annealed Junctions with Advanced CMOS Gate Stacks for 32nm Node: Perspectives on Device Performance and Manufacturability 145
C. Ortolland, T. Noda, T. Chiarella, S. Kubicek, C. Kerner, W. Vandervorst, A. Opdebeeck, C. Vrancken, N. Horiguchi, M. De Potter, M. Aoulaiche, E. Rosseel, S. B. Felch, P. Absil, R. Schreutelkamp, S. Biesemans and T. Hoffmann

19.2 Advanced Junction Profile Design Scheme by Low-Temperature Millisecond Annealing and Co-Implant for High Performance CMOS 147
K. Ikeda, T. Miyashita, T. Kubo, T. Yamamoto, T. Sukegawa, K. Okabe, H. Ohta, Y. S. Kim, H. Nagai, M. Nishikawa, Y. Shimamune, A. Hatada, Y. Hayami, K. Ohkoshi, N. Tamura, K. Sukegawa, H. Kurata, S. Satoh, M. Kase and T. Sugii

19.3 Low Vt Gate-First Al/TaN/[Ir3Si-HfSi2-x]/HfLaON CMOS Using Simple Laser Annealing/Reflection 149
C. C. Liao, A. Chin, N. C. Su, M.-F. Li and S. J. Wang

19.4 Successful Enhancement of Metal Segregation at NiSi/Si Junction through Pre-Amorphization Technique 151
Y. Nishi, Y. Tsuchiya, A. Kinoshita, A. Hokazono and J. Koga

20.1 New Global Shutter CMOS Imager with 2 Transistors per Pixel 153
M. Funaki, T. Shimizu, S. Orihara, H. Kawanaka, M. Kurihara, H. Sato, N. Katsumata, M. Oikawa, J. Higuchi, K. Oe, R. Kuga, K. Maki and T. Nishibata

20.2 35-nm Gate-Length and Ultra Low-Voltage (0.45 V) Operation Bulk Thyristor-SRAM/DRAM (BT-RAM) Cell with Triple Selective Epitaxy Layers (TELs) 155
T. Sugizaki, M. Nakamura, M. Yanagita, M. Shinohara, T. Ikuta, T. Ohchi, K. Kugimiya, S. Kanda, K. Yagami and T. Oda

20.3 Band Offset FinFET-Based URAM (Unified-RAM) Built on SiC for Multi-Functioning NVM and Capacitorless 1T-DRAM 157
J.-W. Han, S.-W. Ryu, S. Kim, C.-J. Kim, J.-H. Ahn, S.-J. Choi, K. J. Choi, B. J. Cho, J. S. Kim, K. H. Kim, G. S. Lee, J. S. Oh, M. H. Song, Y. C. Park, J. W. Kim and Y.-K. Choi

21.2 Integrated Wafer-Scale Growth and Transfer of Directional Carbon Nanotubes and Misaligned-Carbon-Nanotube-Immune Logic Structures 159
N. Patil, A. Lin, E. R. Myers, H.-S. P. Wong and S. Mitra

21.3 Performance Enhancement Schemes Featuring Lattice Mismatched S/D Stressors Concurrently Realized on CMOS Platform: e-SiGeSn S/D for pFETs by Sn+ Implant and SiC S/D for nFETs by C+ Implant 161
G. H. Wang, E.-H. Toh, X. Wang, D. H. L. Seng, S. Tripathy, T. Osipowicz, T. K. Chan, G. Samudra and Y.-C. Yeo

Has the Sun Finally Risen on Photovoltaics?

Mark R. Pinto

Applied Materials
3225 Oakmead Village Drive, PO Box 58039
Santa Clara, CA USA 95052-8039
mark_pinto@amat.com

Abstract

The idea of solar generated electricity dates to discovery of the photovoltaic (PV) effect in 1839 [1] through to the first practical silicon solar cell in 1954 [4]. But even with concerns about oil and the environment, PV currently generates less than 0.1% of the world's electricity. We present here the case that PV is on the verge of becoming a major source of electrical power through a principle similar to that which underlies VLSI – the reduction of unit cost through nanomanufacturing.

Keywords: Photovoltaics, PV, grid parity, nanomanufacturing technology, amorphous silicon, microcrystalline silicon, tandem junction, CdTe, CIGS.

Background

While the first semiconductor [2] and silicon p-n junction [3] solar cells date to before the invention of the transistor, it was not until 1954 – with the advent of the first practical silicon cell, at an efficiency of 6% [4] – that the first solar modules (i.e. "panels") appeared. These were followed soon after with a first field application, the so-called "solar battery" used to power rural phone lines, along with increasingly visible discussion of the practicality of solar-derived energy (see for instance [5]).

Fig. 1 (a) First PV modules using silicon cells, 1954; (b) first commercial application for rural telecommunications, 1955. Source: Bell Laboratories.

However solar technology did not progress nearly as rapidly as the transistor. For nearly 20 years thereafter the main application of solar cells was off-grid power, where a wired utility connection was impractical, expensive or inconvenient – e.g. satellites, remote microwave repeaters, and even hand held calculators. Interest in broader applications can be roughly traced to the oil price spikes of the 1970s where significant solar R&D was invested by multinational firms including most large oil companies. In 1979, U.S. President Carter proposed a goal of achieving 20% of U.S. energy from solar by 2000 and even installed a solar water heater on the White House roof.

From the early 1980s, another period of 20 years followed where the discussion of solar as potential large scale source of energy virtually disappeared – with the notable exceptions of Japan and Germany. (The aforementioned solar water heater on the White House was removed in 1986). Many view the current heightened business in solar as ephemeral, resulting from the latest spike in oil prices, further magnified by concerns on climate change along with questioned government intervention. Has the time finally come for solar really contribute to solving the world's growing energy needs or will it merely be the next dotcom bubble?

Cost Per Watt and Grid Parity

To fully assess the potential of PV, it is essential to look at trended costs compared to other electricity sources. The cost of a solar system includes the production cost of the modules plus the cost of installation. Focusing first on the module cost, figure 2 shows the average selling price of a module per watt peak as a function of cumulative historical production. Watt peak (Wp) is defined as the power generated under direct sunlight at a reference level equivalent to maximum terrestrial sun at a 25°C ambient.

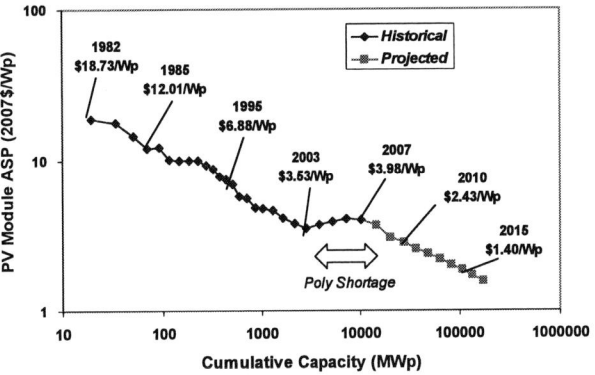

Fig. 2 PV c-Si module average selling price per watt peak over all types of modules used for power applications (data source: [6]). Historical data is dominated by c-Si and strongly affected in 2003-2007 by poly shortages.

Figure 2 is in the form of a traditional learning curve, which highlights the reduction of unit cost with technology progress and manufacturing scale. Because it is difficult to determine an average production cost for the module industry, the average module selling price is used instead which exaggerates effects of supply and demand, especially in the period of polysilicon shortage beginning in 2004. However up to the shortage, per Wp prices had reduced by ~20% per doubling of volume – overall ~5X in real terms over 20 years. Note that a similar learning curve exists for ICs where transistor cost has followed a 28% reduction per doubling. In addition to the expected end to the poly shortage, future average prices should decrease further due to the onset of lower cost thin film technologies (see below).

Of key importance is determining when costs reach market-making inflection points. For ICs, these points have been associated with the progression from mainframes to PCs to cell phones and other consumer devices. For PV, the significant inflection point is "grid parity" – where PV generated electricity costs the same as the utility grid rate. To translate module costs (prices) to kWh energy costs, one needs to include two key factors: the cost of module installation and sunlight intensity, the latter being geography dependent. An approximation for the PV per kWh energy cost is given by:

$$C_{kWh} \approx \kappa \bullet (C_P + C_I) + C_{inv} \qquad (1)$$

where C_P and C_I are module production and installation costs per Wp respectively; κ is a function of peak sunlight hours, overall system loss, efficiency degradation and cost of capital; C_{inv} is the per kWh cost of inverter replacement. From typical ASP and installation data, figure 3 shows kWh equivalent PV electricity prices for a region approximating an average location in California (~1,800 sun hours).

978-1-4244-1802-2/08/$25.00 ©2008 IEEE

With the other assumptions as listed in the caption, the coefficients in equation (1) work out to $\kappa \sim 0.05$Wp/kWh and $C_{inv} \sim \$0.009$/kWh.

There are 3 relevant comparative costs to determine grid parity – peak retail rate (U_{peak}), average retail rate (U_{avg}) and the utility generation rate (U_{gen}). U_{peak} is typically reflective of costs in mid-summer when PV is most effective. U_{avg} is the cost a consumer should typically compare against. U_{gen} is the ultimate cost point – the lowest cost of new generation to a utility. As shown in figure 3, PV is beginning to enter very interesting cost regimes. In fact for large scale installations and with the technology to be described shortly, PV electricity is likely to be generated below U_{avg} by the end of the decade. Note also that the ability of PV to be arbitrarily distributed (unlike a nuclear plant) works in favor of closing the cost gap to U_{gen}. It is further expected that U_{gen} will increase in real terms in the future – in fact, coal prices in Asia were reported to have recently increased 30% year to date, following a doubling in price in 2007 [7].

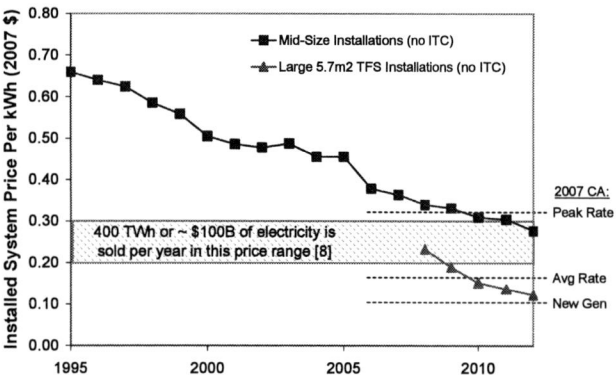

Fig. 3 Equivalent PV per kWh energy cost for a typical location in California (CA). Assumed parameters to convert from underlying module ASPs and installation estimates include: sun hours = 1800, module lifetime and finance term = 25 years, cost of capital = 4%, electrical system loss = 20%, efficiency degradation = 0.5% per year, inverter lifetime = 10 years.

Government incentives, especially in Japan and Germany, have played an important role in both stimulating demand and fostering industry investment to reduce costs. Although they can take various forms, the net effect of incentives is to lower the kWh cost of a PV system such that it becomes more competitive with existing retail rates. For instance if the 30% U.S. Investment Tax Credit (ITC) was applied above, mid-size PV installations become cheaper than U_{peak} in 2006. The increasing worldwide trend for disincentives on carbon producing forms of electricity would raise U_{gen} thereby accelerating PV competitiveness. Even in the absence of incentives, studies predict that PV will reach parity with average retail electricity over most of continental Europe between 2010-2020 [9].

Nanomanufacturing Technologies

Accelerating adoption of PV clearly depends on further reduction of cost per Wp. Historically, the PV industry has typically borrowed older IC technologies while developing few targeted solutions due to a relative lack of investment. However the IC and flat panel LCD markets have supported significant R&D on materials, processes and manufacturing platforms which hold great potential for PV advancements. These applications share a need for very thin films and have required continuously lower manufacturing costs at higher volumes along with stringent reliability specs. Because of these attributes, we use the term "nanomanufacturing technologies" to define the underlying materials, processes and equipment foundation required to produce these kinds of products [10].

A common aspect of nanomanufacturing technology market applications is the focus on growing end demand by reducing unit production cost. It is instructive to decompose unit cost as follows:

$$\frac{Production\ Cost}{Unit} = \frac{Process\ Cost\ /\ Area}{Units\ /\ Area} \qquad (2)$$

For ICs, units are transistors (or bits). Cost per transistor has been reduced by $>10^7$ since its inception primarily by integrating more transistors per unit area (the denominator) but also by not allowing the process cost per unit area (the numerator) to increase substantially through innovations in areas such as process tool productivity.

For nanomanufacturing technology markets other than the IC, the process cost per area numerator in equation (2) is of equal or greater importance to the denominator. As an example, to enable larger and lower cost LCDs, the overall metric of interest is cost per area alone, i.e. the relevant "Unit" is also display area. The dominance of cost per area as a driver suggests the use of large area processes to increase productivity. LCD substrates have hence scaled upwards from <1m^2 (Gen 2) to >5m^2 (Gen 8) over 14 years, leading to a deposition process cost per area reduction of more than 4x – all while maintaining nanoscale film properties, uniformity and throughput.

Applying the cost per unit equation to PV, we replace "Units" in equation (2) with "Watts". Like displays, PV cost per area can be driven by large area nanomanufacturing equipment. Figure 4 shows commercial examples for key PV technologies. Watts per area corresponds directly to conversion efficiency which can be driven by improvements to materials, device structures and even better nanomanufacturing uniformity, to be discussed below.

Fig. 4 Large area PV process equipment: (a) Applied ATON™ PVD for anti-reflective coatings on trays of c-Si wafers, (b) Applied PECVD 5.7 for deposition of thin film a-Si/µSi-based absorber layers.

Note however that while efficiency is a critically important parameter, it has a limited upside (<100%) whereas the cost per area may be driven down by substantially greater factors through large area processing and material innovations. Furthermore large equipment leads to accelerated factory scaling – a single PVD system as in figure 4a can process the equivalent of 100MWp annually or roughly the same output as that of the *entire* PV industry in 1997.

Primary PV Technologies

From the outset of the PV industry, multiple technology paths have been pursued, some of which compete directly while others serve distinct market segments. The following overview gives emphasis to technologies that show the greatest potential for cost per watt economies of scale. For more details, the seminal reference of Green [11] and the most recent PVSEC proceedings [12] are recommended.

A. Crystalline Silicon (c-Si) and III-V PV

PV modules made from silicon substrates – typically either mono-crystalline or multi-crystalline wafers – have dominated the market to date with over 90% of historical production. With module efficiencies between 13-20% (multi-crystalline at the low end, mono-crystalline at the high end) and moderate production cost, c-Si has been especially attractive for the residential market.

A significant component of c-Si process cost per area is the starting material – wafers make up about 40% of the overall module cost even under normal conditions. But because of the rapid growth of the market, raw silicon consumed for PV now exceeds that for ICs, further stressing the importance of reducing starting material cost. As wafers remain the most preferred c-Si implementation (over ribbons and sheets), thin wafer cutting technologies are vital – to minimize the amount of silicon per wafer as well as the loss in cutting. Figure 5a shows the inside of a modern wafer production system, typically resulting in substrates 100-200µm thick.

Conventional c-Si p-n cells are fabricated using a few key steps:

emitter diffusion, surface etch, nitride ARC deposition, interconnect formation, edge delete, test and sort. Significant cost per area improvements can be driven by improving equipment scale, process throughput, factory automation and overall yield. The challenge from a nanomanufacturing technology perspective is to deliver these improvements while also accommodating the thinner wafers together with structural improvements for higher efficiencies. Figure 4a is an illustration of applying scale and throughput to an established PVD process. For interconnect, metal screen printing has emerged as the c-Si method of choice, combining patterning + deposition at resolution on the order of 10 microns, but with similar challenges to realize solutions at high throughput and thin wafers (see figure 5b).

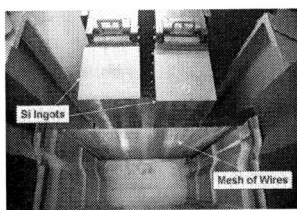

Fig. 5 Process technologies for low cost c-Si PV cells: (a) two multi-crystalline ingots about to be cut into thin wafers by a mesh of wires, (b) laser edge trimming of ultra thin, screen printed solar cell.

To improve overall system costs, c-Si manufacturers continue to focus on efficiency improvement. There are several methods to improve efficiency, beginning with optimization of the standard p-n process. Even tightening the wafer-to-wafer distribution of efficiencies, which has tended to be quite wide for existing PV factories, can increase the average watts per area produced. Beyond improvements in yield and material tuning, a number of structures have been proposed to maximize light absorption and collected carriers; two of the more popular such structures are shown in figure 6 [13, 14]. While improving watts per area, these cells also add process complexity, requiring ongoing control of process cost per area to avoid an overall increase in production cost per watt.

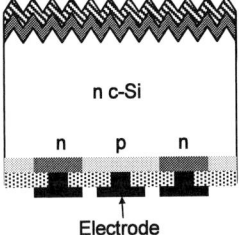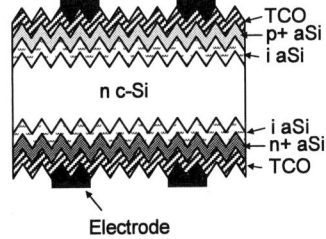

Fig. 6 High efficiency c-Si cells (a) rear point contact cell with gridless front side and reduced recombination through localized p and n contacts [13], (b) HIT cell with a-Si layers on the front and back of a c-Si wafer to improve surface passivation, mechanical stress and the thermal coefficient [14].

Modules made of c-Si require mounting and interconnection of wafers onto a backplane and lamination of a front glass. This step is a substantive component of total production cost (~20%). A completed module typically exhibits a 2-3 point loss in efficiency from the peak single cell, on-wafer values of 16-22%.

Crystalline non-silicon PV cells have also found uses. Of particular note are III-V based multi-junction cells that have reached efficiencies >40% [15]. While ideal for space applications where watts per *gram* is the most important metric, III-V cells lack broad commercial demand due to significantly higher production costs. Nonetheless, they are still being investigated with optical concentrators to attempt to achieve higher collection areas.

B. Thin Film (TF) PV

Thin film (TF) PV modules are produced by depositing a thin absorber layer – most commonly Si, CdTe, $CuInSe_2$ (CIS) or

$CuInGaSe_2$ (CIGS) – onto glass or flexible foils. In comparison to wafers, TF has the potential to lower material use and eliminate module assembly steps. Because of its close proximity to c-Si and its advantages in material abundance, process scalability and component reliability, we begin by considering thin film silicon (TFSi) PV.

The first TFSi cells were made of amorphous silicon (a-Si) in 1976 [16]. The simplest TFSi cell is a single junction a-Si p-i-n, using plasma enhanced chemical vapor deposition (PECVD) of a-Si and a transparent conducting oxide (TCO) as a front contact. These cells typically result in stabilized module efficiencies of ~6%. While much lower than c-Si, they can still yield useful products if manufacturing costs can be made extremely low.

Fig. 7 Large area TFSi production (a) 5.7m² substrate in comparison to previous 1.4m² format, (b) model layout of Applied SunFab™ factory to produce 5.7m² TFSi modules using PECVD tools like those in figure 4b.

TFSi is ideal for applying large area nanomanufacturing technologies to reduce cost. In fact, the primary layers used in a-Si single junction cells – a-Si, TCO and metal – are all used in large flat panel LCD production. By adapting these approaches to produce a-Si cells on 5.7m² glass substrates 4X larger than prior formats (see figure 7a), production costs of $1.20/Wp are now achievable, about half the cost of the cheapest c-Si modules.

The potential of TFSi will depend primarily on how far efficiency can be improved. Several approaches are being pursued, including improving the quality of the silicon towards c-Si on glass [17] or the more widely adopted approach to grow multiple p-i-n junctions with added layers of multi-crystalline silicon (μ-Si) [18] or incorporating Ge [19]. Figure 8 shows the device structure of the tandem junction (TJ) a-Si/μ-Si cell along with a typical spectral response. The TJ attempts to capture more of the photon spectrum and minimize energy losses through the different bandgaps in each layer. Initial TJ cell efficiencies to 15% have been achieved [20], although production modules with stabilized efficiencies >10% have not yet been reported. The primary process challenge is producing uniform micro-crystallinity over full module sizes, and recent lab results suggest this will be demonstrated on 5.7m² substrates at production costs of <$1/Wp within 2 years. As a result TFSi TJ technology on 5.7m² substrates shows great economic promise, as projected in figure 3.

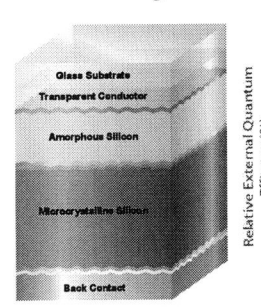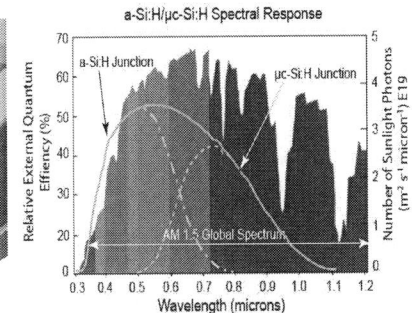

Fig. 8 Thin film tandem junction a-Si/μ-Si PV cells (a) device structure on a glass substrate, (b) spectral response highlighting the advantages of a dual junction, dual bandgap cell.

The non-Si TF absorbers – CdTe and CI(G)S – promise the potential of higher efficiencies than a-Si and with lower costs than c-Si wafers. There are however several significant challenges. First, these materials have been difficult to scale to large sizes at efficiencies close to their ideal values. Second, compounding the cost challenge is the fact that each of these compounds includes one or more elements that are rare (e.g. In and Te), resulting in potentially fundamental raw material supply and cost issues, and some of the constituents are even toxic (e.g. Cd). Third, modules with these materials have tended to require enhanced encapsulants to ensure reliability.

The most significant commercial success to date of the alternate TF materials has been for $0.72m^2$ CdTe modules reported at efficiencies of ~10% at production costs ~$1.25/Wp [21]. Higher efficiency panels have been obtained on CI(G)S but not yet at manufacturable quantities or competitive costs [22]. With the innovations possible on multi-junction TFSi and its inherently better scalability (substrate size and material supply), both CdTe and CI(G)S are being challenged to find paths to efficiencies even closer to c-Si while also needing to realize more complex, multi-element material control on large areas.

The primary application of low cost TF modules is in large commercial PV systems which are not typically constrained by space and are most dependent on module cost. Very large panels have an additional advantage for these systems due to lower hardware and labor requirements especially where mechanized assembly can be used. The resulting installation savings for $5.7m^2$ panels over smaller form factors ($\leq1.4m^2$) has been quantified at >17%, equivalent to a 2-3 point efficiency improvement – i.e. the cost to install a system with 8% efficient $5.7m^2$ panels is equal to or better than 10% $1.4m^2$ panels. This becomes an important added benefit for TFSi.

Fig. 9 High growth and emerging applications for PV (a) c-Si in existing rooftops, (b) solar farms, (c) solar canopy parking lots and (d) large size solar window glass for building integrated photovoltaics (BIPV).

Conclusions and Outlook

The electricity price of PV is nearing parity with retail electricity, first in sun favored regions with high existing utility rates like California and Italy. Rapid progress on thin film technologies is accelerating parity and opening new markets, building on the growth of c-Si PV (see figure 9). Technology roadmaps suggest there are several years of near certain cost reduction while concern about carbon footprint and rising prices for existing energy will continue to be catalysts for PV. This bullish environment has led to high-growth market forecasts as shown in figure 10. The mean of these forecasts implies a module market of ~$20B in 2010.

The key enabler to continued growth is the cost per watt reduction roadmap, and innovations in nanomanufacturing technologies will be a foundation of the future success of PV. Also of critical is the reduction of installation costs through standards, lower cost materials and lower cost (yet more reliable, efficient and intelligent) electronics. And innovations in energy storage, critical to cleaner vehicles, would further magnify the PV application space to non daylight power.

While PV is very close to an elastic commodity – lower price means more demand – there will also undoubtedly be supply-demand imbalances and market cyclicality. However unlike fiber optics and the dotcom era, there is a proven electricity market behind the solar

demand, and one that will continue to grow with the development of emerging global economies – so let's go catch some rays!

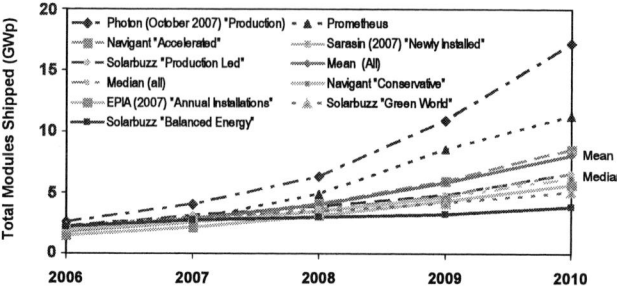

Fig. 10 Market forecasts compiled from industry sources (Jan 2008) project global production range of PV capacity from 3.8 to 17.2GWp in 2010.

Acknowledgements

The author thanks Charlie Gay for his deep insights which helped form the basis of this manuscript. Additional support from Cathy Boone, Pat Lamey, Ian Latchford, Jonathan Pickering, Blair Swezey and Lisong Zhou is also gratefully acknowledged.

References

[1] E. Becquerel, "On electric effects under the influence of solar radiation," *Compt. Rend.*, vol. 9, p. 561, 1839.

[2] C. E. Fritts, "On a New Form of Selenium Photocell", *Proc. AAAS*, vol. 33, p. 97, 1883.

[3] R. S. Ohl, "Light-Sensitive Electric Device," U.S. Patent 2,402,662, filed 1941.

[4] M. Chapin, C. S. Fuller and G. L. Pearson, "A new silicon p-n junction photocell for converting solar radiation into electrical power," *Phys. Rev.*, vol. 96, p. 676, 1954.

[5] F. Daniels, "A Limitless Resource: Solar Energy," *The New York Times*, March 18, 1956.

[6] P. Mints, Principal Analyst, Navigant Consulting PV Services Program, private communication, 2008.

[7] S. Oster and A. Davis, "China Spurs Coal Price Surge," *The Wall Street Journal*, February 12, 2008.

[8] Photon Consulting, *Solar Annual 2007*, Solar Verlag Gmbh, 2007.

[9] G. Van Schaeffer, "PV Cost and Price Development: Learning from the past to think about the future", 20, ECN Policy Studies, Photex Workshop, Brussels, 2003.

[10] M. R. Pinto, "Nanomanufacturing Technology: Exa-Units and Nano-Dollars," *Future Trends in Microelectronics*, p. 154, Wiley, 2007.

[11] M. A. Green, *Solar Cells: Operating Principles, Technology and System Applications*, Prentice-Hall, 1982.

[12] Tech. Digest European PVSEC-21, Milan, Sept 2007; Tech. Digest Intl. PVSEC-17, Fukuoka, Dec 2007.

[13] D. DeCeuster, et.al., "Low cost, high volume production of > 22% efficiency solar cells," Tech. Digest European PVSEC-22, p. 81, 2007.

[14] Y. Tsunomura, et.al., "22% efficiency HIT solar cell," Tech. Digest Intl. PVSEC-17, p. 387, 2007.

[15] R. R. King, et.al. "40% efficient metamorphic GaInP/GaInAs/Ge multijunction solar cells." App. Phys. Lett., 90, 183516 (2007).

[16] D. E. Carlson and C. R. Wronski, "Amorphous silicon solar cell," *App. Phys. Lett.*, vol. 28, p. 671, 1976.

[17] P. A. Basore, "Pilot production of thin-film crystalline on glass modules," Proc. 29th IEEE Photovoltaics Specialists Conf., p. 49, 2002.

[18] J. Meier, et.al., "*On the way towards high efficiency thin film silicon solar cell by the "micromorph" concept*" Proc. Mat. Res. Soc. Symp. Spring Meeting, 3, p. 420, 1996.

[19] J. Yang, A. Banerjee and S. Guha, *Solar Energy Materials and Solar Cells*, Volume 78, Issues 1-4, July 2003, p. 597-612.

[20] S. Fukuda et.al., "High efficiency thin film silicon hybrid cell and module with newly developed interlayer," Tech. Digest European PVSEC-21, p. 1535, 2006.

[21] M. Gloeckler, "First Solar company overview," Tech. Digest Intl. PVSEC-17, p. 166, 2007.

[22] H.-W. Schock, "Chalcopyrite (CIS) based solar cells, developments and production in Europe," Tech. Digest Intl. PVSEC-17, p. 40, 2007.

Silicon Smart Microchips for Intelligent Sensing

M. Ishida, T. Kawano, H. Takao, and K. Sawada,

Department of Electrical and Electronic Engineering,
Toyohashi University of Technology,
Hibarigaoka 1-1, Tempaku-cho, Toyohashi 441-8250, Japan
Ph.:+81-532-44-6745, Fax: +81-532-44-6757, Email: ishida@eee.tut.ac.jp

Abstract

Silicon smart microchips with CMOS/MEMS technology are realized for intelligent sensing. In our developed chips, tow new type sensor chips are presented here. One is Si microprobe electrode array for using in the recording of neurons in the tissue. The probe array can be fabricated on IC chip, using standard IC process followed by a selective Si probe growth. Another one is pH image sensors successfully fabricated by using the CCD/CMOS image sensor technique, and real time imaging of a chemical reaction and pH distribution imaging was demonstrated.

Introduction

Development of silicon sensing devices with Integrated Circuits using MEMS/NEMS technology will realize an ideal sensing chip, which may be called as "Smart micro chips", and results in ideal sensor chips as shown in Fig.1, which have several sensor devices (Multi-modal sensors), signal processing circuits, rf transmitters, and power supply using natural energy. Until now, we developed several smart microchips for Intelligent Sensing as shown in Fig.2.

In this presentation, two smart micro-sensor-chips for monitoring of neural activities and real time monitoring of pH image to biochemical, biomedical and clinical applications are presented in our developed chips. Also, a novel Radio Frequency (RF) induced power supply system for micro sensor nodes is studied. Supplied power in RF-electromagnetic wave is received and rectified in each micro sensor device. Integration of radio frequency transmitter (RF) technology with

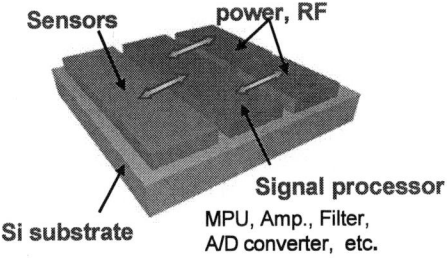

Fig.1. Concept of "Smart Microchips"

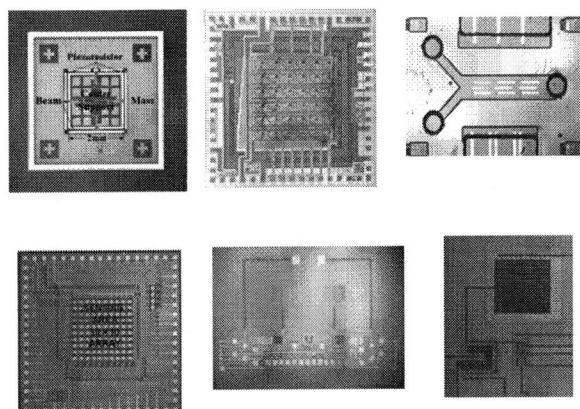

Fig.2. Smart microchips developed in TUT

CMOS/MEMS micro-sensors is required to realize the wireless smart micro-sensors system.

Silicon microprobe array chips

One of our developed smart chips is a smart silicon microprobe array chip to record neural activities for analysis of the mechanism of retina and brain, or for neural interfacing. Our group proposed the microelectrode array chip with extremely-fine, in other words, low-invasive silicon probes of 1-2μm diameter, and fabricated using standard CMOS process followed by the selective vapor-liquid-solid (VLS) growth method. Figure 3 indicates an ideal smart microprobe array chip for this application. Using the chip, the feasibility of in-vitro recording of neural activities in a carp retina and of in-vivo recording of the peripheral nervous activity of a rat was demonstrated.

The study of nervous system is a key problem in recent neuroscience. The analysis of the nervous system requires simultaneous recording as well as stimulation of a large number of neurons in the tissue. Many penetrating microelectrodes have been reported. However, with the previous fabrication technology, diameter of the needle has been of the order of 10^{-2} mm. The electrode probes are too large for electrode penetration due to the small neurons, 5 - 10 μm in diameter.

978-1-4244-1802-2/08/$25.00 ©2008 IEEE

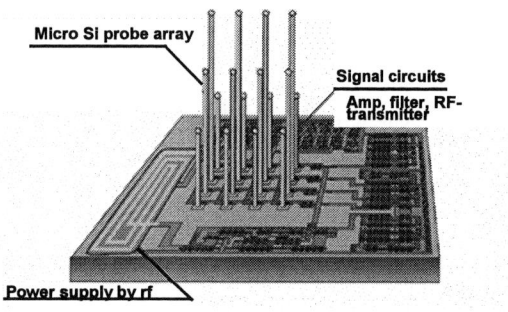

Fig.3. Ideal smart chip for neural recordings.

We have proposed the penetrating microelectrode array by a selective Si growth, based on vapor-liquid-solid (VLS) growth [1,2]. The Si microprobes have been fabricated using standard CMOS process followed by the VLS growth, using catalytic-Au dots and Si_2H_6 gas source at 500 - 700°C. The greatest merit of the device is that the microprobe array can be combined with integrated signal-processors on the same chip. Figure 4 shows a Si microprobe, with 2 μm in diameter and 120 μm in length, fabricated at drain region of the MOSFET. For the neural applications, the electrode device needs to be combined with amplifier, filter, other analog circuits as well as digital circuits, on the same chip, improving the performance.

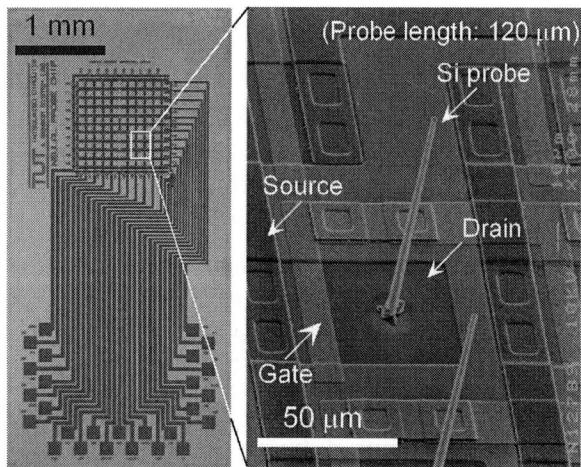

Fig.4. Si microprobe electrodes fabricated at drain region of MOSFET by IC process followed by the selective VLS Si growth.

The diameter of the penetrating needle by the selective VLS growth has been dramatically reduced. In this work, the probe array, each with 2 μm in diameter, and 60 -120μm in length, at 40 μm pitch was designed. Figure 5 shows the process sequence for the Si microprobe electrode array. In this process, Si (111) substrate was used, because the Si probe grows in the direction of <111> by the VLS growth. The Si probe

with a few microns in diameter was controlled with patterned catalytic-Au dots, by photolithography (Fig.5 (1)). Si probes with 2 - 4 μm in diameter were grown from the Au dots with 4 - 10 μm in diameter, using Si_2H_6 gas source with molecular-beam-epitaxy (Fig.5 (2)). Conductance of the probe was controlled by impurity diffusion with phosphorus. The probes were encapsulated with insulating film, 50 nm-thick SiO_2, by oxidation of the probe, except at the tip. The tips of the probes were coated with a biocompatible metal, 30 nm-thick Au (Fig.5 (3)). Finally, wafer dicing and Au wire bonding were carried out (Fig.5 (4)).

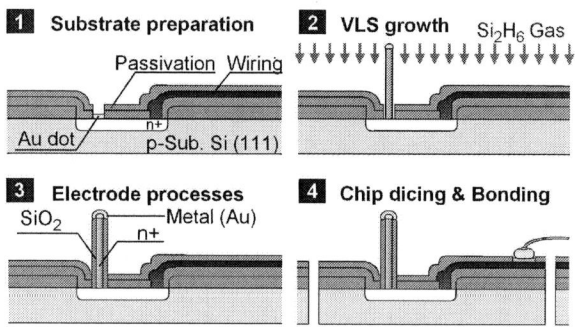

Fig.5. Process flow for the Si microprobe using the VLS growth and electrode processes (doping, encapsulation and tip metal coating).

The resistance of the doped Si microprobe with phosphorous diffusion was about 1kΩ, and the resistance of the insulating layer around the probe was about 100 MΩ. Impedance spectrum of the probe 3.5 μm in diameter and 75 μm in length, measured in saline solution of ringer. The impedance of the probe was less than 500 kΩ at 1 kHz, which is low enough to be used in recording of neural signals.

Recording of neural responses

A retina, isolated from a carp, was placed on the chip with the ganglion cell layer facing down with Si probe penetrated into the retina. The Si probes 3.5 μm in diameter and 75 μm in length was used in the recording. Figure 6 shows the schematic diagram of the experimental setup for the recording of light-evoked retina. During the light stimulation to the retina, repeated full-field flashes were used, with wavelengths varying from 480 nm to 720 nm, at an intensity of $4.8 \times 10^{12.5}$ photons/cm²·s for 0.1 s. Figure 7 shows the neural responses recorded with two channel probes. Under the stimulation conditions, neural responses to the light stimuli were recorded, via the Si probes. Amplitude of the neural responses was ranging from 50 to 80 μV, and the responses depending on the wavelengths of the light stimuli were obtained in the visible range. The recorded neural signals are a typical electroretinogram (ERG) response.

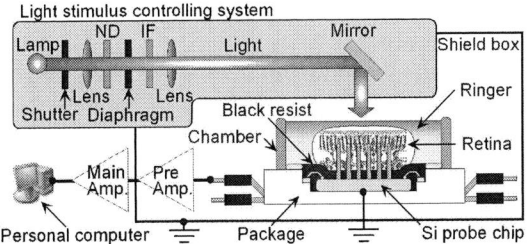

Fig.6. Experimental setup for the recording of light evoked responses of neurons in the retina.

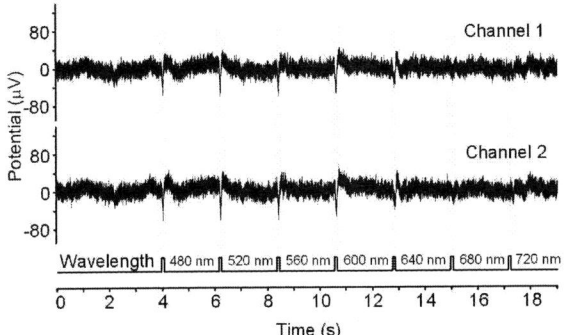

Fig.7. Light-evoked neural responses of a carp retina recorded with two channel Si microprobes. The responses were depending on the wavelengths from 480 nm to 720 nm.

In recent advanced neuroscience, the microprobe array could be applied to minimally-invasive electrode implants, realizing the interfacing between neurons and human artificial limbs, sensory devices, etc.[3] We also presented not only micro-probes but also micro-tubes which may be used for drug delivery on the chip for bio-medical applications. [4]

pH image sensor

32×32 pH image sensors were successfully fabricated by using the CCD/CMOS image sensor technique, and real time imaging of a chemical reaction and pH distribution was carried out for the first time. The pH variations by a chemical reaction are observed by 200 ms step (i.e. 5 flames per sec). The pH image sensor was able to take a pH image of mouse stomach successfully. It means that the novel image sensor can be applied to a biomedical and biochemical field.

Conventionally, pH is measured on the supposition that a solution is a uniform concentration. If the distribution of biochemical phenomena can be obtained as a visible image in real time, the local variations in these will lead to a better understanding. The technique of this pH sensor is based on the principle of a charge-coupled-device (CCD). [5] The cross section and the operation mechanism of the charge-transfer-type pH sensor are shown in Fig.8. The charge-transfer-type pH sensor consists of seven elements: an input diode (ID), an input control gate (ICG), an ion-sensing region, a transfer gate (TG), a floating diffusion (FD) region, a reset switch and a

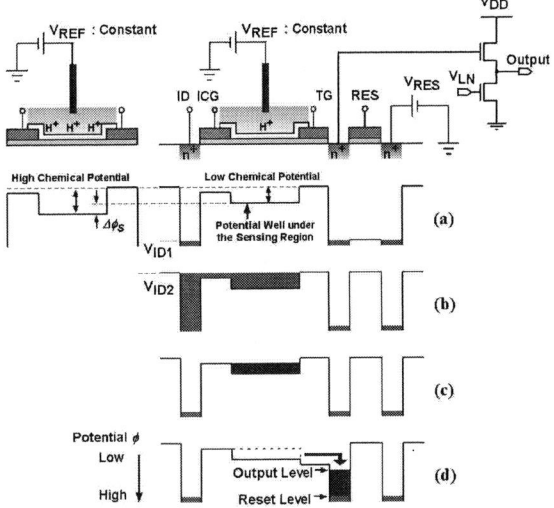

Fig.8. The principle of conversion from pH to charge: (a) The clock cycle is initiated. (b) The input diode is briefly pulsed from V_{ID1} to V_{ID2}. (c) The input diode is kept again V_{ID1}. (d) The transfer gate is turned on and the charge in the sensing region is transferred to the floating diffusion region.

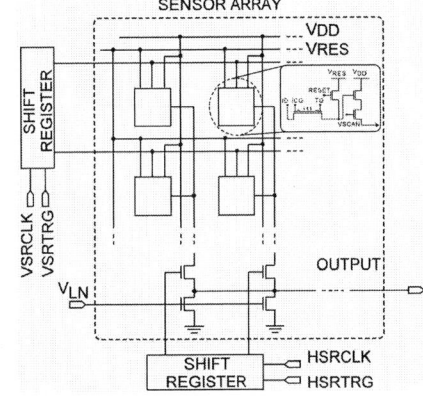

Fig.9. Block diagram of pH image sensor

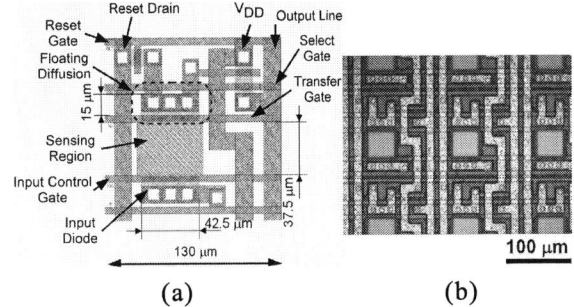

Fig.10. Schematic diagram and photograph of single pixel of pH sensor. (a) Layout. (b) Photograph.

source follower circuit. The ion-sensing region is the thin Si3N4 film that acts as an ion-sensitive membrane. The block diagram of the proposed 32×32 pH image sensor is shown in Fig. 9. The image sensor is integrated with horizontal and vertical shift resistor. The pH image sensors were fabricated by the CCD/CMOS technology.[6] A pixel layout and

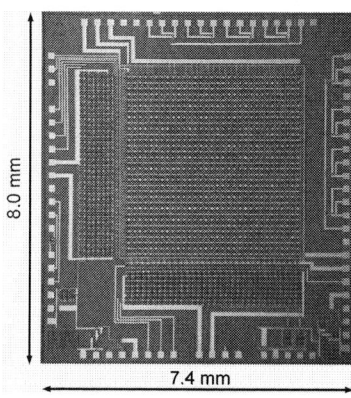

Fig.11. Photograph of pH image sensor

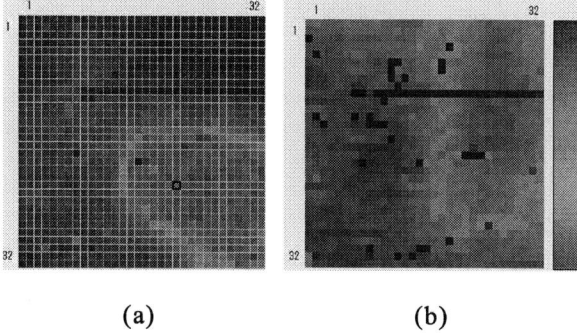

(a)   (b)

Fig.12. The images of 2-D distribution of pH. (a) Steamed SUSHI rice. (b) Mouse stomach.

photograph of the pH sensor is shown in Fig. 10(a), (b). The size of a pixel is 130 μm square, and the sensing region is 40 μm square. Two-dimensional sensor comprising a 32 × 32 pixel array is shown in Fig.11. The chip size is 8.0 mm × 7.4 mm, and the size of sensing area is 4.16 mm × 4.16 mm. An imaging result of pH distribution variation is tried. [7] The images were taken by 200 ms step (i.e. 5 flames per sec). At first, pH 9.2 buffer solution (70 μl) was filled on a pH image sensor, and pH 6.9 buffer solution (70 μl) was dropped in upper side of the chip. The pH 6.9 solution is diffused about 500 μm during 200msec in this condition. Figure 12 (a) shows another pH images of steamed rice of "SUSHI". The rice was dropped in a buffer solution with pH 9.2. It is confirmed a shape of rice by pH in formations. "SUSHI" rice shows about pH 2.5, because vinegar is contained in "SUSHI" rice. Finally we tried to observe a living related material. A stomach of mouse was set on the pH image sensor chip with PBS (Phosphate-Buffered Saline) solution (pH 7.4). The place of the stomach is clearly confirmed and the pH of stomach indicates is acid (about pH 5.5) as shown in Fig.12(b). From these results, the novel image sensor can be applied to a biomedical and biochemical field.

RF transmitters and power circuits for smart chips

Integrated techniques for the RF transmitter by CMOS compatible processes have been successfully developed.[8] After matching by inserting the bonding wire inductor between the on-chip integrated antenna and the VCO output, the measured emission power at distance of 5 m from RF transmitter was -37 dBm (0.2 μW). The integrated RF induced power supply system with a large capacitor of Surface Mount Devices (SMDs) is also developed to be integrated. Deep holes are fabricated by Deep Reactive Ion Etching (DRIE) for SMD mounting. It was confirmed that this system supplies well regulated 4.0V/1mA DC power to the load for 10msec periodically. [9]

Conclusions

Silicon smart microchips for intelligent sensing are discussed with our developed chips of Si microprobe electrode array chip and pH image sensor using CMOS/MEMS technologies. These chips are new type sensing chips and inform more valuable information which is not known until now. In future, these chips could have more functional devices on chips such as rf transmitters and power supply, and be used to many applications.

Acknowledgements

This work was supported in The 21st Century COE Program "Intelligent Human Sensing", JST- CREST, Global COE program "Frontiers of Intelligent Sensing" and a Grant-in-Aid for Scientific Research (A) from the ministry of Education, Culture, Sports, Science and Technology Japan.

References

[1] M. Ishida, et al., "Selective growth of Si wires for intelligent nerve potential sensors using vapor-liquid-solid growth," Proc. 10th Int. Conf. Solid-State Sensors and Actuators (Transducers' 99), 866 (1999).

[2] T. Kawano, et al., "Selective vapor-liquid-solid epitaxial growth of micro-Si probe electrode arrays with on-chip MOSFETs on Si (111) substrates," IEEE Trans. on Electron Devices, **ED-51**, 415 (2004).

[3] M. A. L. Nicolelis, "Actions from thoughts," Nature, **409**, 403 (2001).

[4] K.Kuniharu, T.Kawashima, T.Kawano, H.Takao, S.Sawada, and M.Ishida"Integration of out-of-plane silicon dioxide microtubes, silicon microprobes and on-chip NMOSFETs by selective vapor-liquid –solid growth" J. Micromech. Microeng. **18** (2008) 035033 (9pp)

[5] K. Sawada, S. Mimura, M. Ishida, et al., "Novel CCD based pH Imaging Sensor", IEEE Trans. ED. vol. 46(9), pp.1846-1849,(1999.)

[6] T. Hizawa, K. Sawada, H. Takao, M. Ishida, "Fabrication of a two-dimensional pH image sensor using a charge transfer technique", *Sens. Actuators* B117, pp. 509-515(2006).

[7] T. Hizawa, J. Matsuo, T. Ishida, H. Takao, H. Abe, K. Sawada and M. Ishida "32 X 32 pH image sensors for real time obserbation of biochemical phenomena" TRANSDUCERS.07, Lyon, France, June 10-14, 2007. 2EP2.P

[8] J.W.Kim, H.Takao.K.Sawada and M.Ishida "Integration inductors for RF transmitters in CMOS/MEMS smart microsensor systems" Sensors 2007, 7,1387-1398 (2007).

[9] M. Sudou, H. Takao, K. Sawada, M. Ishida "A novel RF induced power supply system for monolithically integrated Ubiquitous micro sensor nodes" Proc. of Transducers'07, Lyon France, Jun 10-14 pp.907-910 (2007). Sensors & Actuators A: Physics (2007) in press.

FinFET Performance Advantage at 22nm: An AC perspective

M. Guillorn, J. Chang, A. Bryant, N. Fuller, O. Dokumaci, X. Wang, J. Newbury, K. Babich, J. Ott,
B. Haran, R. Yu, C. Lavoie, D. Klaus, Y. Zhang, E. Sikorski, W. Graham, B. To, M. Lofaro, J. Tornello,
D. Koli, B. Yang*, A. Pyzyna, D. Neumeyer, M. Khater, A. Yagishita[†], H. Kawasaki[†] and W. Haensch

IBM Research, *AMD, Inc., [†]Toshiba, Inc, IBM T.J. Watson Research Center, Yorktown Heights, NY

Abstract

At the 22 nm node, we estimate that superior electrostatics and reduced junction capacitance in FinFETs may provide a 13~23% reduction in delay relative to planar FETs. However, this benefit is offset by enhanced gate-to-source/drain capacitance (Cgs) in FinFETs. Here, we measure FinFET C_{gs} capacitance at 22nm-like dimensions and determine that, with optimization, the FinFET capacitance penalty can be limited to <6%, resulting in an overall advantage of up to 17% over a planar technology. (Keywords: FinFET, parasitic capacitance)

Introduction

Continued gate pitch scaling demands gate length scaling, making FinFETs an attractive option for the 22nm technology node. At realistic 22nm node dimensions (Table 1), the heavy channel doping required to control off-state leakage in planar devices at 22nm will present a significant performance penalty [1]. Using TCAD and circuit simulations, we predict that improved short channel effects and reduced junction capacitance (C_j) in FinFETs can be exploited to reduce circuit delay by as much as 13 to 23% compared to a planar, silicon-on-insulator technology with SiON gate dielectric.

Table 1. Relevant dimensions for 22nm technology node

	planar	FinFET	This work
Vdd	.9V	.9V	
Gate Length (Lg)	25 nm	25 nm	≥25 nm
Contact-to-gate space	20 nm	20 nm	≥20 nm
Fin Width (Df)		12 nm	≥9 nm
Fin Height (Hf)		40 nm	≥25 nm
Fin Pitch (FP)		40 nm	≥50 nm

The FinFET advantage will be offset by the fundamental FinFET gate-to-source/drain capacitance (C_{gs}) penalty which arises from the fringing fields between the gate and the top and bottom of the source/drain ("C_{tb}", Fig 1b) [2]. While planar FETs have an analogous fringing capacitance that exists at width boundaries, the penalty from this decreases with larger device width. No such mitigating strategy is available in a FinFET technology with fixed fin height (H_f).

In this study, we use a simple analytical model (Fig 1c) to roughly separate C_{gs} into two components: a planar-like portion (C_{fin}) and the FinFET penalty (C_{tb}).

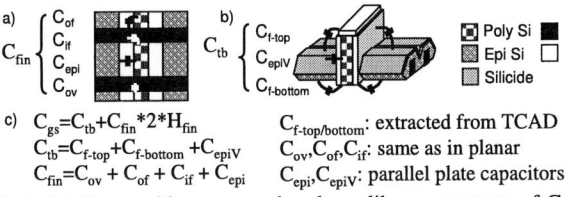

a) $C_{fin} \begin{cases} C_{of} \\ C_{if} \\ C_{epi} \\ C_{ov} \end{cases}$

b) $C_{tb} \begin{cases} C_{f\text{-}top} \\ C_{epiV} \\ C_{f\text{-}bottom} \end{cases}$

■ Poly Si
□ Epi Si
▨ Silicide

c) $C_{gs} = C_{tb} + C_{fin} * 2 * H_{fin}$
$C_{tb} = C_{f\text{-}top} + C_{f\text{-}bottom} + C_{epiV}$
$C_{fin} = C_{ov} + C_{of} + C_{if} + C_{epi}$

$C_{f\text{-}top/bottom}$: extracted from TCAD
C_{ov}, C_{of}, C_{if}: same as in planar
C_{epi}, C_{epiV}: parallel plate capacitors

Fig 1. (a) C_{fin} roughly captures the planar-like components of C_{gs}, while (b) C_{tb} roughly captures the fundamental FinFET C_{gs} penalty. (c) Simple analytical model for C_{gs}.

Device Fabrication

FinFETs were fabricated with fin pitch (FP) down to 50nm and H_f from 25nm to 56nm using the flow in Fig 2. Two contact schemes were explored: via-contacted merged s/d and bar-contacted unmerged s/d regions. Devices have undoped channels and a poly-Si/SiON gate stack resulting in a negative threshold voltage. We measured C_{gs} on large arrays of several thousand fins biased with a V_{gs} of -0.5V.

The topology differences caused by varying H_f impact many subsequent unit processes, resulting in differences in fin thickness (D_{fin}), BOX undercut, and vertical epi growth (epiV). Many of the lithography, deposition, and etch processes are sensitive to pattern density, so changing FP also affects parameters such as D_{fin}, spacer thickness (tsp), and epiV. Structural variables were cataloged using extensive TEM inspection of cross-sections and used in subsequent simulation.

Results

C_{tb} and C_{fin} dependence on parameters such as FP, H_f, and spacer thickness (t_{sp}) was extracted from 3-D TCAD simulations of merged s/d FinFET structures (Fig 3a). The resulting model shows good agreement to sensitivies observed in experimental data (Fig 3b). Note that due to a discrepancy between TCAD predictions and experimental data, a constant 10aF/fin adder has been added to all experimental data presented here to aid comparison to model predictions.

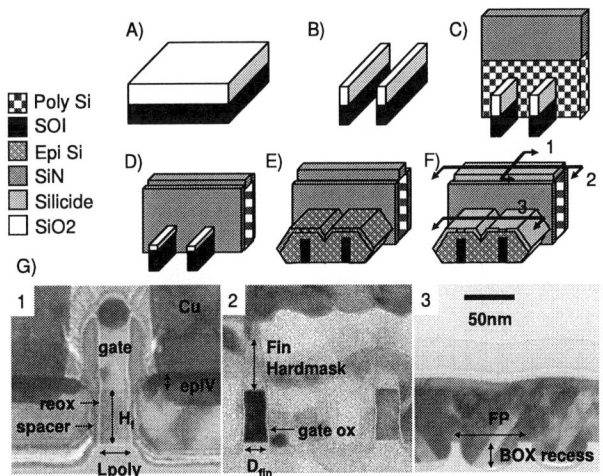

Fig. 2 Process flow: A) SOI thinning and fin oxide hardmask (HM) growth; B) Fin definition: electron beam lithography and RIE; C) Gate stack deposition (1.1nm SiON + polysilicon), gate electron beam lithography and RIE; D) Spacer formation followed by shallow tilted As implant; E) Epitaxial Si growth; F) Deep source/drain As implant, RTA and NiPt silicide formation. Devices are contacted using copper vias and wiring. (G) TEM of device cross-sections.

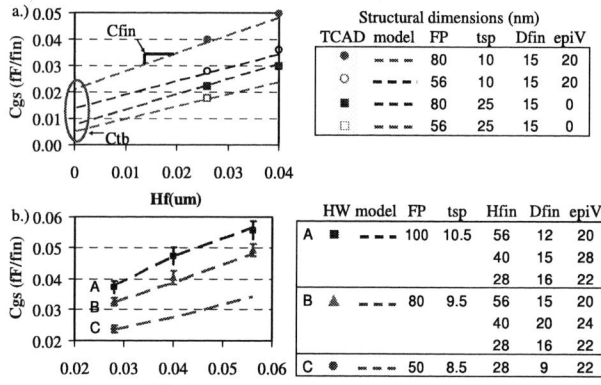

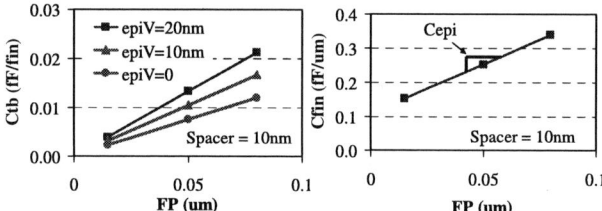

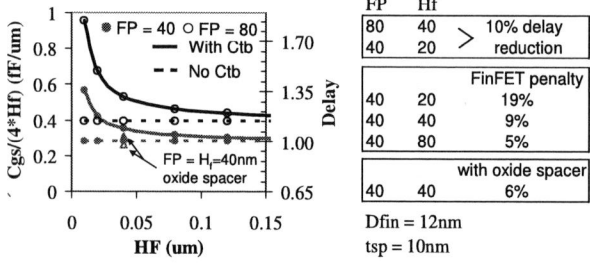

Fig. 3. (a) Extraction of model parameters from TCAD simulation (b) Comparison of model predictions to experimental data.

For 22nm node merged s/d FinFET technology, we predict FP to be the most effective means for C_{gs} reduction. C_{tb} and C_{fin} are both reduced by minimizing FP (Fig 4), predominantly due to a decrease in direct & fringing capacitance to the epi (C_{epi}, C_{epiV}, and $C_{f-top/bottom}$). Even at a fixed FP/H_f ratio, we find that reducing FP is beneficial; for example, going from a FP/H_f of 80/40nm to 40/20nm results in a 0.2fF/μm C_{gs} reduction. Using a CV/I-based model, we estimate this reduction to correspond to a ~10% inverter delay reduction (Fig 5).

Fortunately, minimizing FP is consonant with the need for granularity in device width. Assuming a minimum designed pitch for the 22nm node of 80nm, a FP of 40nm is possible using a pitch split technique (e.g. sidewall image transfer). For a fixed FP, C_{gs} may be further reduced by increasing H_{fin} (Fig 5). For a FP/H_f of 40/40nm, the delay penalty from C_{tb} is estimated to be ~9%. If the spacer is converted from nitride to oxide, the delay penalty falls to ~6%. Assuming a ~13 to 23% delay reduction from superior electrostatics and reduced C_j, FinFETs may therefore provide up to a 17% performance advantage over planar CMOS.

Fig. 4. Model predictions for C_{tb} and C_{fin} sensitivity to FP and epi.

Fig. 5. Estimation of FinFET penalty (With C_{tb}) as compared to planar devices (idealized as "No C_{tb}"). The relative delay normalized to the case of "FP=40nm, no C_{tb}" is given by the 2nd y-axis.

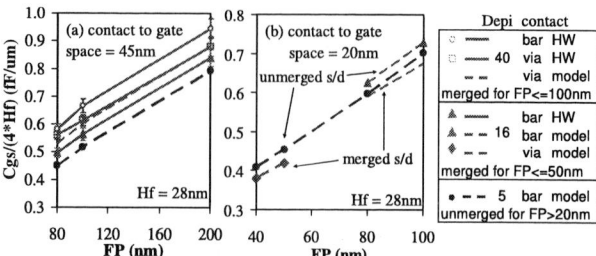

Fig. 6. Source/drain structure tradeoffs (a) at the dimensions of fabricated devices and (b) extrapolated to 22nm-like dimensions. (a) includes experimental data for bar-contacted devices with D_{epi} = 40nm and 16nm as well as extrapolation for D_{epi} = 5nm assuming that C_{gs} varies linearly with D_{epi}. For (b), via and bar contact capacitances are estimated from TCAD simulation.

The preceding analysis assumes a merged s/d structure. Devices built in this manner may be contacted with vias identical to those used in planar technology. However, switching to an unmerged s/d structure by reducing epi thickness (D_{epi}) will reduce C_{gs} since the sensitivity of C_{tb} to FP depends strongly on vertical epi growth. Moreover, the slope of C_{fin} vs. FP is due almost entirely to C_{epi} (Fig 4).

Unmerged s/d regions require either the use of bar contacts or contact vias with a pitch equal to FP. The latter becomes less feasible as FP is reduced. Consequently, we focus on the tradeoff between via-contacted, merged FETs and bar-contacted, unmerged FETs. Experimental data, measured on structures with a contact-to-gate distance of 45nm, reveals lower C_{gs} in unmerged (D_{epi}= 16nm), bar-contacted devices than merged (D_{epi}=40nm), via-contacted devices (Fig 6a). However, when this tradeoff is extrapolated to a contact-to-gate distance of 20nm and FP = 40nm, the merged structure has lower C_{gs}.

Conclusion

We have fabricated FinFET devices at dimensions relevant for the 22nm technology node and extracted capacitance Cgs inherent to this 3D device structure. This capacitance component is a crucial parameter for evaluating FinFETs as an alternative to planar CMOS at the 22 nm technology node. The sensitivity of parasitic C_{gs} in FinFETs to structural variables such as H_f, FP, t_{sp}, and epiV is captured in a simple analytical model calibrated to both TCAD simulations and experimental measurements. Based on this model, a strategy for minimizing FinFET C_{gs} dictates minimizing FP, maximizing H_f, and using an oxide spacer and a via-contacted, merged s/d structure. This optimized structure should limit the FinFET Cgs penalty to ~6%, allowing the retention of up to a 17% FinFET performance advantage over a 22nm planar technology.

Acknowledgments

This work was performed by the Research Alliance Teams at various IBM Research and Development Facilities.

References

[1] W. Haensch, E. J. Nowak, R. H. Dennard, P. M. Solomon, A. Bryant, O. H. Dokumaci, A. Kumar, X. Wang, J. B. Johnson, and M. V. Fischetti, "Silicon CMOS devices beyond scaling," IBM J. Research & Development, vol. 50, pp. 339, 2006.

[2] W. Wu, M. Chan, "Analysis of Geometry-Dependent Parasitics in Multifin Double-Gate FinFETs," IEEE Trans. Electron Devices, vol. 54, pp. 692-698, April, 2007.

Flexible and Robust Capping-Metal Gate Integration Technology enabling multiple-V_T CMOS in MuGFETs

A. Veloso, L. Witters, M. Demand, I. Ferain[1], N. J. Son[2], B. Kaczer, Ph. J. Roussel, E. Simoen, T. Kauerauf, C. Adelmann, S. Brus, O. Richard, H. Bender, T. Conard, R. Vos, R. Rooyackers, S. Van Elshocht, N. Collaert, K. De Meyer[1], S. Biesemans, and M. Jurczak

IMEC, [2]assignee at IMEC from Samsung, Kapeldreef 75, B-3001 Leuven, Belgium; [1]also K. U. Leuven, Belgium
Tel.: +32-16-28 17 28, Fax: +32-16-28 17 06, email: Anabela.Veloso@imec.be

Abstract

We report, for the first time, a comprehensive study on various capping integration options for WF engineering in MuGFET devices with TiN gate electrode: HfSiO/cap/TiN, cap/HfSiO/TiN and HfSiO/TiN/cap/TiN vs. reference deposition sequence HfSiO/TiN (cap=Al$_2$O$_3$ for pmos, and Dy$_2$O$_3$ or La$_2$O$_3$ for nmos). We show that: 1) low-V_T values (<0.3V) are achieved for both nmos and pmos, with excellent process control and device behavior down to Lg≈50nm and W$_{FIN}$≥20nm, for optimized gate stack configurations; 2) inserting a cap layer in-between TiN layers instead of HfSiO/cap/TiN leads to improved mobility, reduced CET without impacting J$_G$, similar noise response and improved BTI behavior, with correction of the abnormal PBTI degradation seen for HfSiO/DyO/TiN. Is also enables simplified and more robust CMOS co-integration of low- and med-V_T devices in the same wafer, avoiding loss in CET and damage of the host dielectric with the cap removal process.

Introduction

FinFET-based multi-gate (MuGFET) devices are one of the most promising candidates for enabling continued MOSFET scaling beyond the 32nm technology node. Due to full depletion of the FINs, V_T tuning options for narrow FINs are limited to workfunction (WF) tuning with metal gate (MG). Medium-V_T values are set by a mid-gap WF MG electrode, and high-V_T (HVT) targets can be achieved by increased FIN doping (N$_a$) only in wider FIN devices. MuGFET viability for low-V_T (LVT) CMOS applications remains a challenge, even if, compared with planar bulk devices, they need smaller WF shifts from mid-gap to reach the low-V_T targets. Recent developments have shown that introducing a cap layer (La$_2$O$_3$ [1], Dy$_2$O$_3$ [2] for nmos; AlO$_x$ [3,4] for pmos) between a high-k host dielectric and MG can successfully tune its WF towards band-edge. In this paper, we propose a practical integration scheme, with cap layers selectively inserted in the flow at different locations in the gate stack, compatible with the MuGFET 3-D architecture and enabling a wide V_T range (LVT-HVT) on the same wafer, needed for flexible circuits operation. Impact on V_T, CET, J$_G$, mobility, device performance, noise and reliability behavior of the various capping options evaluated will be addressed here.

Device fabrication

MuGFET devices were fabricated from a baseline CMOS flow described in [5] on SOI wafers with 65nm Si (H$_{FIN}$) on 145nm BOX, and FINs patterned down to W$_{FIN}$≈20nm. Reference gate stack consists of 5nm plasma enhanced-ALD (PE-ALD) TiN deposited on 2.3nm MOCVD HfSiO + 1nm interfacial oxide layer (IL). Similar high-k, IL and TiN were used in devices with thin (≤1nm) cap layers deposited (Fig.1): a) after IL and before HfSiO; b) immediately after HfSiO; and c) in-between 2 TiN layers depositions such that total TiN thickness=5nm. Dy$_2$O$_3$ by AVD® or La$_2$O$_3$ by ALD were used as cap layers for nmos and Al$_2$O$_3$ by ALD for pmos. A highly-selective (dry+wet) gate etch process is used with negligible Si recess. SEG used in most wafers prior to HDD I/I corresponds to a 750°C, 30min thermal budget. 1050°C spike anneal in N$_2$ is used for junctions activation. After device fabrication, HRTEM (Fig.2) show no contrasting oxide cap layer inside the MG electrode and XPS analysis confirm that the caps diffuse in TiN (DyO results in Fig.3).

Results and discussion

Fig.4 shows an overview of ΔV_T and CET values for various AlO-capped pmos devices. ΔV_T≈100mV is obtained for HfSiO/1nm AlO/TiN at the expense of ≤2Å CET increase. Comparable or slightly larger ΔV_T (≈100-140mV) are obtained for HfSiO/TiN/AlO/TiN stacks (AlO=0.5-1nm) with no penalty in CET. Depositing the AlO cap further away from the dielectric interface (by using a thicker 1st TiN metal layer; 'cap4' in Fig.4) results in smaller ΔV_T (≈100mV) but also in smaller CET (ΔCET≈1.1 Å for HfSiO/2nm TiN/1nm AlO/3nm TiN), interesting for CET scaling. It should be noted that in uncapped samples, smaller CET values are also obtained with thinner PEALD TiN layers (ΔCET≈2Å + ΔWF<0 (more n-type) occur for 5→2nm TiN), indicative of an interaction of the TiN ALD process on the final CET. Fig.5 shows that all AlO-

capped devices have tight V_T distributions (σ≈4-13mV for Lg≥57nm, W$_{FIN}$≈25nm), with V_T roll-offs (Fig.6) comparable to reference. Fig.7 shows there is a small mobility reduction for large (L$_g$=10μm) HfSiO/AlO/TiN devices, though overall very comparable values are obtained for all stacks down to smaller L$_g$.

For nmos, Fig.8 shows ΔVTs and ΔCETs obtained when a DyO or LaO cap layer is deposited prior to or immediately after HfSiO deposition and in the TiN/cap/TiN configuration. Overall, a CET increase is observed for all capped samples, which can be minimized by careful optimization of the layers thicknesses. Similarly to AlO cap in pmos, slightly smaller CETs are obtained with TiN/cap/TiN stacks. However, they correspond to smaller ΔV_Ts compared to stacks with other cap locations, except for LaO. In this case, ΔV_T only decreases when depositing the cap further away from the gate dielectric interface (1st TiN layer=2nm instead of 1nm), suggesting LaO diffuses faster than DyO in TiN. Devices with tight V_T distributions down to Lg≈50nm (Fig.9), and aggressive low-V_T values <0.3V (Figs.9 and 10) are obtained for various optimized capped stacks, with V_T roll-offs comparable to reference. Mobility-wise, Fig.11 shows that depositing the cap layer in-between metal layers is beneficial, with values similar to the reference. ITP characteristics show no degradation, or even a small performance improvement for optimized cap/TiN or TiN/cap/TiN devices (data for DyO-nmos and AlO-pmos in Fig.12). All capped and uncapped devices follow the same J$_G$-CET trend-line, as seen in Fig.13, with an overview of the WF tuning results obtained presented in Fig.14.

Low-frequency noise characterization was done in MuGFETs with 10 FINs, W$_{FIN}$≈25nm and Lg≈1μm. Fig.15 shows the normalized input-referred voltage noise spectral density (S$_{VG}$) values, computed from the measured I$_D$ noise spectral density. A small increase occurs for capped samples, with no difference when the cap is placed in the gate stack just before or within the TiN. Figs.16 and 17 show the results of the reliability evaluation performed. From TBBD (Fig.16), operating voltages >1.0V were extrapolated for a 10 years lifetime, with AlO-capped devices outperforming the reference. NBTI ΔV_T shifts were standard evaluated by monitoring I$_D$ (V$_{G,meas}$ set at ≅V_T) at predetermined 0.5sec interruptions of the stress voltage (Fig.17a), after which a more in-depth study was done, recording short traces of ΔV_T relaxation after various stress conditions and extracting the recoverable R and permanent P components of the BTI degradation [6] (Fig.17b). Results showed comparable behavior for reference and TiN/AlO/TiN devices, and slightly higher degradation (especially in R) for HfSiO/AlO/TiN. Similar method applied to nmos (Fig.17c) showed an abnormal BTI behavior for HfSiO/DyO/TiN devices, as reported in [7]. Interestingly, when inserting the DyO cap in-between the TiN layers, a considerable improvement was obtained, resulting in normal and comparable to the reference (or quantitatively, even slightly better) BTI behavior for optimized HfSiO/TiN/DyO/TiN stacks (1st TiN thickness>1nm).

Combining the benefits of the several cap options explored in this work, a proposed multiple-V_T CMOS integration scheme is illustrated in Fig.18: uncapped med-V_T devices, with cap layers just below or within TiN MG for LVT, and N$_a$↑ in wider FINs for HVT. The different caps locations allow a simplified selective cap removal process [3,8,9] of the nmos (pmos) cap layer from the underlying material in pmos (nmos) areas, less damaging to the host dielectric.

Conclusions

Several capping integration options for WF engineering in MuGFET devices with TiN gate electrode were evaluated, demonstrating the feasibility of a multiple-V_T CMOS integration scheme. Low-V_T capped devices, with excellent process control and device behavior down to Lg≈50nm and W$_{FIN}$≥20nm were fabricated. Insertion of cap layers in-between TiN layers reduces CET with no impact in J$_G$, similar noise response and improved BTI behavior.

References

[1] V. Narayanan et al., VLSI Tech. Dig. 2006, 224; [2] T. Lee et al., EDL 27(8), 640 (2006); [3] H.-S. Jung et al., VLSI Tech. Dig. 2006, 204; [4] H.-S. Jung et al., VLSI Tech. Dig. 2007, 196; [5] Collaert et al., VLSI Tech. Dig. 2005, 108; [6] B. Kaczer et al., IRPS2008 (accepted); [7] R. O'Connor et al., IRPS2008 (accepted); [8] K.L. Lee et al., VLSI Tech. Dig. 2006, 202; [9] H.Y. Yu et al., VLSI Tech. Dig. 2007, 18.

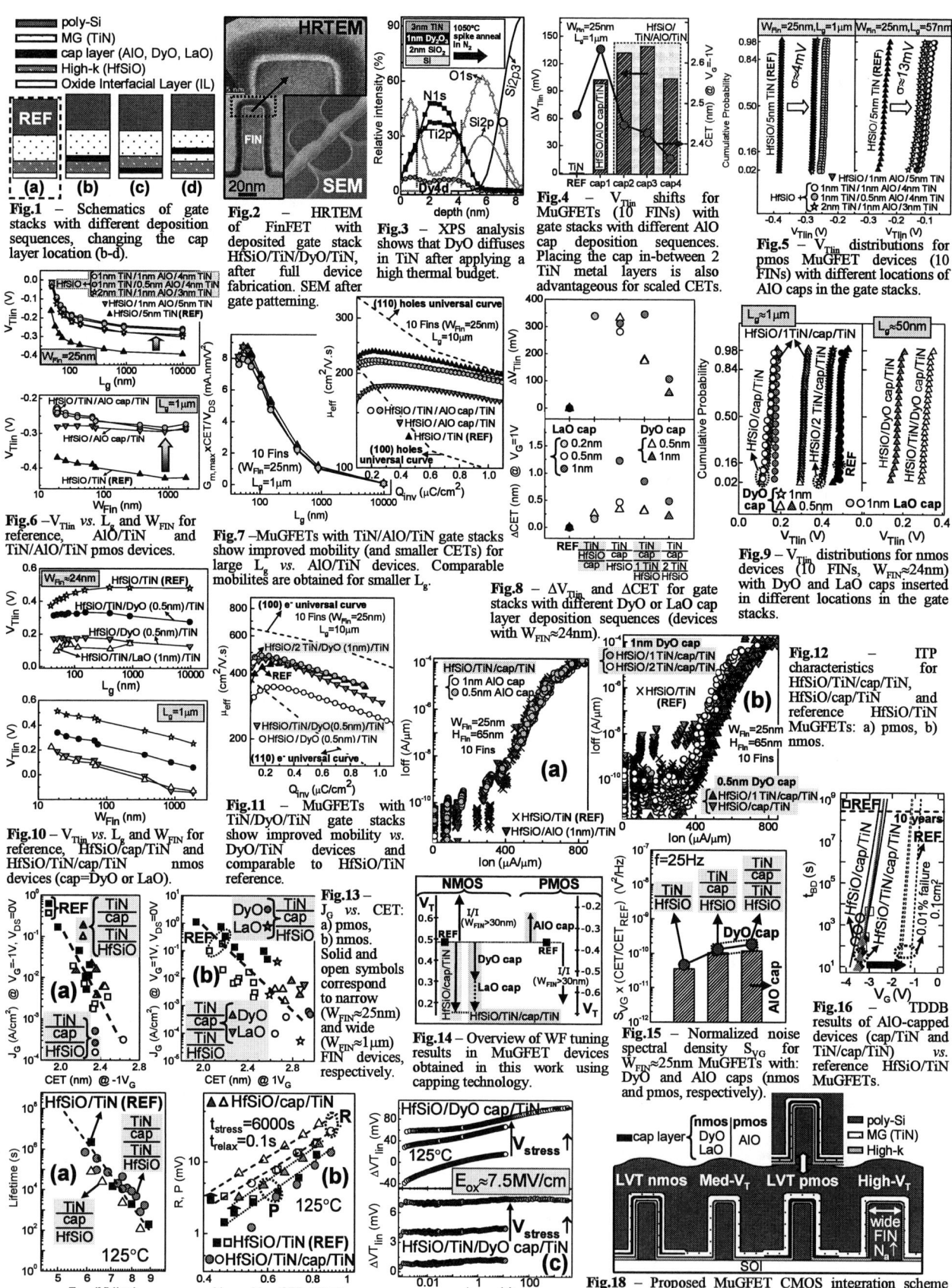

Fig.1 – Schematics of gate stacks with different deposition sequences, changing the cap layer location (b-d).

Fig.2 – HRTEM of FinFET with deposited gate stack HfSiO/TiN/DyO/TiN, after full device fabrication. SEM after gate patterning.

Fig.3 – XPS analysis shows that DyO diffuses in TiN after applying a high thermal budget.

Fig.4 – V_{Tlin} shifts for MuGFETs (10 FINs) with gate stacks with different AlO cap deposition sequences. Placing the cap in-between 2 TiN metal layers is also advantageous for scaled CETs.

Fig.5 – V_{Tlin} distributions for pmos MuGFET devices (10 FINs) with different locations of AlO caps in the gate stacks.

Fig.6 – V_{Tlin} vs. L_g and W_{FIN} for reference, AlO/TiN and TiN/AlO/TiN pmos devices.

Fig.7 – MuGFETs with TiN/AlO/TiN gate stacks show improved mobility (and smaller CETs) for large L_g vs. AlO/TiN devices. Comparable mobilities are obtained for smaller L_g.

Fig.8 – ΔV_{Tlin} and ΔCET for gate stacks with different DyO or LaO cap layer deposition sequences (devices with $W_{FIN}\approx24nm$).

Fig.9 – V_{Tlin} distributions for nmos devices (10 FINs, $W_{FIN}\approx24nm$) with DyO and LaO caps inserted in different locations in the gate stacks.

Fig.10 – V_{Tlin} vs. L_g and W_{FIN} for reference, HfSiO/cap/TiN and HfSiO/TiN/cap/TiN nmos devices (cap=DyO or LaO).

Fig.11 – MuGFETs with TiN/DyO/TiN gate stacks show improved mobility vs. DyO/TiN devices and comparable to HfSiO/TiN reference.

Fig.12 – ITP characteristics for HfSiO/TiN/cap/TiN, HfSiO/cap/TiN and reference HfSiO/TiN MuGFETs: a) pmos, b) nmos.

Fig.13 – J_G vs. CET: a) pmos, b) nmos. Solid and open symbols correspond to narrow ($W_{FIN}\approx25nm$) and wide ($W_{FIN}\approx1\mu m$) FIN devices, respectively.

Fig.14 – Overview of WF tuning results in MuGFET devices obtained in this work using capping technology.

Fig.15 – Normalized noise spectral density S_{VG} for $W_{FIN}\approx25nm$ MuGFETs with: DyO and AlO caps (nmos and pmos, respectively).

Fig.16 – TDDB results of AlO-capped devices (cap/TiN and TiN/cap/TiN) vs. reference HfSiO/TiN MuGFETs.

Fig.17 – NBTI reliability results of AlO-capped devices (cap/TiN and TiN/cap/TiN) vs. reference HfSiO/TiN MuGFETs are shown in a) and b), with the recoverable BTI component (R in b) slightly better for TiN/cap/TiN and reference devices. DyO cap insertion in TiN/cap/TiN can correct an abnormal HfSiO/DyO/TiN PBTI behavior (c).

Fig.18 – Proposed MuGFET CMOS integration scheme allowing multiple-V_T devices (LVT, med-V_T and HVT) in the same wafer. Flexibility in the cap layers location for LVT devices (cap deposition can be done after high-k or in-between TiN layers) avoids the underlying material (high-k or 1st TiN layer) from seeing a cap removal process twice.

Novel integration process and performances analysis of Low STandby Power (LSTP) 3D Multi-Channel CMOSFET (MCFET) on SOI with Metal / High-K Gate stack

E. Bernard[1,2,3], T. Ernst[1], B. Guillaumot[2], N. Vulliet[2], V. Barral[1], V. Maffini-Alvaro[1], F. Andrieu[1], C. Vizioz[1], Y. Campidelli[2], P. Gautier[1], J.M. Hartmann[1], R. Kies[1], V. Delaye[1], F. Aussenac[1], T. Poiroux[1], P. Coronel[2], A. Souifi[3], T. Skotnicki[2] and S. Deleonibus[1]

[1]CEA/LETI, Minatec, 17 rue des Martyrs, 38054 Grenoble, France; [2]STMicroelectronics, 850 rue J. Monnet, 38926 Crolles, France; [3]INL-INSA Lyon, 7 avenue Jean Capelle, 69621 Villeurbanne, France; Tel: 33 4 38 78 92 39, Fax : 33 4 38 78 94 56, emilie.bernard@cea.fr

Abstract

For the first time, ultra low I_{OFF} (16.5pA/µm) and high I_{ON}N,P (2.27mA/µm and 1.32mA/µm) currents are obtained with a Multi-Channel CMOSFET (MCFET) architecture on SOI with a Metal / high-K gate stack. This leads to the best I_{ON}/I_{OFF} ratios ever reported: 1.4×10^8 (0.8×10^8) for 50nm n- (p-) MCFETs. We show, based on specifically developed integration process, characterization methods and analytical modeling, how those performances are obtained thanks to specific 3D MCFET features, in particular, transport properties, saturation regime and electrostatic behavior.

Introduction

The 3D Multi-Channel MOSFET (MCFET) architecture pushes further the scaling limits of the CMOS technology. It offers both a very low I_{OFF} thanks to the multi-gate electrostatic control and a high current drivability due to the 3D vertically stacked channels.

Recently, very high current drivability was achieved in bulk multi-channel devices with TiN/SiO_2 gate stacks and self-aligned Sources/Drains (S/D) [1]. We previously showed that metal / high-K gate stacks led in such architectures to low gate leakage currents and good reliability [2]. However, LSTP targets were not achieved due to specific MCFET issues, in particular 3D series resistances distribution in the S/D [3] and mobility optimization in tunnel-etched channels [2]. We propose here a Silicon-On-Nothing (SON) based [4] multi-channel process (Fig. 1) on SOI with an excellent electrostatic control, optimized gate stack, S/D and transport properties for CMOS Low STandby Power (LSTP) applications.

Device Fabrication

First, a $(25\text{nm Si}/30\text{nm Si}_{0.8}Ge_{0.2}) \times 3$ superlattice was epitaxially grown on top of a SOI substrate thanks to Reduce Pressure – Chemical Vapour Deposition (Fig. 1). The superlattice was then etched (with a gate mask) down to the bottom SOI layer. Crystalline Si S/D were selectively grown on the latter and implanted. Compared to [2], S/D epitaxy shape and implant conditions were optimized in order to reduce the series resistances. The active area was then patterned and etched. The SiGe layers were selectively removed using pure CF_4 at high pressure and low microwave power in a remote plasma tool. The gate stack made of 3nm Atomic Layer Deposition (ALD) HfO_2 + 10nm CVD TiN + 100nm N^+ poly-Silicon layers was then deposited (measured Capacitance Equivalent Thickness in inversion (CET) = 1.95nm). A second gate etch was subsequently carried out followed by the formation of nitride spacers. After the dopant activation anneal, S/D were silicided (NiSi) followed by a standard Back End Of the Line process. A planar FD-SOI device with the same gate stack and silicon thickness was processed as a reference using the technology described in [5].

Results and discussion

Fig. 2 presents a cross-sectional TEM picture of a 5 channels 50nm gate length n-MCFET transistor with the surrounding TiN/HfO_2 gate stack. The five channel device is composed of 2 double gates and a bottom ultra-thin body FD-SOI transistor. The Si channel thicknesses are 10nm. The inset is a HRTEM picture of the device showing the good cristallinity of the Si channels. The use of an optimized ALD process leads to very uniform thicknesses of the HfO_2 and interfacial oxide layers in narrow etched tunnels. I_D-V_G (Fig. 3) and I_D-V_D (Fig. 4) curves of 50nm n- and p-MCFETs show well-behaved characteristics. Ultra low I_{OFF} currents (16.5pA/µm) are obtained leading to excellent I_{ON}/I_{OFF} currents ratios of 1.4×10^8 and 0.8×10^8, exceeding previously reported results for sub-100nm n- and p- LSTP MOSFETs respectively (Fig. 5 and 6). The threshold voltages obtained with TiN/HfO_2 gate stacks are close to the LSTP ITRS requirements. Compared to an optimized planar FD-SOI reference, a gain of 5.5 (4.0) on the drain current is obtained for n-

(p-) MCFET @ V_G–V_T = 0.75V (Fig. 7). The experimental and modeled current gains as a function of the effective gate length are shown in Fig. 8. They increase when the gate length decreases due to a lower mobility for long channel MCFETs (Fig. 13). The expected gain of 5 (5 channels) is exceeded for transistors with small W thanks to the GAA configuration. Those gains are then plotted as a function of the drain voltage (Fig. 9). The gain is much more complex than the simple geometrical gain and depends on several intrinsic parameters, in particular: the drain saturation voltages (V_{Dsat}), specific transport properties and series resistances sharing between the 3D stacked channels [3]. The gain increases with the drain voltage and can be higher than the expected GAA MCFET gain. This is explained quantitatively in our model by a higher V_{Dsat} in 3D MCFETs than in planar FD-SOI devices (inset Fig. 9). Excellent short channel effects are obtained when the channel width W is taken equal to the channel length L (i.e. GAA configuration) (Fig. 10). Sub-threshold Slopes (SS) and Drain-Induced Barrier Lowering (DIBL) values are below 66mV/dec and 21mV/V for 50nm n- and p-MCFETs. The bottom FD-SOI transistor (channel 5) does not degrade the sub-threshold characteristics despite its single gate configuration when the channel width is appropriately scaled (W=L). Moreover, as expected with 5 channels, the drain current is decreased by almost 21% when channel 5 is deactivated (by applying substrate voltage, V_B=-30V) (Fig. 11). The higher mobility obtained for channel 5 compared to the other channels (Fig. 12) is partly compensated by higher series resistances.

To analyse further the gain dependence with gate length, mobility was extracted by split-CV measurements in various gate lengths transistors (Fig. 13) [6]. Mobilities in short channel n-MCFETs are similar to those in optimized planar FD-SOI references (Fig. 13 and 14). However, long channel MCFETs exhibit lower mobilities than in planar FD-SOI devices. To de-correlate the impact of electron scattering (possible interface degradation) and the subbands repopulations (stress, confinement), injection velocities (V_{inj}) and Ballisticity Rates (BR) of various gate length n-MCFET were extracted using the PB method described in [7] (Fig. 15 and 16). V_{inj} values are close to the FD-SOI reference ones, evidencing a similar confinement. The lower BR with increasing L, consistent with the mobility results in Fig. 13, is attributed to a degraded silicon channel surface due to tunnel over-etch in long cavities. However, the BR values are among the best results reported in the literature and indicate no additional scattering in short MCFETs.

Conclusions

We reported ultra low I_{OFF} and high I_{ON} currents in 3D LSTP MCFETs with TiN/HfO_2 gate stacks on SOI. Thanks to optimized S/D, CET and to the GAA configuration, the current gain compared to a planar FD-SOI reference exceeds the expected value of 5. The obtained electrical characteristics, in particular ultra low I_{OFF} currents (16.5pA/µm), make this device very promising for further Low STandby Power CMOS applications.

Acknowledgements

This work has been carried out in the frame of the CEA-LETI / ALLIANCE (ST, NXP and Freescale) collaboration.

References

[1] M. Kim et al., VLSI Tech., pp. 68-69, 2006.
[2] E. Bernard et al., Proc. ESSDERC, pp. 147-150, 2007.
[3] E. Bernard et al., IEEE SOI Conference, pp. 93-94, 2007.
[4] S. Monfray et al., IEDM Tech. Dig., pp. 645-648, 2001.
[5] F. Andrieu et al., IEDM Tech Dig., pp. 641-644, 2006.
[6] K. Romanjek et al., Electron Device Letter, Vol. 25, pp. 583-585, 2004.
[7] V. Barral et al., VLSI Tech., pp. 128-129, 2007.
[8] S.Y. Lee et al., IEEE ICICDT Conference, 2006.

978-1-4244-1802-2/08/$25.00 ©2008 IEEE

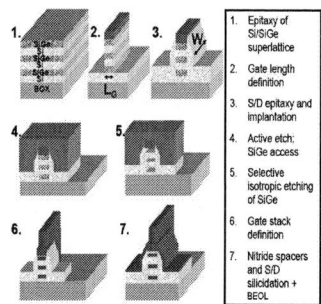

Fig. 1: Main process steps of the MCFET integration.

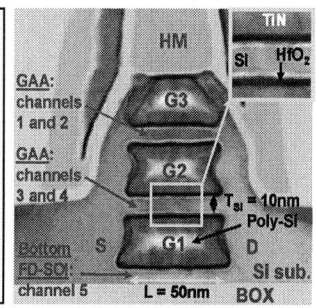

Fig. 2: Cross-section TEM picture of a 50nm gate length n-MCFET transistor. Inset: High-Resolution TEM picture showing a good gate stack conformity.

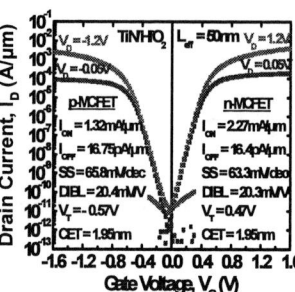

Fig. 3: I_D-V_G characteristics of 50nm gate length n- and p-MCFETs on SOI with TiN/HfO$_2$ gate stack. Ultra low I_{OFF} currents are achieved.

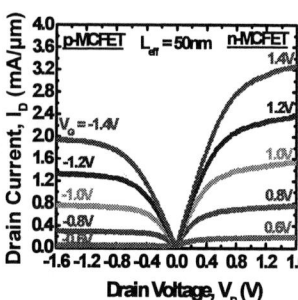

Fig. 4: I_D-V_D characteristics of 50nm gate length n- and p-MCFETs on SOI with TiN/HfO$_2$ gate stack.

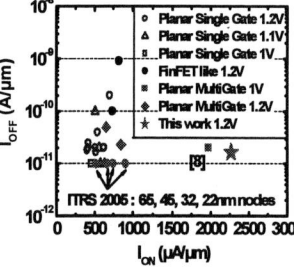

Fig. 5: NMOS LSTP benchmark. Each (unreferenced) point corresponds to an article in the last 5 years VLSI, IEDM and SSDM conferences. The currents are normalized by the top view W.

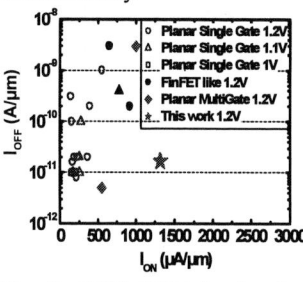

Fig. 6: PMOS LSTP benchmark. Each point corresponds to an article in the last 5 years VLSI, IEDM and SSDM conferences. The currents are normalized by the top view W.

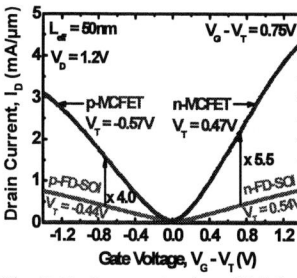

Fig. 7: Drain current gains of 5.5 for n-MOS and 4.0 for p-MOS are obtained @ V_G–V_T = 0.75V and V_D=1.2V for 50nm gate length devices.

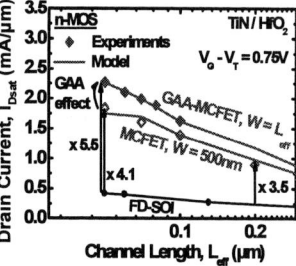

Fig. 8: Experimental and modeled drain currents vs. gate length for n-MCFET compared to a planar n-FD-SOI reference. The GAA configuration enhances the current gain due to lateral conduction.

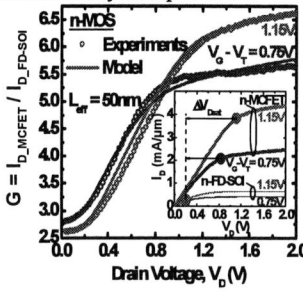

Fig. 9: Drain currents gains for n-MCFET compared to a planar n-FD-SOI reference for various V_G–V_T and analytical modeling. Inset: I_D-V_D of the two architectures showing the V_{Dsat} differences.

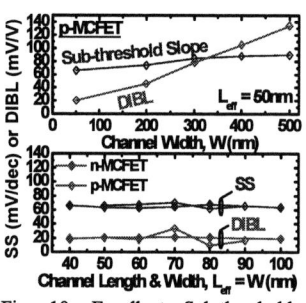

Fig. 10: Excellent Sub-threshold Slope and DIBL values are obtained down to 40nm gate length for n- and p-MCFETs. The channel width scaling has a strong impact on short channel effects control.

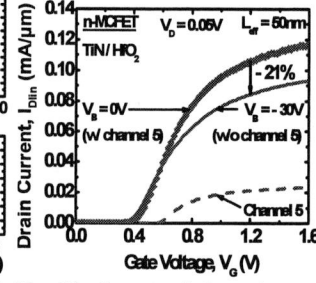

Fig. 11: Impact of the substrate voltage polarisation on the drain current for a 50nm gate length n-MCFET. Bottom FD-SOI transistor (channel 5) is killed @ V_B= -30V. The drain current is enhanced by 21% when channel 5 is activated.

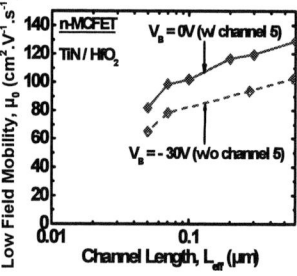

Fig. 12: Comparison of the extracted low field mobility μ_0 on SOI for MCFET when channel 5 is turned ON and OFF. Its low field mobility is higher than in other channels.

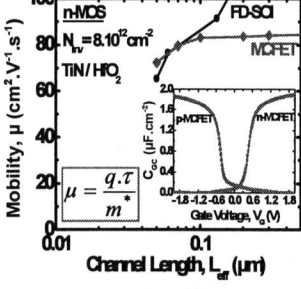

Fig. 13: Extracted mobility on SOI (N_{inv}= 8.10^{12}cm^{-2}) for MCFET and FD-SOI architectures. Values are similar for short channel devices (sub-100nm) but are degraded in long channel MCFETs. Inset: C-V measurements for CET and Q_{inv} extraction. (In the equation, m* is the effective conduction mass.)

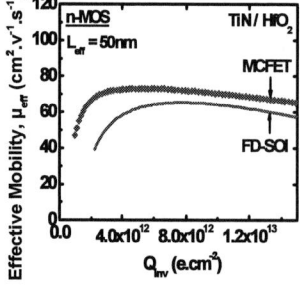

Fig. 14: Electron effective mobility for 50nm MCFET and FD-SOI architectures. A slightly higher mobility is obtained in the 50nm n-MCFET @ high effective field.

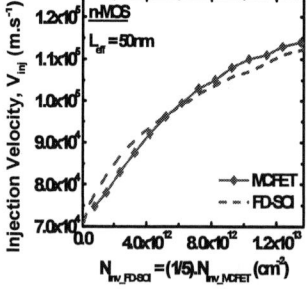

Fig. 15: V_{inj} vs. N_{inv} for n-MCFET and FD-SOI reference. The injection velocity accounts for the subband populations in the Si film (confinement). The values are similar for the two architectures.

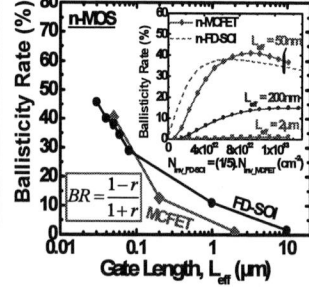

Fig. 16: Ballisticity rate vs. gate length for n-MCFET and FD-SOI reference (N_{inv_FD-SOI} = N_{inv_MCFET}/5 = 8.10^{12}cm^{-2}). Very good ballisticity rates are obtained. Inset: Ballisticity Rate vs. N_{inv}. (In the equation, r is the backscattering coefficient.)

14

Three-Dimensional Stress Engineering in FinFETs for Mobility/On-Current Enhancement and Gate Current Reduction

Masumi Saitoh, Akio Kaneko[1], Kimitoshi Okano[2], Tomoko Kinoshita[3],
Satoshi Inaba[2], Yoshiaki Toyoshima[2] and Ken Uchida

Advanced LSI Technology Laboratory, Corporate R&D Center, [1]Process & Manufacturing Center, [2]Center for Semiconductor Research & Development, [3]System LSI Division, Semiconductor Company, Toshiba Corporation, 8 Shinsugita-cho, Isogo-ku, Yokohama 235-8522, Japan, Tel: +81-45-776-5959, Fax: +81-45-776-4113, E-mail: masumi.saitoh@toshiba.co.jp

Abstract

In this paper, the first systematic study of uniaxial stress effects on mobility (μ)/on-current (I_{on}) enhancement and gate current (I_g) reduction in FinFETs is described. We demonstrate for the first time that I_g of (110) side-surface pFinFETs is largely reduced by longitudinal compressive stress due to out-of-plane mass increase. (110) n/pFinFETs are superior to (100) FinFETs in terms of higher μ/I_{on} enhancement ratio by longitudinal strain and comparable/ higher short-channel I_{dsat}. Three-dimensional stress design in FinFETs including transverse and vertical stresses is proposed based on the understanding of stress effects beyond bulk piezoresistance.

1. Introduction

FinFET is one of the most promising device structures for 32 nm node and beyond. To enhance the current drive of FinFETs, application of the same local strain techniques as planar FETs have been reported [1,2]. However, accurate evaluation of strain effects on μ, I_{on} and I_g in FinFETs beyond the framework of bulk piezoresistance has not been performed. Also, stress design for three-dimensional FinFET structures has not been explored.

In this work, μ/I_{on} enhancement and I_g reduction in FinFETs by uniaxial stress from all the directions are systematically investigated. (110) and (100) side-surface FinFETs are compared from the viewpoints of stress effects and short-channel I_{dsat}.

2. Stress-Induced Gate Current Reduction in (110) FinFETs

In this study, SOI FinFETs made with sidewall pattern transfer process [3] are used. Side surfaces of FinFETs are (110) or (100) plane. Fin width ranges from 15 nm to 60 nm, and fin height is 50 nm. In-plane uniaxial stresses (longitudinal (long.) or transverse (trans.), tensile (tens.) or compressive (comp.)) are mechanically applied to FinFETs using the bending apparatus.

I_g modulation by strain in (100)-plane FETs has been reported [4], but there has been no report on (110) FETs. In this work, I_g change in (110) side-surface FinFETs and (110) planar FETs by long. uniaxial stresses is examined for the first time. Fig. 1 shows I_g-V_g of (110) FinFETs with 1.3-nm-thick SiON. While I_g of nFET does not change by tensile strain, I_g of pFET is reduced by compressive strain, which also enhances μ. I_g modulation by strain in (100) and (110) planar FETs is systematically shown in Fig. 2. Largest I_g reduction by comp. strain is observed in (110) pFETs. From linear extrapolation, one-order reduction of I_g by 1.3% comp. strain is expected.

To clarify the origin of I_g reduction in (110) pFETs, the change of out-of-plane mass m_z by strain is discussed. Fig. 3 shows V_{th} reduction by strain. While ΔV_{th} in (100) pFETs is independent of N_{sub}, ΔV_{th} in (110) pFETs increases with N_{sub}. Large V_{th} reduction under strong carrier confinement indicates the lowering of ground subband energy by strain. This is the first evidence of m_z increase by comp. strain in (110) pFETs, which was only predicted by theory [4].

I_g can be expressed as the sum of $qNTf$ for each subband (N: carrier density, T: transmission probability, f: impact frequency on the interface)[5]. m_z increase by strain leads to subband energy lowering and reduction in both T and f, and thus I_g is reduced. Fig. 4 shows $\Delta I_g/I_g$ calculated assuming m_z increase extracted from ΔV_{th} in Fig. 3. Agreement of experiment and calculation indicates large I_g reduction in (110) pFETs actually arises from m_z increase by strain.

3. Mobility Enhancement by Strain in FinFETs

μ of FinFETs with 10 fin channels (SOI doping (N_{SOI}): 1e15 cm[-3]) is measured with split-CV technique. Fig. 5 shows measured μ-E_{eff} in (110) and (100) n/pFETs. Although the same trend as bulk planar FETs is obtained as a whole, (110) pFinFETs show relatively low μ at low E_{eff} due to Coulomb scattering by interface states.

A. Longitudinal Stress Effects ~ Comparison with Planar FETs ~

Fig. 6 shows μ enhancement by longitudinal tensile strain ($\Delta\mu/\mu$) as a function of E_{eff} in planar nFETs and nFinFETs. $\Delta\mu/\mu$ of FinFETs is comparable to planar FETs, indicating the same long. stress as planar FETs is effective also in nFinFETs. Also, $\Delta\mu/\mu$ of (110) FETs increases with E_{eff} and surpasses (100) FETs at high E_{eff}. In (110) nFETs, conductivity mass m_c of 2-fold valleys is reduced by <110> tensile strain. Since occupancy of 2-fold valleys increases with E_{eff} due to band non-parabolicity [6], $\Delta\mu/\mu$ also increases with E_{eff}. As a result of high $\Delta\mu/\mu$ in high E_{eff}, $\Delta\mu/\mu$ further increases in highly-doped (110) FinFETs required for aggressive scaling (Fig. 7).

Fig. 8 shows $\Delta\mu/\mu$ by long. comp. strain in pFETs. $\Delta\mu/\mu$ of (100) pFinFETs is negligibly low and $\Delta\mu/\mu$ of (110) pFinFETs is lower than planar FETs at low E_{eff}. Since $\Delta\mu/\mu$ of (110) pFETs stems from not only m_c reduction, but also subband splitting [7], Coulomb scattering due to interface states (elastic scattering) degrades $\Delta\mu/\mu$ of (110) pFinFETs. Therefore, improvement of interface quality in (110) pFinFETs is essential for both higher μ and higher $\Delta\mu/\mu$.

B. Three-Dimensional Stress Engineering in FinFETs

Fig. 9 shows $\Delta\mu/\mu$ by transverse strain (fin width direction) in FinFETs. Transverse strain is also effective for μ enhancement, though little attention has been paid to the width direction so far. Trans. tensile stress increases μ of both (110) nFinFET and pFinFET, but comp. stress should be applied for (100) nFinFET. Fig. 10 shows $\Delta\mu/\mu$ by transverse strain in planar FETs. This result can be regarded as $\Delta\mu/\mu$ by vertical strain (fin height direction) in FinFETs. Vertical compressive stress is most effective among all the stress directions for both (110) nFinFET and (100) pFinFET.

Table I lists piezoresistance π measured with FinFETs and bulk π. π of (110) nFETs surpasses bulk values in all the directions. This results from high $\Delta\mu/\mu$ under strong carrier confinement in (110) nFETs. Trans. comp. stress and vertical comp. stress are highly effective for (100) nFinFET and pFinFET, respectively.

In addition to extrinsic stress effects, intrinsic stress inside fin channel is analyzed by nano-beam diffraction (NBD). Fig. 11 shows measured transverse strain at each position. Although there exist no intentional stressors, 0.3% compressive strain is induced along the width direction for gate dopants of both P and B. Vertical strain is negligibly small. One possible origin of trans. comp. strain is volume expansion of surrounding poly-Si gate [8]. In (100) nFETs, trans. comp. stress is effective for μ increase (Table I). Hence, boost-up of trans. stress (i.e. by highly-doped gate) is desirable. Optimum stress design in (100) and (110) n/pFinFETs is summarized in Fig. 12.

4. Comparison of Stressed Short-L_g (110) and (100) FinFETs

Fig. 13 shows I_{dsat}-L_{eff} in (110) and (100) nFinFETs. Large I_{dsat} difference at long L_{eff} caused by μ difference disappears at short L_{eff}, possibly due to strong velocity saturation in nFETs [9]. I_{dsat} increase by strain in 50-nm-long nFinFETs is shown in Fig. 14. $\Delta I_{dsat}/I_{dsat}$ of (110) FETs exceeds (100) FETs at high E_{eff} in the same way as $\Delta\mu/\mu$ in Fig. 6, while I_{dsat} is comparable in both devices. Fig. 15 shows I_{dsat}-L_{eff} in pFETs. I_{dsat} of short-L_{eff} (110) pFETs is still slightly higher than (100) pFETs. Since $\Delta\mu/\mu$ in (100) pFETs by long. strain is much smaller than (110) pFETs (Fig. 8), (110) n/pFinFETs are superior to (100) FETs in terms of both I_{dsat} and $\Delta I_{dsat}/I_{dsat}$.

5. Conclusions

The effects of uniaxial stress on μ/I_{dsat} increase and I_g reduction in FinFETs are systematically studied. (110) FinFETs have the advantages such as I_g reduction by comp. stress (pFETs), higher $\Delta\mu/\mu$ by long. stress, and comparable/higher I_{dsat} over (100) FinFETs. 3D stress engineering including transverse and vertical stresses is crucial for realizing ultimately-high performance FinFETs.

978-1-4244-1802-2/08/$25.00 ©2008 IEEE

References [1] N. Collaert et al., EDL **26**, 820 (2005). [2] J. Kavalieros et al., VLSI2006, p.62. [3] A. Kaneko et al., IEDM2005, p.863. [4] S. E. Thompson et al., ED **53**, 1010 (2006). [5] K.-N. Yang et al., ED **47**, 2161 (2000). [6] K. Uchida et al., IEDM2006, p.711. [7] M. Saitoh et al., IEDM2007, p.1019. [8] Y. Mishima et al., JJAP **46**, 943 (2007). [9] H. R. Harris et al., IEDM2007, p.57.

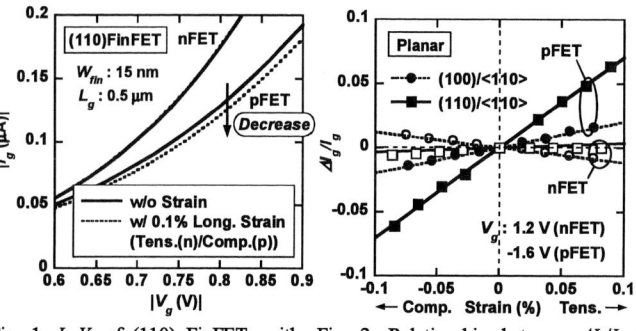
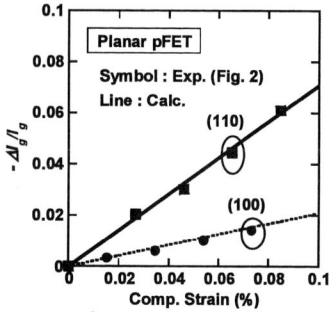

Fig. 1. I_g-V_g of (110) FinFETs with and without strain. Gate insulator is 1.3-nm-thick SiON. While I_g of nFET does not change by tens. strain, I_g of pFET is reduced by comp. strain.

Fig. 2. Relationship between $\Delta I_g/I_g$ and long. strain in planar (100) and (110) n/pFETs (L_g : 1 μm). The largest I_g modulation (7% reduction for 0.1% strain) is obtained in (110) pFETs.

Fig. 3. Measured V_{th} reduction (ΔV_{th}) by long. comp. strain in low-N_{sub} and high-N_{sub} planar pFETs. ΔV_{th} in (110) pFETs increases with N_{sub}, while ΔV_{th} in (100) pFETs is constant.

Fig. 4. $\Delta I_g/I_g$ by comp. strain in pFETs calculated assuming m_z increase by strain extracted from ΔV_{th} in Fig. 3. Calculation agrees well with the experiment.

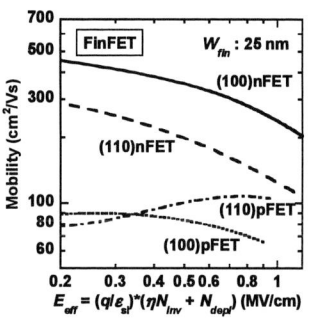
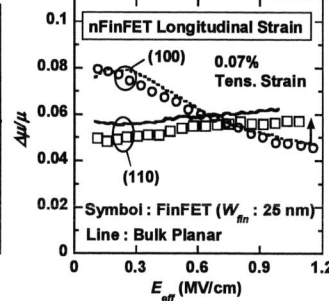

Fig. 5. Measured μ-E_{eff} in (110) and (100) n/pFinFETs (poly/SiO2(2.5 nm), L_g : 10 μm). The same trend as bulk planar FETs is obtained as a whole.

Fig. 6. $\Delta\mu/\mu$ by long. tens. strain in nFinFETs and planar nFETs. Strain estimation in FinFETs is corrected considering the effect of BOX.

Fig. 7. $\Delta\mu/\mu$ by long. tens. strain in nFinFETs with high and low N_{SOI}. $\Delta\mu/\mu$ of high-N_{SOI} (110)/(100) FinFETs is higher/lower than low-N_{SOI} FETs.

Fig. 8. $\Delta\mu/\mu$ by long. comp. strain in pFinFETs and planar pFETs. $\Delta\mu/\mu$ of (100) pFinFETs is negligibly low in the whole E_{eff} range.

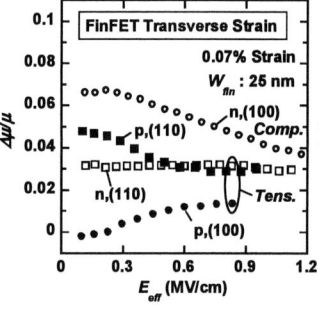
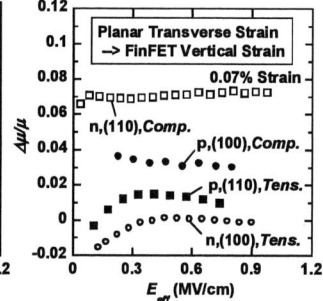
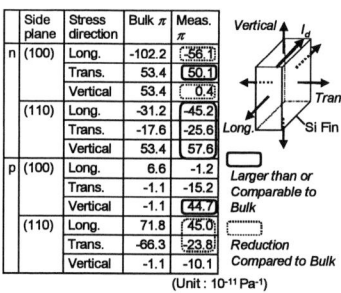

Side plane	Stress direction	Bulk π	Meas. π
n (100)	Long.	-102.2	-56.1
	Trans.	53.4	50.1
	Vertical	53.4	0.4
(110)	Long.	-31.2	-45.2
	Trans.	-17.6	-25.6
	Vertical	53.4	57.6
p (100)	Long.	6.6	-1.2
	Trans.	-1.1	-15.2
	Vertical	-1.1	44.7
(110)	Long.	71.8	45.0
	Trans.	-66.3	-23.8
	Vertical	-1.1	-10.1

(Unit: 10^{-11} Pa^{-1})

Fig. 9. $\Delta\mu/\mu$ by transverse strain in FinFETs. Transverse tensile stress increases μ of both (110) nFinFET and pFinFET, but compressive stress should be applied for (100) nFinFET.

Fig. 10. $\Delta\mu/\mu$ by transverse strain in planar FETs. This can be regarded as $\Delta\mu/\mu$ by vertical strain in FinFETs. Vertical comp. stress is effective in (110) nFinFET and (100) pFinFET.

Table I. π measured with FinFETs (E_{eff}: 0.8 MV/cm) and bulk π in each stress direction. π for (110) nFinFETs surpass bulk values in all the directions. Inset figure explains stress directions.

Fig. 11. Cross-sectional TEM images of 15-nm-wide and 200-nm-wide fins (Top) and trans. comp. strain at each position (indicated by white circles) measured with NBD (Bottom).

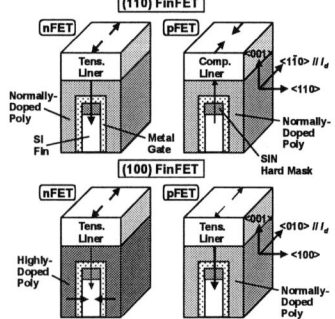
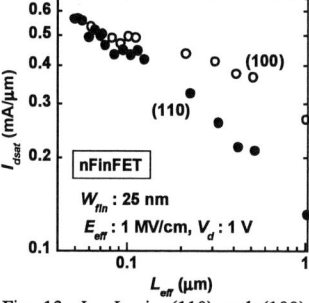
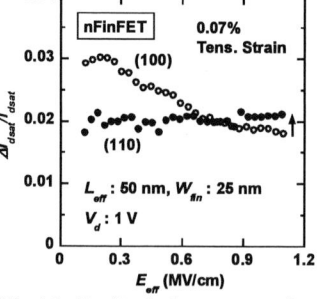
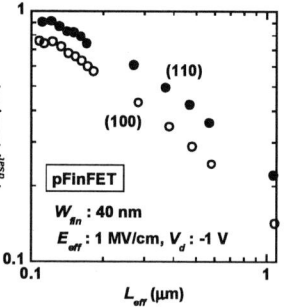

Fig. 12. Proposed 3D stress design in (110) and (100) n/pFinFETs. Trans. comp. stress by highly-doped poly is desirable in (100) nFinFETs.

Fig. 13. I_{dsat}-L_{eff} in (110) and (100) nFinFETs. Large I_{dsat} difference between (110) and (100) nFETs at long L_{eff} disappears at short L_{eff}.

Fig. 14. $\Delta I_{dsat}/I_{dsat}$ by long. tens. strain in 50-nm-long nFinFETs. $\Delta I_{dsat}/I_{dsat}$ of (110) FETs exceeds (100) FETs at high E_{eff} in the same way as $\Delta\mu/\mu$.

Fig. 15. I_{dsat}-L_{eff} in (110) and (100) pFinFETs. I_{dsat} of short-L_{eff} (110) pFETs is still slightly higher than (100) pFETs.

Experimental Study of Single Source-Heterojunction MOS Transistors (SHOTs) under Quasi-Ballistic Transport

T. Mizuno[1,3], Y. Moriyama[2], T. Tezuka[2], N. Sugiyama[2], and S. Takagi[1,4]

[1]MIRAI-AIST, [2]MIRAI-ASET, 1 Komukai Toshiba-cho, Saiwai-ku, Kawasaki, Japan 212-8582
[3]Kanagawa University, 2946, Tsuchiya, Hiratsuka 259-1293, Japan   (mizuno@info.kanagawa-u.ac.jp)
[4]The University of Tokyo, 7-3-1 Hongo, Bunkyo-ku, Tokyo 113-8656, Japan

I. Introduction

In order to breakthrough the carrier velocity saturation limited by thermal velocity or Fermi velocity at the source edge in quasi-ballistic transistors [1,2], we have proposed and experimentally demonstrated high velocity electron injection into the channel by the excess kinetic energy corresponding to the conduction band offset ΔE_C at the source heterojunction, called the source heterojunction MOS transistors (SHOTs) [3]. In an ideal case of SHOTs, the injected electron velocity v can be increased by the ΔE_C, that is, $v=(2\Delta E_C/m^*)^{1/2}$, where m^* is the effective mass of inversion electrons. In SHOTs, it is necessary to optimize the lateral position of the source heterojunction near the source pn junction [4]. On the other hand, it has been reported [5] that the scattering of carriers even near the drain regions also causes the reduction of drain currents in quasi-ballistic transistors. Therefore, the energy spike at the drain heterojunction should be reduced to realize high performance SHOTs.

In this study, we have experimentally demonstrated the high performance of the single source heterojunction MOSFETs (SHOTs) without the energy spike of the drain heterojuction. We have studied the influence of the drain energy spike on the MOSFET performance. Moreover, we have clarified the physical mechanism for the high velocity electron injection in SHOTs, through the lattice temperature dependence of the electron velocity, comparing with that of strained-SOIs (SSOIs).

II. Design Concept and Experimental for SHOTs

Our simulation results indicated that the ΔE_C value, the source heterojunction position of $(Y_H - Y_J)$, and the graded-heterojunction structures are key factors to enhance the drain current drive of SHOTs shown in Fig.1 (a) over a wide range of drain bias [4]. In this work, we have realized the single source heterojunction by using oblique n^+ ion implantation for the source/drain regions. After pattering the CVD-SiO$_2$/poly-Si gate electrodes, Fig.2 (a) shows that Ge ion implantation into strained-Si layers was used to form relaxed-SiGe source layers on strained-SOI MOS structures [3]. Next, as shown in Fig.2(b), oblique As$^+$ ions with an incident angle θ (5°≤θ≤9°) were implanted to form the single source heterojunction outside the source n^+ region by controlling the position of the source Y_J and the drain heterojunction inside the drain n^+ region. The Ge content of SGOI substrates, the gate oxide thickness, the SGOI thickness, the strained-Si thickness were taken to be 28%, 5nm, 50nm, and 15nm, respectively. The minimum channel length was 100nm. As a reference, we have also fabricated SSOIs by using the same processes.

As shown in Fig.1(b), by exchanging the source and the drain electrodes of SHOTs, the single drain heterojunction MOSFETs (DHOTs) with no source band-offset and with the drain energy spike can be realized. Using the DHOTs, we have also studied the influence of the drain energy spike on the electron transport.

In this study, we have evaluated the electron velocity v from the intrinsic transconductance G_{MI}, that is, $v= G_{MI}/WC_{OX}$, where W and C_{OX} are the channel width and the gate capacitance, respectively. In addition, $G_{MI}=G_M/(1-G_MR_P/2)$, where G_M is a measured transconductance, and R_P is the parasitic resistance of both the source and the drain regions which was measured by a test structure. Taking the effects of R_P into account, effective gate voltage V_{GEFF} and drain voltages V_{DEFF} were used in this study. Namely, $V_{GEFF}= V_G-I_DR_P/2$ and $V_{DEFF} = V_D-I_DR_P$, where I_D is the drain current.

III. Results and Discussions

A. Optimization of SHOT Structures

In order to determine the Y_J as a function of θ, we measured both the gate to source capacitance C_{GS} and the gate to drain capacitance C_{GD}. Fig.3 shows the capacitance data at θ of 9°. The fact that the minimum C_{GS} (C_{GS}^{min}) is much lower than the minimum C_{GD} (C_{GD}^{min}), is the direct proof of shorter Y_J in the source n^+ region than in the drain region. As a result, SHOTs with shorter Y_J exhibit higher V_{TH}, because the source heterojunction exists outside the source n^+ region. ΔE_C is roughly estimated from the V_{TH} shift value to be 75meV. Using the C_{GS}^{min} value, the Y_J value can be evaluated from $Y_J=C_{GS}^{min}/WC_{OX}$. Fig.4 shows the source Ge profile evaluated by

EDX and the source Y_J values of SHOTs in 7°≤θ≤9° and SSOIs, resulting in Y_H of about 65nm. When θ=9°, the Y_J at the source side is longer than Y_H. However, the source Y_J value increases with decreasing θ, and so $(Y_H-Y_J)<0$, when θ≤8°. The drain Y_J is almost the same as the source Y_J of SSOIs.

Fig.5 shows the G_{MI} characteristics of SHOTs with the source (Y_H-Y_J) of 16nm, DHOTs with the drain (Y_H-Y_J) of 16nm, and SSOIs at lattice temperatures T of 300K and 25K. The G_{MI} of SHOTs is enhanced at both temperatures, while the G_{MI} of DHOTs is almost the same as that of SSOIs, irrespective of T.

Fig.6 shows the v at room temperature obtained by the maximum G_{MI} of SHOTs, DHOTs, and SSOIs as a function of (Y_H-Y_J). The v values of SHOTs are enhanced at larger (Y_H-Y_J), against those of both DHOTs and SSOIs, but rapidly decreases, when the (Y_H-Y_J) decreases, that is, the position of the source heterojunction moves toward the higher concentration region of the n^+ region. This is due to the reduction of the velocity of electron injection into the channels [4]. The v values of DHOTs are almost the same as the v of SSOIs.

Fig.7 shows the V_{DEFF} dependence of v of SHOTs and SSOIs. In the optimized source (Y_H-Y_J) value of 16nm, v in $V_{DEFF}>0.6V$ is enhanced against SSOIs. However, the SHOT with (Y_H-Y_J) of only 1nm exhibits the drastic v reduction in $V_{DEFF}<1V$, which is due to higher barrier height at ΔE_C edge under lower V_{DEFF} conditions [3]. Consequently, by optimizing the position of the source heterojunction of (Y_H-Y_J), the v of SHOTs can be enhanced over a wide range of V_{DEFF}.

B. Temperature Dependence of Injected Electron Velocity

Fig.8 shows the T dependence of v of SHOTs, DHOTs, and SSOIs. The v values of SHOTs keep increasing with decreasing T, and the v enhancement in SHOTs is observed in a whole range of T. On the other hand, DHOTs have almost the same v values as those of SSOIs. Therefore, in long L_{EFF} of 100nm, the drain energy spike does not affect the electron transport in the whole range of T. Fig.9 also shows the V_{DEFF} dependence of v of SHOTs and SSOIs at various T. When $V_{DEFF}>0.6V$, the V_{DEFF} dependence of v becomes weaker in lower T.

The v enhancement factors of SHOTs against DHOTs and SSOIs, as shown in Fig.10, gradually increase with decreasing T in proportion to $T^{-0.02}$. On the contrary, the simulation results without considering tunneling effects at the source heterojunction show that the v enhancement of SHOTs rapidly decreases with decreasing T. This suggests that the electron injection into the channels is caused mainly by the tunneling effects, not by the thermal activation process at the source heterojunction.

The relationship between the v values and low field effective electron mobility μ_{EFF} at various T is examined to evaluate the increase in the electron injection velocity due to ΔE_C. Fig.11 shows that the values of v in all the devices have a power law of the μ_{EFF} ($v \propto \mu_{EFF}^{0.34}$), indicating that the v increase can be explained by the phonon scattering reduction effects of electrons in lower T. The v gap between SHOTs and other devices shown as the double arrow is attributable to the excess kinetic energy ΔE_K in SHOTs introduced by ΔE_C. Here, $\Delta E_K \equiv 1/2 \cdot m^*(v_{SH}^2-v_{SS}^2)$, where v_{SH} and v_{SS} are electron velocities of SHOTs and SSOIs, respectively. Using $v \propto \mu_{EFF}^\alpha$ in various V_{DEFF} (the power coefficient α is a function of V_{DEFF}), we can obtain the enhancement factor of the velocity injection efficiency $\Delta E_K/\Delta E_C$. Fig.12 shows the $\Delta E_K/\Delta E_C$ vs. μ_{EFF}. It is noted that $\Delta E_K/\Delta E_C$ increases with increasing μ_{EFF} by reducing the scattering rate of electrons in lower T. Moreover, $\Delta E_K/\Delta E_C$ at various V_{DEFF} becomes the same higher value at μ_{EFF} of around 3000cm^2/Vs, although the maximum $\Delta E_K/\Delta E_C$ value is only 2% in L_{EFF} of 100nm. The $\Delta E_K/\Delta E_C$ value is expected to be much increased under quasi-ballistic transport, resulting in the realization of much higher injection velocity in SHOTs.

IV. Summary

We have experimentally studied the single source heterojunction MOSFETs (SHOTs) without the drain energy spike by optimizing the source heterostructures. The electron velocity enhancement of SHOTs against DHOTs and SSOIs slightly increases with decreasing the lattice temperatures. It has been experimentally demonstrated that the enhancement of the velocity injection efficiency associated with ΔE_C becomes higher by reducing the scattering of electrons in lower temperature. Consequently, much higher carrier velocity is expected to be realized in SHOTs under quasi-ballistic transport, where the scattering rate of carriers is significantly suppressed.

Acknowledgements: We would like to thank Dr. M.Hirose and Dr. T.Kanayama for continuous supports. This work was supported by NEDO.

References: [1]K.Natori, J.Appl.Phys., **76**, 4879 (1994). [2]S.Takagi, VLSI Symp., p.115 (2003). [3]T.Mizuno et al., IEEE Trans., **ED-52**, 2690 (2005). [4]T.Mizuno et al., IEEE Trans., **ED-54**, 2598 (2007). [5] H.Tsuchiya et al., SSDM, p.350 (2006).

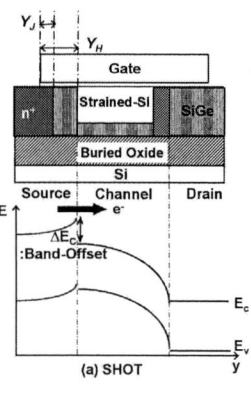

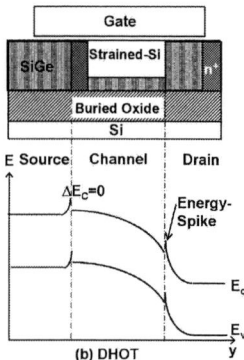

Fig.1 Schematic cross section of (a) source-(SHOT) and drain-heterojunction MOSFETs (DHOT). Y_H and Y_J mean the lateral length of SiGe layer and n^+ region, respectively.

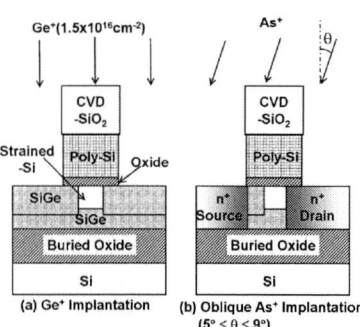

Fig.2 Process steps for fabricating the single source-heterojunction transistors. (a) Ge ion implantation process for forming the source SiGe layers. (b) Inclined As$^+$ ion implantation for asymmetrical source/drain n^+ regions, where $5° \leq \theta \leq 9°$. The source Y_J is expected to be changed roughly by about 29nm ($\equiv T_G \cdot (\tan 9° - \tan 5°)$), where T_G is a total thickness of the gate electrode.

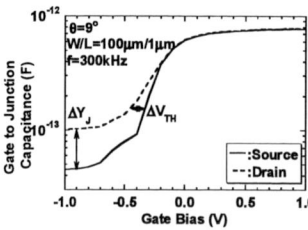

Fig.3 Gate to source (solid line) and drain capacitance (dashed line), where $\theta=9°$, $L=1\mu m$, and $W=100\mu m$.

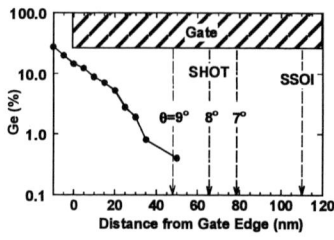

Fig.4 Surface Ge atom profile from the source to the channel regions evaluated by EDX analysis and the broken lines show the location of the lateral n^+ pn junction of SHOTs and SSOIs evaluated by the C_{GS}.

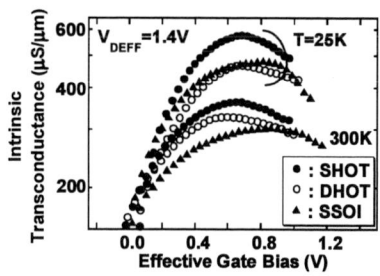

Fig.5 Intrinsic transconductance characteristics at 300K and 25K of SHOT with ($Y_H - Y_J$) of 49nm (closed circles), DHOT (open circles), and SSOIs (triangles), where $L_{EFF}=100nm$ and $V_{DEFF}=1.4V$.

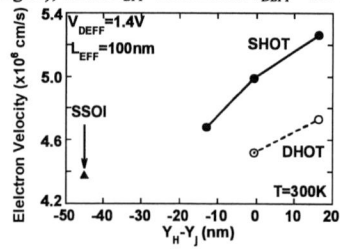

Fig.6 Electron velocity evaluated by the maximum G_{MI} of SHOT (closed circles), DHOT (open circles), and SSOIs (triangle) as a function of ($Y_H - Y_J$), where $T=300K$, $L_{EFF}=100nm$ and $V_{DEFF}=1.4V$.

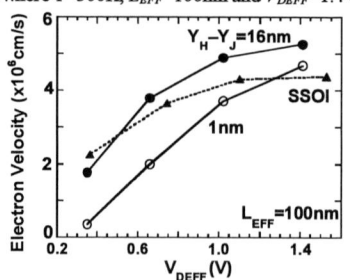

Fig.7 V_{DEFF} dependence of electron velocity obtained by the maximum G_{MI} of SHOT with (Y_H-Y_J) of 16nm (closed circles) and 1nm (open circles), and SSOI (triangles), where $T=300K$ and $L_{EFF}=100nm$.

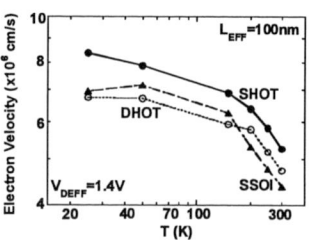

Fig.8 Electron velocity of SHOT with Y_J =49nm (closed circles), DHOT (open circles), and SSOIs (triangle), where, $L_{EFF}=100nm$ and $V_{DEFF}=1.4V$.

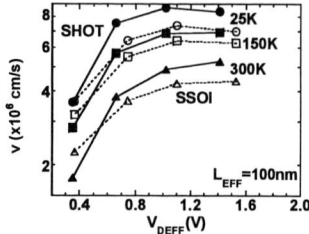

Fig.9 Electron velocity of SHOTs (closed characters) and SSOIs (open characters) vs. V_{DEFF} in various temperatures. Circles, squares, and triangles show the data at T of 25K, 150K, and 300K, respectively.

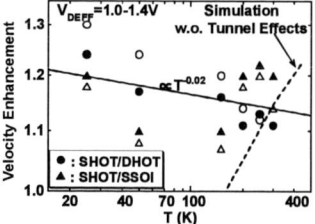

Fig.10 Temperature dependence of electron velocity enhancement of SHOTs against DHOTs (circles) and SSOIs (triangles). Closed and open characters show the data at V_{DEFF} of 1.0 and 1.4V, respectively. The dashed line indicates the simulated velocity enhancement of SHOTs against SSOIs without considering the tunneling effects at the ΔE_C edge.

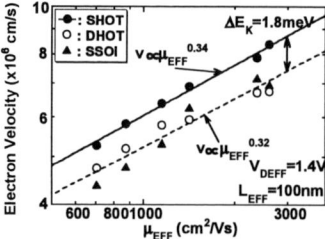

Fig.11 Electron velocity of SHOTs (closed circles), DHOTs (open circles), and SSOIs (triangles) vs. effective electron mobility at various temperatures, where $V_{DEFF}=1.4V$. Bold arrow shows ΔE_K ($\equiv m^* \cdot \Delta v^2/2$).

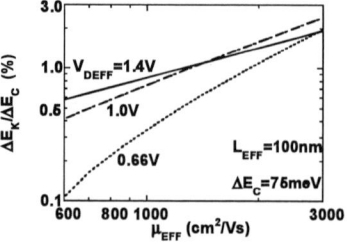

Fig.12 $\Delta E_K/\Delta E_C$ vs. μ_{EFF} for various V_{DEFF}.

18

Advanced DSS MOSFET Technology for Ultrahigh Performance Applications

M. Awano, H. Onoda, K. Miyashita, K. Adachi*, Y. Kawase**, K. Miyano**, H. Yoshimura and T. Nakayama

System LSI Division, *Center for Semiconductor R&D, **Process and Manufacturing Engineering Center,
Semiconductor Company, Toshiba Corporation, 8, Shinsugita-Cho, Isogo-Ku, Yokohama 235-8522, Japan
Phone: +81-45-776-5803, Fax: +81-45-776-4111, E-mail: awano@semicon.toshiba.co.jp

Abstract

Dopant segregated Schottky MOSFET (DSS FET) is one of the key technologies which can improve the MOSFET performance thanks to reduction of external resistance and increase of carrier injection velocity. We have found that both laser spike annealing (LSA) and fluorine co-implant can reduce external resistance furthermore, which leads to boost drive currents of DSS FETs by 7% respectively. By optimizing these technologies, high drive currents of 1310 μA/μm and 1080 μA/μm at Ioff of 100 nA/μm are achieved at 1.0V and 0.9V respectively, without use of high-k/metal gate.

Introduction

For high performance MOSFETs at 45nm node and beyond, metal gate and high-k gate dielectrics are booked to be utilized since they enable to thin gate dielectric thickness without increasing gate leakage current. Excessive reduction of gate dielectric thickness, however, causes gate capacitance increase, which will degrade the performance in some circuits. Therefore, on-current increase without gate capacitance increase is actually demanded. Although there are few choices of high performance nFET technologies to adopt compared to pFET booster technologies, recently DSS FETs have become spotlighted among others as nFET booster technology [1-3]. This enhanced performance derives from increased carrier injection velocity [1] and decreased external resistance [2].

The drive currents of DSS FETs in the previous reports are, however, actually inferior to those with metal gate/high-k reported in [4], [5] due to difference of gate dielectric thickness at inversion.

In this work, we present a novel booster technology which combines DSS FETs with LSA and co-implant, of which performance is on a par with performance with metal gate/high-k for the first time.

Device Fabrication

The process flow of DSS transistors in this experiment is shown in Fig. 1. Poly-Si gates with 1.2 nm thick SiON gate dielectric are utilized in this experiment. The following processes are newly added into DSS transistor flow utilized in [3]. LSA is incorporated just before spike RTA in order to form sharper DSS junction. In addition, co-implants of carbon, nitrogen, or fluorine are implemented just after halo implants and dopant segregation implants respectively. Cross sectional TEM image of DSS nFET shown in Fig. 2 indicates decently-formed nickel silicide adjacent to offset spacer edge, as well as proximal tensile stress liner.

Results and Discussion

Fig. 3 shows comparison of Ion-Ioff characteristics of DSS FETs with and without LSA as well as conventional FETs (Conv.). DSS FETs show Ion of 1100 μA/μm at Ioff=100 nA/μm and Vdd=1.0V, while Conv. FETs show 1030 μA/μm at the same condition, which means Ion improvement by 7%. Moreover, DSS FETs combined with LSA show Ion of 1180 μA/μm, which stands for additional 7% Ion improvement by LSA. Vtsat vs. Lg curves for these three devices are shown in Fig. 4. DSS FETs with LSA have similar immunity against short channel effect (SCE) to DSS FETs without LSA, although Conv. FETs have worse immunity. In Fig. 5, off-current (Ioff) at Lg = 40nm is plotted against off-state gate-to-drain leakage current (Igd) normalized by on-state gate leakage current (Ig_on). DSS FETs with LSA are found to have larger S/D overlap without degrading Ioff compared to DSS FETs without LSA. These results imply that LSA generates more abrupt DSS junction horizontally although horizontal junction "depth" becomes deeper.

Cg-Vg curves of these two devices are plotted in Fig. 6, where Tinv of both devices are 1.9 nm, which means gate poly depletion

effect is unchanged by LSA, probably because spike RTA followed by LSA limits dopant activation in poly-Si gates. On the other hand, on-state resistance at linear region (Ron) vs. Lg plots in Fig. 7 clearly demonstrate 60 ohm*μm reduction of external resistance by LSA, keeping channel mobility. This reduction, which can explain Ion improvement, is probably because of the reduction of contact resistance at DSS junction.

Next, another improvement factor by co-implantation is shown. Bar graph of Fig. 8 compares drive currents at Ioff = 100 nA/μm of DSS FETs with LSA and co-implants by carbon, nitrogen, or fluorine. These co-implant technologies are clearly seen to modify the FET's performance. DSS FETs with LSA and fluorine co-implant show Ion of 1265 μA/μm at Ioff of 100 nA/μm and Vdd of 1.0V, which gives additional 7% Ion increase from DSS FETs with LSA, while carbon co-implant degrades the performance. According to Vtsat vs. Lg plots in Fig. 9, carbon or nitrogen slightly exacerbate SCE. Fluorine co-implant, however, does not degrade it at all. From Ron vs. Lg plots in Fig. 10, only fluorine co-implant is found to improve external resistance with 5%, which is the source of Ion improvement. This beneficial trend of DSS FETs with LSA and fluorine co-implant is maintained until narrow gate pitches, shown in Fig. 11, while Conv. FETs are supposed to require gate spacer structure change for narrow gate pitches.

One of concerns in DSS FETs might be hot carrier injection. As shown in Fig. 12, DSS itself does not cause substrate current (Isub) increase. Although LSA surely increases Isub due to more abrupt junction, fluorine co-implant mitigates Isub increase. As a result, DSS with LSA and fluorine co-implant satisfy 10 years lifetime against hot carrier injection in Fig. 13.

Finally, the best record of Ion vs. Ioff characteristics with poly-Si/SiON gate electrodes are demonstrated by optimizing these technologies shown in Fig 14. High drive currents of 1310 μA/μm and 1080 μA/μm at Ioff of 100 nA/μm are achieved at 1.0V and 0.9V respectively. These DSS FETs exhibit Id-Vg curves with S-slope of 92.5 mV/dec (Fig. 15), and decent Id-Vd curves (Fig. 16).

Conclusion

We found that LSA and fluorine co-implant enhance the drive current of DSS FETs by 7% respectively. These effects are mainly driven by reduction of external resistance, especially at DSS junction. We confirmed that this beneficial trend is maintained with narrow gate pitch, which can support the scalability. In addtition, this technology is verified to guarantee hot carrier reliability specs. Finally, we optimized these technologies and the best-record drive currents of 1310 μA/μm and 1080 μA/μm at Ioff = 100 nA/μm are achieved at 1.0V and 0.9V respectively. This enhanced DSS technology with LSA and fluorine co-implant is a promising candidate for 32 nm node and beyond, for it has additional advantages about cost effectiveness and compatibility with metal gate/high-k process.

Acknowledgments

The authors would like to thank T. Yoshida, T. Komoda and T. Sanuki for their continuous supports and discussion.

References

[1] A. Kinoshita et al., IEDM Tech. Dig., p. 79, 2006.
[2] T. Kinoshita et al., IEDM Tech. Dig., p. 71, 2006.
[3] H. Onoda et al., Symp. VLSI Tech. Dig., p. 76, 2007.
[4] K. Mistry et al., IEDM Tech. Dig., p. 247, 2007.
[5] S. Mayuzumi et al., IEDM Tech. Dig., p. 293, 2007.

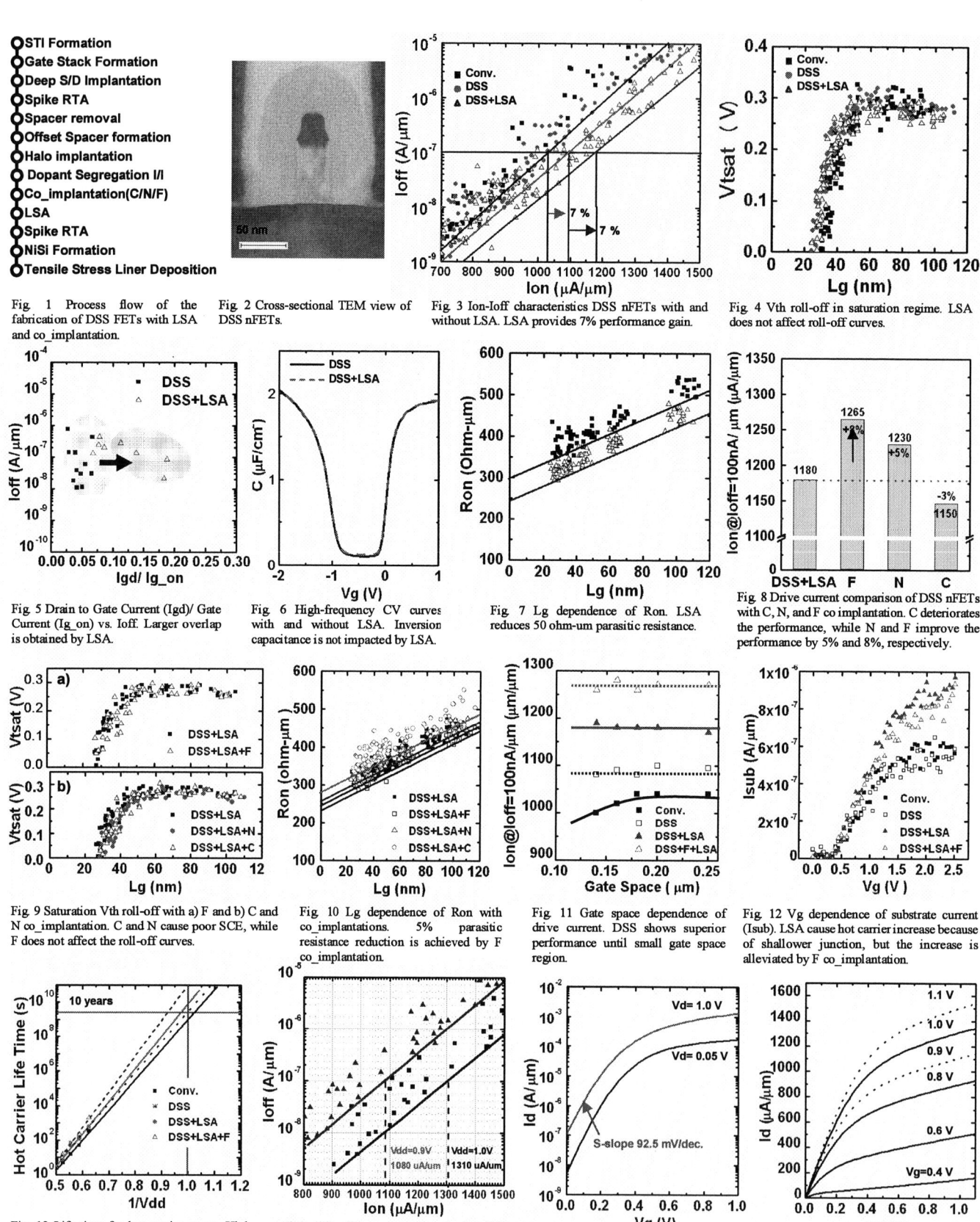

- STI Formation
- Gate Stack Formation
- Deep S/D Implantation
- Spike RTA
- Spacer removal
- Offset Spacer formation
- Halo implantation
- Dopant Segregation I/I
- Co_implantation(C/N/F)
- LSA
- Spike RTA
- NiSi Formation
- Tensile Stress Liner Deposition

Fig. 1 Process flow of the fabrication of DSS FETs with LSA and co_implantation.

Fig. 2 Cross-sectional TEM view of DSS nFETs.

Fig. 3 Ion-Ioff characteristics DSS nFETs with and without LSA. LSA provides 7% performance gain.

Fig. 4 Vth roll-off in saturation regime. LSA does not affect roll-off curves.

Fig. 5 Drain to Gate Current (Igd)/ Gate Current (Ig_on) vs. Ioff. Larger overlap is obtained by LSA.

Fig. 6 High-frequency CV curves with and without LSA. Inversion capacitance is not impacted by LSA.

Fig. 7 Lg dependence of Ron. LSA reduces 50 ohm-um parasitic resistance.

Fig. 8 Drive current comparison of DSS nFETs with C, N, and F co_implantation. C deteriorates the performance, while N and F improve the performance by 5% and 8%, respectively.

Fig. 9 Saturation Vth roll-off with a) F and b) C and N co_implantation. C and N cause poor SCE, while F does not affect the roll-off curves.

Fig. 10 Lg dependence of Ron with co_implantations. 5% parasitic resistance reduction is achieved by F co_implantation.

Fig. 11 Gate space dependence of drive current. DSS shows superior performance until small gate space region.

Fig. 12 Vg dependence of substrate current (Isub). LSA cause hot carrier increase because of shallower junction, but the increase is alleviated by F co_implantation.

Fig. 13 Life time for hot carrier stress. Higher Life time is extrapolated by F co_implantation. Also, DSS does not deteriorate hot carrier reliability. (Definition of hot carrier life time is

Fig. 14 Best record of Ion-Ioff characteristics. High drive current of 1310 uA/um and 1080 uA/um are achieved at 1V and 0.9V, respectively.

Fig. 15 Id-Vg characteristics under optimized condition.

Fig. 16 Id-Vd characteristics under optimized condition.

A New Source/Drain Germanium-Enrichment Process Comprising Ge Deposition and Laser-Induced Local Melting and Recrystallization for P-FET Performance Enhancement

Fangyue Liu[1,2], Hoong-Shing Wong[1], Kah-Wee Ang[2], Ming Zhu[1], Xincai Wang[3], Doreen Mei-Ying Lai[4], Poh-Chong Lim[4], Ben Lian Huat Tan[2], S. Tripathy[4], Sue-Ann Oh[4], Ganesh S. Samudra[1], N. Balasubramanian[2], and Yee-Chia Yeo[1].

[1]Silicon Nano Device Lab., Dept. of Electrical and Computer Engineering, National University of Singapore (NUS), 117576, Singapore.
[2]Institute of Microelectronics, 11 Science Park Road, 117685 Singapore.
[3]Singapore Institute of Manufacturing Technology (SIMTECH), 71 Nanyang Drive, 638075, Singapore.
[4]Institute of Materials Research & Engineering, 3 Research Link, 117602, Singapore.
Phone: +65 6516-2298, Fax: +65 6779-1103, E-mail: yeo@ieee.org

ABSTRACT

We report for the first time a new process technology for boosting the Ge content in SiGe source/drain (S/D) stressors to increase strain and performance levels in p-FETs. By laser-induced local melting and inter-mixing of an amorphous Ge layer with an underlying $Si_{0.8}Ge_{0.2}$ S/D region, a graded SiGe S/D stressor is formed upon recrystallization. Peak Ge content in the graded SiGe S/D is doubled over the as-grown film. Raman analysis confirmed the retention of high S/D strain levels due to the rapid non-equilibrium recrystallization process. The new process technology developed here employs several simple additional steps, including amorphous Ge deposition and laser anneal (LA). For a p-FET with Ge enriched S/D, 21% and 12% I_{Dsat} enhancement at a fixed I_{OFF} of 2×10^{-8} A/μm is observed over control p-FETs with $Si_{0.8}Ge_{0.2}$ S/D formed by RTA and LA, respectively.

INTRODUCTION

Increasing the Ge content of the embedded SiGe (eSiGe) Source/Drain (S/D) stressors improves p-FET performance [1-4] (Fig. 1). However, challenges are faced in managing strain relaxation and defect formation in an epi-SiGe S/D formed using a conventional process. Scaling the Ge content in S/D is difficult without new process capabilities. Non-equilibrium processes, such as laser anneal (LA), has never been exploited for defect/strain management in SiGe S/D formation to achieve much higher Ge content.

In this work, we present the first demonstration of a new process technology that incorporates additional Ge in SiGe S/D by using laser anneal. We exploit the non-equilibrium attributes of LA, i.e. superfast heating (~ns) and rapid recrystallization (at a rate of ~ 1 m/s), to form high quality and highly strained SiGe S/D. LA also suppresses nucleation and growth of the misfit dislocations, therefore enabling us to achieve a high Ge content in SiGe S/D for enhanced performance benefits. A graded eSiGe S/D with a peak Ge content of 40% is obtained. We integrated this newly developed process technology in SOI p-FETs to achieve significant drive current enhancement.

NEW S/D GERMANIUM ENRICHMENT PROCESS

Fig. 2 (a) shows the key process steps for forming a p-FET with enriched Ge content in SiGe S/D stressor. 8-inch silicon-on-insulator (SOI) substrates with a Si thickness of ~50 nm were used for this technology demonstration. After LOCOS formation, threshold voltage V_T adjust implant was performed prior to poly-Si/SiO$_2$ gate stack formation. A SiO$_2$ hardmask over the gate stack serves as a protection layer during subsequent pulsed laser annealing. S/D extension and a slim SiN spacer were formed followed by S/D recess etch and selective epitaxy of $Si_{0.8}Ge_{0.2}$ using ultra-high vacuum chemical vapor deposition. Deep S/D implant was performed. On two wafers, the S/D dopants were activated by either rapid thermal anneal (RTA) or pulsed laser anneal (LA). The LA employs 23ns pulses (1Hz) with wavelength λ of 248 nm and was done in N$_2$ ambient. On one wafer in which Ge content in the S/D is to be increased, 10 nm amorphous Ge (α-Ge) was deposited followed by LA to achieve local melting and intermixing of Ge with SiGe S/D. The fast cooling and re-growth of the molten layer forms a graded SiGe S/D with high crystalline quality and high Ge content. We developed a selective etching process employing H$_2$O$_2$ solution to remove the remaining Ge and $Si_{1-x}Ge_x$ (x > 60%) over gate and spacer. Fig. 2(d) shows a completed SOI p-FET with high Ge content in S/D formed by the new S/D Ge-enrichment process. EDS analysis shows a peak Ge content of 40%, 2 times higher than that in the as-grown $Si_{0.8}Ge_{0.2}$ S/D.

RESULTS AND DISCUSSION

A. Material Analysis

The Ge depth profile obtained by SIMS (Fig. 3) indicates the formation of a graded SiGe S/D, and is consistent with EDS analysis. Several factors contributed to the non-uniform distribution of Ge. After melting by laser, the liquid-solid interface or melting front moves up during recrystallization. Due to the partition ratio of Ge [5] and strain effect, the SiGe solid tends to be progressively enriched with Ge, thus giving a graded SiGe S/D. High resolution XRD in Fig. 4 shows well-defined SiGe peaks and indicates the high-quality of single crystalline SiGe epi-layer formed (Fig. 4). Of all 3 device splits, laser-annealed α-Ge-on-SiGe shows the largest shift of the SiGe satellite peak from the Si peak and has the highest substitutional Ge content. Raman shifts for Si-Si peaks in SiGe film are shown in Fig. 5. The extent of blue shifts is proportional to compressive strain in SiGe. Very high strain of -1.1% was obtained in the laser-annealed α-Ge-on-SiGe film, as compared to a mere -0.6% strain for rapid-thermal-annealed $Si_{0.8}Ge_{0.2}$ and -0.7% strain for laser-annealed $Si_{0.8}Ge_{0.2}$. The high strain levels in the new Ge enriched SiGe S/D stressor shall next be exploited for p-FET application.

B. Electrical Characteristics

Fig. 6 shows the I_D-V_G characteristics of closely matched p-FETs with gate length L_G of 120 nm, showing comparable I_{OFF}, subthreshold swing, and drain-induced barrier lowering DIBL. Two devices with S/D activated by RTA only or LA only have $Si_{0.8}Ge_{0.2}$ S/D, while the device with laser-annealed α-Ge/SiGe has Ge-enriched graded SiGe S/D. The strained p-FET with Ge-enriched graded SiGe S/D shows significantly higher saturation drain current I_{Dsat} than the controls (Fig. 7). The LA control also shows I_{Dsat} enhancement over the RTA device due to higher S/D dopant activation and lower S/D sheet resistance of S/D (Fig. 8). Sheet resistance of the Ge-enriched graded SiGe S/D is comparable to that of the laser-annealed $Si_{0.8}Ge_{0.2}$ S/D, for the same laser anneal conditions. Hence, the I_{Dsat} enhancement for the p-FET with Ge-enriched graded SiGe S/D over the laser-annealed p-FET with $Si_{0.8}Ge_{0.2}$ S/D is due to higher strain level. Fig. 9 shows comparable gate leakage current of the devices, indicating no gate dielectric degradation due to LA. Fig. 10 shows the I_{Dsat} of devices of different gate lengths. In general, increasing I_{Dsat} enhancement was observed with decreasing L_G due to the enhanced strain effects in devices with smaller L_G. The I_{OFF}-I_{Dsat} for p-FETs with L_G in the range of 120-150 nm is plotted in Fig. 11. The Ge-enriched graded SiGe S/D gives an I_{Dsat} enhancement of 12% compared to LA control, and 21% improvement compared to RTA devices. For $I_{D,lin}$, a higher enhancement of 21% and 32% was obtained for the Ge enriched p-FET compared to LA and RTA control, respectively (Fig. 12). Linear peak transconductance G_m shows same level of enhancement over the LA control device (Fig. 13). I_{Dsat} performance is also compared at a given DIBL (Fig. 14). At a DIBL of 0.07V/V, p-FETs with Ge-enriched graded SiGe S/D demonstrate ~14% improvement over the LA control p-FET.

CONCLUSION

A new process to increase the Ge content in the SiGe S/D was demonstrated. This enables the formation of p-FETs with Ge-enriched graded SiGe S/D, giving significant I_{Dsat} enhancement of about ~14% over control strained p-FETs with similar short-channel effects and S/D sheet resistance. This new technology is promising for further strain enhancement for 22 nm technology node and beyond.

Acknowledgment. This work was supported by the Nanoelectronics Research Program, Agency for Science, Technology & Research, Singapore.

REFERENCES

[1] K. Mistry et al., Symp. VLSI Tech. Dig., pp.50, 2004.
[2] P. Verheyen et al., IEDM Tech. Dig., pp.886 , 2005.
[3] R. A.Donaton et al., IEDM Tech. Dig.., pp.1, 2006.
[4] S. K.H. Fung et al., IEDM Tech. Dig.., pp.1035, 2007.
[5] D. P. Brunco et al., J. Appl. Phys. 78, pp. 1575, 1995.

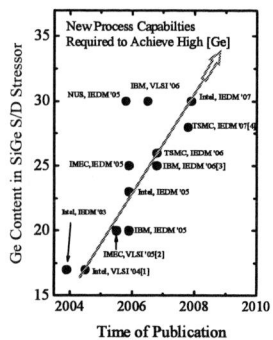

(a) Process Flow

- Gate Stack Formation
- SDE Implant and Spacer Formation
- Recess Etch and SEG Si$_{0.8}$Ge$_{0.2}$ S/D
- Deep S/D Implant and RTA
- Splits on Amorphous Ge Deposition
  Ge-Enrich S/D: 10 nm Ge Deposition
  Controls: No Ge Deposition
- Shallow S/D Implant
- Laser anneal or RTA
- Selective Ge removal [H$_2$O$_2$, 10 s] for p-FET with Ge Deposition
- Al contact formation

(b) New S/D Ge Enrichment Process
Deposit Ge + S/D Implant + Laser Anneal

(c) Selective Ge Removal

(d) TEM of Ge-Enriched SiGe S/D

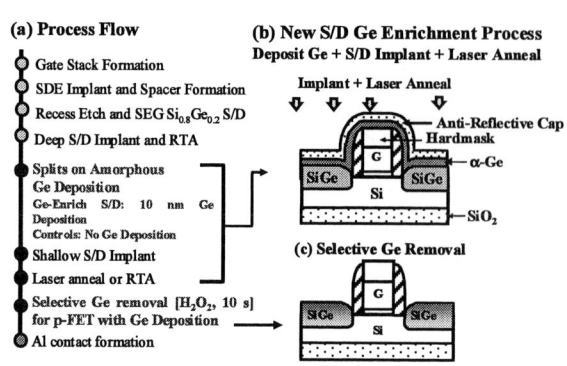

Fig. 1. Higher Ge content in SiGe S/D of p-FET enhances strain effects. This work presents a new process to achieve high Ge content in SiGe S/D.

Fig. 2. (a) Key process steps for forming a p-FET with enriched Ge content in SiGe S/D stressor. The novel Ge enrichment process involves (b) deposition of α-Ge (10 nm), laser anneal, and (c) selective removal of unreacted Ge to form the complete device. (d) TEM picture of fabricated p-FET with Ge-enriched SiGe S/D, showing the achievement of up to 40% Ge concentration in the S/D.

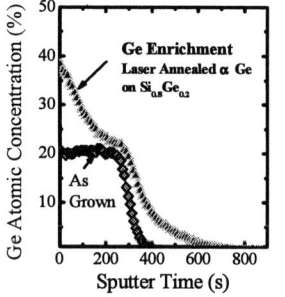

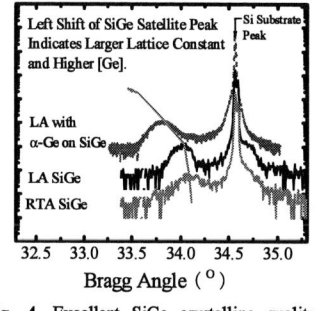

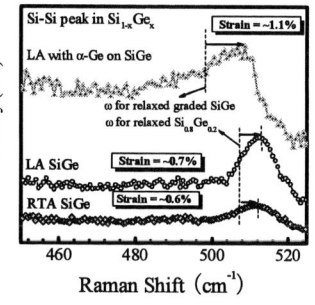

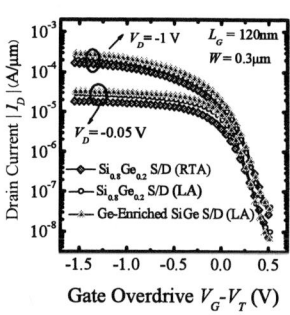

Fig. 3. SIMS analysis reveals pronounced Ge enrichment, and Ge content as high as 40% was achieved in the SiGe S/D stressors.

Fig. 4. Excellent SiGe crystalline quality revealed by well defined peaks in high resolution X-ray diffraction (HRXRD). Laser annealed α-Ge on SiGe gives the highest substitutional Ge content for strain engineering.

Fig. 5. UV Raman spectra showing that Ge enriched SiGe layer formed by laser anneal of α-Ge on SiGe has the highest compressive strain of 1.1%

Fig. 6. I_D-V_G characteristics of p-FETs with S/D formed by RTA only, LA only, and LA of α-Ge/SiGe. All p-FETs show comparable I_{OFF}, sub-threshold swing, and DIBL.

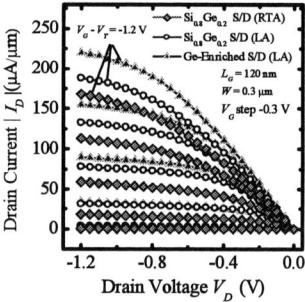

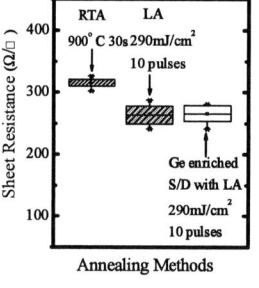

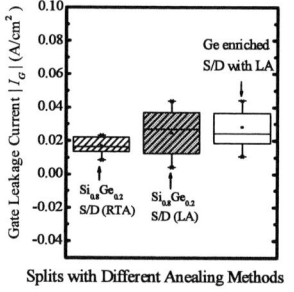

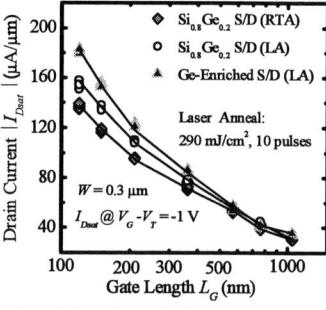

Fig. 7. The p-FET with Ge-enriched graded SiGe S/D formed by laser-anneal of α-Ge/SiGe gives significant I_{Dsat} enhancement over the control p-FETs.

Fig. 8. LA at 290 mJ/cm^2 achieved lower sheet resistance compared to RTA. All laser-annealed samples have comparable sheet resistance.

Fig. 9. Comparable gate leakage current for LA and RTA devices confirms that the gate dielectric integrity is similar for all p-FETs.

Fig. 10. P-FETs with high Ge S/D content formed by laser induced intermixing and regrowth show the largest I_{Dsat} gain over control devices. Higher current gain is observed at smaller L_G.

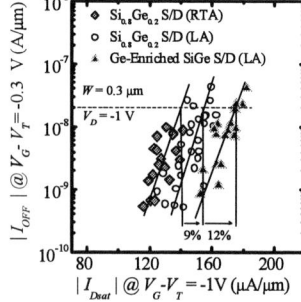

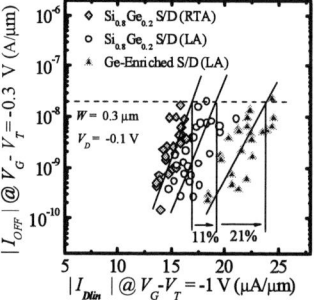

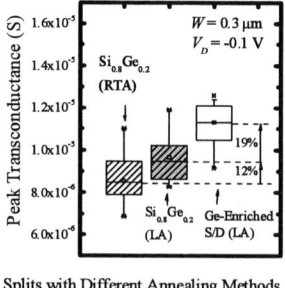

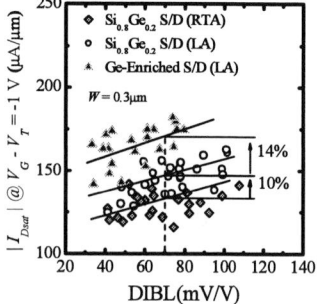

Fig. 11. At I_{OFF} of 20nA/μm, p-FETs with Ge enriched S/D show an I_{Dsat} enhancement of 21% and 12% compared to RTA and LA p-FETs, respectively.

Fig. 12. $I_{D,lin}$ at a given I_{OFF} shows an enhancement of 32 % and 21 % in p-FETs with enriched Ge at S/D compared to RTA and LA control p-FETs, respectively.

Fig. 13. Ge enriched S/D p-FETs has a peak transconductance improvement of 31% and 19%, respectively, over the RTA and LA controls.

Fig. 14. Excellent I_{Dsat} improvement is demonstrated by the novel S/D Ge-enrichment process at a comparable DIBL of 70 mV/V.

Novel and Cost-Efficient Single Metallic Silicide Integration Solution with Dual Schottky-Barrier Achieved by Aluminum Inter-diffusion for FinFET CMOS Technology with Enhanced Performance

Rinus Tek-Po Lee, Alvin Tian-Yi Koh, Wei-Wei Fang, Kian-Ming Tan, Andy Eu-Jin Lim, Tsung-Yang Liow
Chow Shue-Yin[+], Anna M. Yong[+], Hoong Shing Wong, Guo-Qiang Lo*, Ganesh S. Samudra, Dong-Zhi Chi[+] and Yee-Chia Yeo.

Silicon Nano Device Lab., Dept. of Electrical and Computer Engineering, National University of Singapore, 117576,
[+]Institute of Materials Research and Engineering, Singapore, *Institute of Microelectronics, Singapore
Phone: +65 516-2298, Fax: +65 6779-1103, E-mail: yeo@ieee.org

Abstract

We have developed a novel and cost-efficient silicide integration solution to achieve a hole barrier height of 215 meV and electron barrier height of 665 meV simultaneously with a single metallic silicide based on aluminum inter-diffusion. It is proposed that aluminum diffuses into PtSi and forms an alloy, which lowers the electron barrier height of PtSi due to a change in the intrinsic PtSi workfunction. Additionally, we have integrated platinum germanosilicide with an ultra-low hole barrier height of 215 meV in P-FinFETs to provide a 21% enhancement in drive current performance, which is attributed to the 20 % reduction in series resistance. We have also ascertained the compatibility of PtSiGe with laser thermal annealing for further performance enhancement.

Will series resistance be a showstopper for FinFETs?

The FinFET architecture is the most promising transistor structure to address SCE and leakage issues in scaled CMOS devices. However, the use of narrow fins for effective gate electrostatic control exacerbates the already critical issue of increasing series resistance for aggressively scaled devices [1]. It was reported that (110) PFETs suffers from higher S/D series resistances compared to conventional (100) PFETs [2]. This will further aggravate the series resistance issue in FinFETs due to the (110) fin sidewall surfaces. Various groups have proposed innovative S/D solutions to alleviate the impact of series resistance on device performance. However, integration issues of process complexity, and cost efficiency remains a key consideration as these solutions require NFETs and PFETs to be optimized independently for enhanced device performance.

In this work, we propose a novel, cost-efficient and highly manufacturable single metallic silicide integration solution based on Al inter-diffusion to realize low series resistances in both N and P-FinFETs simultaneously to enhance device performance without added process complexity. In addition, we demonstrate the feasibility of integrating platinum germanosilicides (PtSiGe) into P-FinFETs for series resistance engineering. Lastly, we report the integration of PtSiGe into the most aggressively scaled laser thermal anneal FinFET fabricated to date for further performance enhancement.

Optimization of Platinum Germanosilicide for P-FinFETs

10 nm Pt films were deposited on epitaxially grown $Si_{0.74}Ge_{0.26}$ film [Fig 1(a)]. Our results show that the formation of platinum germanosilicide (PtSiGe) does not disrupt the strain-state/crystalline quality of SiGe [Fig. 1(a)-(c)]. It was observed in Fig 2 that sheet resistance of NiSiGe remains constant from 400 - 600 °C. At temperatures greater than 600 °C, sheet resistance of NiSiGe degrades drastically due to agglomeration. Sheet resistances were found to improve gradually with higher annealing temperatures for PtSiGe and remain stable. Additionally, it can be seen in Fig. 2 that the stress in PtSiGe increases from ~ 0.3 GPa to ~ 1.2 GPa after annealing at temperatures greater than 600 °C. TEM images confirm the enhanced morphological stability of PtSiGe in agreement with sheet resistance measurements. Voids and "spikes" were observed to form in NiSiGe and penetrate into the substrate when annealed at 700 °C [Fig. 3]. We demonstrated that PtSiGe is chemically stable in dilute aqua regia for up to 10 minutes to allow the selective removal of unreacted Pt for a self-aligned germanosilicidation process [Fig. 4]. An ultra-low hole barrier of 215 meV for PtSiGe on p-Si was extracted from the Richardson plot using the thermionic-emission model.

Device Integration, Results and Discussion

To verify the feasibility of forming a low barrier PtSiGe for improved P-FinFET performance, we integrated the PtSiGe process into a standard P-FinFET process flow and compared the device performances in a statistical manner. Fig. 5 compares the $I_{OFF} - I_{Dsat}$ characteristics of P-FinFETs with PtSiGe and NiSiGe at 100 nA/μm and shows an I_{Dsat} enhancement of 21 % over the control P-FinFETs with NiSiGe. In a comparison of I_{OFF} versus I_{Dlin}, a higher enhancement of 30 % was also obtained for P-FinFETs with PtSiGe. Device performances compared at a fixed DIBL of 60 mV/V and SS of 100 mV/dec. show I_{Dsat} enhancement of 34% and 37%, respectively. Fig. 9 shows that series resistance is reduced by 20 % for FinFETs integrated with PtSiGe, which contributes to the I_{Dsat} performance gain observed in FinFETs with PtSiGe.

Novel and Cost Efficient Single Metallic Silicide Integration Solution

As shown in Fig. 10. the co-integration of Pt based silicides on NFETs will degrade contact resistivity due to the large electron barrier of Pt based silicides to N^+/P junctions. For a simplified and cost-efficient single silicide integration solution, we introduce the concept of Al inter-diffusion. The incorporation of Al into PtSi based on inter-diffusion is effective in modulating the electron barrier height of PtSi towards the conduction band [Fig. 11(a)]. The distributions of electron barrier for PtSi and PtSi(Al) are plotted in Fig. 11(b) and shows an average 207 meV shift towards the conduction band. A larger shift can be obtained with the use of lower work function materials such as Yb or Er. SIMS analysis in Fig. 12. shows that Al is distributed homogenously within the bulk PtSi layer. We postulate that the effective electron barrier modulation observed with Al inter-diffusion can be attributed to the alloying of Al with PtSi. This results in a change in the intrinsic workfunction of PtSi. This will minimize the impact of degraded contact resistivity on NFET performance. With our optimized silicidation process, we are able to demonstrate superior device characteristics for both NFETs and PFETs with a single metallic silicide [Fig. 13]. The superior PFET performance is attributed to the lowering of series resistance. On the other hand, the appreciable drive current enhancement of NFETs is attributed to the high stress PtSi(Al) films [Fig. 2] when subjected to a 600 °C anneal for the inter-diffusion process. This gives rise to increase longitudinal tensile channel stress, which is beneficial for NFET performance. Fig. 14 shows the junction leakage for NiSi and PtSi(Al). Although there is an appreciable increase in leakage but it is within an acceptable range that can be further optimized with junction design. As shown in Fig. 15, our proposed single metallic silicide integration solution is simple and cost efficient, which will allow device engineers to independently optimize transistor performance without the use of a costly and complicated dual silicide integration approach.

In addition, we have also fabricated the most aggressively scaled laser thermal annealed (LTA) FinFET transistor to date with a gate length of 95 nm [Fig 16]. The tight distributions of subthreshold swing and DIBL indicate that the fabricated LTA FinFETs possess excellent device characteristics [Fig. 17]. The further integration of PtSiGe with LTA provides superior drive current performance compared to LTA P-FinFETs with NiSiGe and demonstrates the compatibility of PtSiGe to LTA.

Conclusion

We have demonstrated a novel and cost efficient single metallic silicide integration solution with Al inter-diffusion to independently optimize NFETs and PFETs performance. This leads to an improvement in contact resistance and will allow further device scaling. Our results also showed that silicide-induced stress can potentially have a synergistic effect when used in conjunction with series resistance engineering to provide further performance enhancement. Additionally, the integration of PtSiGe into P-FinFETs provides a 21% enhancement in drive current performance compared to NiSiGe due to the 20 % reduction in series resistance. We have also investigated the compatibility of PtSiGe with LTA and fabricated the most aggressively scaled LTA FinFET to date.

References

[1] A. Dixit, et al., IEEE TED pp. 1132-1140, 2005, [2] B.Yang, Symp. VLSI pp. 126-127, 2007

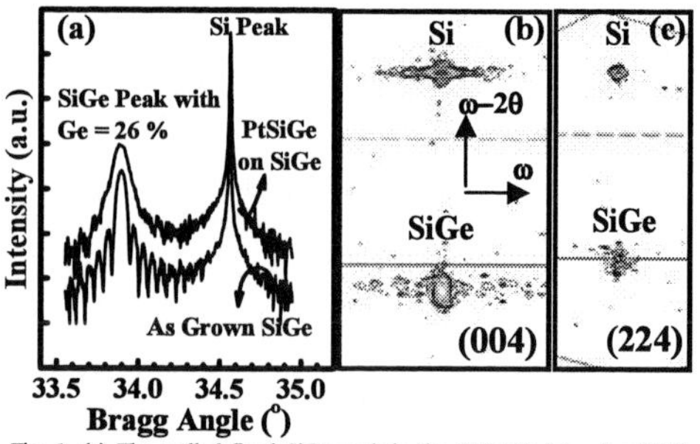

Fig. 1. (a) The well defined SiGe peak in the HRXRD data after PtSiGe formation indicates retention of the high crystalline quality of the as-grown SiGe film. (b) and (c) (004) and (224) reciprocal space maps of the SiGe layers indicate that SiGe remains fully strained after PtSiGe formation.

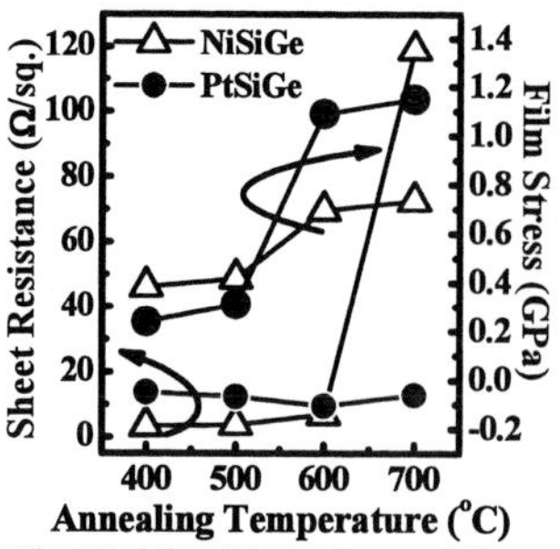

Fig. 2. Evolution of sheet resistance and film stress with temperature. The dramatic increase in sheet resistance for NiSiGe is attributed to film agglomeration.

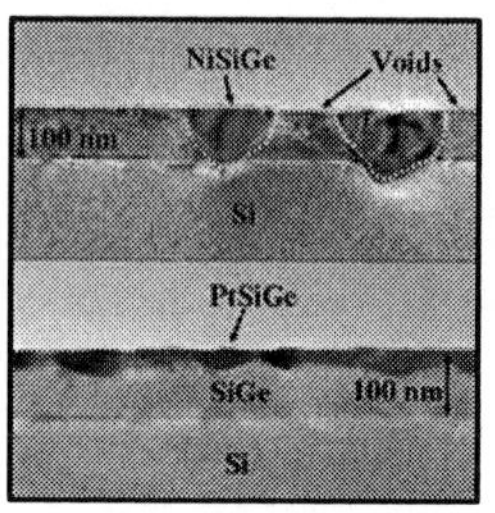

Fig. 3. Improved PtSiGe morphological stability is obvious in the TEM images for NiSiGe and PtSiGe films formed at 700 °C. Formation of voids and the presence of NiSiGe "spikes" into the Si substrate

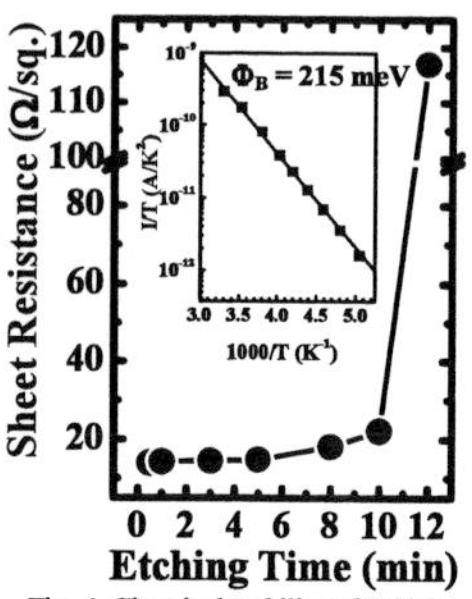

Fig. 4. Chemical stability of PtSiGe films in aqua regia with time. Inset shows the Richardson plot for the extraction of PtSiGe Φ_B=215 meV.

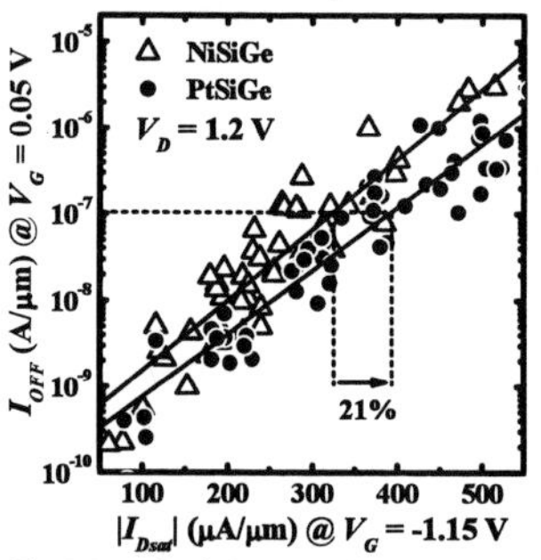

Fig. 5. I_{OFF} - $|I_{Dsat}|$ plot comparing FinFETs with NiSiGe and PtSiGe S/D. A 21 % I_{Dsat} enhancement at $I_{OFF} = 1 \times 10^{-7}$ A/μm was observed for FinFETs with PtSiGe S/D.

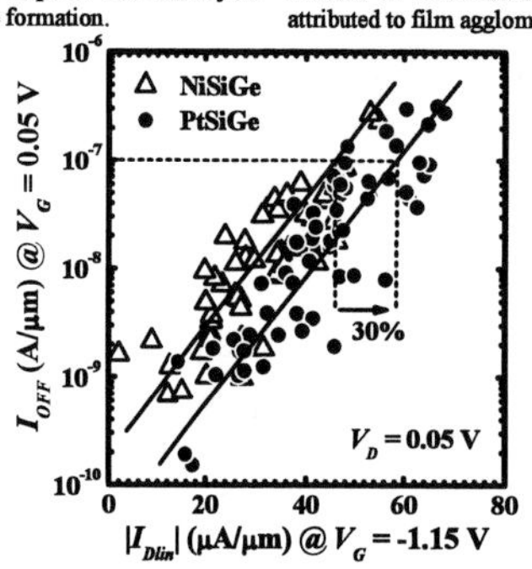

Fig. 6. Comparison of I_{OFF} versus I_{Dlin} shows an enhancement of 41 % at $I_{OFF} = 1 \times 10^{-7}$ A/μm for FinFETs with PtSiGe S/D compared to NiSiGe

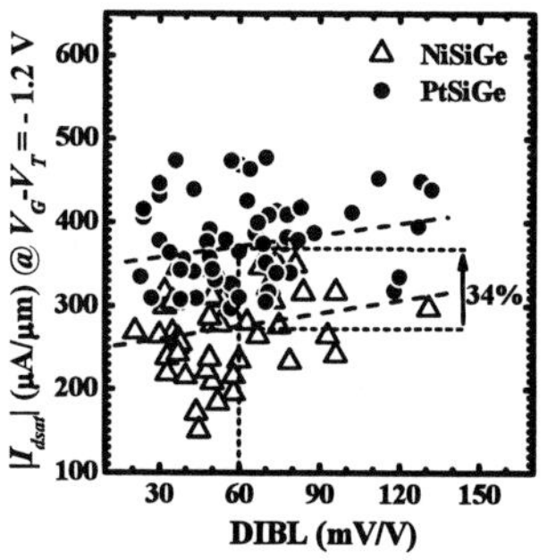

Fig. 7. Comparison of FinFETs with NiSiGe S/D and PtSiGe S/D at a fixed DIBL of 60 mV/V shows I_{Dsat} enhancement of 34 % for PtSiGe S/D.

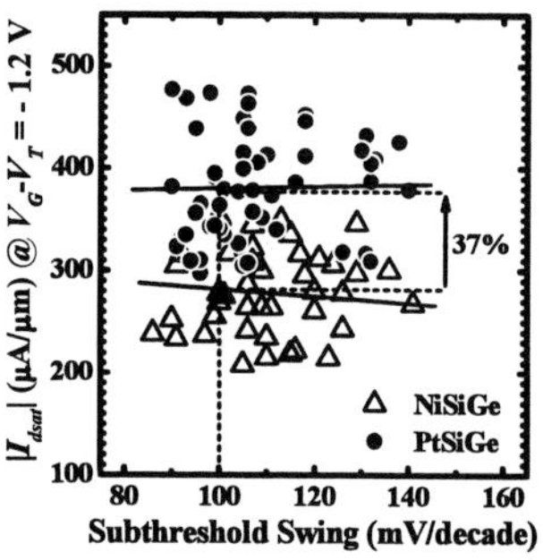

Fig. 8. At a fixed subthreshold swing of 100 mV/dec., FinFETs with PtSiGe S/D shows I_{Dsat} enhancement of 34 % compared to NiSiGe S/D.

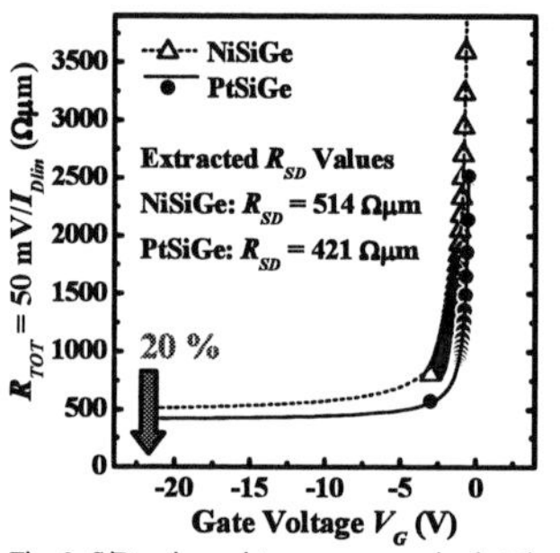

Fig. 9. S/D series resistance was examined at the asymptotic behavior of the R_{TOT} curve at V_G = -20 V. PtSiGe S/D shows a 20 % reduction in R_{SD}.

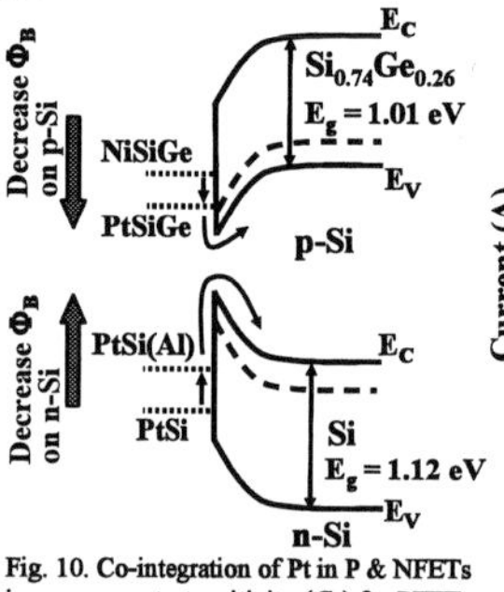

Fig. 10. Co-integration of Pt in P & NFETs improves contact resistivity (C_R) for PFETs but degrades C_R for NFETs. Interdiffusion of Al into PtSi is proposed to improve the contact resistivty of PtSi for NFETs.

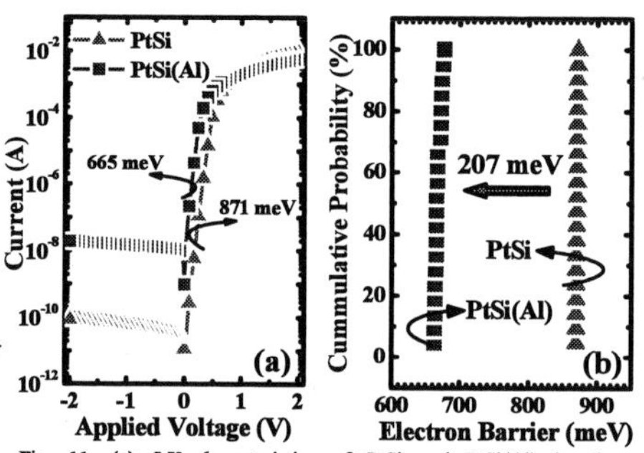

Fig. 11. (a) I-V characteristics of PtSi and PtSi(Al) junctions. Enhancement in the reverse leakage current is due to Φ_B reduction with Al interdiffusion in PtSi. (b) Electron barrier height comparison between PtSi and PtSi(Al) shows an effective 207 meV reduction of PtSi electron barrier height.

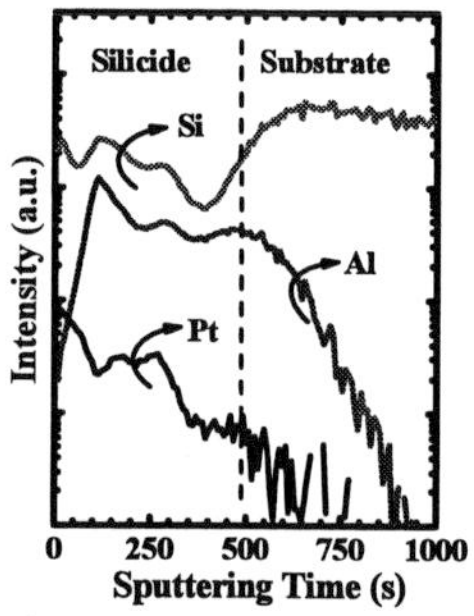

Fig. 12. SIMS analysis reveals that Al is distributed homogenously within the PtSi film after 600 °C RTA.

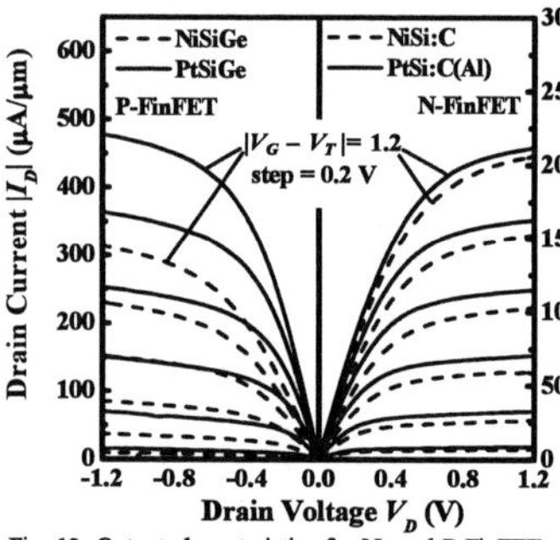

Fig. 13. Output characteristics for N- and P-FinFETs with L_G = 250 and 70 nm, respectively, integrated with our optimized silicidation process.

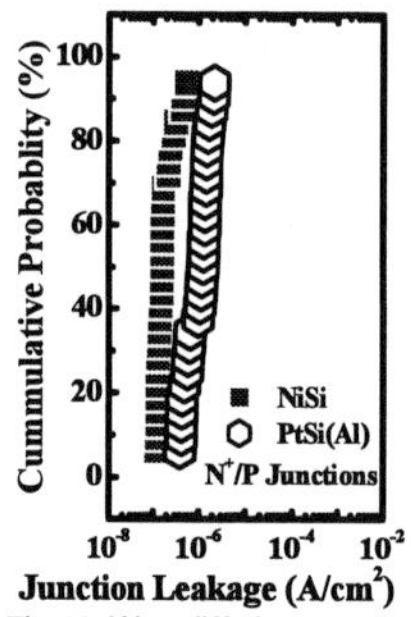

Fig. 14. Al interdiffusion causes a slight leakage increase but it is within the acceptable range.

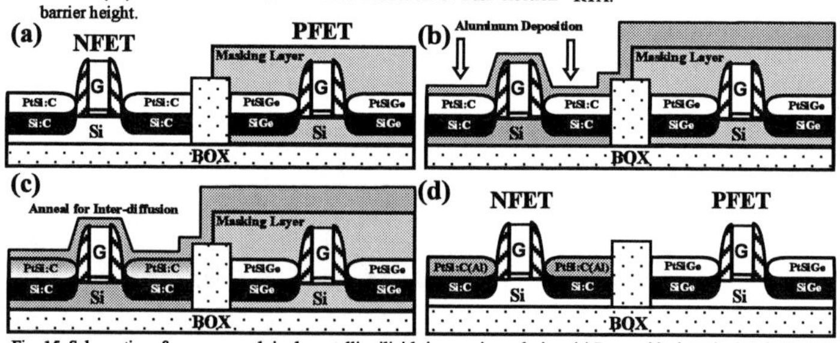

Fig. 15. Schematics of our proposed single metallic silicide integration solution. (a) Pt metal is deposited on both N & PFETs and anneal to form Pt metal silicides (i.e. PtSiGe and/or PtSi and/or PtSi:C). (b) A masking layer is deposited on the PFETs followed by Al metal deposition on the entire wafer. (c) A 600 °C anneal is performed for Al interdiffusion in PtSi and/or PtSi:C to modulate the electron barrier height of PtSi. (d) Removal of excess Al with dilute HF acid completes the integration process for a single metallic silicide with dual Schottky-barrier height.

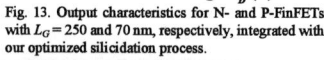

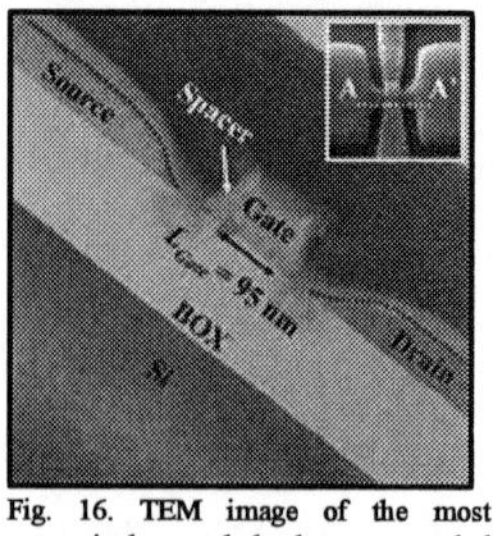

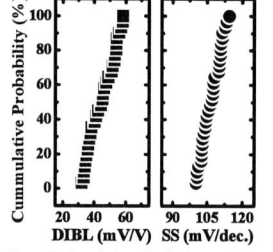

Fig. 16. TEM image of the most aggressively scaled laser annealed FinFET with L_G = 95 nm fabricated to date.

Fig. 17. Cumulative distributions of SS and DIBL for the laser anneal FinFETs.

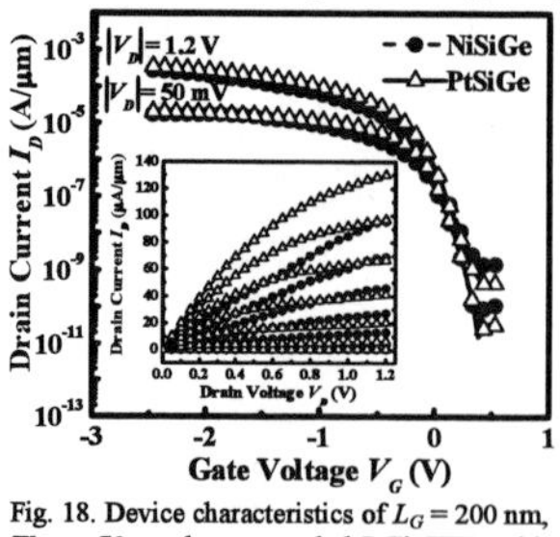

Fig. 18. Device characteristics of L_G = 200 nm, W_{Fin} = 70 nm laser annealed P-FinFETs with NiSiGe and PtSiGe S/D.

	Single Silicide Process	Dual Silicide Process	This Work
Independent Optimization for N & PFETs	✗	✓	✓
Cost Effectiveness	✓	✗	✓
Device Performance	✗	✓	✓
Process Simplicity	✓	✗	✓

Table 1. Comparison of integration challenges and options for conventional approaches and our proposed single metallic silicide integration solution to form self-aligned metallic silicides on the S/D regions to achieve enhanced device performance.

Experimental Study of Mobility in [110]- and [100]-Directed Multiple Silicon Nanowire GAA MOSFETs on (100) SOI

Jiezhi Chen, Takuya Saraya, Kousuke Miyaji, Ken Shimizu, and Toshiro Hiramoto

Institute of Industrial Science, University of Tokyo, 4-6-1 Komaba, Meguro-ku, Tokyo 153-8505, Japan

Tel: +81-3-5452-6264, Fax: +81-3-5452-6265, e-mail: chen@nano.iis.u.tokyo.ac.jp

Abstract

Experimental investigations of silicon nanowire mobility characteristics on (100) SOI as shrinking nanowire width to sub-10nm are reported. Accurate mobility estimations by advanced split CV method for 50~1000 nanowires are performed. For the first time, electron and hole mobility in [100]-directed nanowires are studied and compared with [110] nanowires. It is shown that both electron and hole mobility decreases monotonically and electron mobility of [100]-directed nanowire tends to be comparable to that of [110]-directed nanowire as decreasing nanowire width.

Introduction

Nanowire (NW) MOSFETs, as one of promising candidates for future VLSI technologies with high performance, have attracted much attention recently [1-3]. However, the mobility characteristics of NW have not been studied systematically due to the limited fabrication technologies. The gate-channel capacitance C_{gc} of NW-MOSFETs is ultra-small and it is quite difficult to extract the mobility with split CV method.

In this paper, aiming at elucidating transport in nanowire MOSFETs in detail, accurate estimations of carrier mobility of [110] and [100]-directed multiple NWs on (100) SOI are carried out. It is found that [100] mobility approaches to [110] as decreasing nanowire width. The possible physical mechanisms of this mobility behavior are also discussed.

Device Fabrication

The device fabrication starts with thinning (100) SOI to desired thickness. When NWs patterns are defined by EB lithography and followed by RIE, NW width is further narrowed using SC1 solution. Then, BHF is used to over-etch the BOX layer under NWs for gate-all-around (GAA) structure. The final NW width W_{NW} after gate oxidation is from 7nm~48 nm, with gate oxidation thickness around 13.7 nm on (100)-surface while 21.6nm on (110)-surface. The SOI thickness (T_{SOI}, NW height) and NW width are estimated by ellipsometry and SEM observation. Fig. 1(a) shows the schematic view of fabricated NWs-MOSFETs. Fig. 1(b) shows SEM image of NW cross section after gate poly-Si deposition, where the NW height is around 22nm and the width is around 30nm. Ultra-thin-body (UTB) MOSFETs are also fabricated for reference.

Device Measurement

Fig. 2 shows I_d-V_g characteristics of [100]/(100) NW MOSFETs with 1000 nanowires (T_{SOI}=22nm and L_g=3μm). The drain current decreases as NWs turn to be narrower. Fig. 3 (a) and (b) show C_{gc} of devices with L_g=3μm and 4μm, where W_{NW} is 7nm~48nm. Furthermore, C_{gc} of NW with different W_{NW} is shown in Fig. 3 (c). Comparing the calculation results with rectangular and cylinder shape, the narrower NWs turns to be cylinder shape instead of rectangular shape. Especially, we observed an additional C_{gc} reduction at W_{NW}=7nm, which may originate from volume inversion [2]. Thanks to GAA structure and surface roughness control, sub-threshold characteristics show ideal performance in the whole NWs width range, as shown in Fig. 4. Fig. 5(a) shows experimental V_{th} shift of nFET

and pFET. Fig. 5(b) shows subband calculations of electron and hole with EMA and K.P method, which agree with experimental data. Fig. 6 shows I_d and C_{gc} of 48nm NWs at Vg=3 V, changing NW number and gate length (L_g=3μm, 4μm). The linear relationships along with NW number can be observed clearly.

Mobility Estimations and Discussion

Accurate mobility estimations are performed by using the double L_m method [4], where two gate lengths are designed to remove the effect of parasitic capacitance and resistance, as shown in Fig. 7. Here, the average surface carrier density is calculated by $N_{inv}= N_l/(2W_{NW}+2T_{SOI})$, where N_l means the line inversion carrier density obtained from CV measurement.

Mobility in [100]-directed NW nFETs with two different T_{SOI} are shown in Figs. 8 and 9. In the wider-limit, mobility of NW with T_{SOI}=22nm converges to thick UTB mobility, and mobility of T_{SOI}=30nm NW is even higher than that of UTB at high N_{inv}. Although rough W_{NW} estimations will cause N_{inv} definition errors, the mobility degradation tendency with decreasing W_{NW} is clear. Devices with thinner T_{SOI} have smaller mobility.

Mobility in [110]-directed NW nFETs with T_{SOI} =18nm is shown in Fig. 10. Because electron mobility on (110) surface is lower than that of (100) surface, [110] NW has smaller mobility than [100] NW with W_{NW}=48nm. However, this degraded mobility in [110] NW diminishes and is comparable to that of [100] NW as decreasing W_{NW}, as shown in Fig. 11.

As reported in UTB nFET, due to the subband modulations, a (100) double-gate (DG) nFET has smaller mobility than a (100) single-gate (SG) nFET, while (110) DG nFET has larger mobility than (110) SG [5]. Please note that the [110]-directed NW nFET is in form of DG nFET on (110) surface, and [100] NW nFET is in form of DG on (100) surface. Therefore, comparable mobility in [110] and [100] NW nFETs can be explained by the subband modulation effect on (110) surface.

Fig. 11 shows the extracted [110] and [100] hole mobility in NW pFET with W_{NW} from 15~40nm. Hole mobility decreases monotonically as decreasing W_{NW}. It is also found that [100] NW has lower mobility than [110] NWs.

Conclusions

The electron and hole mobility in [110] and [100] oriented NWs on (100) SOI has been studied systematically with W_{NW} down to sub-10nm. The mobility of [110] and [100] NWs decreases monotonically as narrowing NW width. It is found that electron mobility of [100] and [110] NWs tend to be comparable as decreasing W_{NW} due to subband modulations in DG nFET.

Acknowledgement

This work was partly supported by Nanoelectronics Project by METI, Japan.

References

[1]T. Tezuka et al, IEDM, p.887, 2007. [2]S.D. Suk et al, IEDM, p.891, 2007. [3]M.J.H. van Dal et al, VLSI, p. 110, 2007. [4]A. Toriumi et al, IEDM, p.671, 2006. [5]G. Tsutsui et al, IEDM, p.729, 2005.

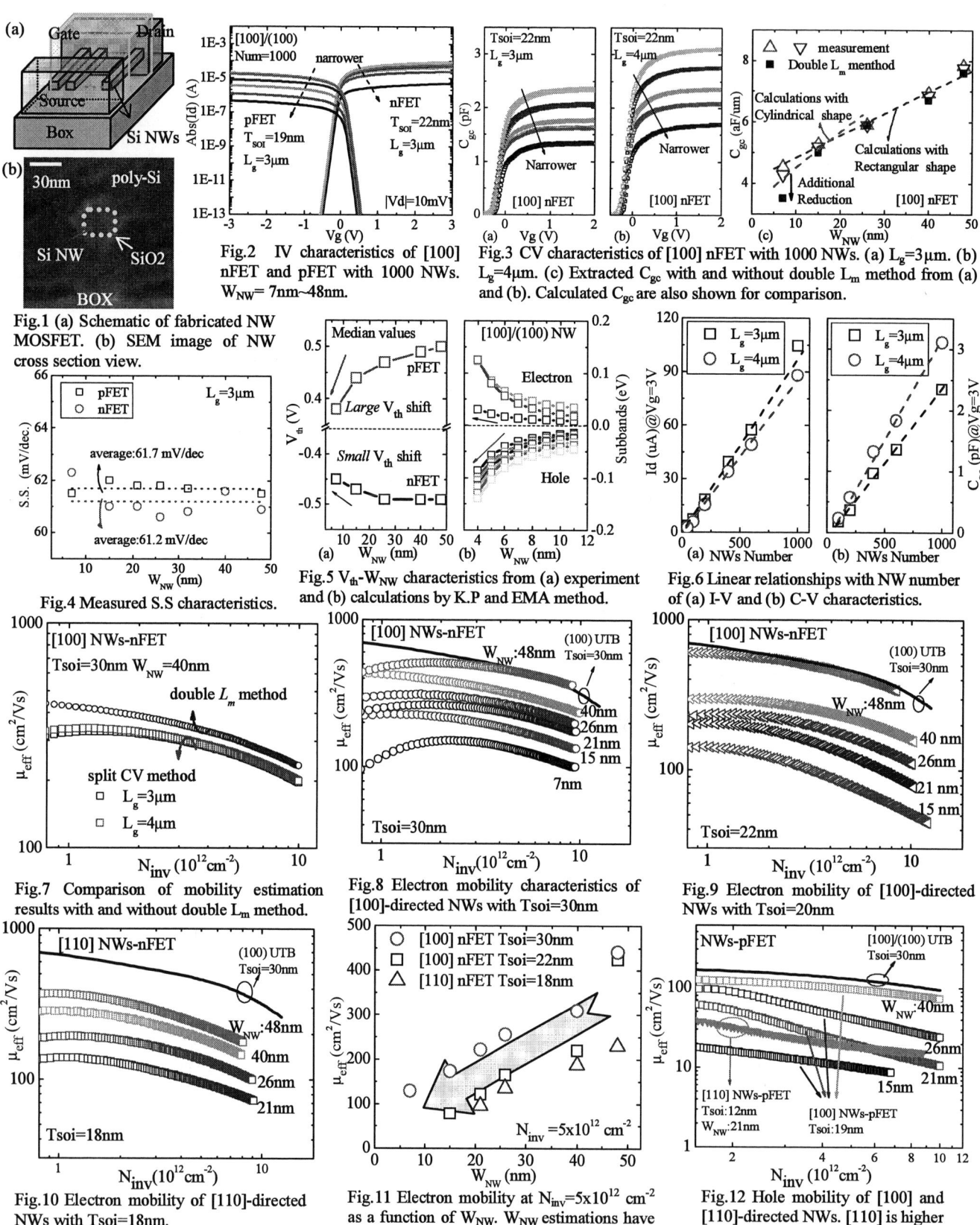

Fig.1 (a) Schematic of fabricated NW MOSFET. (b) SEM image of NW cross section view.

Fig.2 IV characteristics of [100] nFET and pFET with 1000 NWs. W_{NW}= 7nm~48nm.

Fig.3 CV characteristics of [100] nFET with 1000 NWs. (a) L_g=3μm. (b) L_g=4μm. (c) Extracted C_{gc} with and without double L_m method from (a) and (b). Calculated C_{gc} are also shown for comparison.

Fig.4 Measured S.S characteristics.

Fig.5 V_{th}-W_{NW} characteristics from (a) experiment and (b) calculations by K.P and EMA method.

Fig.6 Linear relationships with NW number of (a) I-V and (b) C-V characteristics.

Fig.7 Comparison of mobility estimation results with and without double L_m method.

Fig.8 Electron mobility characteristics of [100]-directed NWs with Tsoi=30nm

Fig.9 Electron mobility of [100]-directed NWs with Tsoi=20nm

Fig.10 Electron mobility of [110]-directed NWs with Tsoi=18nm.

Fig.11 Electron mobility at N_{inv}=5x10^{12} cm^{-2} as a function of W_{NW}. W_{NW} estimations have 5nm error by SEM observations.

Fig.12 Hole mobility of [100] and [110]-directed NWs. [110] is higher than [100] even in narrower NW.

Performance Breakthrough in 8 nm Gate Length Gate-All-Around Nanowire Transistors using Metallic Nanowire Contacts

Y. Jiang[1,2], T. Y. Liow[1,2], N. Singh[1], L.H. Tan[1], G. Q. Lo[1], D. S. H. Chan[2] and D. L. Kwong[1]

[1]Institute of Microelectronics, 11 Science Park Road, Science Park-II, 117685 Singapore.
[2]Silicon Nano Device Lab., Dept. of Electrical and Computer Engineering, National University of Singapore
Phone: +65 67705705, Fax: +65 67731914, E-mail: logq@ime.a-star.edu.sg

Abstract

Parasitic S/D resistances in extremely scaled GAA nanowire devices can pathologically limit the device drive current performance. We demonstrate for the first time, that S/D extension dopant profile engineering together with successful integration of low resistivity metallic nanowire contacts greatly reduces parasitic resistances. This allows 8 nm gate length GAA nanowire devices in this work to attain record-high drive currents of 3740 $\mu A/\mu m$.

Introduction

Gate-All-Around (GAA) silicon nanowire (SNW) transistors is one of the most promising candidates for further scaling requirement owing to its superior short channel immunity. [1],[2]. While channel control improves as channel dimension shrinks, high series resistance arises due to the narrow nanowire source and drain (S/D) regions, which pathologically limits the device drive current performance [3]. In ultra-short channel nanowire transistors, the adjoining source and drain regions (doped semiconductor nanowires of comparatively large lengths) contribute significantly to the S/D series resistances, forming a large fraction of the total resistance. In order to obtain high performance in ultra-short channel nanowire devices, S/D extension dopant profile and silicide engineering in the S/D extension regions need to be optimized, not unlike that for SOI FinFETs [4].

In this work, the successful formation of low resistivity metallic nanowires contacts to nanowire transistors has been demonstrated for the first time. We report greatly enhanced drive current performance in 8 nm gate GAA nanowire transistors with optimized NiSi nanowire contact and S/D extension dopant engineering.

Experiment

N-channel GAA SiNW devices were fabricated using a CMOS-compatible process. The key steps are summarized in the process flow schematic as shown in Fig. 1. Nanowires are formed on thin SOI wafers using deep UV lithography, trimming and etching. Nanowire diameters were then further trimmed down to ~10 nm using thermal oxidation. Poly-silicon gates down to 8 nm were defined using a combination of photoresist, gate hardmask and gate trimming. Following deep S/D implantation and activation, low resistivity NiSi nanowire contacts to the nanowire S/D extensions were formed using an optimized salicidation process. This was followed by backend contact and metallization processes. Fig. 2(a) shows an SEM image of a released nanowire, while Fig. 2(b) shows a nanowire transistor prior to S/D salicidation. A cross-section TEM of the 10 nm nanowire channel is shown in Fig. 3, showing the gate-all-around structure of the nanowire device. Fig. 4 shows a cross-section TEM to illustrate the gate length of these devices.

Results and Discussion

The salicidation process of nanowire transistors is a non-trivial process. Deposition of too thick a Ni film can result in device shorts due to excessive lateral silicide encroachment. At the same time, both (sheet) resistivity and contact resistivity should be as low as possible. For Ni silicide, resistivity is a function of its thickness [5]. Ultra-thin Ni films of 1 to 4 nm were deposited on thin SOI samples in a Canon Anelva PVD system and annealed in a single wafer rapid thermal furnace to form Ni silicide films. A 2nd run was performed to confirm the repeatability of the results. The sheet resistance and resistivity values are summarized in Fig. 5(a) and Fig. 5(b) respectively. It was observed that the Ni silicide films corresponding to 4 nm Ni have resistivity values (~27 μOhm.cm) which are close to that of thick NiSi films. For silicide films corresponding to 1 to 3nm of Ni, the resistivity value for all three thicknesses was about (~70 μOhm.cm). This represents significant deviation from bulk values. SEM observations and EDX analyses did not reveal any signs of silicide film agglomeration. Hence, the occurrence of this phenomenon can possibly be due to the formation of higher resistivity Ni_3Si_2 phase. It should be noted that such resistivity values are still much less than that of highly doped Si (Fig. 5(b)). Moreover, calculations show that even 4.4 nm of NiSi can reduce the series resistance contribution from 50 nm long nanowire S/D regions to less than 17.5 Ohm.μm. per side, which is still acceptable for nanowire diameter scaling to 5 nm.

Besides sheet resistivity, contact resistivity is also an important parameter that requires careful optimization due to the extremely small contact area between the nanowire S/D extensions and the silicide. A high dopant concentration near the interface to the silicide is desirable, since this increases the tunneling current across the Schottky barrier, and will thus result in low contact resistivity. On the other hand, it is difficult to achieve highly doped nanowire S/D extensions without compromising short channel effects. For the devices in this work, the As+ SDE implant was implanted after deposition of an offset SiO_2 liner for a slight gate-

underlap. P+ implant was employed after the formation of nitride spacers. Fig. 6 shows the simulated active dopant concentrations for 10 nm nanowire devices after S/D activation. Implanting P+ at a dose of 5 x 10^15 cm^-2 results in significantly higher dopant concentration near the spacer edge than that obtained with implanting at a dose of 1 x 10^15 cm^-2. This results in high dopant concentration at the semiconductor region near the silicide interface, which is crucial for ensuring a low contact resistivity.

The device splits are shown in Table I. All splits received the same SDE implants. After nitride spacer formation, both the Control and split A received 1 x 10^15 cm^-2 P+ implants while split B received 5 x 10^15 cm^-2 P+ implants. All splits underwent the same 950°C spike anneal. Subsequently, both splits A & B were silicided using 4 nm of deposited Ni. The choice of this Ni thickness is thin enough to avoid "over-silicidation" of the nanowire S/D regions, which would otherwise result in excessive lateral silicide encroachment (Excessive silicide encroachment can lead to Schottky transistor-like device behaviour or device shorts.). The I_{Off}-I_{On} plot is shown in Fig. 7. At a fixed I_{Off} of 100 nA/μm, split A shows a 640% enhancement in I_{On} from 400 $\mu A/\mu m$ to 2560 $\mu A/\mu m$. This highlights the importance of silicidation for improving nanowire drive current performance. Split B shows a further 43% enhancement in drive current than split A, giving a very high drive current of 3670 $\mu A/\mu m$ for n-channel nanowire transistors at the same I_{Off}. Fig. 8 plots the series resistance R_{SD} fitting at V_G=10V for the various splits. The inset plots R_{Tot} against gate voltage with an extroplation to 10V. At low drain bias and high gate bias, the channel resistance diminishes asymptotically and R_{Tot} approaches the value of the parasitic S/D series resistances. It is observed that silicidation improves the series resistance tremendously. The series resistance of split B is also drastically improved over that of split A, which indicates that high S/D extension dopant concentration indeed helps to reduce the contact resistivity by thinning the Schottky barrier and lowering the effective Schottky barrier height at the n+-Si-NiSi junction. This series resistance reduction gives rise to the 43% further enhancement in I_{On} for split B.

It is possible that the high deep S/D doping in split B can affect short channel control due to increased dopant diffusion into the channel regions. Fig. 9 plots I_{On} against subthreshold swing, which gives the dependence of I_{On} on the gate length. The enhancement in I_{On} with decreasing gate length is obvious in split B, as the total resistance is no longer dominated by parasitic series resistances. On the other extreme, I_{On} in the control does not show a significant dependence on gate length as the current is suppressed due its high parasitic resistances. Fig. 10 plots I_{Off} against $V_{T,sat}$, which compares the short channel matching for the various splits. Excellent matching at short gate lengths is observed, which confirms that no short channel degradation occurred in splits A and B compared to control. Fig. 11 compares the transfer characteristics of a split B device against a control device (L_G=8 nm). Excellent DIBL and SS of ~22 mV/V and ~75 mV/dec were obtained owing to the GAA device architecture and gate-underlapped S/D extension doping. Without comprising short channel performance, successful integration of low resistivity metallic contacts on GAA nanowire devices results in a record-high drive current of 3740 $\mu A/\mu m$ at a V_G - V_T = 1 V. This is achieved at a relatively thick oxide thickness of 4 nm, which will translate to reduced gate delay. Table II summarizes and compares the key device parameters of this work with other prior work. Devices from this work exceed the performance of prior work by a significant amount, since the limit imposed by parasitic series resistances has been lifted.

Conclusions

Nickel silicide resistivity values at ultra-thin thicknesses deviate significantly from its bulk value. The formation of low resistivity metallic nanowire contacts to nanowire transistors requires process optimization, and has been successfully demonstrated in this work. The record-high absolute device performance obtained strongly suggests that low resistivity metallic nanowire contact technology is of great importance in realizing state-of-the-art nanowire devices.

Reference

[1]. K. H.Yeo et al., *IEDM Tech. Dig.* , 2006, pp.539-542
[2]. N. Singh et al., *IEDM. Tech. Dig.*, 2006, pp.547-550
[3]. J. Kedzierski et al., *TED.*,2003, pp. 952-958
[4]. M. C. Öztürk et al., Symp. VLSI Tech., 2005, pp. 194-195
[5]. R. Mukai et al., *Thin Solid Film*, 1995, pp. 567-572
[6]. H. Lee at al., *VLSI Tech. Symp.*, 2006, pp.58-59
[7]. F. -L. Yang et al., *VLSI Tech. Symp.*, 2004, pp. 196-197
[8]. Y. Tian et al., *IEDM Tech.Dig.*, 2007, pp. 895-898

Fig. 1 Process Key Steps:

- Nanowire formation and release
- Gate stack formation & trimming
- LDD implantation and spacer formation
- **Deep S/D implantation (Optimization)**
- **Nanowire Silicidation (Thin Ni silicidation)**
- Metalization and backend process

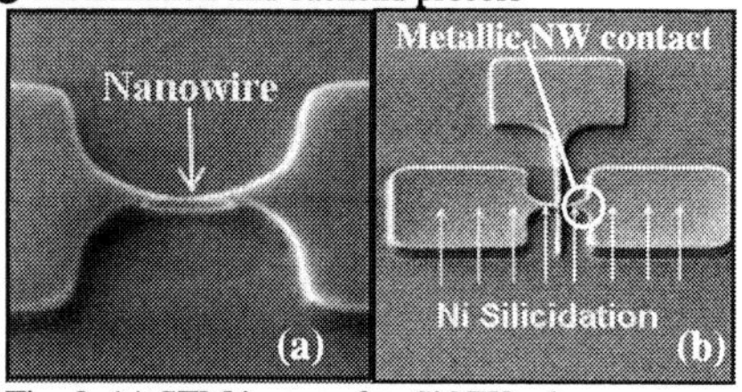

Fig. 2 (a) SEM image after Si NW release (b) SEM image after gate etch. Arrows indicate nanowire and pad Silicidation.

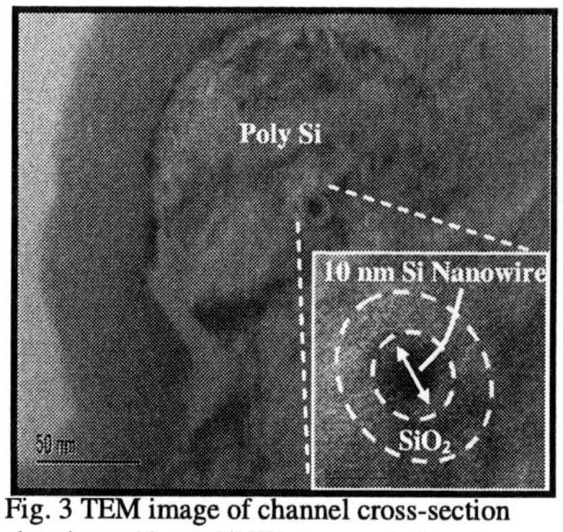

Fig. 3 TEM image of channel cross-section showing a 10 nm SiNW surrounded by 4 nm oxide and followed by poly silicon.

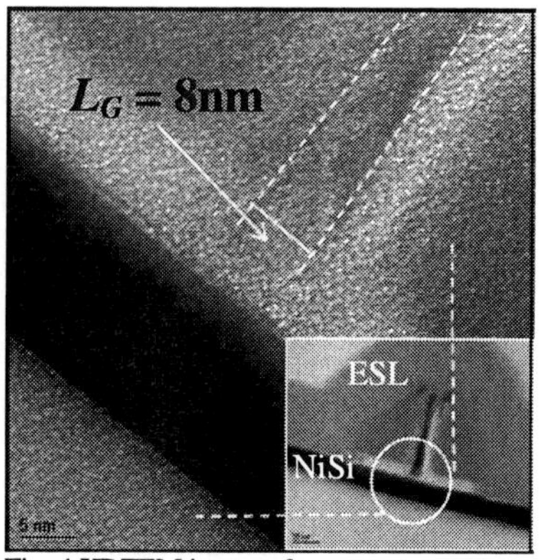

Fig. 4 HRTEM image of gate structure cross-section, gate length is ~8nm. Inset shows zoom out view of gate structure, with ESL as etching stop layer.

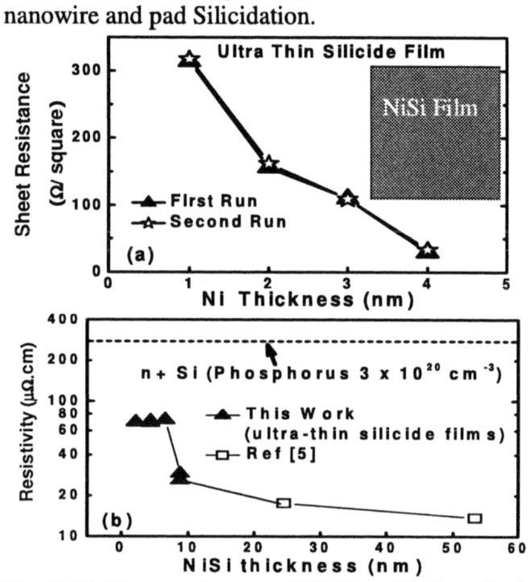

Fig. 5 (a) Sheet resistance of Ni Si vs. Ni deposited thickness. Inset shows SEM image of NiSi film. (b) NiSi Resistivity vs NiSi thickness.

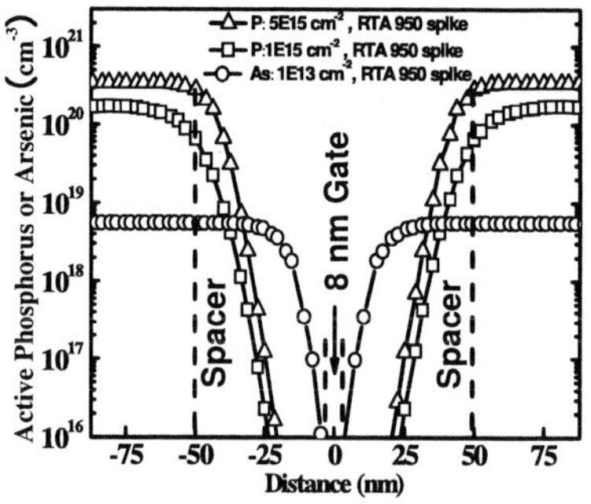

Fig. 6 Active dopant profile with different implant condition. High dopant concentration underneath the spacers is essential for low silicide contact resistivity.

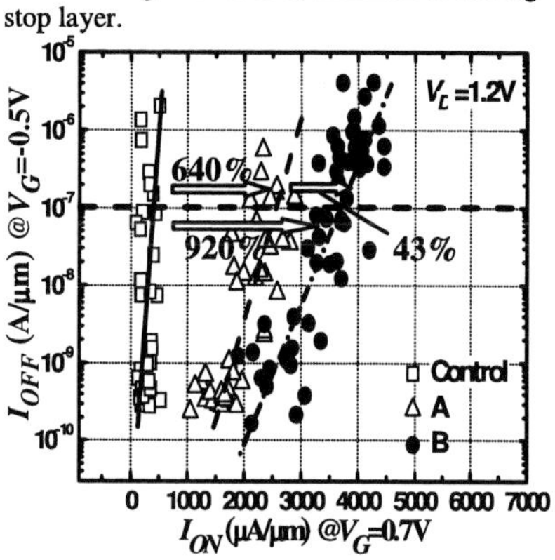

Fig. 7 Comparison of I_{off} versus I_{on}. Reduction in extension and contact resistance results in 43% enhancement of B over A.

Table I: Split table for this work

samples	P dose (cm^{-3})	Silicidation
Control	$1E^{15}$	×
A	$1E^{15}$	4 nm Ni
B	$5E^{15}$	4 nm Ni

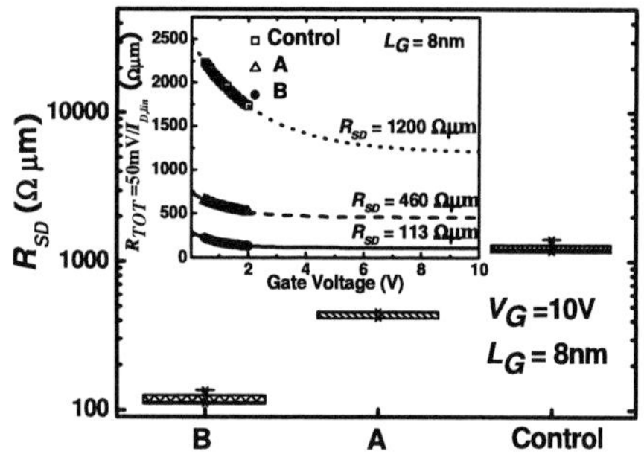

Fig. 8 Series Resistance for three samples at L_G = 8nm. Inset shows R_{total} versus V_G. Series resistance of sample B is 113Ω μm at V_G = 10V.

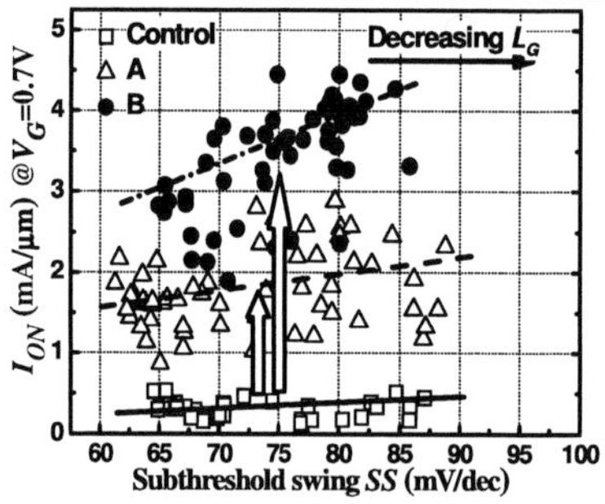

Fig. 9 Comparison of I_{oN} versus SS. Sample B shows large improvement over control and sample A.

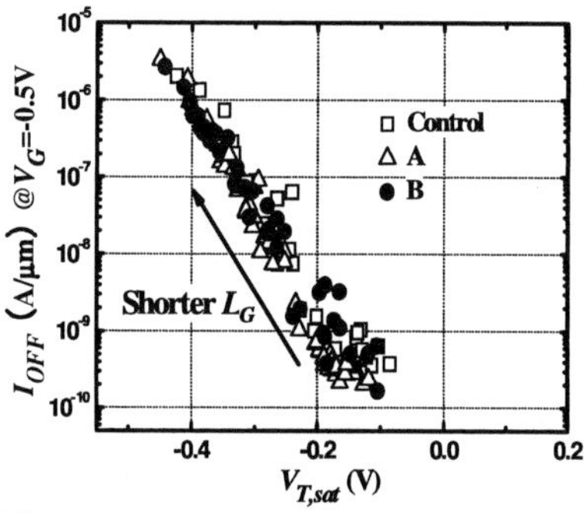

Fig. 10 Comparison of I_{OFF} versus $V_{T,sat}$. No short channel degradation has been found on sample A and B compared to control.

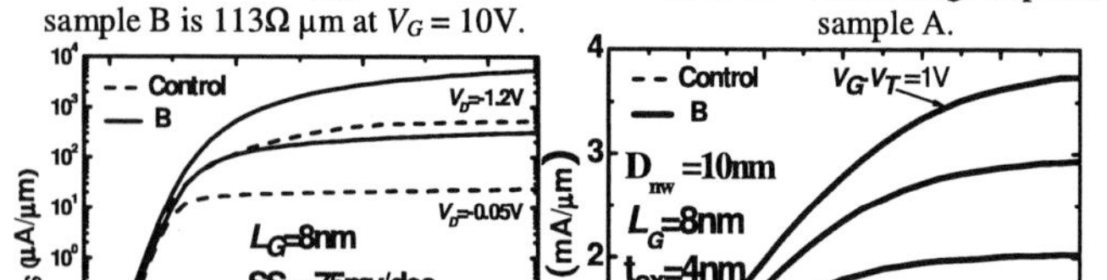

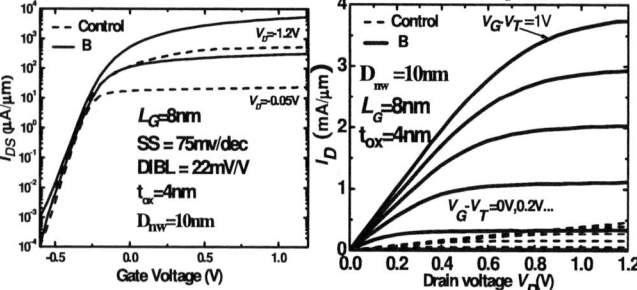

Fig.11 (a)(b)showing DC characteristics comparison between control and Sample B. Large enhancement has been obtained on sample B. Drive current is 3740μA/μm@V_G-V_T=1V.

Table II: Comparison of SNWTs using CMOS compatible process

	This work	Ref[1]	Ref[2]	Ref[6]	Ref[7]	Ref[8]
NW size radius (nm)	5	4	1.5	W=3, H=14	5	~5
Gate structure	GAA	GAA	GAA	GAA	Ωshape	GAA
Gate type	Poly Si	TiN	Poly Si	Poly Si	Poly Si	Poly Si
Dielectric	SiO2	SiO2	SiO2	HfO2	SiO2	SiO2
Lg(nm)	8	15	350	5	10	130
tox/EOT (nm)	4	3.5	4	1.2	1.9	5
Vdd (V)	1.2	1.0	1.2	1.0	1	1.5
Ion (μA/μm)	3740	1440	2400	497	522	1039
Normalization	Diameter	Diameter	Diameter	Perimeter	---	Diameter
SS (mV/Dec)	75	72	60	208	75	72-74
DIBL (mV/V)	22	50	6	230	80	4~12
Ion/Ioff	$>10^7$	10^6	10^6	5×10^2	10^5	$>10^8$

5 nm Gate Length Nanowire-FETs and Planar UTB-FETs with Pure Germanium Source/Drain Stressors and Laser-Free <u>Melt</u>-<u>E</u>nhanced <u>D</u>opant (MeltED) Diffusion and Activation Technique

Tsung-Yang Liow[1,2], Kian-Ming Tan[1], Rinus T. P. Lee[1], Ming Zhu[1], Ben L.-H. Tan[2],
Ganesh S. Samudra[1], N. Balasubramanian[2], and Yee-Chia Yeo[1]

[1]Silicon Nano Device Lab., Dept. of Electrical and Computer Engineering, National University of Singapore (NUS), 117576 Singapore.
[2]Institute of Microelectronics, 11 Science Park Road, 117685 Singapore.
Phone: +65 6516-2298, Fax: +65 6779-1103, E-mail: yeo@ieee.org

Abstract

We report the first demonstration of pure Ge source/drain (S/D) stressors (unembedded) on the ultra-narrow or ultra-thin Si S/D regions of Nanowire-FETs and UTB-FETs, compressively straining the channels to provide up to ~100% I_{Dsat} enhancement. Devices with 5 nm gate lengths were fabricated. In addition, we report a novel Melt-Enhanced Dopant (MeltED) diffusion and activation technique to form embedded Ge S/D stressor in the S/D regions of nanowire-FETs, boosting the channel strain even further, and achieving ~125% I_{Dsat} enhancement.

Introduction

Embedded SiGe source and drain (S/D) stressors induce uniaxial compressive channel stress, improve hole mobility, and enhance I_{Dsat} performance in p-channel transistors. There have been a few reports on integrating embedded SiGe stressors (with low Ge %) in multiple-gate transistors [1],[2]. Integrating embedded SiGe stressors with Ultra-Thin Body (UTB) SOI planar or even nanowire transistors is extremely challenging, since there is very limited margin for S/D recess etch prior to SiGe epitaxy. With reduced body thickness, lattice strain coupling from the S/D stressors is reduced. No successful attempt has ever been reported on forming Ge S/D stressors or even embedding Ge S/D stressors in nanowire-FETs or UTB-FETs.

In this work, we demonstrate for the first time, the successful integration of pure Ge S/D stressors with planar UTB-FETs and nanowire-FETs, resulting in very dramatic performance enhancement. The effect of substrate compliance in reducing defect density and increasing induced channel strain is also exploited. We further show that such highly-strained Ge S/D devices can be further improved by employing a laser-free MeltED diffusion and activation technique which uniformly dopes the Ge S/D regions. This technique simultaneously forms embedded SiGe stressors with very high Ge content (up to ~85%), effecting further strain-induced hole mobility enhancement as a result of improved lattice strain coupling.

Device Fabrication

Multiple-gate ultra-thin body transistors with 30-80 nm widths were fabricated on 8 nm thick (100) surface SOI wafers. Fig.1 summarizes the key process steps. The gate etch process was tuned to produce bottom-tapered gates for small physical gate lengths L_G. A translucent bottom-tapered gate running over a nanowire is shown in Fig. 2(a), the gate length of which is obtained from TEM to be 5 nm. To induce high compressive channel strain, lattice-mismatched Ge stressors were epitaxially grown in the S/D regions of 3 device splits. Excellent selectivity was achieved over SiO_2 and SiN, as shown in Fig. 3.

Two of these device splits have thick Ge S/D stressor films: 'Thick Ge S/D', which is unembedded, and 'Ge S/D (MeltED)' which is embedded with a new Laser-Free Melt-Enhanced Dopant (MeltED) Diffusion and Activation Technique to be described later. One split ('Thin Ge S/D') has a thin Ge S/D stressor, the thickness of which is 50% that of the other two splits. For the splits with thick Ge stressors, one split (Ge S/D (MeltED)) was capped with SiO_2 and subjected to a 950°C RTA during S/D activation. This melted the Ge film which immediately re-crystallized upon cooling, using the underlying Si as a seed. The other two Ge splits and the Si control were also capped with SiO_2 but underwent a non-melt 900°C RTA during S/D activation.

Results and Discussion

A. Ge S/D stressor technology for UTB planar and nanowire transistors

Ge S/D stressors (Thick Ge) result in 80% I_{Dsat} enhancement (Fig. 4) and 135% I_{Dlin} enhancement (Fig. 5) over Control devices with raised Si S/D at a fixed I_{Off} of 1 ×10⁻⁷ A/µm. Device width W is 80 nm. At a fixed DIBL of 75 mV/V, 105% enhancement in peak linear transconductance G_m was obtained. Such a large G_m enhancement indicates large hole mobility enhancement due to compressive channel strain. G_m enhancement is more pronounced at larger DIBL values (shorter L_G), in agreement with typical observations for other local S/D stressor technologies. While Ge films more than a few nanometers thick tend to relax significantly when grown on bulk Si, substrate compliance in ultra-thin SOI possibly allows for greater elastic strain accommodation due to the enhanced viscous flow of the underlying SiO_2 under large amounts of shear stress [3]. Furthermore, narrow structures tend to allow elastic lateral expansion of the Ge film [4]. TEM analyses (Fig. 7) showed that Ge grown on a wider SOI pattern had a much higher defect density compared to a narrower nanowire-like pattern, which has few observable defects. Fig. 8 shows the dependence of I_{Dsat} (extracted from I_{Off}-I_{Dsat} plots) on device width. For the control device with raised Si S/D, the increase in I_{Dsat} with smaller W can be attributed to the non-negligible contribution of sidewall conduction in the 8 nm SOI. A trend of

increasing strain-induced enhancement with decreasing W is observed. This correlates well with the reduced defect density in narrow patterns. For W = 30 nm, 96% I_{Dsat} enhancement was obtained for the Thick Ge S/D split, with the Thin Ge S/D split not too far behind (77%). This is especially significant, considering that the Thin Ge split only has half the Ge S/D stressor thickness. It also suggests that for devices with very narrow widths, e.g. nanowire devices, a thin Ge S/D stressor may be sufficient to induce large channel strain for significant performance benefits.

B. MeltED Diffusion and Activation Technique & Stressor Embedding

Key steps for forming MeltED or embedded Ge S/D stressors is shown in Fig. 9. When Ge is melted at temperatures exceeding 938°C, extremely fast liquid-state diffusion of dopants allow for rapid and uniform distribution of dopants throughout the Ge S/D stressor. During this process, Si and Ge inter-diffusion also occurs at the heterointerface, resulting in "embedding" of the Ge stressor for improved channel stress coupling effects. Fig. 10 shows the sheet resistance values of films with 2 different Ge thicknesses. The films were doped near the surface with low energy 5keV BF_2^+ implantation of identical dose. After capping and undergoing the MeltED anneal, the sheet resistances were measured and plotted with calculated resistivity values in Fig. 10. The identical resistivity values clearly indicate uniform distribution and incorporation of B in Ge. Hence, the results prove that this technique can facilitate uniform dopant diffusion and incorporation regardless of Ge thickness or S/D geometries.

Compared to Ge S/D devices without Ge melting (referred to as Thick Ge S/D Stressors earlier in Fig. 4), devices with MeltED or embedded Ge S/D exhibit a further 10% I_{Dsat} enhancement (Fig. 11) as well as a further 15% peak G_m improvement (Fig. 12). Since MeltED Ge S/D devices employ a slightly higher activation thermal budget, one might question if the enhancement is simply due to a decrease in effective channel lengths or S/D series resistances. Fig. 13 plots the cumulative distributions of SS, R_{tot}, and $G_{m,max}$ for 2 sets of devices with similar mask gate lengths. It is quite clear that non-melted Ge and MeltED Ge devices have comparable short channel control and S/D series resistances (estimated from R_{tot} at high gate overdrive). I_D-V_G characteristics in Fig. 14 show comparable SS and DIBL between a MeltED Ge S/D and a Si S/D control device. G_m is clearly enhanced by ~22%. The MeltED Ge S/D device shows a dramatic 120% I_{Dsat} enhancement over the Si S/D control device (Fig. 15). On average, the I_{Dsat} enhancement at an I_{Off} of 1 × 10⁻⁷ A/µm obtained with the MeltED Ge technique is close to 100%.

Since the lattice mismatch between Ge and Si is 4%, the substrate compliance effect of narrow and thin SOI lines may not be able to completely suppress the formation of dislocations. It is important that such defects are formed outside the S/D depletion regions during transistor operation. Fig. 16 compares the junction leakage currents of Si Control, non-melted Ge S/D and MeltED Ge S/D devices. It is found that the leakage currents are quite comparable and are of a low magnitude.

Energy dispersive spectrometry (EDS) of a MeltED Ge S/D nanowire's S/D reveals uniform ~85% Ge concentration. It is postulated that as Ge melts, Si also dissolves in the molten Ge. In a Si/Ge core/shell nanowire S/D region, the small volume of Si can completely dissolve in the relatively much larger surrounding volume of molten Ge, leading to the formation of a uniform SiGe alloy upon re-crystallization. In this case, the seed for re-crystallization is located underneath the gate spacers. As a result, fully embedded Ge-rich SiGe stressors are formed in the S/D regions of nanowire devices. I_D-V_G transfer characteristics in Fig. 18 show comparable SS and DIBL for MeltED Ge and non-melted Ge S/D nanowire devices. Such embedded stressors result in a further 16% I_{Dsat} enhancement in the MeltED Ge S/D nanowire device, improving its drive current to 609 µA/µm (Fig. 19). Note that such performance levels were achieved without S/D metallization.

Conclusions

Pure Ge S/D stressors induce large compressive channel stress, enhancing hole mobility greatly. Substrate compliance effects result in larger I_{Dsat} enhancement for narrow width nanowire devices compared to wider width devices. By employing a Ge melting technique, dopant activation and stressor embedding was accomplished in a single process step, achieving I_{Dsat} enhancement of ~100% and ~125% for planar UTB-FETs and nanowire-FETs respectively.

References

[1] P. Verheyen et al., Symp. VLSI Tech., pp. 194-195, 2005.
[2] J. Kavalieros et al., Symp. VLSI Tech., pp. 62-63, 2006.
[3] F. Liu et al., Phys. Rev. Lett., Vol 89, No. 13, pp. 136101, Sep. 2002.
[4] D. Zubia et al., J. Vac. Sci. Technol. B, Vol. 18, No. 6, Nov./Dec. 2000.

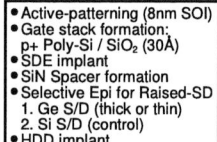

- Active-patterning (8nm SOI)
- Gate stack formation:
  p+ Poly-Si / SiO₂ (30Å)
- SDE implant
- SiN Spacer formation
- Selective Epi for Raised-SD
  1. Ge S/D (thick or thin)
  2. Si S/D (control)
- HDD implant
- SiO₂ capping
- RTA activation
- Ni silicidation for Si control
- Contact and metallization

Fig. 1. Key steps in process flow including 5 nm gate definition and selective Ge epitaxy in S/D.

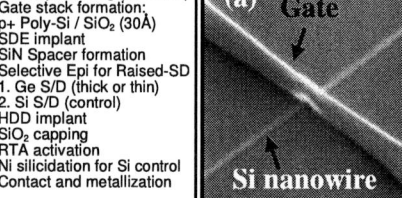

Fig. 2. (a) Tilted-SEM image of a UTB nanowire device's translucent gate. Etch process was designed to form tapered gates which is narrowest at the bottom. (b) TEM showing the gate length at the gate bottom.

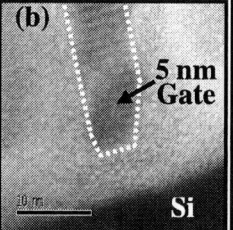

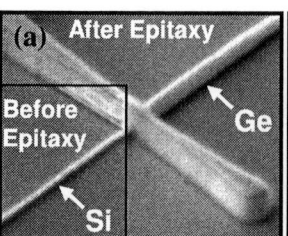

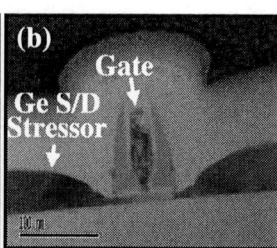

Fig. 3. (a) Tilted-SEM image of a UTB nanowire device before (inset) and after selective Ge epitaxy. Excellent selectivity to SiN and Ge SiO₂ was achieved. (b) TEM showing the tapered gate profile and Ge raised S/D regions on a UTB device with 8 nm body thickness.

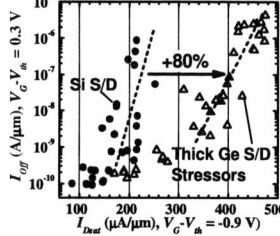

Fig. 4. Ge S/D stressors result in 80% I_{Dsat} enhancement at a fixed I_{Off} of 1×10⁻⁷A/μm. Ge stressors grown on S/D are unembedded. Raised Si S/D were used for all control devices.

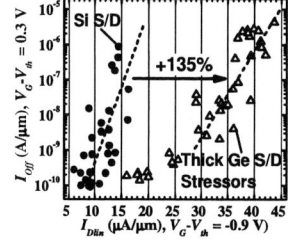

Fig. 5. At I_{Off} of 1×10⁻⁷A/μm, I_{Dlin} enhancement of 135% is larger than I_{Dsat} enhancement, as I_{Dlin} is less limited by series resistance.

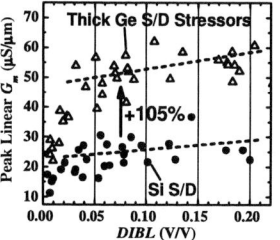

Fig. 6. Hole mobility is very significantly enhanced by large compressive stress due to Ge S/D stressors. Peak transconductance is increased by 105% at a DIBL of 75 mV/V.

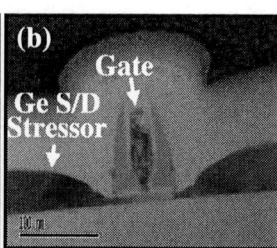

Fig. 7. Defect density is lower in nanowire-like regions than in wide active patterns, suggesting lesser strain relaxation in narrower structures due to substrate compliance effect.

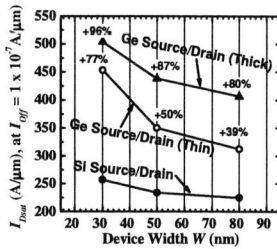

Fig. 8. I_{Dsat} enhancement (indicated in %) for 2 different Ge S/D stressor thicknesses, compared to a control with raised Si S/D. UTB-FETs with small width W have larger enhancement. For W of 30 nm, I_{Dsat} enhancement due to Thin Ge S/D stressor approaches that of Thick Ge S/D.

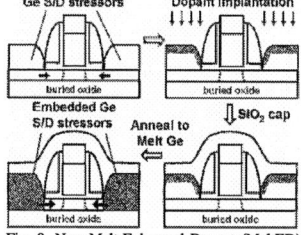

Fig. 9. New Melt-Enhanced Dopant (MeltED) diffusion and activation process. After shallow S/D implant and SiO₂ capping, a 950°C spike anneal melts the Ge-rich region, and achieves these key objectives: Interface inter-diffusion *embeds* the Ge stressor; Dopant diffuses, redistributes uniformly, and is substitutionally incorporated as Ge recrystallizes.

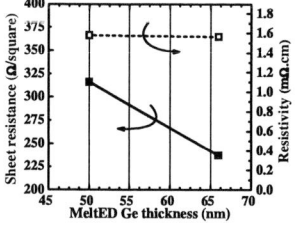

Fig. 10 Sheet resistance measurements of MeltED Ge with 2 thicknesses. Both received surface BF₂⁺ implants only. Both films have near identical resistivity values, which confirms that Boron is uniformly diffused in the liquid Ge and is substitutionally incorporated as Ge re-crystallizes.

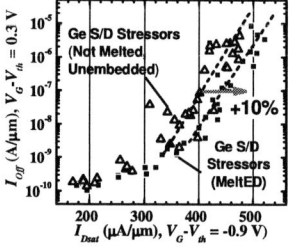

Fig. 11 MeltED Ge S/D stressors are embedded, and gives a further 10% I_{Dsat} enhancement at I_{Off} of 1×10⁻⁷A/μm, as compared to the unembedded Ge S/D stressors (also plotted in Fig. 4). Greater strain effects come with S/D stressor embedding.

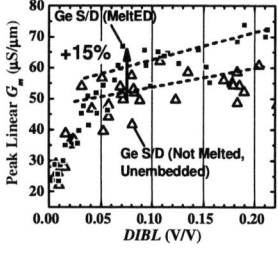

Fig. 12. An additional 15% enhancement in peak transconductance was obtained at a DIBL of 75 mV/V as a result of increased channel strain.

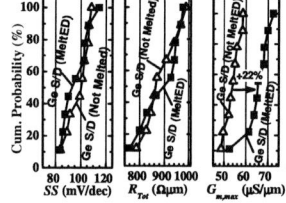

Fig. 13. P-FETs with MeltED/Embedded Ge S/D have 22% higher $G_{m,max}$ than those with unembedded Ge S/D (not melted). All p-FETs have the same short channel control and L_G (35nm). R_{Tot} at high gate overdrive (V_G-V_{th} = 2.5V) estimates R_{SD} to be comparable.

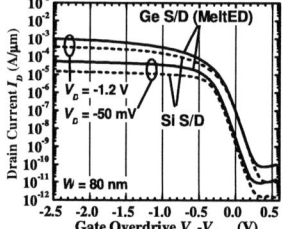

Fig. 14. I_D-V_G plot showing comparable DIBL and SS for a p-FET with Raised Si Control and a p-FET with Embedded Ge S/D (MeltED).

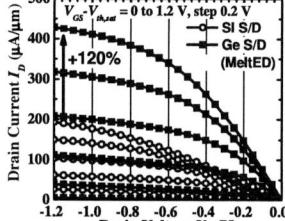

Fig. 15. I_D-V_D plot of same pair of devices in Fig. 14. Embedded Ge S/D (MeltED) gives a 120% I_{Dsat} enhancement over a p-FET with Si S/D.

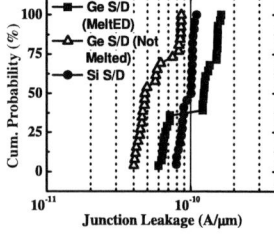

Fig. 16. Ge S/D stressors does not significantly impact junction leakage, indicating that defects are well-confined outside the extension regions.

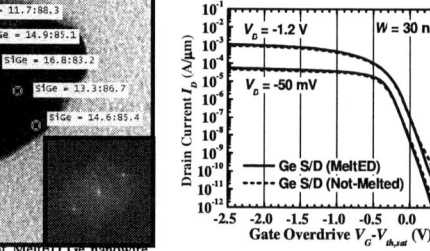

Fig. 17 EDS analysis of MeltED Ge nanowire S/D shows uniform ~85% Ge concentration from top to bottom. Reciprocal space diffractogram (inset) shows single crystallinity.

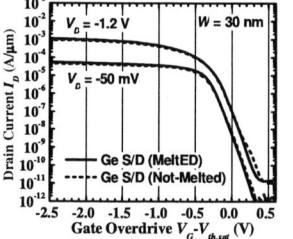

Fig. 18. I_D-V_G transfer characteristics showing a pair of nanowire p-FETs with Ge S/D with comparable DIBL and SS.

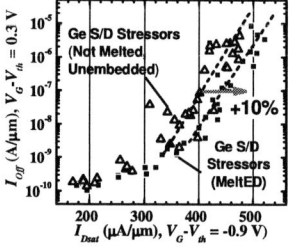

Fig. 19. I_D-V_D plot of same pair of Nanowire p-FETs in Fig. 18. Embedded Ge S/D (MeltED) shows a further 16% I_{Dsat} enhancement over unembedded Ge S/D.

30

TSNWFET for SRAM cell application: Performance Variation and Process Dependency

Sung Dae Suk, Yun Young Yeoh, Ming Li, Kyoung Hwan Yeo, Sung-Han Kim*,
Dong-Won Kim, Donggun Park, and Won-Seoung Lee

Advanced Technology Development Team 1, MTT2 Team*, R&D Center, Samsung Electronics Co.
San 24, Nongseo-Dong, Kiheung-Ku, Yongin-City, Kyoungi-Do, 449-711, KOREA
Phone: +82-31-209-6668, Fax:+82-31-209-3274, E-mail: sd1.suk@samsung.com

Abstract

I_{ON} is increased about 25 % with the width/height (W/H) of 12/24 nm nanowire (NW) in comparison with the W/H of 12/12 nm at V_G-V_{TH}=1 V. With these results, we have successfully fabricated NW SRAM arrays with the W/H of 5/15 nm and L_G of 40 nm for the first time. Static noise margin (SNM) of 325 mV is achieved at V_D=1 V. NW height and gate oxide thickness dependency of n-ch twin silicon nanowire MOSFET (TSNWFET) on device variations is investigated. Line edge roughness and size variation are more critical than random dopant fluctuation in TSNWFET.

Introduction

As technology is developed into the deep nano-era, not only performance but device variations such as random dopant fluctuation (RDF) and line edge roughness (LER) become one of the severe issues for the further CMOS applications [1]. Normally, channel width and C_{OX} are more flexible parameters than V_G, V_{TH} and V_D to improve I_{ON}. In this paper, we firstly handled the two parameters of H and gate oxide thickness (T_{OX}) of TSNWFETs [2] to enhance the performance and analyze the variations. Additionally, reliabilities are investigated with the various NW sizes. Finally, SRAM arrays with TSNWFETs are successfully fabricated and evaluated with the butterfly curves by using the tall NW.

Electrical Result

Fig. 1 and Fig. 2 show the process flow and the split groups of NW size and T_{OX} for investigating the device performance as well variations and reliability. Fig. 3 -5 show I-V curves, V_{TH}/I_{ON} and I_{ON}/I_{OFF} correlations, and normalized I_{ON} by circumference as a function of H. Size variation of NW is minimized to about 1 nm in a wafer by subtly controlled process. 25 % improvement of the measured I_{ON} is observed as H increases 2 times while 45 % improvement of it as T_{OX} decreases from 2.5 nm to 1.3 nm at W of 12 nm, V_G-V_{TH}=1 V and I_{OFF} of around 10^{-11} A. 3 times higher H at W of 5 nm also causes 33 % I_{ON} enhancement as well. However, normalized I_{ON} decreases with taller NW contrary to the measured I_{ON} trend. This means that the current density of round-shape NW is enhanced in comparison with that of elliptical one though of measured I_{ON} reduction. Fig. 6 (a) shows V_{TH}'s as a function of L_G. C, D and E groups have the immunity to roll-off until L_G of 30 nm but A and B groups suffer 50mV reduction of V_{TH}. D group (W/H = 5/5 nm) also shows 0.1 V higher V_{TH} than other groups due to quantum confinement effect while quantum confinement is not effective on group E because of higher H. Fig. 7 (a) shows the drain induced barrier lowering (DIBL) trends along to L_G. DIBLs are measured to be 50mV/V at 30 nm L_G and 2.5 nm T_{OX} groups. DIBL is only improved with T_{OX} of 1.3 nm.

Fig. 8 demonstrates 'Pelgrom' plots which represent the normalized transistor variations [3]. Gate-all-around (GAA) TSNWFET shows at least 2 times smaller variation while 3.5 times smaller width than planar's due to no channel implantation and good gate controllability [4]. As W/H or T_{OX} decrease, the V_{TH} variation is much suppressed shown in C and D groups. In order to know the detailed factors affected to variations on TSNWFET, V_{TH} and DIBL are analyzed with standard deviations (STD). V_{TH} and DIBL differences of forward and reverse sweep measurement (ΔV_{TH} and $\Delta DIBL$) are also examined as a function of L_G to investigate the asymmetry of S/D in NW by S/D dopants shown in Fig. 6 (b) and Fig. 7 (b). STDs of ΔV_{TH} show similar values with various NW sizes at long L_G of 114 nm shown in Fig. 6 (b)

so that size variation at S/D ends of NW is negligible. Basically, STD of V_{TH} with group D (W/H of 5/5 nm) is larger than that with group A (W/H of 12/12 nm) as shown in Fig. 6 (a). However, STD of ΔV_{TH} indicates opposite trend that larger NW shows larger variation between A and D groups at L_G of 30 nm in Fig. 6 (b). In Fig. 7 (b), STD of $\Delta DIBL$ and relative variation of $\Delta DIBL$ are also reduced as NW size decreases from group A to group D at L_G of 30 nm. In other words, group D has better gate controllability than group A. From these results, it is concluded that though the asymmetry of S/D with smaller NW is gradually enlarged by dopants from S/D due to decrease in existence of dopants and increase in randomness by discreteness of few dopants in NW, influence of S/D dopants on V_{TH} and DIBL variations is reduced with smaller NW. V_{TH} variations in TSNWFET mainly come from LER including NW shape and size.

Fig. 9 shows S.Swing as a function of L_G for all groups. S.Swing is improved about 5 mV/dec. with thinner T_{OX} of 1.3 nm due to increase in C_{OX}. Fig. 10 shows GIDL graphs for H split groups and gate leakage curves along to T_{OX}. Because of small cross-sectional area of the NW with lower dopant diffusivity into the channel, the GIDLs are shown similar values. The gate leakage with 1.3 nm SiO_2 shows 2 order higher with 8×10^{11} A compared to 2.5 nm SiO_2. Fig. 11 shows hot carrier lifetime with the various NW sizes. 10 year lifetime is guaranteed at 1.35 V with 5 nm NW while at 1.49 V with 12 nm NW. From these data, the larger and taller NW show better hot carrier lifetime because of smaller vertical electric field. Additionally, NBTI characteristics at 125 °C are investigated in p-ch TSNWFET along to d_{NW} of W/H=1. When d_{NW} decreases, V_{TH} degradation and its STDs are enlarged due to increase in vertical field and influence of surface conditions to channel.

Slightly taller NW channel is beneficial to achieve high measured I_{ON} with proper V_{TH} for high performance application and improve hot carrier lifetime compared to round-shape NW at same NW width and similar I_{OFF}. In case of variation, GAA NW is more affected by LER or size variation than by RDF as the NW size decreases. Based on these reasons, Fig. 13 shows top view SEM and cross-sectional TEM images of SRAM arrays with TSNWFETs having W/H of 5/15 nm NW, T_{OX} of 2.5 nm and L_G of 40 nm. Fig. 14 shows I-V curves of pull up (PU) and pull down (PD) transistors. PU and PD transistors are well operated with good SCE immunity and V_{TH} variation of 0.12V. Typically, PMOS current is larger than NMOS due to the compressive stress by SiGe under S/D [5][6]. By these I-V curves, static noise margin (SNM) is obtained 325 mV at V_D=1 V and 278 mV at V_D=0.8 V shown in Fig. 15. The minimum SNM is achieved 262 mV at V_D=1 V.

Conclusion

Slightly taller NW can become a good candidate for SRAM with considering of V_{TH} control, high measured I_{ON} at the same NW width and better reliability, while round-shape NW can be still effective to achieve the high performance with lower CV/I. Variations are also extremely suppressed with GAA TSNWFET compared to planar MOSFET. Therefore, 40 nm SRAM arrays with selected NW size of W/H = 5/15 nm are successfully fabricated and SNM of about 325 mV at V_D=1 V is achieved with the V_{TH} range of 0.12 V. From these investigations and results, GAA TSNWFET seems to be a solution for future memory and logic devices

Reference

[1] A. Asenov, *Symp. on VLSI Tech*, p. 86, 2007 [2] S. D. Suk *et al.*, *IEDM Tech*, p.717, 2005. [3] M. J. M. Pelgrom *et al.*, *IEDM Tech*, p. 915, 1998. [4] K. J. Kuhn, *IEDM Tech*, p. 471, 2007. [5] M. Li *et al.*, *IEDM Tech*, p. 899, 2007. [6] S. D. Suk *et al.*, *IEDM Tech*, p.891, 2007

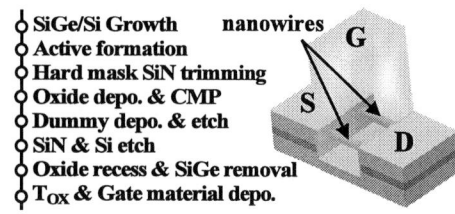

- SiGe/Si Growth nanowires
- Active formation
- Hard mask SiN trimming
- Oxide depo. & CMP
- Dummy depo. & etch
- SiN & Si etch
- Oxide recess & SiGe removal
- T_{OX} & Gate material depo.

Fig. 1. Process flow and schematic view of TSNWFET.

Split Group	d_{NW}		Tox (nm)	Cross-Sectional NW
	W (nm)	H (nm)		
A	12	12	2.5	
B	12	24	2.5	W/H (nm) 12/12
C	12	12	1.3	
D	5	5	2.5	
E	5	15	2.5	12/24

Fig. 2. NW size and T_{OX} are split for investigating the performance and reliability.

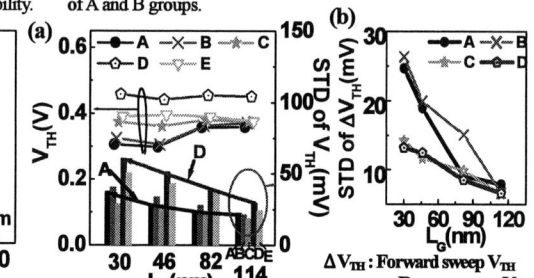

Fig. 3. I_D-V_G curves of A,B, C and D groups and I_D-V_D curves of A and B groups.

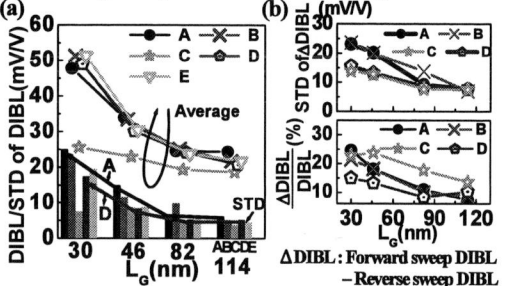

Fig. 4. As H increases 2 times at W of 12 nm NW, I_{ON} is enhanced 25 % at V_G-V_T of 1 V. But normalized I_{ON} by circumference decreases with taller NW. Round-shape NW helps to enhance the current density, while taller NW helps to improve the absolute measured I_{ON} at the same W.

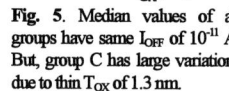

Fig. 5. Median values of all groups have same I_{OFF} of 10^{-11} A. But, group C has large variations due to thin T_{OX} of 1.3 nm.

Fig. 6. STD of V_{TH} with small NW is larger than that with large one. But, STD of ΔV_{TH} shows reverse trend as L_G decreases. This result implies that asymmetry of S/D by random positioned dopants in NW is relatively not much affected to V_{TH} variation with smaller d_{NW}.

Fig. 7. DIBLs and STDs of DIBL are improved as T_{OX} decreases. STD of ΔDIBL indicate better gate controllability with C and D groups. Relatively large variations of ΔDIBL with A and C groups compared to group D implies that DIBL variation for larger NW has affected more than that for smaller one by asymmetric variation of S/D from RDF.

Fig. 8. GAA NW FET has more immune to V_{TH} variations than planar MOSFET due to no channel implantation and good gate controllability. As d_{NW} or T_{OX} decreases, the slopes lean more as shown in C and D groups due to increase in gate controllability compared with group A.

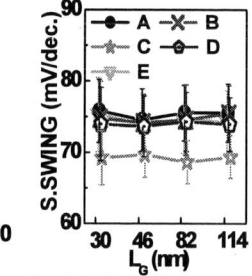

Fig. 9. S.Swing is improved about 5 mV/dec. when T_{OX} changes from 2.5 nm to 1.3 nm due to increase in C_{OX}.

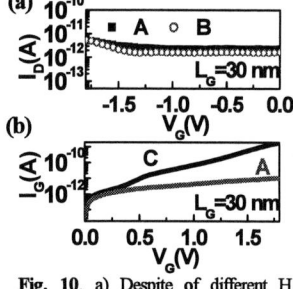

Fig. 10. a) Despite of different H, GIDLs show similar value.
b) Gate leakage current of 1.3 nm SiO_2 is 8×10^{-11} A at V_G=1 V.

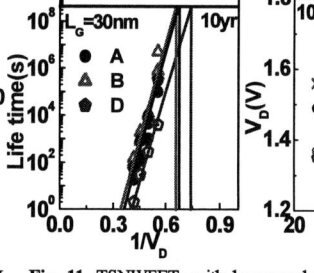

Fig. 11. TSNWFETs with larger and taller NW show better hot carrier lifetime due to smaller vertical electric field. 10 year lifetime is guaranteed at 1.35 V with the W/H of 5/5 nm NW while at 1.49 V with the W/H of 12/12 nm NW.

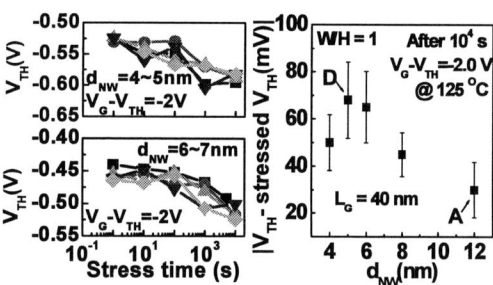

Fig. 12. As d_{NW} is reduced, ΔV_{TH} and STDs of ΔV_{TH} are enlarged due to relatively increase in vertical field and influence of surface conditions to channel.

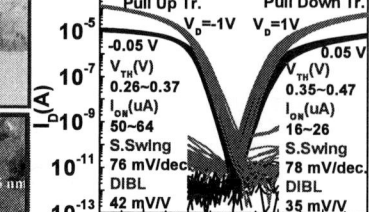

Fig. 13. Top view SEM and cross-sectional TEM images of SRAM arrays with TSNWFETs. For the high I_{ON} at the same W and proper V_{TH} of high performance, TSNWFET having W/H of 5X15 nm, T_{OX} of 2.5 nm and L_G of 40 nm is fabricated.

Fig. 14. PU and PD transistors are well operated with good SCE immunity. The variation of V_{TH} is about 0.12 V. Due to the compressive stress by SiGe, PMOS current is larger than NMOS.

Fig. 15. Static Noise Margin (SNM) is achieved 325 mV at V_D=1 V. L_G is 40 nm.

Novel V_{th} Tuning Process for HfO$_2$ CMOS with Oxygen-doped TaC$_x$

W. Mizubayashi[1], K. Akiyama[2], W. Wang[1], M. Ikeda[2], K. Iwamoto[2], Y. Kamimuta[2], A. Hirano[2],
H. Ota[1], T. Nabatame[2], and A. Toriumi[1, 3]

[1] MIRAI-ASRC, AIST, Tsukuba 305-8569, Japan
Tel: +81-29-849-1649, Fax: +81-29-849-1529, E-mail: w.mizubayashi@aist.go.jp
[2] MIRAI-ASET, Tsukuba 305-8569, Japan, [3] The University of Tokyo, Tokyo 113-8656, Japan

Abstract

We have investigated effects of the oxygen doping into TaC$_x$ on the effective work function ($\Phi_{m,eff}$) in TaC$_x$/SiO$_2$/Si and TaC$_x$/HfO$_2$/Si gate stacks. It has been found for the first time that the threshold voltage (V_{th}) is tunable within 0.5~0.6V for HfO$_2$ MOSFETs by adjusting the oxygen content within 0~12 at. % in TaC$_x$. Furthermore, it has been shown that unknown oxygen content in TaC$_x$ gates is a possible origin of scattering among the $\Phi_{m,eff}$ data reported.

1. Introduction

TaC$_x$ with a sufficient thermal stability has been reported for a promising gate electrode of high-k MOSFETs [1-10]. The effective work function ($\Phi_{m,eff}$) values on SiO$_2$, HfSiO(N) and HfO$_2$, however, have been very scattered over the range 4.18 to 4.9 eV [1-8], as shown in **Fig. 1**. This fact is still controversial and has made the TaC$_x$ technology difficult for practical applications. On the other hand, nitrogen and oxygen doped TaCN [8] and TaCON [9] gate electrodes have been reported for achieving the low V_{th} in CMOS. It has been clearly shown that nitrogen in the TaCN has an effect of the positive $\Phi_{m,eff}$ shift [10], a role of oxygen in TaC$_x$, however, has not been understood yet.

In this paper, effects of oxygen doping on the $\Phi_{m,eff}$ in TaC$_x$ have been intensively studied, and it is demonstrated for the first time that the V_{th} tuning for HfO$_2$ MOSFETs by adjusting the oxygen content in TaC$_x$. This understanding also suggests a possible reason why TaC$_x$ $\Phi_{m,eff}$ values are widely scattered in literatures.

2. Experimental

n-MOS capacitors, n- and p-MOSFETs with the TaC$_x$ gate electrodes on SiO$_2$ and HfO$_2$ were fabricated by the gate last process. 2~10 nm-thick SiO$_2$ were thermally grown at 1000 $^{\circ}$C in O$_2$, while 1~8 nm-thick HfO$_2$ were deposited on 2 and 4 nm-thick SiO$_2$ by ALD. PDA was performed at 800 $^{\circ}$C in N$_2$. 5.1~18.4 nm-thick TaC$_x$ films were deposited by PVD, followed by a-Si deposition using PVD. The post gate electrode deposition annealing at 600 $^{\circ}$C was performed, and then the poly-Si layers were removed. Finally, FGA was performed at 400°C in H$_2$. The oxygen content in TaC$_x$ was controlled by low temperature oxidation at 400 $^{\circ}$C in 20 % O$_2$. As shown in **Fig. 2**, since the oxygen content in TaC$_x$ films monotonically increases with thinning the TaC$_x$ thickness, it is easy to adjust the oxygen content in TaC$_x$. We estimated $\Phi_{m,eff}$ of TaC$_x$ on SiO$_2$ from an extrapolation in the linear relationship between V_{FB} and SiO$_2$ thickness of MOS capacitors.

3. Results & Discussion

V_{FB} of MOS capacitors with TaC$_x$ on SiO$_2$ and HfO$_2$ was estimated as a function of EOT by changing the doped oxygen content in TaC$_x$. **Fig. 3** shows that the V_{FB} of TaC$_x$/SiO$_2$/Si is little dependent on EOT, while it is sensitive to the oxygen content in TaC$_x$. The relationship between $\Phi_{m,eff}$ of TaC$_x$ on SiO$_2$ and the oxygen content in TaC$_x$ is shown in **Fig. 4**. $\Phi_{m,eff}$ of TaC$_x$ is by 0.4 eV variable on SiO$_2$ by changing the oxygen content from 0 to 12 at. % in TaC$_x$, while no resistivity increase of TaC$_x$ is observed in spite of the oxygen doping. It is conjectured that scattered values of $\Phi_{m,eff}$ in TaC$_x$ reported in literatures are related to the uncontrolled oxygen content included in TaC$_x$. **Figs. 5** and **6** show C-V characteristics and V_{FB} vs. EOT$_{HfO2}$ for TaC$_x$/HfO$_2$ gate stack n-MOS capacitors. The V_{FB} of TaC$_x$ on HfO$_2$ as well as on SiO$_2$ becomes large with increasing the oxygen content in TaC$_x$. As shown in **Figs. 6 and 7**, crystallinity of TaC$_x$ on HfO$_2$ is not directly associated with the behavior of V_{FB} induced by oxygen incorporation in TaC$_x$. In addition, for the oxygen contents in TaC$_x$ higher than 6 at. %, an additional positive V_{FB} shift with EOT$_{HfO2}$ decrease is observed. Above results are summarized as, (1) $\Phi_{m,eff}$ of TaC$_x$ depends on doped oxygen content, (2) $\Phi_{m,eff}$ on HfO$_2$ is different from that on SiO$_2$, and (3) the V_{FB} shift on HfO$_2$ with oxygen doped TaC$_x$ is further enhanced with decreasing HfO$_2$ thickness. It is noted that without defining the oxygen content of TaC$_x$ a reproducible V_{FB} value of TaC$_x$/HfO$_2$/Si MOS capacitors cannot be expected.

The V_{th} tuning technique by utilizing the oxygen doped TaC$_x$ can be applied to HfO$_2$ CMOS. For both n- and p-MOSFETs with 12 at. % oxygen doped TaC$_x$/HfO$_2$(2 nm), no gate depletion is observed (**Fig. 8**). **Figs. 9** and **10** show I$_D$-V$_G$ characteristics and V_{th}, respectively, for both n- and p-MOSFETs with TaC$_x$/HfO$_2$/SiO$_2$/Si as a parameter of the oxygen content in TaC$_x$. It is demonstrated that the V_{th} for HfO$_2$ n- and p-MOSFETs is controllable within 0.5~0.6V by changing the oxygen content (0~12 at. %) in TaC$_x$. Both electron and hole mobilities at 0.8 MV/cm are about ~75% and ~100% of the universal curves, respectively, regardless of the oxygen content in the TaC$_x$ as shown in **Fig. 11**. It is also shown that NBTI behavior is almost unchanged by the oxygen incorporation in TaC$_x$ (**Fig. 12**). These results indicate that TaC$_x$ with a controlled oxygen content can effectively adjust V_{th} in n- and p-MOSFETs without causing any significant degradation of transport properties and reliability.

4. Conclusion

We demonstrated the V_{th} tunability by 0.5~0.6V for both HfO$_2$ n- and p- MOSFETs by increasing the oxygen content up to 12at. % in TaC$_x$. In addition, it has been pointed out that uncontrolled oxygen content in TaC$_x$ is a possible origin of scattering among the $\Phi_{m,eff}$ values reported. Thus, it should be emphasized that the oxygen content in TaC$_x$ should be rigidly controlled for the TaC$_x$/high-k/Si gate stack application.

Acknowledgement This work was supported by NEDO.

References

[1] R. Ichihara et al., SSDM, 2005, p. 850.
[2] W. S. Hwang et al., VLSI, 2007, p. 156.
[3] W. Mizubayashi et al., ISCSI-V, 2007, p. 219.
[4] Y. T. Hou et al., IEDM, 2005, p. 35.
[5] R. Mitsuhashi et al., ISAGST, 2007.
[6] L.-Å. Ragnarsson et al., EDL, 28, 486 (2007).
[7] J. K. Schaeffer et al., IEDM, 2004, p. 287.
[8] V. S. Chang et al., IEDM, 2007, p. 535.
[9] S. Kubicek et al., IEDM, 2007, p. 49.
[10] J. K. Schaeffer et al., J. Appl. Phys., 101, 014503 (2007).

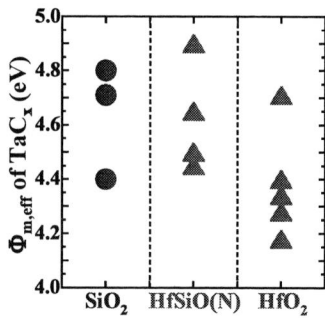

Fig. 1. Comparison of the reported data (Ref. 1-8) of the effective work function ($\Phi_{m,eff}$) of the TaC$_x$ gate electrode on the gate dielectrics such as SiO$_2$, HfSiO(N), and HfO$_2$. The $\Phi_{m,eff}$ is scattered over the range of 4.18~4.9 eV.

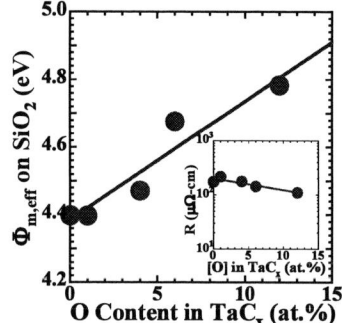

Fig. 2. Relationship between oxygen content in TaC$_x$ and TaC$_x$ thickness. The oxygen content in TaC$_x$ is estimated by the inset HR-RBS depth profiles and increases with thinning TaC$_x$ thickness.

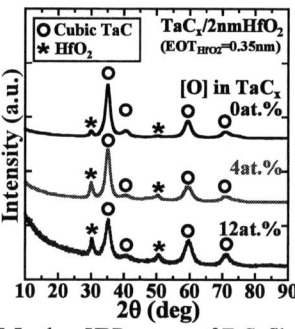

Fig. 3. V_{FB} versus EOT for TaC$_x$/SiO$_2$ MOS capacitors. V_{FB} shifts to the positive direction with increasing the oxygen content in TaC$_x$. The Q_{ss} shows a low value of 2~8x10^{10}cm^{-2} in all cases.

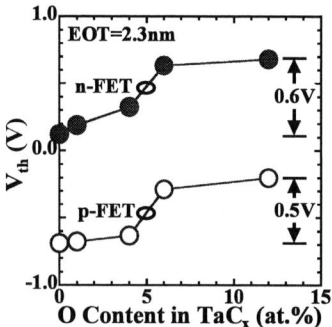

Fig. 4. Relationship between $\Phi_{m,eff}$ of the TaC$_x$ gate electrode on SiO$_2$ and the oxygen content in TaC$_x$. The inset shows the resistivity of TaC$_x$ films against [O] in TaC$_x$.

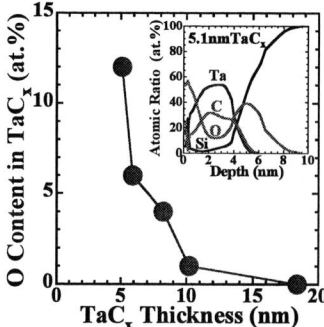

Fig. 5. High frequency C-V characteristics of HfO$_2$(3nm)/SiO$_2$(4nm) MOS capacitors with the TaC$_x$ gate electrode.

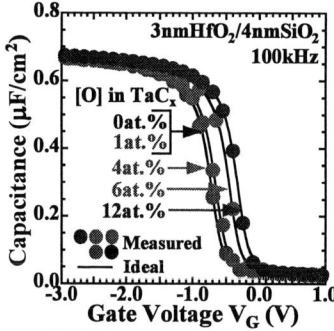

Fig. 6. V_{FB} vs. EOT$_{HfO2}$ for TaC$_x$/HfO$_2$/SiO$_2$ MOS capacitors. V_{FB} on HfO$_2$ shifts to the positive direction with increasing the oxygen content in TaC$_x$.

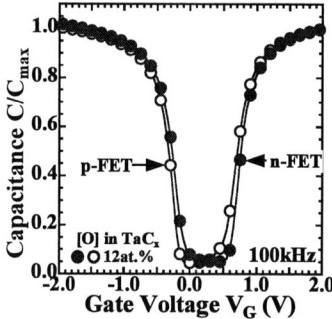

Fig. 7. In-plane XRD patterns of TaC$_x$ films on HfO$_2$(2nm)/SiO$_2$(4nm) structures. The TaC$_x$ films consist of the cubic structure in all cases, regardless of the oxygen content in TaC$_x$.

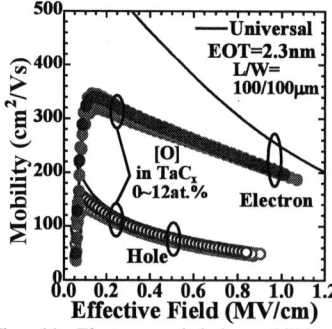

Fig. 8. High frequency C-V characteristics of HfO$_2$(2nm)/SiO$_2$(2nm) n- and p-MOSFETs with the 12 at. % oxygen doped TaC$_x$ gate electrode. The gate depletions for n- and p-MOSFETs are not observed.

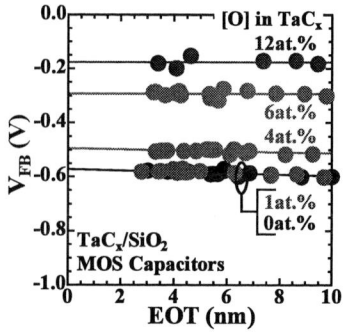

Fig. 9. I_D-V_G characteristics of HfO$_2$(2nm) /SiO$_2$(2nm) n- and p-MOSFETs with the TaC$_x$ gate electrodes. The V_{th} shifts to the positive direction with the increase of oxygen content in TaC$_x$ for n- and p-MOSFETs.

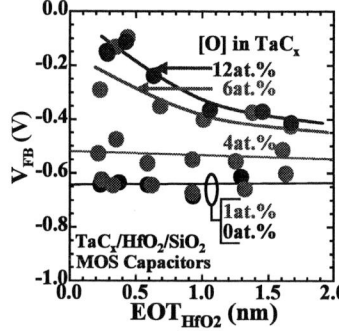

Fig. 10. V_{th} of HfO$_2$(2nm)/SiO$_2$(2nm) MOSFETs with the TaC$_x$ gate electrode plotted as a function of oxygen content in TaC$_x$. The V_{th} can be controlled in the wide range of 0.6 and 0.5V for n- and p-MOSFETs, respectively

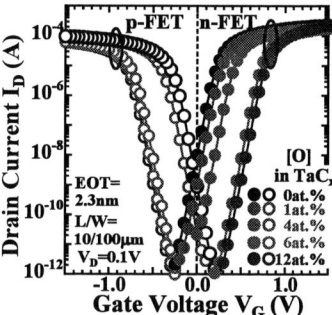

Fig. 11. Electron and hole mobilities vs. effective field for HfO$_2$(2nm)/SiO$_2$(2nm) n- and p-MOSFETs with the TaC$_x$ gate electrode. No degradation of electron and hole mobilities is observed, regardless of the oxygen content in TaC$_x$.

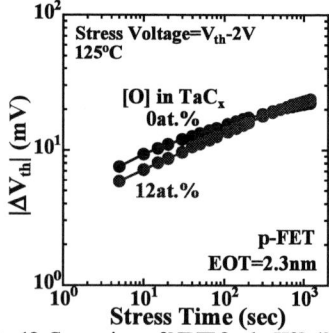

Fig. 12. Comparison of NBTI for the HfO$_2$(2nm) /SiO$_2$(2nm) gate stack p-MOSFETs with and without containing oxygen in TaC$_x$. NBTI behavior is almost unchanged by the oxygen incorporation in TaC$_x$.

34

Novel Process To Pattern Selectively Dual Dielectric Capping Layers Using Soft-Mask Only

T. Schram, S. Kubicek, E. Rohr, S. Brus, C. Vrancken, S.-Z. Chang[1], V.S. Chang[1], R. Mitsuhashi[2], Y. Okuno[2], A. Akheyar[3], H.-J. Cho[4], J.C. Hooker[5], V. Paraschiv, R. Vos, F. Sebai, M. Ercken, P. Kelkar, A. Delabie, C. Adelmann, T. Witters, L-A. Ragnarsson, C. Kerner, T. Chiarella, M. Aoulaiche[†], Moon-Ju Cho[†], T. Kauerauf[†], K.De Meyer[†], A. Lauwers, T. Hoffmann, P.P. Absil and S. Biesemans.

IMEC, assignee at IMEC from [1]TSMC, [2]Matsushita, [3]Infineon, [4]Samsung, [5]NXP and [†]IMEC and KULeuven

IMEC vzw. kapeldreef 75, B-3001 Leuven, Belgium

Phone: +32 16 28 85 54 Fax: +32 16 28 17 06 E-mail: Tom.Schram@imec.be

Abstract

We are reporting for the first time on the use of simple resist-based selective high-k dielectric capping removal processes of La_2O_3, Dy_2O_3 and Al_2O_3 on both HfSiO(N) and SiO_2 to fabricate functional HK/MG CMOS ring oscillators with 40% fewer process steps compared to our previous report [1]. Both selective high-k removal (using wet chemistries) and resist strip processes (using NMP and APM) have been characterized physically and electrically indicating no major impact on Vt, EOT, Jg, mobility and gate dielectric integrity (PBTI, TDDB and charge pumping).

Introduction

Low-V_T HK/MG CMOS has been recently successfully demonstrated with both "gate-first" [1,2] and "gate-last" [3] integration schemes. While the "gate-last" is now introduced in production for high performance products, the "gate-first" option remains attractive for low-cost applications if its complexity can be reduced to the standard poly-Si/SiON CMOS process flow. Moreover the "gate-first" scheme is more flexible since it allows post gate formation high thermal processing as it may be required for certain applications such as embedded DRAM circuits. Among the different possibilities for "gate-first" HK/MG low-V_T CMOS, we have previously reported on our dual metal dual dielectric process flow (DMDD) using mostly hard-masks to pattern selectively nMOS and pMOS regions [2]. In this work, we report on the use of soft-mask processes and wet removal chemistries to simplify the process complexity to Single Metal Dual Dielectric (SMDD) providing several advantages (Table 1). First, it significantly reduces the number of process step by 40% (6 steps). It also allows simpler gate etch profile control since the same metal is used for both n- and p-MOS areas. Finally it has better scaling opportunity since it does not create a recess in the STI are as produced at the n-p boundary in the DMDD approach [2]. The main SMDD concern resides in the reduced range of achievable V_T since it is now entirely based on the use of capping layers [4,5]. The SMDD process flow is described in Fig.1. Note that capping layers may be located above or below the bulk dielectric. We discuss in details the processes used, their selectivities as well as their impact on the key electrical parameters, and validate the CMOS integration with functional ring oscillators.

Process Description

A. Photo-lithography on high-k dielectric capping layers

In order to develop a simple patterning strategy we have opted to use resist soft-mask directly on the dielectric capping to selectively remove it from the complementary areas (La_2O_3 / Dy_2O_3 from pMOS and Al_2O_3 from nMOS). The initial process consisted of a standard 248nm photo resist directly applied on the capping layers but resulted in poor adhesion. Different 'priming' conditions have been attempted but did not improve the adhesion as confirmed by the small contact angles measured for on Dy_2O_3 and Al_2O_3 (Table 2). This adhesion problem can be circumvented with the use of a wet developable BARC layer (WetBARC) available from Brewer Science, Inc.. Fig.2 illustrates the superior adhesion and sharper patterning achieved with WetBARC.

B. Selective removal of high-k dielectric capping layers

The high-k wet capping removal required for the proposed process flow must be resist-compatible, highly selective to the underlying layer (SiO_2 or HfSiON for capping "below" or "above" respectively) and damage-less to the dielectric stack. Table 3 summarizes the chemistries used and further details can be found in [6]. Diluted HCl (dHCl) is the chemistry of choice for La_2O_3 and Dy_2O_3 removal satisfying the required selectivities (Fig.3). For Al_2O_3 we have opted to use extended resist development (TMAH-based) since it was found to remove also Al_2O_3 although dHCl would also be suitable. We have tested the robustness of this selective etch by exposing wafers to extended dHCl etch times for the case of Dy_2O_3 and Al_2O_3.

C. Resist strip with exposed bulk and capping high-k dielectrics

Once the high-k capping has been selectively removed the photo resist must be stripped without impacting further the exposed materials. The resist strip (NMP-based) and post-cleans (APM-based) process details are provided in Table 3. It is worth noting that a less aggressive post-clean is applied in the case of Al_2O_3 to minimize its removal (etch rate in APM is 2.4 nm/min @65C and only 0.65 nm/min @ RT).

Results and Discussion

The capping material removal efficiency for La_2O_3 and Dy_2O_3 has been quantified by TXRF measurement (Fig.4) showing residual La and Dy when removed from Hf-based dielectric but below detection limit when from removed SiO_2 and SiON. Although La shows a 10x reduction in residual compared to Dy for the "above–HK" case, it may be indicative that an intermixed layer is formed at the bulk/cap interface. This was verified electrically in transistors showing a residual V_T increase of 35mV after AlO and 120mV after DyO removal on top of HfSiO, compared to the uncap reference (Fig.5). In case of removal on the interfacial SiO_2 the residual V_T difference is below 30mV. Fig.6 shows that in case the cap is not removed a shift up to 500mV and 250mV can be obtained with La_2O_3 and Al_2O_3 cap, respectively. Fig.7 shows the impact of the resist strip on the $V_{T,lin}$, EOT and mobility for AlO and DyO. It can be clearly observed that exposing the high-k to soft resist strip processes is not significantly impacting the key device parameters. The impact on gate leakage is shown in Fig.8. The impact on TDDB is illustrated in Fig.9. As the progressive wear-out at low biasing voltages is very slow, the real lifetime of the devices will be much longer. These measurements show that the TDDB behavior HfSiO with removed Al_2O_3 layers matches that of virgin uncapped HfSiO. The strip process also does not affect the predicted lifetime. Fig.10 and 11 show the charge pumping and EOT corrected PBTI predictions. Both figures illustrate that the resist and cap layer removal do not aversely affect the trapping and PBTI behavior of the gate stack.

We also integrated these processes into the SMDD process flow of Fig.1 resulting in working ring oscillators as shown in Fig. 12.

Conclusions

We have successfully demonstrated the use of soft-mask processes to pattern selective dual HK layers (SMDD) without significant negative impact to the key electrical device parameters. This provides a 40% reduction in process complexity over our previous report on DMDD process flow [1].

References

[1] M. Chudzik et al., Symp. VLSI Technology, p.194, 2007. [2] S. Kubicek et al., IEDM Tech. Dig. pp. 49-52, 2007. [3] K.Misry et al., IEDM Tech. Dig., p.247, 2007. [4] H.-S. Jung et al., Symp. VLSI Technology, p232-233, 2005. [5] H. Alshareef et al., APL, Vol.89, 2006. [6] Vos et al, ECS Transactions 11(4), 275-283, 2007.

	DMDD	SMDD
# of extra process steps compared to conventional SiO₂/Si case	15	9
# of extra mask steps	2	2
Gate etch aspects	Different metals etched at the same time	Only one metal etched
V_T tuning flexibility	cap layer and metal	Cap layer only
Gate dielectric integrity	Gate dielectric not touched be removal processes	Gate dielectric only exposed to wet chemistries

Table 1: comparison of the advantages of DMDD and SMDD high-k metal gate integration.

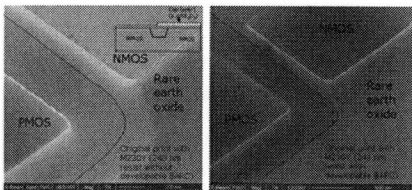

Figure 2: N-P MOS boundary after etching and resist removal using the resist only process and the optimized wetBARC based process.

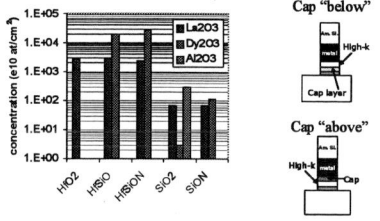

Figure 4: residual atom concentration of La, Dy and Al after cap layer removal, as measured by TXRF (detection limit 100e10 at/cm²)

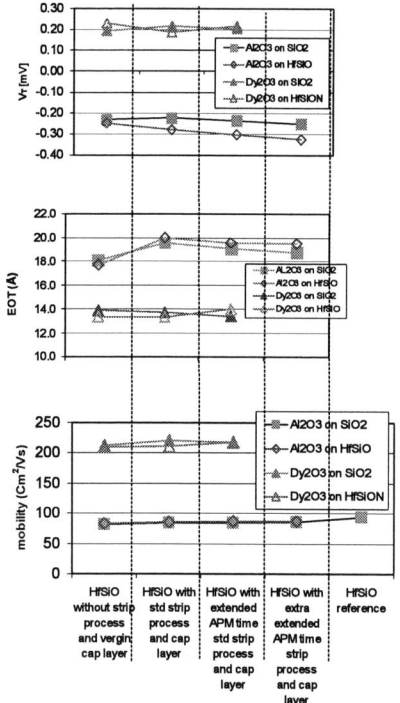

Figure 7: V_T, EOT and mobility of stripped and virgin capped high-k for the different strip processes used.

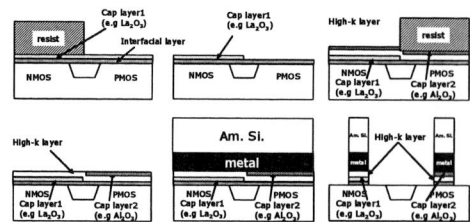

Figure 1: SMDD CMOS process flow, with the nMOS cap located below and pMOS cap located above the bulk high-k dielectric.

	Resist	Cap removal	Resist strip	Post-clean
Dy₂O₃	248 nm	dilute HCl	Wet organic strip (NMP/AEE)	APM at 65C
La₂O₃	Wet BARC + 248 nm	Dilute HCl	Wet organic strip (NMP/AEE)	APM at 65C
Al₂O₃	248 nm	During litho development step (1 min ~3.5 % TMAH) (ER 2.4 nm/min)	Wet organic strip (NMP/AEE)	APM at RT (with some sacrificial Al₂O₃ removal)

Table 3: Processes used to selectively remove the used cap layers to high-k dielectrics and the subsequent strips.

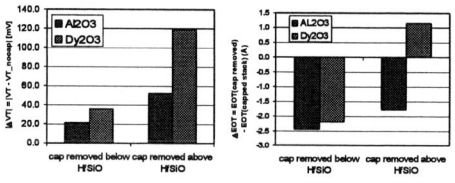

Figure 5: Absolute value of the residual V_T and EOT difference due to the selective cap removal of Dy₂O₃ or Al₂O₃. The V_T shift is referenced to the same gatestack without cap layer.

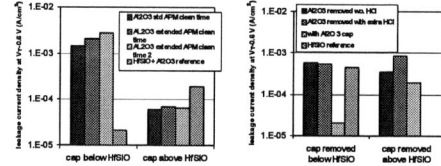

Figure 8: leakage comparison between (left) Al₂O₃ capped HfSiO subjected to strip and without strip. (right) reference HfSiO and Al₂O₃ capped HfSiO after cap removal.

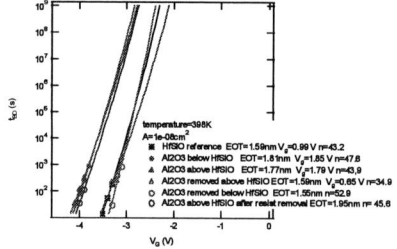

Figure 9: TDDB (soft breakdown) data for stripped Al₂O₃ capped HfSiO and selectively etched Al₂O₃ layers on top of HfSiO.

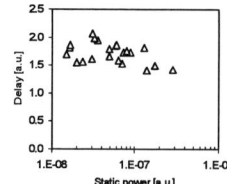

Figure 12: Ring oscillator data of a SMDD MIPS circuit prepared according to the method described in Fig.1.

	Contact Angle On Dy₂O₃	Contact Angle On Al₂O₃
Unprimed:	33.5⁰	29.6⁰
HMDS Primed 130C, 60sec	48.3⁰	33.9⁰
HMDS Primed 160C, 60sec	48⁰	NA
HMDS Primed 150C,120sec	42.6⁰	33.4⁰

Table 2: contact angle measurements on differently primed Dy₂O₃ [NMOS shift] or Al₂O₃ wafers. A contact angle of at least 60 deg is needed for good resist adhesion.

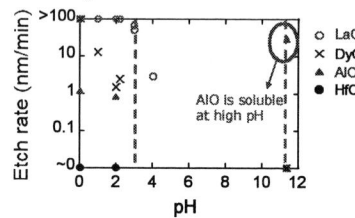

Figure 3: etch rate of the different used cap layers and high-k as function of pH.

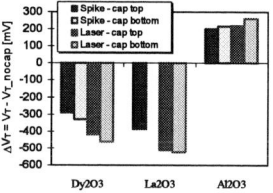

Figure 6: effective V_T shift obtained from a stack consisting of an oxide interfacial layer with a thin cap layer (Al₂O₃ [PMOS shift], La₂O₃ or Dy₂O₃ [NMOS shift]) located above and below HfSiO.

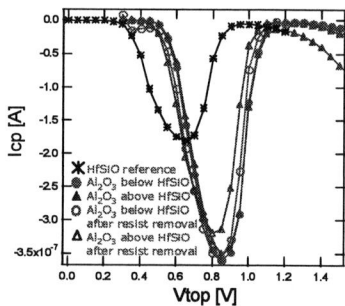

Figure 10: charge pumping data on stripped and bare Al₂O₃ capped HfSO, as compared to that of uncapped HfSiO.

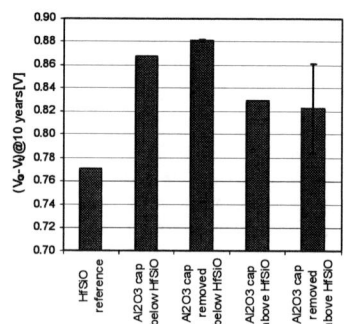

Figure 11: PBTI data on stripped and bare Al₂O₃ capped HfSO, as compared to that of uncapped HfSiO.

Single metal/single dielectric gate stack realizing triple effective workfunction for embedded memory application

Kenzo Manabe, Koji Masuzaki, *Takashi Ogura, Takashi Nakagawa, Motofumi Saitoh, Hiroshi Sunamura, Toru Tatsumi and Hirohito Watanabe

Device Platforms Research Laboratories, NEC Corporation
*Process Technology Division, NEC Electronics Corporation
1120, Shimokuzawa, Sagamihara, Kanagawa 229-1198, Japan
Phone: +81-42-771-2294, Fax: +81-42-771-2481, E-mail: k-manabe@ce.jp.nec.com

Abstract

We demonstrate midgap and band-edge effective workfunctions (EWFs) control with simple metal gate process scheme (single metal gate/single gate dielectric), using impurity-segregated $NiSi_2$/SiON structure for embedded memory application. The application of midgap and band-edge EWF enables us to lower power consumption in SRAM and logic devices by 30% and 15% compared to poly-Si devices, respectively, due to reduced channel impurity concentration, suppressed gate depletion and high carrier mobility. These results show that $NiSi_2$/SiON stack is one of the most promising candidates for future system on chip (SoC) devices with embedded memory.

1. Introduction

To improve device performance and reduce power consumption in future system on chip (SoC) devices with embedded memory, gate electrodes with not only band-edge effective workfunction (EWF) but also midgap EWF are needed because excess channel doping (N_{ch}) would be required to obtain high threshold voltage (V_{th}) in memory cell transistors with band-edge EWF electrodes. Such doping degrades carrier mobility and increases gate induced drain leakage (GIDL) and intrinsic V_{th} fluctuation [1]. If we can tailor the EWF to midgap and band-edge using *single metal gate/single gate dielectric* (Fig. 1), we can realize high performance logic and low power memory transistors with low V_{th} fluctuation at a minimal cost. NiSi fully-silicided (FUSI) metal gate electrode has a midgap EWF and its EWF on SiON can be controlled by doping into poly-Si prior to silicidation [2]. The increased V_{th} shift (ΔV_{th}) is also reported for P-doped $NiSi_2$ and B-doped NiSi by post-doping [3]. However, ΔV_{th} in *single* Ni-FUSI is not sufficient for logic CMOS devices. Moreover, we must solve process problems such as electrode peeling and device destruction induced by volume expansion of FUSI electrode [4]. In this work, we searched for a way to obtain midgap and band-edge EWFs with simple metal gate process and investigated the impacts of $NiSi_2$ on MOSFET for embedded memory application.

2. Process improvement and EWF of $NiSi_2$-FUSI

By using $NiSi_2$/SiON structure, we can overcome issues in doped Ni-FUSI such as 1) peeling, 2) volume expansion and 3) insufficient ΔV_{th}. Adhesion strength of Ni-FUSI on SiON increased with increasing Si content in electrode and peeling did not occur for $NiSi_2$ until Si-substrate destruction (Fig. 2). Volume expansion was suppressed with increasing Si content in FUSI and did not exist for $NiSi_2$ (Fig. 3). These results show that the use of $NiSi_2$ electrode successfully overcomes peeling and volume expansion problems

We have found that the EWFs for doped Ni-FUSI approach Si band-edge with increasing Si content in electrodes (Fig. 4). Band-edge EWFs (4.1 eV and 5 eV) are obtained for As-doped and B-doped $NiSi_2$ while the EWF for undoped $NiSi_2$ is at midgap (4.6 eV). These results show that we can obtain band-edge and midgap EWFs in *single metal gate* ($NiSi_2$) and *single gate dielectric* (SiON) *scheme* simply by pre-doping. Figure 5 shows that shift in EWF by impurity (ΔEWF) for $NiSi_2$ is larger than that for NiSi at the same dose. The dose dependence of ΔEWF for $NiSi_2$ shows saturation behavior in the dose range of > 4×10^{15} cm^{-2}. This EWF saturation tendency is useful because we can suppress V_{th} fluctuation induced by dose fluctuation. These results show that triple EWF for embedded memory devices is obtained using single metal ($NiSi_2$) without process issues.

3. Device performance for $NiSi_2$-FUSI CMOSFET

To investigate the impacts of doped $NiSi_2$ on logic (low-V_{th}) device performance, CMOSFET was fabricated using the process flow as shown in Fig. 6. Low V_{th} of ± 0.2 V was achieved by the band-edge EWFs of As-doped and B-doepd $NiSi_2$ (Fig. 7). In these MOSFETs, mobility degradation was not observed (Fig. 8). I_{ON} for $NiSi_2$ NMOSFET increased by 11% compared to that for poly-Si FET due to elimination of poly-Si depletion (Fig. 9). Moreover, the combination of $NiSi_2$ and thinner gate oxide (1.5 nm) resulted in $I_{ON} = 1.02$ mA/μm at $I_{OFF} = 100$ nA/μm at V_D of 1.2 V without any mobility enhancement techniques. Figure 10 indicates that gate leakage current (I_G) for $NiSi_2$ MOSFET was 1.5-2.5 orders of magnitude lower than that for poly-Si MOSFET. Figure 11 shows that V_{th} fluctuation decreased with increasing pre-dope dose into poly-Si gate because of saturated ΔEWF at higher dose (Fig. 5).

To study the impacts of $NiSi_2$ on memory (high-V_{th}) FET, we adjusted N_{ch} so that V_{th} for $NiSi_2$ FET is similar to that for poly-Si FET (Fig. 12). Since the EWF for $NiSi_2$ is inside Si bandgap (Fig. 4), we can reduce N_{ch} for $NiSi_2$ compared to that for poly-Si. As a result, I_{OFF} for $NiSi_2$ FET decreased by 60% compared to that for poly-Si FET due to suppression of GIDL (Fig.12). Moreover, I_{ON} increased (+20%) by using $NiSi_2$ (Fig.13) since the use of $NiSi_2$ eliminated poly-Si depletion (+10%) and mobility was enhanced by 10% due to lower N_{ch} (Fig. 14).

Figure 15 shows that 0.66 μm^2 SRAM cell successfully operated at $V_{DD} = 0.8$ V. In this SRAM, cell current (IBL) for $NiSi_2$ device was higher than that for poly-Si device at the same static noise margin (SNM) when $V_{DD} = 1.2$ V (Fig. 16) because I_{ON} for $NiSi_2$ device was enhanced by 20% compared to poly-Si device (Fig. 13). The IBL-SNM trend (Fig. 16) for $NiSi_2$ device with $V_{DD} = 1$ V was similar to that for poly-Si with $V_{DD} = 1.2$ V, which indicates that 20% reduction of V_{DD} can be achieved without sacrificing device performance by using $NiSi_2$. Long term (10 years) reliability was obtained with respect to BTI characteristic (Fig. 17).

The improvement in device performance by the use of $NiSi_2$ is summarized in TABLE I. The enhancement of I_{ON} by 11% and 20% under the same V_{DD} is obtained compared to poly-Si for logic and memory FETs, respectively, due to eliminated depletion and reduced N_{ch}. This means that active power consumption in logic and memory can be reduced by 15% and 30% retaining device performance because we can reduce V_{DD} by 10 and 20%, respectively. Our results show that the improvement of device performance and reduction of power consumption can be achieved for logic and embedded memory devices by using $NiSi_2$ on SiON in *single metal gate/single gate dielectric scheme*.

4. Conclusion

Triple EWF (midgap and band-edge) was achieved by using doped $NiSi_2$ electrodes for embedded memory devices. Therefore, the application of $NiSi_2$ enables us to improve device performance and to reduce power consumption because of reduced N_{ch} and high mobility. We believe that $NiSi_2$ is one of the most promising metal gate candidates especially for future SoC devices with embedded memory.

References

[1] K. Takeuchi, T. Tatsumi and A. Furukawa, IEDM p.841 (1997)
[2] J. Kedzierski et al., IEDM 2003, p.315 (2003).
[3] Y. Tsuchiya et al., IEDM 2006, p.231 (2006).
[4] M. Saitoh et al., Symp. on VLSI Tech. p.162 (2006).

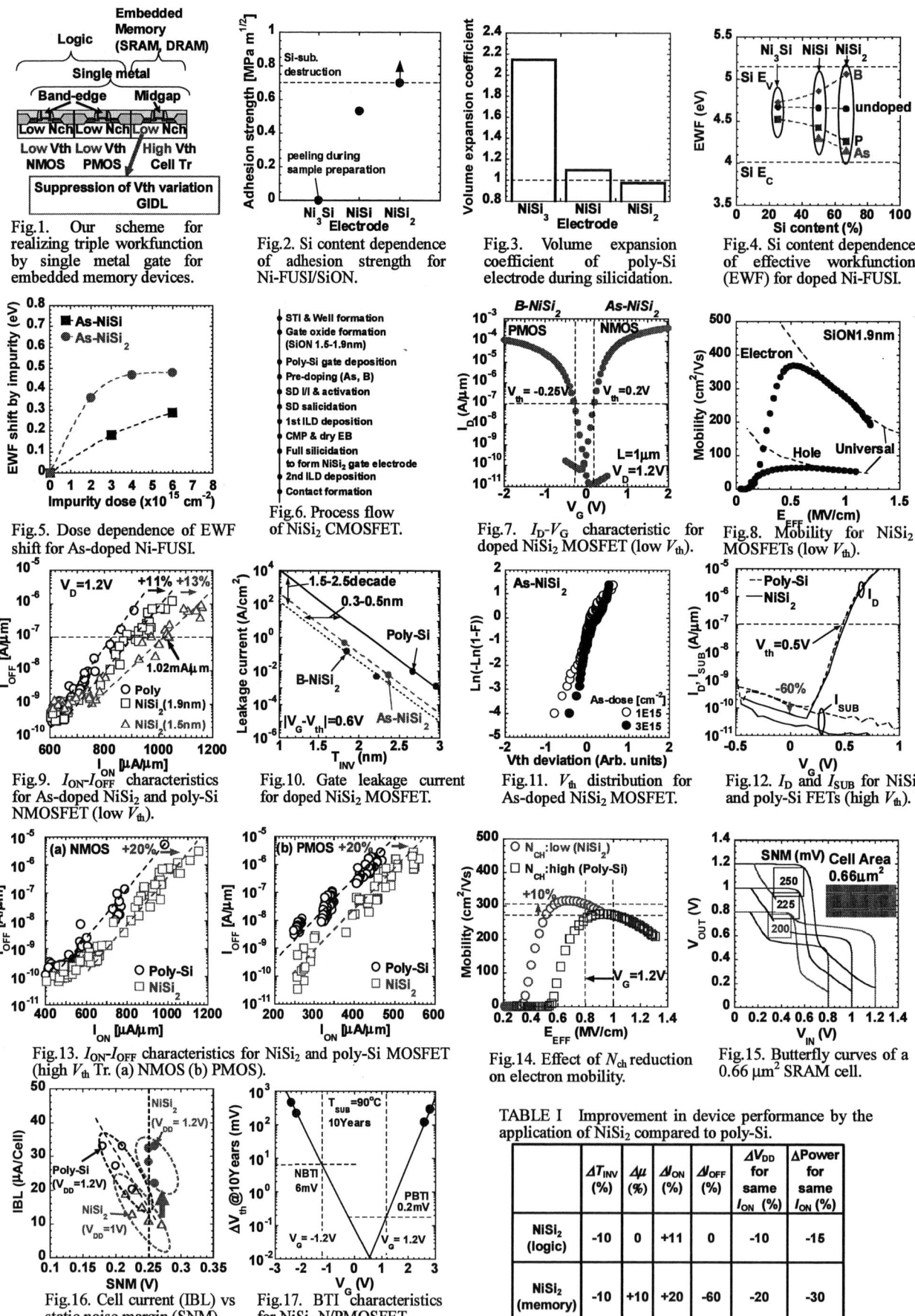

Fig. 1. Our scheme for realizing triple workfunction by single metal gate for embedded memory devices.

Fig. 2. Si content dependence of adhesion strength for Ni-FUSI/SiON.

Fig. 3. Volume expansion coefficient of poly-Si electrode during silicidation.

Fig. 4. Si content dependence of effective workfunction (EWF) for doped Ni-FUSI.

Fig. 5. Dose dependence of EWF shift for As-doped Ni-FUSI.

Fig. 6. Process flow of $NiSi_2$ CMOSFET.

Fig. 7. I_D-V_G characteristic for doped $NiSi_2$ MOSFET (low V_{th}).

Fig. 8. Mobility for $NiSi_2$ MOSFETs (low V_{th}).

Fig. 9. I_{ON}-I_{OFF} characteristics for As-doped $NiSi_2$ and poly-Si NMOSFET (low V_{th}).

Fig. 10. Gate leakage current for doped $NiSi_2$ MOSFET.

Fig. 11. V_{th} distribution for As-doped $NiSi_2$ MOSFET.

Fig. 12. I_D and I_{SUB} for $NiSi_2$ and poly-Si FETs (high V_{th}).

Fig. 13. I_{ON}-I_{OFF} characteristics for $NiSi_2$ and poly-Si MOSFET (high V_{th} Tr. (a) NMOS (b) PMOS).

Fig. 14. Effect of N_{ch} reduction on electron mobility.

Fig. 15. Butterfly curves of a 0.66 μm^2 SRAM cell.

Fig. 16. Cell current (IBL) vs static noise margin (SNM).

Fig. 17. BTI characteristics for $NiSi_2$ N/PMOSFET.

TABLE I Improvement in device performance by the application of $NiSi_2$ compared to poly-Si.

	ΔT_{INV} (%)	$\Delta \mu$ (%)	ΔI_{ON} (%)	ΔI_{OFF} (%)	ΔV_{DD} for same I_{ON} (%)	ΔPower for same I_{ON} (%)
$NiSi_2$ (logic)	-10	0	+11	0	-10	-15
$NiSi_2$ (memory)	-10	+10	+20	-60	-20	-30

Improved FET characteristics by laminate design optimization of metal gates
- Guidelines for optimizing metal gate stack structure -

M. Kadoshima[1], T. Matsuki[1], N. Mise[1], M. Sato[1], M. Hayashi[1], T. Aminaka[1], E. Kurosawa[1], M. Kitajima[1],
S. Miyazaki[2], K. Shiraishi[3], T. Chikyo[4], K. Yamada[5], T. Aoyama[1], Y. Nara[1], Y. Ohji[1]

[1]Semiconductor Leading Edge Technologies Inc. (Selete), 16-1 Onogawa, Tsukuba, Ibaraki 305-8569, Japan,
[2]Hiroshima University, [3]University of Tsukuba, [4]National Institute for Material Science, [5]Waseda University.
Phone: +81-298-49-1472, Fax: +81-298-49-1186, E-mail: kadoshima.masaru@selete.co.jp

Abstract

A laminate design technology of metal gates is proposed to improve FET characteristics regardless of EOT and gate dielectric material. The laminated metal gate structures are basically composed of low-R_s(sheet resistance) metal/ WF(work-function)-lowering layer/ WFM(WF determining metal). A thin WFM (~2 nm) laminated by the Si-based WF-lowering layer such as poly-Si or TaSiN brings an additional benefit of dramatic improvements in mobility and PBTI in nFETs. A thick WFM (~10 nm) suppresses the WF-lowering in pFETs. The concept of the laminate design is indispensable for improving the performance in CMOSFETs.

Introduction

High V_t, low mobility and low reliability are still major issues for realizing gate-first metal/high-k CMOSFETs. Recently, these have been improving by optimizing a layered structure of high-k and another material incorporation into high-k on the basis of enormous findings about high-k [1-4]. It has been commonly believed that V_t, namely V_{FB}, and mobility are identically determined for a fixed material combination between metal gate and high-k under same device fabrication process. However, anomalous V_{FB} shifts and electron mobility (μ_e) improvement have been reported by thinning CVD-TiN or TaN and by stacking CVD-TiN with PVD-TiN as work-function determining metal (WFM) in a gate electrode [5-7]. Such impacts of metal gate structures on FET characteristics have hardly been studied comprehensively. In this study, we have systematically investigated V_{FB} and mobility behaviors by changing gate electrode structures, especially as a function of the WFM thickness. Based on the results, the guidelines are proposed for optimum metal gate electrode structures in CMOSFETs.

Experimental

HfSiON n and pFETs with poly-Si/WFM-TiN and W/WFM-TiN [8] gates were investigated. TiN was selected as WFM because of the high process-friendliness. The TiN films as WFM were deposited on HfSiON by reactive sputtering. Poly-Si and W are deposited on WFM-TiN by CVD and sputtering, respectively. The TiN thickness was set from 0 to 30 nm. HfSiON(2.5 nm)/SiO2(1 nm) was used as a gate dielectric. Spike annealing at 1000°C was basically carried out after the gate electrode formation.

Results and Discussion

1. Variation of FET characteristics by thinning WFM

The C-V curve in HfSiON nFETs with poly-Si/WFM-TiN(2 nm) is negatively shifted as compared to that with poly-Si/WFM-TiN(10 nm) (Fig. 1). The shift anomalously occurs without decreasing accumulation and inversion capacitances. V_{FB} negatively shifts by about 300 mV without an EOT increase when the WFM-TiN thickness is reduced from 30 to 2 nm (Fig. 2). The negative V_{FB} shift by thinning WFM is advantageous to obtain low V_t in nFETs.

μ_e is dramatically improved by thinning WFM along with the negative V_{FB} shift (Fig. 3). High μ_e near the universal curve has been also reported when ~1 nm-thick TaN and 2.3 nm-thick ALD-TiN are used as thin WFM in a poly-Si/WFM gate [6,9]. Temperature dependent components of μ_e at 0.8 MV/cm were extracted according to Matthiessen's rule (Fig. 4). The components are almost identical irrespective of the laminated structure in the metal gates (Fig. 5). Therefore, the μ_e improvement by thinning WFM is not mainly due to decrease of remote phonon scattering but decrease of remote coulomb scattering. Judging from nano-beam diffraction, negligible stress is induced from the metal gates of poly-Si/WFM-TiN(2 and 10 nm) and W/WFM-TiN(2 nm) into the Si channel.

The thin WFM-TiN (2 nm) is more advantageous to PBTI suppression in HfSiON nFETs than the thick WFM-TiN(10 nm) in a low stress region (Fig. 6). Therefore, we should intentionally utilize the thinning of WFM with the negative V_{FB} shift in nFETs for obtaining not only low V_t, but also high μ_e and high reliability.

2. Understanding of anomalous V_{FB} behavior by thinning WFM

V_{FB} is negatively shifted by thinning WFM in SiO2 nFETs with n$^+$poly-Si/WFM-TiN gates as well as in HfSiON nFETs (Fig. 7). On the other hand, no V_{FB} shift occurs in HfSiON nFETs with W/WFM-TiN gates. These strongly indicate that the V_{FB} shift mainly originates in the gate electrode and the poly-Si layer on WFM plays an important role in the shift. The V_{FB} shift is not due to dopants in the poly-Si layer because similar negative V_{FB} shifts are observed in both HfSiON nFETs with n$^+$poly-Si/WFM gates and HfSiON pFETs with p$^+$poly-Si/WFM-TiN gates (Figs. 3 and 13).

The extent of the V_{FB} shift in SiO2 nFETs is slightly larger than that in HfSiON nFETs (Fig. 8). The V_{FB} shifts induced by thinning WFM is not the negative V_{FB} shift observed in high-k FETs after high temperature annealing (so-called "pinning" [10]) but rather due to a WF modulation of WFM-TiN.

Oxidation of Si proceeds in the Si/TiN(2 nm) interface after 1000°C, as compared to that in the Si/TiN(10 nm) interface (Fig. 9). It is revealed that the amounts of TiON or TiO$_x$ in poly-Si/WFM-TiN(2 nm) are less than those in W/WFM-TiN(2 nm) after the annealing by using substrate-side XPS (Fig. 10). No peak of TiSi or Ti with lower WF than TiN is observed. Thus, it is quite likely that very thin TiON (TiO$_x$) in a WFM-TiN/dielectric interface has fixed charges or dipoles [11] that modulate WF (Fig. 2) and μ_e (Fig. 3). WF of WFM-TiN is lowered and μ_e is improved when the amounts of TiON (TiO$_x$) are reduced by the oxidation of Si on thin WFM-TiN(~2 nm) (Fig. 9). A similar effect was reported for a reactive Ti layer in a TiN/Ti/W/WFM-TiN gate [12].

V_{FB} is also negatively shifted by thinning WFM in a low-temperature process below 600°C in HfSiON nFETs with poly-Si/WFM-TiN gates (Fig. 11). Insertion of TaSiN between W and thin WFM-TiN leads to a negative V_{FB} shift which originates in Si of the TaSiN layer (Fig. 11). μ_e is improved by thinning WFM along with the V_{FB} shifts (Fig. 12). Thus, the Si-containing layer on thin WFM-TiN acts as a contributor to WF-lowering and μ_e improvement.

Hole mobility (μ_h) is also improved by thinning WFM in pFETs with p$^+$poly-Si/WFM-TiN gates, but V_{FB} is simultaneously decreased, namely |V_t| is increased (Fig. 13). Since μ_h and NBTI are relatively insensitive to the WFM-TiN thickness (Figs. 13 and 14), high V_{FB}'s should be kept by using a thick WFM-TiN (~10 nm) with little penalty of μ_h for realizing low |V_t| in pFETs.

3. Optimum laminate design of metal gates

Finally, guidelines for optimum laminate design of metal gates are summarized (Fig. 15). The metal gate structures are basically composed of low-R_s(sheet resistance) metal/ Si-containing layer/ WFM. The Si-containing layer on a thin WFM (~2 nm) acts as a WF-lowering layer to reduce V_t (V_{FB}) and to improve μ_e and PBTI in nFETs. A thick WFM (~10 nm) suppresses the WF-lowering in pFETs. The laminate design is widely applicable even when EOT becomes thinner and high-k is incorporated by La, Al, etc. [1-4] in future CMOS, because the effect is supposedly induced by interaction between the Si-containing layer and WFM.

Conclusion

Guidelines for optimizing laminate design of a metal gate have been developed on the basis of the systematic study. A thin WFM (~2 nm) laminated by a Si-containing layer is effective to lower WF and to improve μ_e and PBTI in nFETs. A thick WFM (~10 nm) is used to avoid the WF-lowering in pFETs. The laminate design is necessary for getting the best performance out of metal gate FETs.

References [1] V. Narayanan *et al.*, VLSI2006, p. 224. [2] M. Kadoshima *et. al.*, J. Appl. Phys., **99**, 054506 (2006). [3] P. Sivasubramani *et al.*, VLSI2007, p. 68. [4] Y. Kamimuta *et al.*, IEDM2007, p. 341. [5] J. Widiez *et al.*, Proc. IEEE Int. SOI Conf., 2005, p. 30. [6] S. K. Han *et al.*, IEDM2006, p. 621. [7] Y. Nishida *et al.*, VLSI2007, p. 214. [8] F. Ootsuka *et al.*, SSDM2006, p. 1116. [9] Z. Zhang *et al.*, IEEE EDL, **27**, 185 (2006). [10] M. Kadoshima *et al.*, VLSI2007, p. 66. [11] A. Uedono *et al.*, Jpn. J. Appl. Phys., **46**, 3214 (2007). [12] Y. Akasaka *et al.*, VLSI2006, p. 206.

978-1-4244-1802-2/08/$25.00 ©2008 IEEE

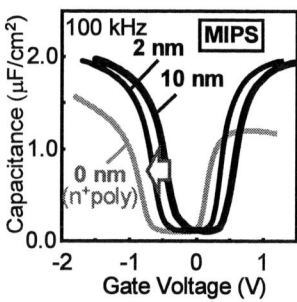

Fig. 1 C-V curves of HfSiON nFETs with poly-Si/WFM-TiN gates. The curve is anomalously shifted with no gate depletion by changing the WFM-TiN thickness.

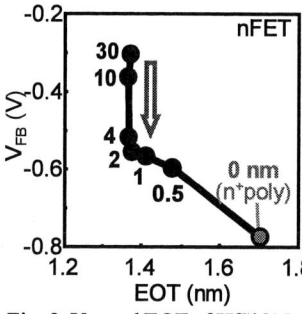

Fig. 2 V_{FB} and EOT of HfSiON nFETs as a function of the WFM-TiN thickness. Labels denote the WFM-TiN thickness in poly-Si/WFM-TiN gate electrodes.

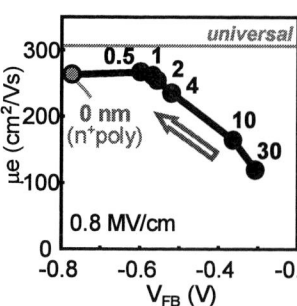

Fig. 3 μ_e at 0.8 MV/cm and V_{FB} as a function of the WFM-TiN thickness. It is likely that μ_e directly correlates with V_{FB}.

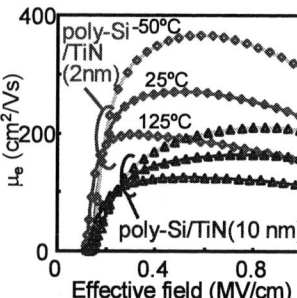

Fig. 4 μ_e in HfSiON nFETs with poly-Si/WFM-TiN as a function of the measured temperature. The WFM-TiN thicknesses are 2 (diamonds) and 10 nm (triangles).

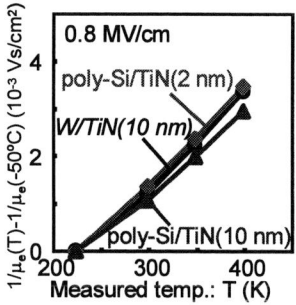

Fig. 5 Temperature dependent components of $1/\mu_e$ at 0.8MV/cm extracted by Matthiessen's rule. The components are almost identical regardless of gate stacks .

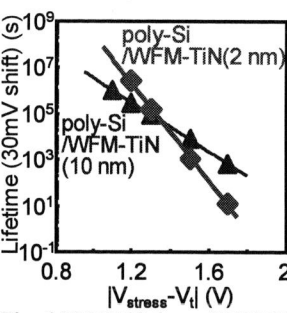

Fig. 6 PBTI Lifetime of HfSiON nFETs. PBTI for TiN(2 nm) is more suppressed than that for TiN(10 nm) in poly-Si/WFM-TiN gates in a low stress region.

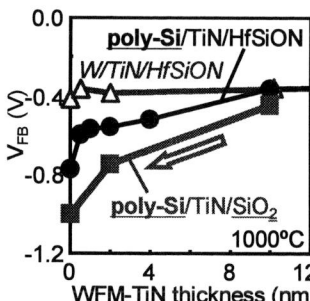

Fig. 7 Change in V_{FB} by thinning WFM-TiN in various gate stacks. The V_{FB} decrease is not observed in W/WFM-TiN/HfSiON, but in poly-Si/WFM-TiN/SiO$_2$(3 nm).

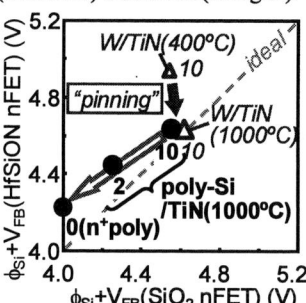

Fig. 8 $\phi_{Si}+V_{FB}$(HfSiON nFET) vs. $\phi_{Si}+V_{FB}$(SiO$_2$ nFET). $\phi_{Si}+V_{FB}$ means effective WF without excluding effects of fixed charges. ϕ_{Si} denotes Fermi-level of Si-sub.

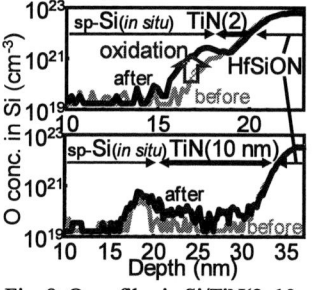

Fig. 9 O profiles in Si/TiN(2, 10 nm)/HfSiON before (gray) and after 1000°C (black). Oxidation of Si proceeds in the Si/TiN interface for Si/TiN(2 nm) (upper graph) as compared to that for Si/TiN(10 nm) (lower).

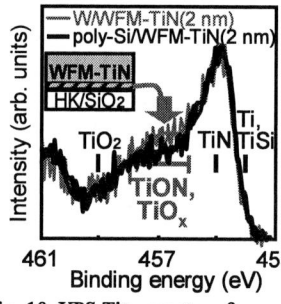

Fig. 10 XPS Ti2p spectra of poly-Si and W/TiN(2 nm)/HfSiON after 1000°C measured from substrate side after Si-sub. removal. A peak of TiON(TiO$_x$) is lowered for poly-Si/TiN(2 nm).

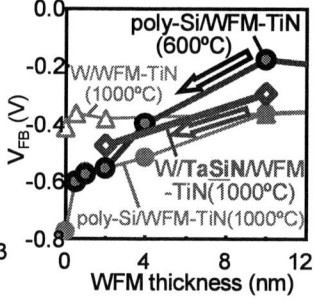

Fig. 11 V_{FB} of HfSiON nFET vs. WFM-TiN thickness. V_{FB} is decreased by thinning WFM in poly-Si/WFM-TiN(600°C process) and W/**TaSiN**/WFM-TiN gates.

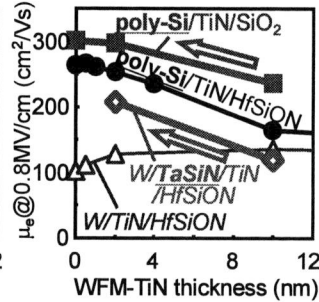

Fig. 12 μ_e vs. WFM-TiN thickness. μ_e improvements are observed whenever a Si-containing layer is directly laminated on WFM-TiN irrespective of gate dielectrics.

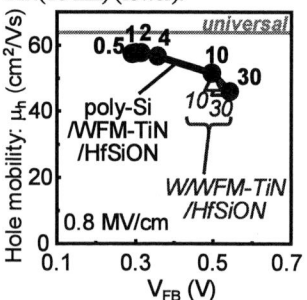

Fig. 13 μ_h and V_{FB} in HfSiON pFETs as a function of the WFM-TiN thickness. For low $|V_t|$ pFETs, high V_{FB} should be kept by using a thick WFM-TiN (~10 nm) with little penalty of μ_h because μ_h is less sensitive to the WFM-TiN thickness than μ_e.

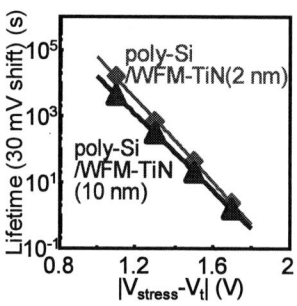

Fig. 14 NBTI Lifetime of HfSiON pFETs. NBTI is almost independent of the TiN thickness in poly-Si/WFM-TiN, unlike PBTI in Fig. 6.

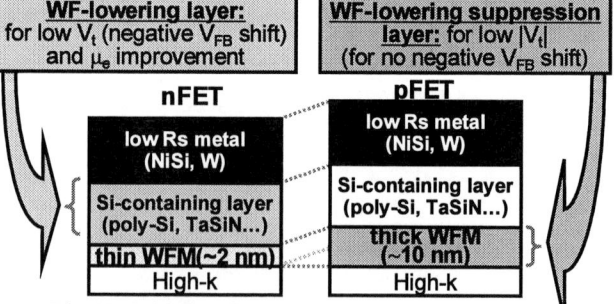

Fig. 15 Guidelines for optimum metal gate structures. A Si-containing layer such as poly-Si and TaSiN on thin WFM (0~4 nm) acts as a WF-lowering layer in nFETs. A thick WFM (~10 nm) should be used in pFETs for low $|V_t|$, namely to suppress the WF-lowering.

40

Fundamentals and extraction of velocity saturation in sub-100nm (110)-Si and (100)-Ge

L.Pantisano, L.Trojman*, J.Mitard, B. DeJaeger, S.Severi, G.Eneman,
G.Crupi[+], T.Hoffmann, I.Ferain*, M.Meuris, M.Heyns*

IMEC Kapeldreef 75 Leuven (Belgium), also *KU Leuven, [+]Univ. Messina, email pantisan@imec.be

Abstract
A novel RFCV-technique is applied to *directly* quantify the short channel devices at high V_{ds}, enabling parameter extraction like velocity saturation and critical field. This technique is applied to benchmark Si (110) and Si(100) as well as Ge devices. Similarities and crucial differences between short channel parameters in Si and Ge are discussed.

Introduction: mobility, velocity and short channel devices
Novel dielectric materials on Si or alternative substrates (Ge, GaAs,...) are typically benchmarked by using the long channel mobility u_{eff} in linear regime (V_{ds}=50mV) and the I_{ON}-I_{OFF} for short channel devices ($V_{ds} \geq 1V$) in saturation conditions. For any gate length I_{ON}/W=$v*Q_{inv}$, where Q_{inv} is the charge in the inversion layer and v the carrier velocity. High mobility u_{eff} is desirable since u_{eff} and v (i.e., I_{ON}) are linked, at least at low lateral field E_{lat} where $v \sim u_{eff}*E_{lat}$. However when E_{lat}>$E_{critical}$ the v starts to saturate [1,2] and u_{eff} is no longer a good metric for performances. More in general short channel parameters like metallurgical length (L_{met}), v_{sat} and mobility u_{eff} **at operating conditions ($V_{ds} \geq 1$)** would better describe the device physics and enable a *proper comparison between different materials*: this fundamental link is still missing for novel substrate materials like Si-(110) and Ge.

We previously reported [3] a technique to extract the short channel electrostatics (L_{met} and series resistance R_{series}) and mobility down to 50nm lengths. This paper expands on the previous methodology and uses an original RFCV technique to extract saturation velocities and critical fields at high V_{ds} for short channel devices. The *fundamental link* between v_{sat} and u_{eff} for 1.2nm EOT MOSFETs on Si ((100) and (110)) and Germanium is reported and benchmarked.

Device fabrication
Conventional CMOS65 with 1.2nm SiON/poly-Si down to L_{poly}=50nm were considered as reference. Using the same doping as CMOS65, N- and PFET were fabricated on the *same* wafer with the *same* dielectric and gate electrode (HfSiON/TiN) with L_{gate}~70nm on both Si-(100) and Si(110) wafers [4]. Finally planar Ge pFET were considered [5] with L_{gate} down to 120nm with Si-passivation layer and HfO_2/TaN gate stacks. Note that these devices were not optimized for best performance.

Short channel parameters at high V_{ds}
Key information on Ge short channel parameters can be extracted from CV measurements, as in fig.1-3. Following a procedure developed for standard poly-Si/SiON [3], good short channel behavior can be demonstrated for our Ge samples (fig.1) with high mobility (fig.2). Similar high mobility (200cm²/Vs) to the ones of fig.2 was already reported [4] for (110)-Si pFET. Note in Fig.3 the lower R_{series} for Ge devices and a similar mobility between pFETs on Ge(100) and Si(110).

Once the basic short channel properties are known (i.e., fig.1-3), the transport properties at high V_{ds} can be done, as discussed in fig.4-6. The inversion capacitance C_{inv} can be determined for all V_{ds} conditions (fig.4) and all the L_{met} (fig.5) of interest using the procedure of Fig.6. Note in fig.5 that the C_{inv} is L_{met} dependent due to charge sharing. Since this is important only for L_{met}<100nm, in the following C_{inv} will be assumed L-independent. After R_{series} normalization, the $u_{eff} = I_{ds}L_{met}/V_{ds}WQ_{inv}$

and $v_{sat} = I_{ds}/WQ_{inv}$, where Q_{inv} is the inversion charge (i.e., integration of C_{inv} in fig.4-5). Contrary to previous v_{sat}-extraction proposed [2], [6], this technique is *direct* and enables a *simultaneous* evaluation of both u_{eff} and v_{sat} in any bias condition (i.e., V_{gs}, V_{ds}).

Velocity saturation in Si (110) and Ge
In fig.7 the v_{sat} vs E_{lat} is compared for HfSiON on Si(100) and Si(110). Note v_{sat} (110)pFET ~ v_{sat} (100)nFET. Since the doping and the gate stack are the same, the v_{sat} increase is due to the Si orientation itself. Fig.8 also demonstrates that Ge features a 2x higher v compared to reference pFET on (100).

In the following we approximated E_{lat}=V_{ds}/L_{met}. This is somewhat a crude approximation, as confirmed by process and device calibrated TCAD simulation (not shown) where the point-to-point field along the channel can increase up to 10x close to the drain (especially in long channels). However it can be shown that the v_{sat}-E_{lat} trend shown in fig.7 for the SiON/poly-Si is correct and thus the v_{sat} and E_{lat} should be considered as mean values.

Mobility and critical field – a perspective
For the first time the u_{eff} vs. E_{lat} in fig.9 and 10 shows fundamental similarities and differences between Si and Ge. Similar to Si [1,7], also for Ge the critical field $E_{critical}$ concept is applicable. At low E_{lat} the u_{eff} is constant (i.e., Ohm's law), while at E_{lat}>$E_{critical}$ u_{eff} decreases (i.e., v starts to saturate – see fig.7,8). An important difference between the high mobility Si (i.e., (110)-pFET) and Ge samples is that for E_{lat}>$E_{critical}$ u_{eff} *decreases faster* for Ge compared to Si.

A combined analysis of fig.7-10 demonstrates the advantage of these novel techniques for benchmarking high mobility substrates. When the pFET on Si (110) and Ge are considered (see fig.11), the I_{ON} on these devices should be *similar* (for a given CET) since v_{sat} is the same. However when the ON-OFF switching is concerned (see fig.12), the u_{eff} (Ge) decreasing faster with E_{lat} (compared to Si) implying that the velocity saturates at a lower E_{lat} thus yielding a *slower* switching speed.

While it's not clear whether these properties of Ge pFETs are linked to the fabrication process (possibly related to the S/D region germanidation) or a fundamental limit of the Ge itself, these pFET in Si-(110) seem superior to these Ge samples.

Summary and Conclusions
A direct extraction of velocity and mobility at high V_{ds} for sub-100nm MOSFETs has been demonstrated using a modified RFCV technique. This technique enables the benchmarking of very different MOSFET technologies (Ge and Si(110)) with respect of carrier velocity and mobility. Saturation velocities are very similar for pFET Ge and pFET Si (110), suggesting similar I_{ON} drivability (2x compared to pFET on Si(100)). However Ge samples seem to reach velocity saturation at lower lateral fields thus possibly limiting the switching speed.

References
[1] Taur, Ning, Fundamentals of Modern VLSI devices, Cambridge Press 1998; [2] Sodini et al TED 1984; [3] Severi et al, TED 2007; [4] L.Trojman et al, INFOS 2007; [5] G.Nicholas, TED 2007 p.2503; [6] Lochtfeld, Antoniadis EDL 2001 p95; [7] Coen, Muller, Sol. State El. **23** p35-40; [8] E. San Andrés, EDL 2006;

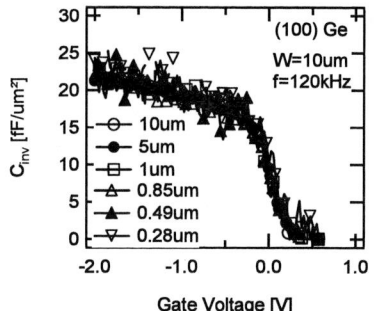

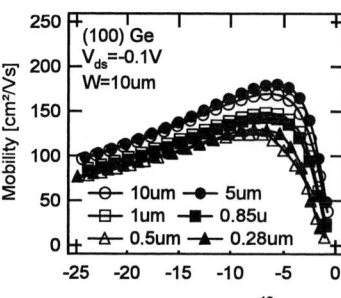

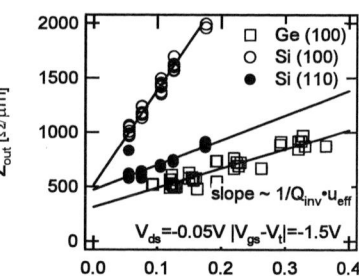

Fig.1: splitCV for Ge samples. L_{met} extraction done taking C_{par} @ $V=V_{fb}$ [ref].

Fig.2: Mobility extracted for all Ge devices. u_{peak} decreases in short L, similar to Si [ref].

Fig.3: The intercept of output resistance vs L_{gate} yields the R_{series}. Note the similar slope for Ge and Si (110) pFET, indicating a similar u_{eff}

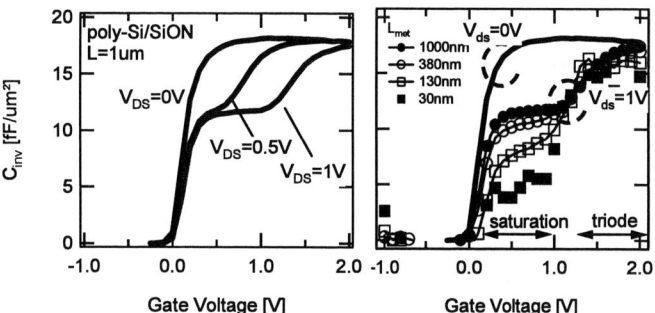

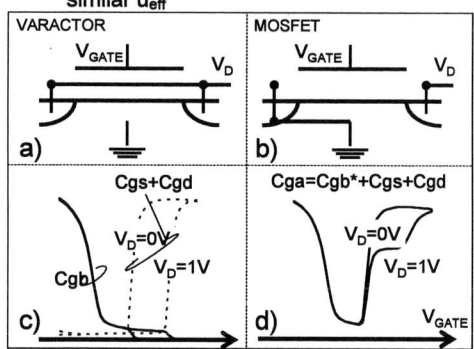

Fig.4: inversion capacitance for several V_{ds} (L=1um). A network analyzer is needed to measure the cross coupling at several (V_{gs}, V_{ds}). frequency = 100MHz

Fig.5: inversion capacitance for several L_{met}. L_{met} was extracted as in [3]. Note in triode region C_{inv} is L-independent, while in saturation is L-dependent (charge sharing).

Fig.6: device connections considered for extraction with S/D connected together (a) or separately (b). Connection in (a) is used for conventional RFCV extraction (see [8] and yields the gate-bulk (C_{gb}) and gate-source-drain ($C_{gs}+C_{gd}$) coupling, while connection as in (b) yields information on the total gate capacitance $C_{ga}=C_{gb}+C_{gs}+C_{gd}$. Note that b) is the electrostatic configuration used in I_{ON}-I_{OFF} measurements. Capacitances measured with $V_D>0$ bias are shown in panel c) and d) respectively. C_{gb}^* for data in panel d) are estimated from panel c) by averaging the C_{gb} at 0 (i.e., source) and operating conditions (i.e., drain). $C_{inv}=C_{ga}-C_{gb}^*$

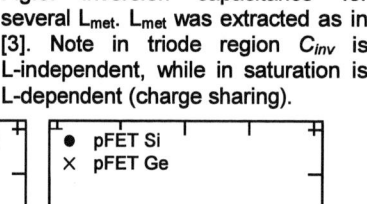

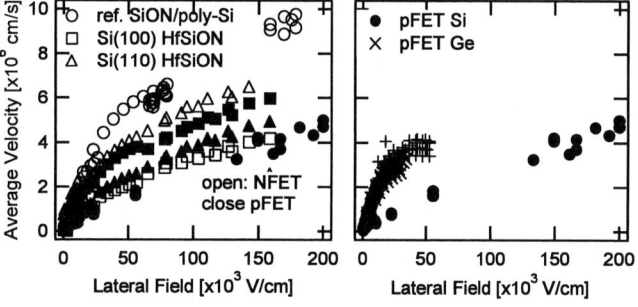

Fig.7: saturation velocity vs. lateral field for n- and pFET HfSiON on Si (110) and (100) for L_{met} down to 70nm

Fig.8: similar results for Ge pFET.

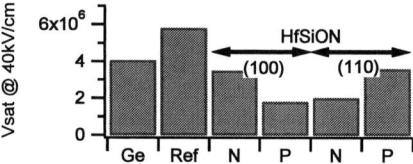

Fig.11: summary of short channel v_{sat} among different dielectric, Si orientations and Ge.

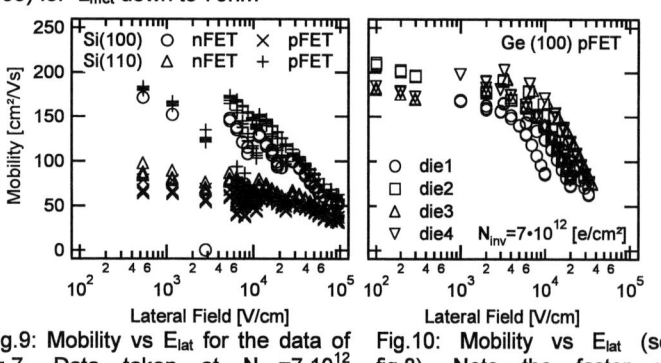

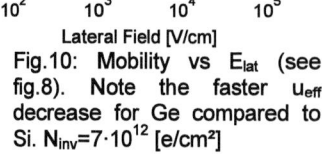

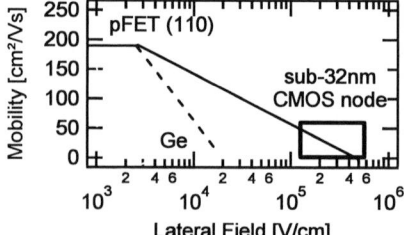

Fig.9: Mobility vs E_{lat} for the data of fig.7. Data taken at $N_{inv}=7\cdot10^{12}$ [e/cm²]. Note that the nFET (100) and pFET (110) and nFET (110) and pFET (100) show similar u_{eff}-E_{lat} behavior, respectively.

Fig.10: Mobility vs E_{lat} (see fig.8). Note the faster u_{eff} decrease for Ge compared to Si. $N_{inv}=7\cdot10^{12}$ [e/cm²]

Fig.12: (extrapolation from fig.9) mobility drops rapidly to 0 in Ge samples compared to Si ones. This indicates that the vsat is reached at lower E_{lat}.

42

Fermi-Level Depinning in Metal/Ge Schottky Junction and Its Application to Metal Source/Drain Ge NMOSFET

Masaharu Kobayashi, Atsuhiro Kinoshita, Krishna Saraswat, H. -S. Philip Wong and Yoshio Nishi

Department of Electrical Engineering, Stanford University, 420 Via Palou Mall, Stanford, CA, 94305, Email: masaharu@stanford.edu

Abstract

We successfully demonstrated Schottky barrier height modulation in metal/Ge Schottky junction by inserting an ultrathin interfacial SiN layer. The SiN layer suppressed strong Fermi level pinning in metal/Ge junction, which resulted in effective control of the Schottky barrier height. We systematically investigated its physics, for the first time, and almost zero Schottky barrier height was successfully obtained for electrons. We applied this technology to metal source/drain Ge NMOSFET and achieved low source/drain resistance.

Introduction

Non-silicon channel material, especially Ge, is one of the key technology boosters to enhance device performance. There are demonstrations of superior performance in Ge PMOSFETs to Si [1-3]. Ge NMOSFETs are, however, still not superior to Si [4-7]. A major obstacle for Ge NMOSFETs is the large source/drain (S/D) resistance due to poor dopant incorporation into Ge. Although metal S/D is a possible candidate to reduce S/D resistance, strong surface Fermi level pinning of Ge [9,10] results in high Schottky barrier height for electrons (Φ_{BNeff}) with typical germanides, such as NiGe, TiGe, CoGe [8]. To mitigate this problem, reduction of Φ_{BNeff} is necessary to achieve low S/D resistance in Ge NMOSFETs.

In this paper, we have successfully demonstrated and investigated Φ_{BNeff} modulation by suppressing Fermi level pinning experimentally and systematically, for the first time. Finally we applied this technology to metal S/D Ge NMOSFET.

Schottky Barrier Height Modulation: Fundamentals

The technologies for mitigating Fermi level pinning are firstly proposed for Si by Connely et al. [11]. Then Takahashi et al. applied a similar technology to Ge [7]. The main concept is illustrated in Fig. 1. The free-electron wavefunction penetrates into the semiconductor, which generates metal-induced gap-state (MIGS) and pins the Fermi level [12]. Since Ge has smaller bandgap and higher dielectric constant than Si, the strong Fermi level pinning occurs near charge neutrality level (E_{CNL}) which is close to the valence band, and thus high Φ_{BNeff} is known to be problematic. An ultrathin insulator inserted between metal/Ge interface blocks the free state penetration into Ge and equivalently reduces MIGS density. As a result, Fermi level pinning is released and Φ_{BNeff} becomes low.

In this study, SiN is chosen as the interfacial insulator to prevent the interfacial GeOx formation during fabrication process and attain precisely controlled thickness. SiN is deposited by well-controlled sputter system and metal/SiN/Ge Schottky diode is fabricated. Fig. 2 is a cross sectional TEM image of metal(Al)/SiN/Ge. As SiN thickness (t_{SiN}) increases, the contact resistance (R_c) decreases due to Φ_{BNeff} reduction firstly, and then, R_c becomes large because of additional tunnel resistance (Fig. 3). Therefore, t_{SiN} should be carefully optimized to minimize R_c. This metal/SiN/Ge Schottky diode is characterized in terms of current-voltage (I-V) characteristics and contact resistance (R_c) by four-point probe measurement [11] shown in Fig. 4. The Φ_{BNeff} extraction procedure in Fig. 5 is based on the thermionic emission model with ideality factor (n) which describes the first-order Schottky effect [12].

Fig. 6 shows I-V characteristics for Al/SiN/n-Ge Schottky diode with different t_{SiN}. In order to observe Φ_{BNeff} modulation clearly, Ge doping concentration is chosen as low as 1×10^{15}cm^{-3} so that current transport is dominated by thermionic emission [13]. The saturation current shows maximum value at optimal t_{SiN}, which indicates that Φ_{BNeff} is successfully modulated with t_{SiN}. Figure 7, the linear scale plot of Fig. 6(b), shows apparent change in the transport mechanisms from rectifying, ohmic, and to symmetric tunneling. Fig. 8 shows R_c as a function of t_{SiN}. Minimum R_c is observed in a manner consistent with Fig. 5. It should be noted that Ge requires thicker t_{SiN} of 2nm than Si which requires $t_{SiN}<$1nm [11] to sufficiently block the wavefunction penetration. Figure 9(a) shows experimental and

calculated n as a function of t_{SiN}. The quite good agreement is an evidence of ohmic contact in the optimum point of t_{SiN} at 2nm. Figure 9(b) shows n difference (Δn) between 313K and 273K. The larger temperature dependence in Δn observed in thicker t_{SiN} indicates that additional transport, maybe Poole-Frenkel mechanism, occurs in thick t_{SiN} samples. The additional transport does not affect the value of the R_c at the optimal point.

Fig. 10 and 11 show the I-V characteristics and R_c of Er/SiN/Ge Schottky diode. Fermi level is strongly pinned even in a low workfunction (WF) metal, Er without SiN, so that the reverse bias current is small [10]. On the other hand, the reverse bias current is maximized and R_c shows minimum with optimal t_{SiN}. Fig. 12 shows the collection of Φ_{BNeff} as a function of t_{SiN} for selected metals. It is clearly found that the initial Φ_{BNeff} of low WF metals, such as Al, Ti and Er, is very close to the E_{CNL}, however the Φ_{BNeff} significantly decreases with increase in t_{SiN} due to Fermi level depinning.

Fig. 13 shows the relationship between R_c and Φ_{BNeff}. An exponential correlation is observed, which indicates Φ_{BNeff} modulation effectively changes R_c. Fig. 14 shows the relationship between Φ_{BNeff} and metal WF. Pinning factor ($S = d\Phi_{BNeff}/dX_M$, X_M: electronegativity of the metal) indicates how much Φ_{BNeff} can be modulated by changing the metal WF. While S is close to zero in samples without SiN [9,10], samples with SiN show $S \sim 0.3$, which is close to 0.4 calculated by Monch's empirical model: $S^{-1}-1 = 0.1 (\varepsilon_\infty - 1)^2$. This significant improvement of S suggests that Φ_{BNeff} can be controlled by the interfacial insulator, not by Ge.

Considering the result in Fig. 14 and higher reverse current than forward bias current at optimal t_{SiN}, Er could have zero or even negative Φ_{BNeff}. This suggests one can potentially enhance the injection velocity (v_{inj}) by introducing a large enhancement bandoffset [14] from low WF metal source to Ge channel, as illustrated in Fig. 15, in the ballistic transport regime.

Novel Metal S/D Ge NMOSFET Implementation

The metal/SiN/Ge structure allows us to implement metal S/D Ge NMOSFET by newly established low Φ_{BNeff} metal S/D. Fig. 16 shows the schematic of fabricated metal S/D Ge NMOSFET and Fig. 17 shows the main fabrication process flow. The gate first process with thermal GeON interfacial layer is used and metal gate and S/D are deposited in a self-aligned manner. Fig. 18 shows the Al-source/SiN/p-Ge substrate junction characteristics. Off-state leakage by hole injection and junction leakage is well suppressed due to successful Φ_{BNeff} modulation. The offstate ambipoloar leakage suppression, important for low bandgap materials, is another advantage of this technology. Fig. 19 demonstrates I_d-V_d characteristics of Ge NMOSFET with L_g of 1.5um. No significant S/D resistance is observed.

Conclusions

Fermi level depinning in metal/Ge Schottky junction was experimentally and systematically investigated. An ultrathin SiN successfully released Fermi level pinning and achieved very low Φ_{BNeff}. The metal S/D Ge NMOSFET with low S/D resistance and low leakage current was successfully demonstrated. We also proposed high v_{inj} from a low WF metal source. This novel junction technology is feasibly integrated with high-k/metal gate and 3D-IC technologies which require low thermal budget process.

Acknowledgement

The authors thank to Prof. Shin-ichi Kobayashi for TEM image. This work was supported by Semiconductor Research Corporation (SRC).

References

[1] Y. Kamata et al., IEDM2005, p.441[2] P. Zimmerman et al., IEDM2006, p.655. [3] T. Yamamoto et al., IEDM2007, p.1041. [4] C. O. Chui et al., IEDM2003, p.437 [5] H. Shang et al., EDL, 25, 135, 2004. [6] D. Kuzum et al., IEDM2007, p.723. [7] T. Takahashi et al., IEDM2007, p.697. [8] D. Han et al., Microelectron. Eng., 82, 93,2005, [9] A. Dimoulas et al., APL, 89, 252110, 2006 [10] T. Nishimura et al., SSDM2006 p. 400 [11] D. Connelly et al., IEEE Nanotechnol., 3, 98, 2004 [12] W Monch, "Electronics Properties of Semiconductor Interfaces", Springer, 2004 [13] S. Sze, "Physics of Semiconductor Devices", Whiley, 1981 [14] T. Mizuno et al., VLSI2004, p.202

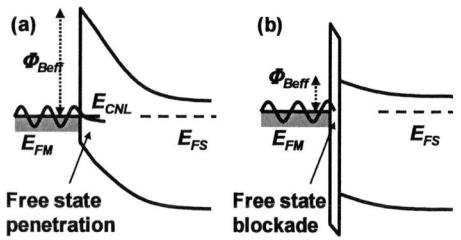

Figure. 1 Schematic band diagram of (a) metal/Ge junction, where free electron wavefunction penetrates into Ge and (b) metal/insulator/Ge junction, where free electron wavefunction penetration is blocked.

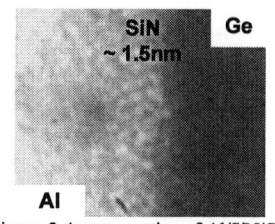

Figure. 2 A cross section of Al/SiN/Ge. An ultrathin SiN is formed.

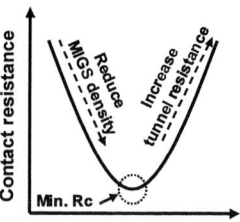

Figure. 3 A schematic of the relationship between R_c and t_{SiN} [11]. There is a minimum R_c at a optimal t_{SiN}

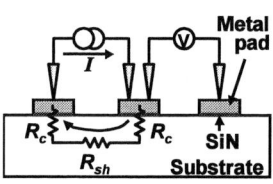

Figure. 4 Measurement set up for R_c. Four-point probe method is used to detect voltage drop at a single contact by forcing current.

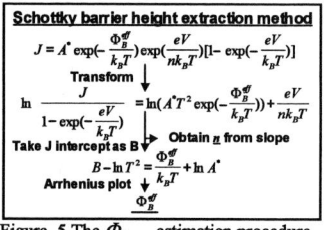

Figure. 5 The Φ_{BNeff} estimation procedure. The thermionic emission model with ideality factor is used. Measurement is done between 313K and 243K. ideality factor (n) is also extracted in this process.

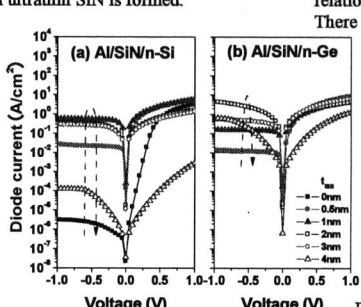

Figure. 6 IV characteristics in Al/SiN/(a)Si and (b)Ge with different t_{SiN}. The reverse current is maximized at optimal t_{SiN}.

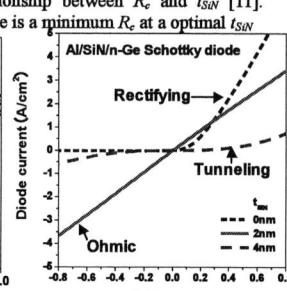

Figure. 7 Linear scale plots of Figure 6(b). Transport mechanism clearly changes from rectifying to ohmic and then tunneling.

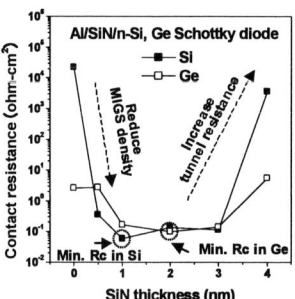

Figure. 8 R_c in Al/SiN/Si, Ge as a function of t_{SiN}. R_c is minimum at optimal t_{SiN}=2nm for Ge due to the balance between two mechanisms.

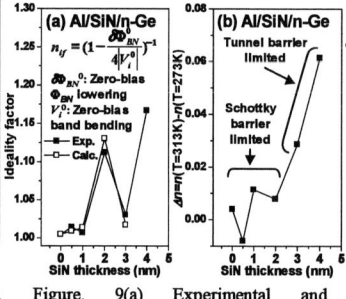

Figure. 9(a) Experimental and calculated ideality factors as a function of t_{SiN} at 293K. (b) Ideality factor difference between 313K and 273K.

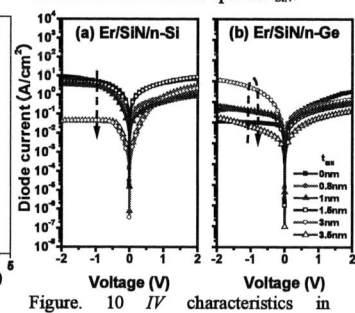

Figure. 10 IV characteristics in Er/SiN/(a)Si and (b)Ge with different t_{SiN}. The reverse current is maximized at optimal t_{SiN}=3nm in Ge.

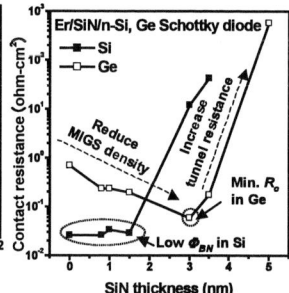

Figure. 11 R_c in Er/SiN/Ge as a function of t_{SiN}. R_c is minimum at a optimal t_{SiN}=3nm for Ge due to the balance between two mechanisms.

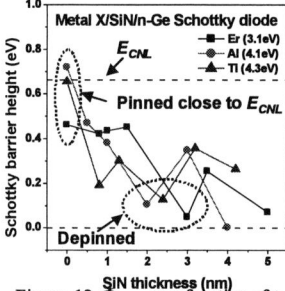

Figure. 12 Φ_{BNeff} as a function of t_{SiN} in selected low WF metal/SiN/Ge. Low WF metals have almost ideal Φ_{BNeff} with optimal t_{SiN}.

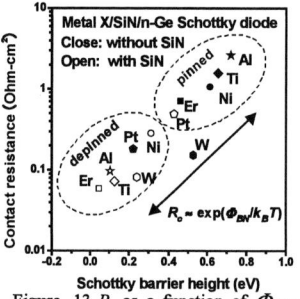

Figure. 13 R_c as a function of Φ_{BNeff} with different WF metals. Linear correlation is held. Φ_{BNeff} lowering due to E_F depinning reduces R_c.

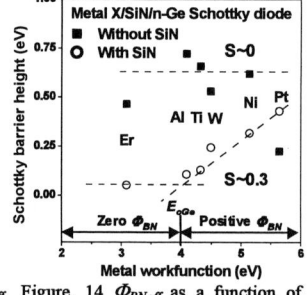

Figure. 14 Φ_{BNeff} as a function of metal WF. S=0.3 in metal/SiN/Ge, consistent with calculated value 0.4 based on analytical model.

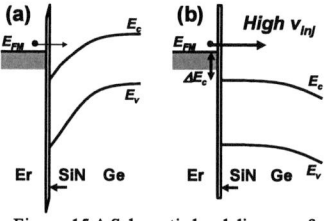

Figure. 15 A Schematic band diagram of Er/SiN/Ge Schottky source in Ge NMOSFET in (a) off and (b) on states. Large band offset possibly enables high injection velocity into channel [14].

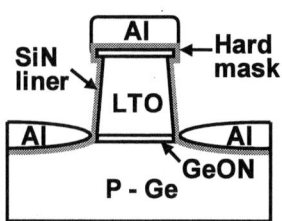

Figure. 16 A schematic of novel metal gate and metal S/D Ge NMOSFET. GeON is used for the interfacial layer. Al is used for a low WF metal.

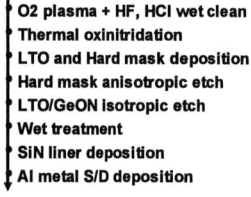

Figure. 17 Fabricaion process flow of the metal S/D Ge NMOSFET with GeON gate first process. Al is used for low WF metal.

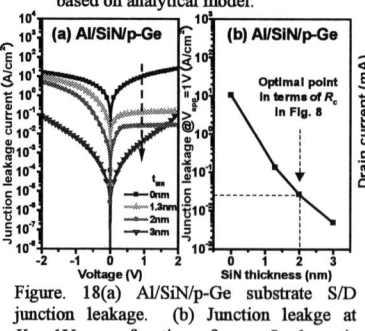

Figure. 18(a) Al/SiN/p-Ge substrate S/D junction leakage. (b) Junction leakge at V_{app}=1V as a function of t_{SiN}. Leakage is well suppressed at t_{SiN}=2nm which is the optimal point in terms of R_c in Fig. 8.

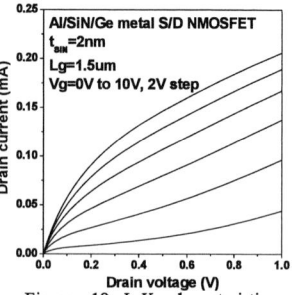

Figure. 19 I_d-V_d characteristics of metal S/D Ge NMOSFET with t_{SiN}=2nm. Parasitic S/D series resistances are effectively reduced.

The effects of Ge composition and Si cap thickness on hot carrier reliability of Si/Si$_{1-x}$Ge$_x$/Si p-MOSFETs with high-K/metal gate

W.-Y Loh[*+], P.Majhi[1], S.-H. Lee[a], J.-W. Oh[*], B. Sassman[*], C. Young[*], G. Bersuker[*], B.-J. Cho[b], C.-S. Park[*],C.-Y. Kang[*], P.Kirsch[2], B.-H. Lee[2], H.R.Harris[3], H-H.Tseng[*], R.Jammy[2]

*SEMTECH, 2706 Montopolis Drive, Austin, TX 78741, USA, [1]Intel assignee, [2]IBM assignee, [3]AMD assignee, [a]Univ. of Texas, [b]KAIST, Daejeon, Korea; [+]email:wei-yip.loh@sematech.org

Abstract

We report on new observations of hot carrier (HC) degradation in strained Si/Si$_{1-x}$Ge$_x$(x =0.2 to 0.5) p-MOSFETs. By using low voltage current-voltage measurement coupled with carrier separation, we are able, *for the first time*, to easily distinguish the energy distribution of the interface traps. High-K dielectrics on SiGe p-channel show higher interface traps generation located close to conduction band under channel hot carrier stressing and uniform interface trap under drain avalanche hot carrier stressing, both of which can be mitigated by increasing Ge% in the Si/SiGe channel. Detailed study on Si capping layer ($\leq$ 20Å) shows good immunity against Drain Avalanche Hot Carrier but is degraded under Channel Hot Carrier stressing. The results suggest that *higher Ge%* and *thinner Si cap* is preferably for hot carrier reliability for low voltage application with 10yrs lifetime at operating voltage of -0.85 V.

Introduction

High mobility Si/SiGe/Si hetero-structure is one of the key areas of research for improving device performance. While the material and device characteristics of SiGe channel has been extensively studied [1], the reliability of high-K on Si/SiGe/Si is still unclear. Si/SiGe/Si heterostructures exhibit significantly higher interface traps [2] and impact ionization [3] compared to Si. Higher gate leakage and additional interfaces hinder measurements of channel-dielectric interface traps using conventional charge pumping and gated diode technique and may require elaborate low temperature measurements [4]. In this work, we revert to interface trap assisted tunneling [5] as a direct and simple measure of the interface traps coupled with carrier separation to delineate the gate leakage component. Hot carrier reliability of high-K dielectrics on SiGe p-channel MOSFET with gate length of L$_g$ = 100 ~ 200 nm are presented.

Device Fabrication

Devices were fabricated using standard CMOS process flow on 8" p-type Si (100). Epitaxial layers of Si$_{1-x}$Ge$_x$ with thickness 100 to 300 Å with a variation of Ge concentration from 20%~ 50%, and different Si capping layer of 0Å, 14Å and 20Å were sequentially deposited in UHV epitaxial-chamber. Gate stack consists of TaN (1000 Å)/HfO$_2$ (40Å) with EOT of 15Å and interface trap density of < 8x10^{11} /cm^2.

Results And Discussions

Fig. 1 shows the transconductance of the SiGe$_x$ (x=20 to 50%) with different Si cap thickness and the high-resolution TEM image of the fabricated devices. Good device characteristics have been achieved for 100nm devices as shown in Fig. 2. Statistical gate leakage shows Si/SiGe/Si channel have lower leakage compared to Si channel with gate leakage reducing with Ge% (Fig. 3). Using carrier separation [6] (Fig. 4), the hole current (J$_s$) and electron current (J$_W$) can be clearly differentiated. Both electron and hole current comprise of a direct tunneling component J$_{DT}$ and a interface-trap assisted tunneling component J$_{ITAT}$ as shown in Fig. 4. Within the Si/SiGe forbidden bandgap, trap-assisted tunneling (ITAT) is the main component and can be used as a monitor for interface traps generation [5]. By applying carrier separation and low voltage measurement, we are able to distinguish the interface traps in terms of region (channel or overlap) and energy level of the traps (closer to conduction or valence band). Increasing Ge% decreases hole current, I$_{SD}$ (Fig. 5) consistent with a larger hole barrier $\phi_{B,hole}$ and larger out-plane mass due to biaxial compressive strain in the SiGe layer [1,7,10]. Fig. 6 shows that the effect of Si cap on gate leakage results in higher electron leakage but

does not affect hole current. The differences in leakage current cannot be explained by lower interface traps as charge pumping results shows otherwise. Under high gate and drain bias, impact ionization in SiGe channel is increased with increasing Ge% as shown in Fig. 7 and 8.

Hot carrier (HC) stressing is carried out under two different conditions: (a) Channel Hot Carrier (CHC) stress with V$_d$ = V$_g$-V$_{th}$ and (b) Drain Avalanche Hot Carrier (DAHC) stress with maximum substrate hole current. Under CHC stress (Fig. 10), SiGe channel shows similar increase in hole J$_{ITAT,h}$ (= I$_d$) as Si channel but order of magnitude increase in electron J$_{ITAT,e}$ (=I$_W$)(Fig. 11). Since both electrons and holes are injected into the SiGe-high-K interface, the damage caused by CHC comprises of damages from both holes and electrons. In contrast, under DAHC, hot hole is predominant due to maximum hole generation in the drain region (Fig. 12). The results suggest that electrons tend to cause interface damage close to the conduction band while holes create interface trap uniformly distributed across the energy band in SiGe channel. In both stressing conditions, increasing Ge% is able to suppress both the electron and hole ITAT, thus reducing the V$_{th}$ shift. To avoid Ge out-diffusion, Si cap (~5nm) has been used to improve device reliability [8]. Figs. 13 and 14 show the effects of Si cap on SiGe channel under both CHC and DAHC stressing. While Si cap suppress interface traps formation under DAHC (Fig. 13), V$_{th}$ degradation is actually enhanced under CHC stressing. In all the above cases, increased interface trap generation close to either conduction or valence band can be explained by the reduction in band off-set to the dielectric resulting in increased trap sampling at Fermi level[9]. The results suggest that increasing Ge% in Si/SiGe/Si would be preferential in terms of hot carrier performance due to reduced energy bandgap and hence lower N$_{IT}$.

Lifetime projection of SiGe p-channel devices under hot carrier stress is shown in Fig. 16. Compared to Si, SiGe shows generally higher degradation (Fig.15) under DAHC stressing which can be attributed to higher impact ionization (Fig. 7). Increasing Ge% actually helps to mitigate degradation under CHC conditions due probably to suppression of hole ITAT. On the other hand, SiGe devices suffers higher degradation under DAHC due to anomalous increase in ITAT which results in high V$_{th}$ shift under high field stress. Lifetime for SiGe p-channel devices appears to be limited under DAHC stressing and a 10-yr lifetime at -0.85V is predicted for SiGe$_{0.2}$ p-channel devices.

Conclusion

Si/SiGe/Si p-channel MOSFET with various Si cap thickness (0 ~ 20Å) have been studied in terms of hot carrier reliability. It was observed that SiGe have degraded hot carrier immunity compared to Si due to higher interface traps, evidenced by the enhanced ITAT. However, higher Ge composition is able to suppress gate leakage and hot carrier degradation due to enhanced hole barrier. Thinner Si capping layer is preferably to mitigate the enhanced interface trap generation under CHC stress. SiGe p-channel shows higher field dependency due to enhanced impact ionization and may be more suitable for low voltage application (|V$_g$|$\leq$0.85V).

References

[1] D. J. Paul, *et al.*, *SST*, 19, 2004. [2] P.-J. Tzeng *et al.*, *IEEE TDMR*, p. 168, 2005. [3] N.S. Waldron et al., *IEDM Tech Dig*, 813, 2003. [4] T. Tsuchiya et al., *IRPS*, p. 449, 2004. [5] A. Ghetti, *et al.*, *IEDM Tech Dig.*, p. 731, 1999. [6] H. Guan *et al.*, *IEEE TED.*, 1608 2000. [7] C. W. Liu *et al.*, *IEEECDM*, 21, 2005. [8] G. Kumar *et al.*, *IEEE TED*, 1142, 2006. [9] Mathew *et al.*, *IEEE EDL*, p. 173, 1999. [10] R. People et al., *APL.*, 48, 1986. [11] W. Zhao *et al.*,*IEEE EDL*, 26, 2005.

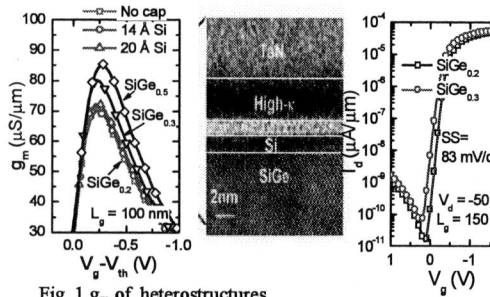

Fig. 1 g_m of heterostructures Si/SiGe/Si p-MOSFET with high-K/TaN gate stack and different Ge% and Si cap thickness. Inset shows the TEM image of the channel region

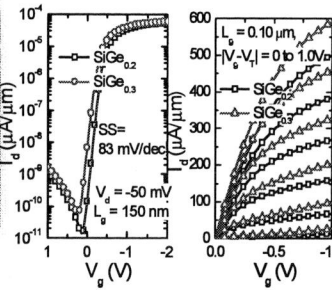

Fig. 2. I_d-V_g and I_d-V_d characteristics of short channel p-MOSFET with different composition of SiGe for 20% and 30% Ge.

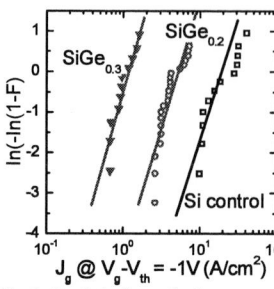

Fig. 3. Statistical gate leakage current at $|V_g$-$V_{th}| = 1$V. Increasing Ge% suppresses gate leakage compared to Si control.

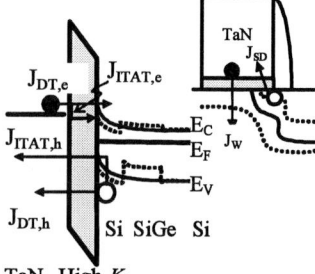

Fig. 4. Energy band alignment of SiGe p-channel with Si cap under inversion conditions [10] and associated leakage current. Inset shows the carrier separation under inversion conditions.

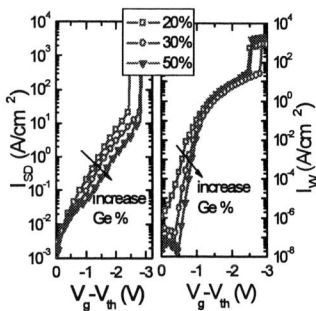

Fig. 5. Source/drain and well leakage component for different SiGe composiiton. Increasing Ge% leads to lower hole current I_{SD} due to enhanced compressive strain, and hence increase in $\phi_{B,hole}$ and m_{eff} [11].

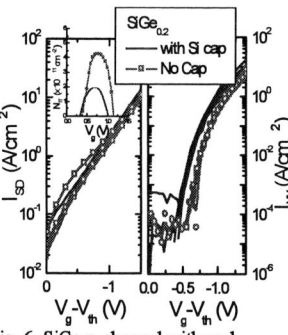

Fig. 6. SiGe p-channel with and without cap shows no impact on I_{SD} while electron current I_W is increased. Charge pumping shows higher N_{IT} for SiGe without Si cap.

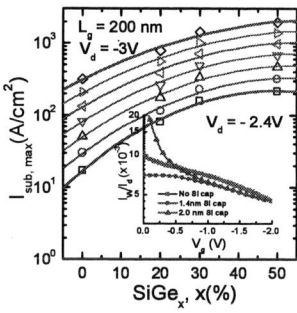

Fig. 7. Max I_{sub} for different Ge% in Si/SiGe/Si p-channel . SiGe p-channel devices show increased impact ionisation. Inset shows that Si cap suppress impact ionisation.

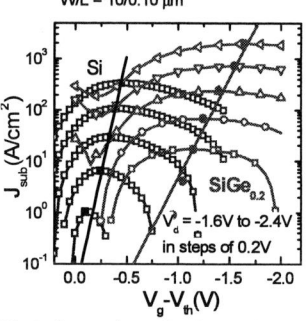

Fig.8. Comparison of substrate hole current under high V_d bias for Si and SiGe channel. SiGe channel shows higher substrate current I_{sub} occurring at higher gate bias

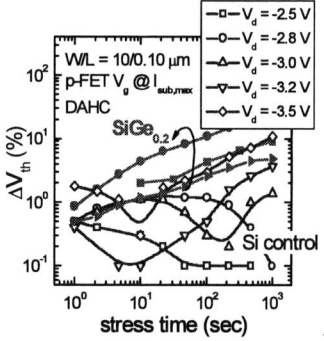

Fig. 9. Relative change in V_{th} for both SiGe$_{0.2}$ and Si p-MSOFET stressed under drain avalanche hot carrier injection with max I_{sub} conditions .

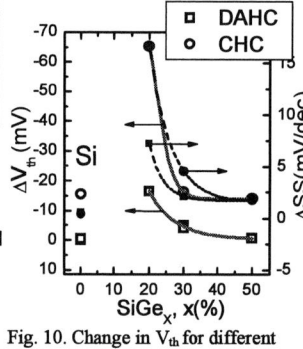

Fig. 10. Change in V_{th} for different Ge% under both DAHC and CHC conditions. Also shown is the subthreshold slope degradation which correlate with the change in V_{th}.

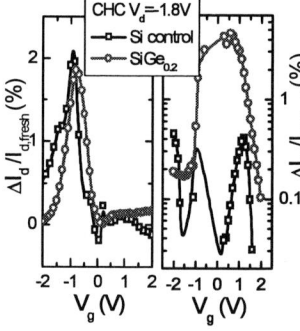

Fig. 11. ITAT of post-stressed SiGe$_{0.2}$ and Si p-MOSFET. Devices stressed under CHC stress for 1000s. SiGe device show anomalous increase in I_W

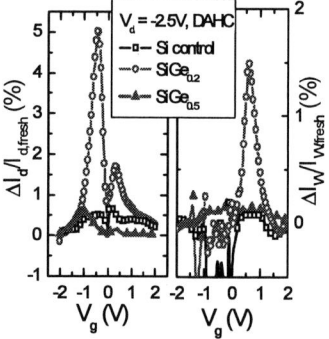

Fig. 12. ITAT of SiGe$_{0.2}$ and Si p-MOSFET stressed under DAHC conditions at V_d = -2.5V for 1000s.

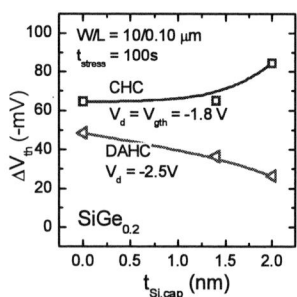

Fig. 13. Change in p-MSOFET V_{th} under both CHC and DAHC for SiGe$_{0.2}$ channel with different Si cap thickness

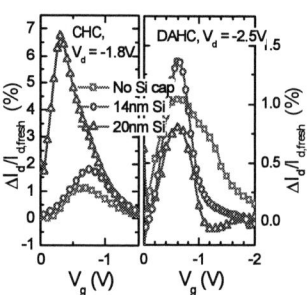

Fig. 14. Well and source ITAT leakage component for SiGe$_{0.2}$ p-MOSFET with different Si cap thickness under CHC and DAHC stress.

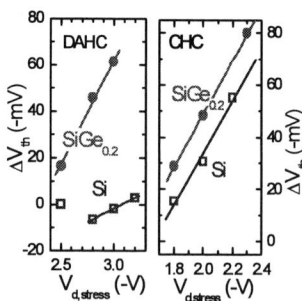

Fig. 15. Voltage dependency of HC degradation in Si and SiGe (20%). HC stress performed for t = 100sec.

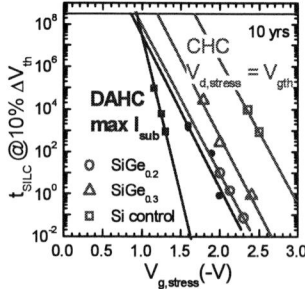

Fig. 16. Projected lifetime for SiGe and Si p-MOSFET under CHC (V_d = V_{gth}) and DAHC (max I_{sub}) conditions.

46

Impact of Source-to-Channel Carrier Injection Properties on Device Performance of Sub-100nm Metal Source/Drain Ge-pMOSFETs

H. Takeda[1], T. Yamamoto[2], T. Ikezawa[3], M. Kawada[3], S. Takagi[4,5] and M. Hane[1]

[1]Device Platforms Research Laboratories, NEC Corporation,
[2]MIRAI-ASET, [3]NEC Informatec Systems, Ltd., [4]MIRAI-AIST, [5]The University of Tokyo
1120 Shimokuzawa, Sagamihara, Kanagawa 229-1198, Japan
Phone: +81-42-771-0797, Fax: +81-42-771-0886, E-mail: h-takeda@im.jp.nec.com

Abstract

Sub-100nm metal source/drain (MSD) Ge-pMOSFETs are successfully fabricated and the device performance is analyzed from the aspect of source-to-channel carrier injection properties. Our full-band based device simulator is able to reproduce the experimental device characteristics, revealing that the low source-to-channel injected carrier density (N_s) of MSD Ge-devices could limit their source-drain current. In deep sub-100nm region, the quasi-ballistic transport nature tends to reduce the carrier velocity advantage of Ge to Si, while Ge-devices rather exhibit higher drain current than Si-ones with sufficiently high N_s condition. This N_s increase is a key to develop high-performance Ge-pMOSFETs.

Keywords: Ge, MOSFET, metal source/drain, Schottky, strain

Introduction

Ge-channel MOSFETs have been attractive for future high speed and low power CMOS devices. Recently, sub-100nm high mobility Ge-pMOSFETs with NiGe metal source/drain (MSD) was presented by one of authors [1] (Fig. 1). The NiGe MSD has a small Schottky barrier (~60meV) to contribute low S/D resistance for p-type Ge-devices. Comprehensive analysis on the sub-100nm MSD Ge-pMOSFETs in terms of the carrier transport characteristics is necessary to explore further scaling benefit. We, therefore, performed predictive carrier transport simulation considering full-band structure and carrier injection from the MSD, and discussed detailed on-current characteristics under near-ballistic condition with an emphasis on the quantitative comparison to the MSD Si-devices.

Analysis of Fabricated MSD Ge-pMOSFETs

Our full-band Monte Carlo device simulation [2] has considered vertical quantization effects, phonon and surface roughness scattering effects [3,4]. Carrier density at the SD/channel interface is calculated by the local density of states (DOS) with a Fermi distribution function of the thermal equilibrium MSD. Tunneling carrier injection through the Schottky barrier of the MSD is also considered based on the WKB approximation.

First, we have confirmed the accuracy of our simulation by means of the reproduction of the experimental results. Surface roughness parameters were found to be critical for reproducing experimentally obtained inversion layer hole mobility (Fig. 2). As for device characteristics, our device simulator enables to reproduce experimental I-V curves of L_{ch}=80nm devices quite well without any further parameter adjustment (Fig. 3). Thus, we used our device simulator for analyzing further aggressively scaled situations by changing device parameters, such as L_g, t_{ox}, and Schottky barrier height ($V_{Schottky}$). For instance, t_{ox}=6.3nm used in the device shown in Fig. 1 gave us rather small I_{on} (but barely approaching a half of the ITRS specification CV/I_{on} (Fig. 4)). We are able to assess the effects of t_{ox} or $V_{Schottky}$ reduction as shown in Figs. 5 and 6 by using our simulation code.

Comparative Analysis between Ge- and Si-pMOSFET

We have found that, under the same bias conditions, sub-100nm MSD Ge-pMOSFET hardly outperforms Si-ones if the same MSD ($V_{Schottky}$=60meV) structure could be achieved for Si-pMOSFETs. Figure 6 shows I_d-V_g characteristics of L_{ch}=80nm MSD Ge- and Si-pMOSFETs with the same bias and MSD conditions. At the on-states, the drain current of MSD Ge-device (I_d^{Ge}) is slightly smaller than that of Si-device (I_d^{Si}) because of the lower DOS of Ge [5]. Similarly, the subthreshold slope for Ge shows slight degradation to Si-devices reflecting higher Ge dielectric constant. These tendencies were found to be maintained even for t_{ox}=2.0nm devices.

Moreover, calculated I_d^{Ge} values become rather smaller than I_d^{Si} at deep sub-100nm region while the I_d^{Ge} values for relatively longer channels are comparable to I_d^{Si}, supposing the same MSD structure for Si-devices (Fig. 7). This can be explained in terms of carrier velocity and sheet density. At the source/channel interface, hole sheet density of Ge-device (N_s^{Ge}) is only about half of that in Si (N_s^{Si}) because of the lower DOS of Ge with the same MSD (Fig. 8(a)).

In the case of L_{ch}>100nm devices, the net carrier velocity across the source/channel interface of Ge (v_s^{Ge}) reaches up to twice of the Si-case (v_s^{Si}) (Fig. 8(b)) since the back-scattering rate ($I_{back}/I_{forward}$) in the Ge-channel becomes sufficiently low reflecting the Schottky barrier rectification nature as show in Fig. 9. Thereby, lower N_s^{Ge} effect is to be reconciled by this higher v_s^{Ge} for long channel devices.

In deep sub-100nm region, however, even Si-cases exhibit quasi-ballistic nature due to the strong longitudinal electric field of MSD structure, thus the v_s^{Si} tends to increase being closer to v_s^{Ge} (Fig. 8(b)), which leads to larger I_d^{Si} than I_d^{Ge} since N_s^{Si} is always twice greater than N_s^{Ge} (Fig. 8(a)). This quasi-ballistic nature also tends to make v_s as close as the source-to-channel injection velocity (v_{inj}) regardless of Ge- or Si-devices. As show in Fig. 10, at L_{ch}=30nm, v_{inj}^{Ge} becomes no more than 1.2 times (=0.95/0.81) of v_{inj}^{Si} with the same MSD structure (same Fermi distribution), since significant amount of low-velocity carriers near the energy-band bottom regulate dominant portion of the velocity distribution even for the Ge-cases. Such an argument might be overlooked if one would restrict merely the low field mobility analysis where high-velocity minor carriers near the edge of the momentum distribution play a role for determining the mobility, but not for the v_{inj} (Fig. 10).

Effects of Channel Injection Improvement

In order to make the most of the higher v_s in Ge-pMOSFETs, it is required to improve the source-to-channel hole injection and increase N_s, for example with highly doped source/drain diffusion layers. If sufficiently high N_s is injected in the Ge-channel, the higher v_s^{Ge} enhances I_d compared to Si-pMOSFETs (Fig. 11), while I_d^{Si} somewhat approaches to I_d^{Ge} at deep sub-100nm devices because of their high ballisticity.

Our simulation results show that compressive uniaxial stress significantly enhances I_d for <110>-channel Ge-pMOSFETs as shown in Fig. 12 since the v_{inj} is increased by the strain induced band curvature change but only when sufficiently high N_s is achieved. Compressive strain also reduces DOS for Ge-MOSFET, thus it could degrade I_d^{Ge} owing to the fixed Fermi level of the MSD.

Conclusion

Highly accurate device simulation has revealed that the lower carrier density at the source/channel interface of MSD structure could diminish the drain current superiority for the sub-100nm MSD Ge-pMOSFETs despite of the higher hole velocity. Ge-pMOSFETs are expected to exhibit higher drain current than Si if they have sufficiently large number of source-to-channel injected carriers. Under such constraint, strained channel technologies are still efficient for increasing the drain current even with the reduced DOS by the strain.

Acknowledgment: This work was partly supported by NEDO.

References: [1] T. Yamamoto et al., IEDM Tech. Dig., 1041 (2007). [2] H. Takeda et al., Proc. of SISPAD, 25 (2007). [3] M.V. Fischetti et al., JAP **80**, 2234 (1996). [4] T. Ando et al., Rev. Mod. Phys. **54**, 437 (1982). [5] S. Takagi et al., SSDM Ext. Abst., 1056 (2006).

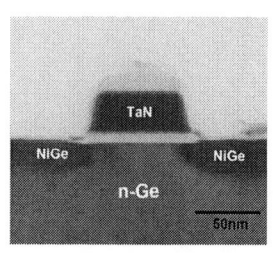

Figure 1: TEM image of the fabricated Ge pMOSFET with NiGe metal source/drain (MSD) [1]. The effective oxide thickness is 6.3nm and the junction depth is 15nm.

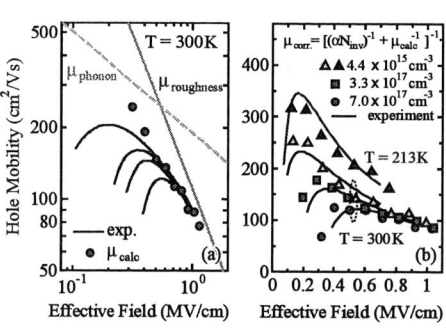

Figure 2: (a) Hole mobility of Ge inversion layer as a function of effective field, that is 1.8 higher than hole mobility of Si [1]. (b) Hole mobility assuming ionized impurity scattering by Matthiessen's rule.

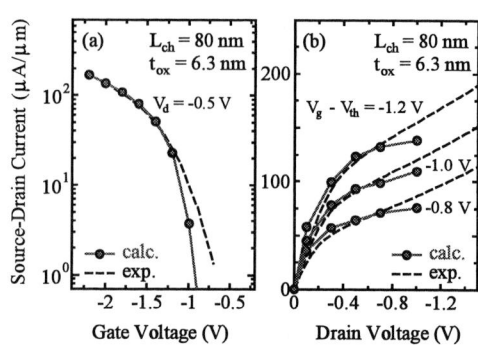

Figure 3: (a) I_d-V_g and (b) I_d-V_d characteristics of L_{ch}= 80nm MSD Ge-pMOSFET with 3.3×10^{17}cm^{-3} channel. Fabricated device has significant recombination leakage current.

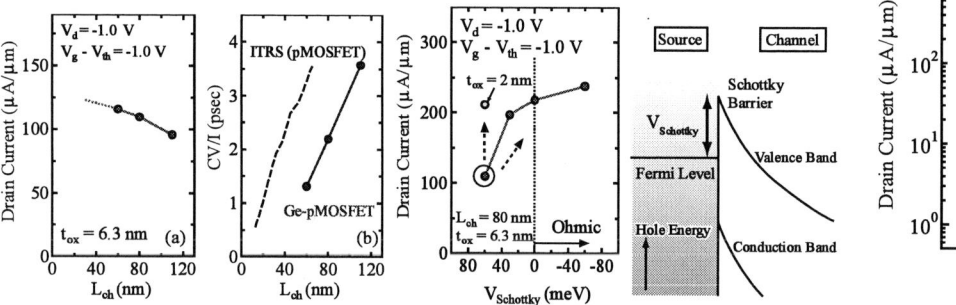

Figure 4: (a) I_{on}-L_{ch} characteristics and (b) CV/I_{on} versus L_{ch}. L_{ch} is defined by the distance between MSD junctions.

Figure 5: Schottky barrier height dependency of I_{on} for L_{ch}=80nm MSD Ge-pMOSFET. Thinned t_{ox}(=2nm) also enhances I_{on}.

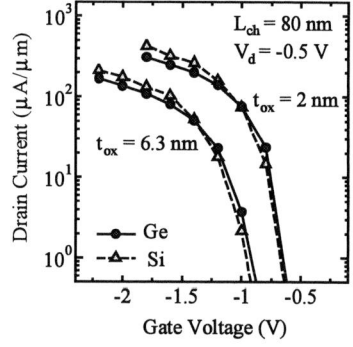

Figure 6: I_d-V_g characteristics of L_{ch}=80nm Ge- and Si-pMOSFET under the same bias conditions (V_d=-0.5 V).

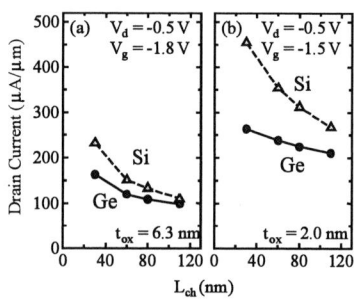

Figure 7: I_d-L_{ch} characteristics of Ge- and Si-pMOSFETs with (a) t_{ox}=6.3nm and (b) t_{ox}=2.0nm, under the same set of bias conditions.

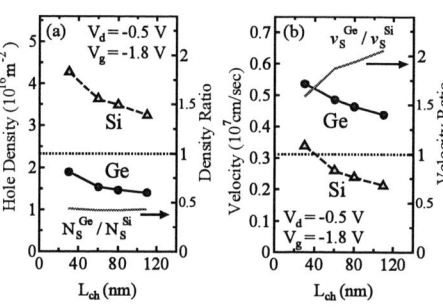

Figure 8: L_{ch} dependencies of (a) hole density (N_s) and (b) hole velocity (v_s) at the source/channel interface. Density ratio (N_s^{Ge}/N_s^{Si}) and velocity ratio (v_s^{Ge}/v_s^{Si}) are also plotted.

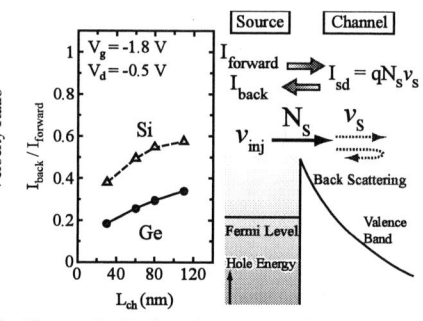

Figure 9: Ratio of $I_{forward}$ and I_{back} at the source/channel interface as a back-scattering rate of MSD Ge- and Si-pMOSFETs.

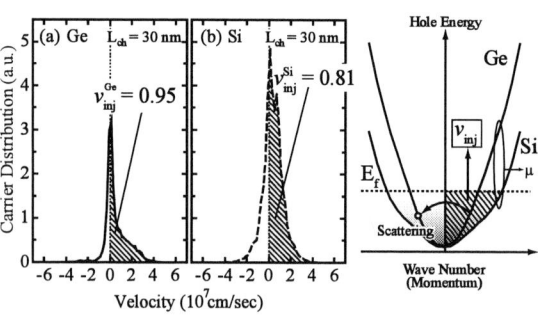

Figure 10: Carrier velocity distribution at the source/channel interface of MSD Ge- and Si-pMOSFET with L_{ch}=30nm. Injection velocity, v_{inj}, represents the average velocity of positive velocity carriers.

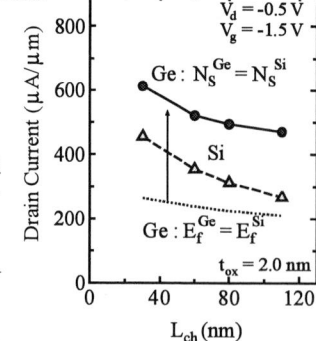

Figure 11: Drain current of MSD Ge-pMOSFETs with t_{ox}=2nm assuming the same N_s as Si-ones.

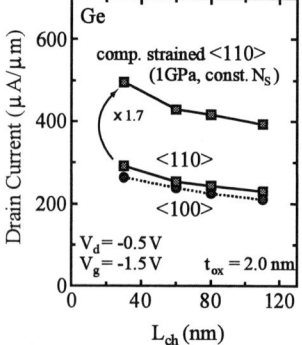

Figure 12: Drain current of strained and unstrained <110>-ch Ge-pMOSFETs assuming the same N_s as unstrained one.

Low V_T Metal-gate/High-k nMOSFETs — PBTI dependence and V_T Tune-ability on La/Dy-capping layer locations and Laser annealing conditions

S. Z. Chang[1], T. Y. Hoffmann, H. Y. Yu[2], M. Aoulaiche[3], E. Rohr, C. Adelmann, B. Kaczer, A. Delabie, P. Favia, S. Van Elshocht, S.Kubicek, T. Scharm,
T. Witters, L. -A. Ragnarsson, X. P. Wang, H. -J. Cho[4], M. Mueller[5], T. Chiarella, P. Absil, and S. Biesemans

IMEC, [1]TSMC, [2]also EEE, NTU, Singapore, [3]also EE of Katholieke Universiteit Leuven, [4]Samsung, [5]NXP-TSMC Research Center
Kapeldreef 75, B-3001, Leuven, Belgium. TEL: +32-16281145; Fax: +32-16281706; email: szchang@tsmc.com, changs@imec.be

Abstract

This paper provides a comprehensive study of the abnormal PBTI behaviors recently observed in La/Dy-capped high-k films in low-V_T nMOSFETs. We found that process details in thermal budget (or dielectric intermixing) and oxygen content of the metal trigger the onset of these abnormalities. The ΔV_T relaxation during the PBTI recovery period induced by bulk trapping/de-trapping is believed to be oxygen vacancies related, and can be suppressed either by reducing dielectric intermixing with lower laser anneal powers (La above or below HK), or by increasing the oxygen concentration, i.e., TaCNO metal electrode instead of TaCN (La above HK). Putting La below HK can result in a similar V_T tune-ability with less thermal budget for intermixing with the IL (with superior PBTI), without loss of current drive-ability. We propose Ta$_2$C/HK/LaO/IL + LLP anneals as an optimum nFETs stack configuration for practical CMOS integration.

Introduction

Hf-based dielectrics are considered the most promising high-k candidate. Gate stacks consisting of deposited metals like Ta or Ti, and doped high-k with rare-earth element via dielectric cappings (LaO, DyO, etc.) has been demonstrated as a practical solution to achieve low V_T nMOSFETs [1-3] for advanced technologies. However, the fast transient charging effect owing to bulk trapping in high-k dielectrics is one of the major reliability challenges [4]. Furthermore, abnormal V_T shifts during the PBTI (Positive Bias Temperature Instability) stress resulting from La/Dy incorporation are also observed recently [5, 6]. Considering the ultra-shallow junction (USJ) requirement, the application of laser anneal for metal-gate/high-k CMOS fabrication is reported [1]. In this work, by varying La/Dy-capping layer positions relative to the host high-k dielectrics (above or below) and with Ta$_2$C/TaCNO metal electrodes, the V_T tune-ability and PBTI implication on nMOSFETs at different thermal budgets through 1150-1350°C S/D laser annealing are investigated, while an optimum stack is suggested.

Device Fabrication and PBTI setting

The high-k gate stack formation starts from an IL of 1nm ISSG SiO$_2$, followed by a host dielectric of 1.8nm HfSiO$_x$ with 60% [Hf] (MOCVD), or 1.5nm HfO$_2$ (ALD), and a capping layer of 0.7nm LaO (ALD) or DyO (AVD), putting below or above HK. Then a 10nm Ta$_2$C (PVD) or TaCNO (MOCVD) electrode and a 100nm Poly-Si cap layer are deposited. S/D were activated via <u>L</u>ow (1150°C), <u>M</u>edium (1250°C), and <u>H</u>igh (1350°C) <u>L</u>aser <u>P</u>ower anneals, or spike anneals (1035°C). PBTI is measured at 110°C by using sense-and-measure technique [7]. PBTI measuring with a 100s recovery time after each stress is also implemented for dielectric trapping/de-trapping investigation.

Results: Device Performance

Fig.1 shows nMOSFETs V_T vs. LaO capping layer locations at different laser-anneal conditions with Ta$_2$C electrode. ~500mV V_T lowering is obtained by adding LaO either above or below HfSiO$_x$. It is observed La-above HK is more sensitive to laser powers, and LLP still results in ~150mV higher in V_T. (Similar trend in DyO capped case). Comparable transistor driving performance between LaO-bottom cap with LLP/MLP anneal and LaO-top cap with MLP anneal is observed (**Fig. 2**). **Figs. 3, 4** are J_G vs. EOT, and mobility vs. T_{inv}, for LaO (DyO) top/bottom cap with Ta$_2$C and TaCNO electrodes, respectively. Ta$_2$C exhibits better EOT scalability.

PBTI: La/Dy- locations and Laser anneal powers

Fig. 5 illustrates the PBTI measuring methodology, and **Figs. 6-8** plot the PBTI induced V_T shifts vs. stress times. A stress-field dependent two polarities V_T shift is observed, either putting LaO (**Fig.6**) or DyO (**Fig.7**) above HfSiO$_x$ with Ta$_2$C electrode, or putting LaO below HfSiO$_x$ with both Ta$_2$C and TaCNO electrodes (**Fig. 8**).

This phenomenon was also reported on FUSI/Dy-silicate stack [6], and can be explained by the competition between electron de-trapping (dominate at low-stress field) and electron trapping/defect generation (dominate at high-stress field). The ΔV_T relaxation during PBTI recovery periods (100s) is also examined, and **Fig.9** shows typical relaxation curves of SiON, HfSiO$_x$ and Dy-capped SiON. It is interesting that Dy-silicate exhibits a different relaxation behavior and also a negatively larger ΔV_T than HfSiO$_x$ and SiON. Under low-, medium- and high- laser-power anneals ΔV_T relaxation of Ta$_2$C/HfSiO$_x$ stack with LaO-above (**Fig. 10**) and LaO-below (**Fig.11**) HfSiO$_x$ were checked. Meanwhile, ΔV_T relaxation of Dy/La-bottom capping with HfSiO$_x$/HfO$_2$ dielectrics and Ta$_2$C/TaCNO electrodes under MLP anneals is also shown in **Fig. 12**. Typical La/Dy-silicate relaxation curves are observed in general, but LLP anneal exhibits different relaxation behaviors, for both top/bottom LaO-capping: ΔV_T follows HfSiO$_x$-like recovery behavior initially and then changes to the silicate-like behaviors gradually at longer stress times. Same behavior is seen on HfO$_2$/DyO-bottom cap (**Fig. 12**). We believe that electron de-trapping from bulk traps (generated by dielectric intermixing) at PBTI stress periods and then trapping back at the recovery periods dominates the relaxation behaviors. Insufficient intermixing at low power anneal (less trap generation) not only reduces ΔV_T relaxation amplitude but also makes relaxation of both host dielectrics and silicate seen simultaneously. The less intermixing property of DyO with HfO$_2$ as compared to SiO$_2$ also leads to the co-existence of V_T relaxation from HfO$_2$ and silicates, even with MLP anneal (**Fig.12**). Looking back at **Fig. 1**, V_T tune-ability in LaO-bottom approach is less sensitive to laser powers, indicating LaO/SiO$_2$ bottom interface plays an important role for V_T tuning [8]. On the other hand, top LaO-capping needs higher thermal budget to drive La to the SiO$_2$ interface for V_T lowering, at an expense of more bulk traps in La-silicate due to the dielectric intermixing process.

Impacts of Metal Electrodes and Oxygen Vacancies

XTEM images (**Fig. 13**) and/or EELS/EDS analysis (**Fig. 14**) indicates LaO/HfSiO$_x$ intermixing, and also interactions between dielectrics and electrodes (Ta$_2$C [9] or TaCNO [**Fig. 14**]). Moreover, image contrast analysis identifies a less-oxygen layer (~1nm) at the bottom of TaCNO electrode. Possibly there is oxygen incorporation from TaCNO into dielectrics during the intermixing process, and it results in the worse EOT scalability of TaCNO than Ta$_2$C (**Figs. 3, 4**). Interestingly, both normal PBTI induced ΔV_T (**Fig.15**) and HfSiO$_x$-like ΔV_T relaxation (**Fig. 16**) are observed in the TaCNO/LaO/HfSiO$_x$ gate stack. We thus believe the bulk traps generated from dielectric intermixing are probably related to the oxygen vacancies incorporation [10] and TaCNO providing oxygen thus suppressing bulk trapping generation in La/Dy-silicates. It is more evident when placing cap layers above HK. ΔV_T extracted from the PBTI recovery periods (**Fig.17**) show smaller ΔV_T on Ta$_2$C + LLP (less intermixing) and TaCNO + HLP (TaCNO provides "O"). Schematic diagrams illustrating these phenomena are shown in **Fig. 18**.

Conclusions

The process details in thermal budget (or dielectric intermixing) and oxygen content trigger the onset of PBTI abnormalities. Putting La below HK can result in a similar V_T tune-ability with less thermal budget for intermixing with the IL (less PBTI impact), without loss of current drive-ability. We propose Ta$_2$C/HK/LaO/IL + LLP anneals as an optimum nFETs stack configuration for practical CMOS integration.

References [1] S. Kubicek et al., IEDM, p.49 (2007); [2] H. Y. Yu et al., Tech. VLSI, p.18 (2007); [3] V. Narayanan et al., Tech. VLSI, 22.2 (2006); [4] B. H. Lee et al., IEDM, p.859 (2004); [5] B. J. O' Sullivan et al., SSDM, p.372 (2007); [6] S.Z. Chang et al., submitted to EDL; [7]. B. Kaczer et al., Tech. IRPS, p. 381 (2005); [8]. P. Sivasubramani et al., Tech. VLSI, p.68 (2007); [9] L. -A. Ragnarsson et al., EDL, Vol. 28, NO.6, p. 486 (2007) ; [10]. C. Shen et al., IEDM, p. 733 (2004)

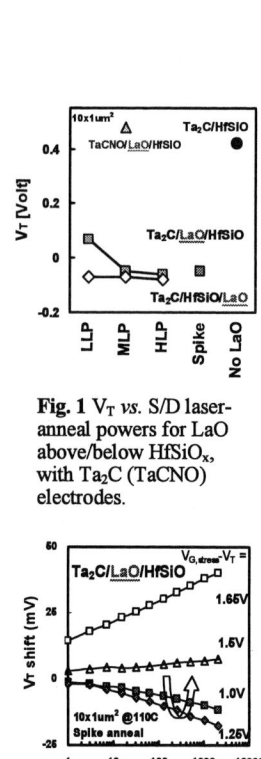

Fig. 1 V_T vs. S/D laser-anneal powers for LaO above/below HfSiO$_x$, with Ta$_2$C (TaCNO) electrodes.

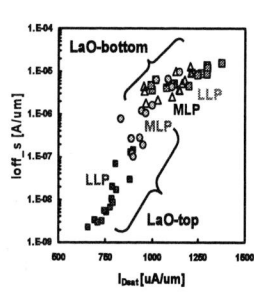

Fig. 2 Ion-Ioff of Ta$_2$C/HfSiO gate stack with LaO above/below HfSiO$_x$, under low and medium laser power anneals.

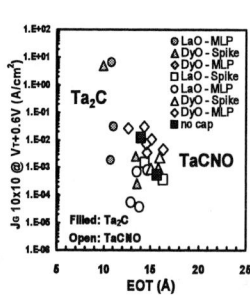

Fig. 3 J$_G$ vs. EOT for LaO (DyO) top/bottom capping with Ta$_2$C and TaCNO electrodes. Ta$_2$C exhibits better EOT scalability.

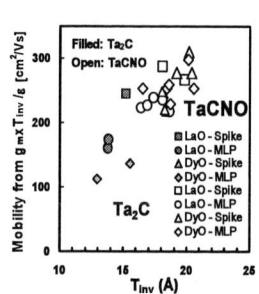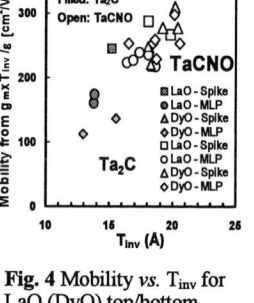

Fig. 4 Mobility vs. T$_{inv}$ for LaO (DyO) top/bottom capping with Ta$_2$C and TaCNO electrodes.

Fig. 5 PBTI measurement setting. Both sense-and-measure, and with 100s recovery time after each stress, are implemented.

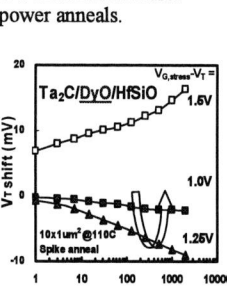

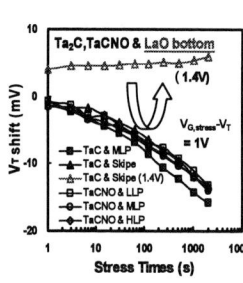

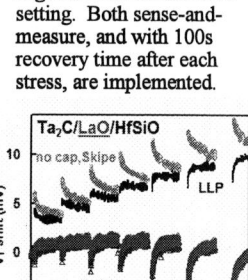

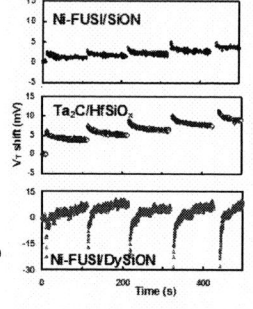

Fig. 6 Stress-field dependent two-polarity PBTI V_T shifts are observed in the Ta$_2$C and LaO top-capping n-FETs.

Fig. 7 Abnormal PBTI induced V_T shift is also observed in DyO top-capping n-FETs.

Fig. 8 Put LaO below HK also shows abnormal ΔV_T under PBTI stress, either with laser or spike anneal.

Fig. 9 Typical ΔV_T relaxation curves during the recovery periods of PBTI stress. HfSiO & SiON have similar behaviors

Fig. 10 (LaO-top) ΔV_T relaxation vs. time during stress recovery. LLP shows different relaxation curves than MLP /HLP anneals

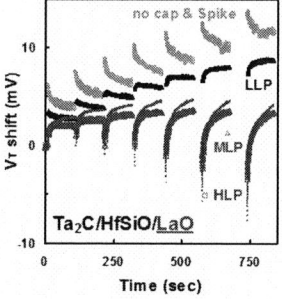

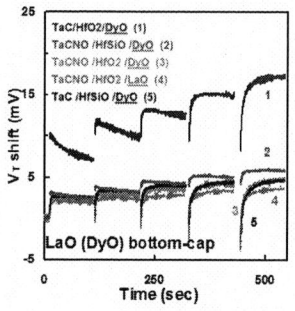

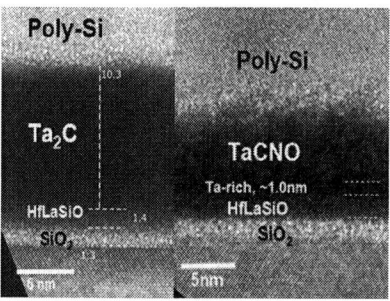

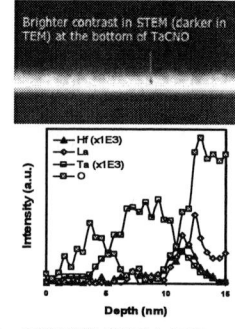

Fig.11 (LaO-bottom) ΔV_T relaxation vs. time. All measurements keep V_G-V_{Stress}= 1.25V (Figs. 10-12, 16).

Fig.12 (LaO/DyO-bottom) ΔV_T relaxation vs. time, with both Ta$_2$C and TaCNO electrodes. Medium power anneals are implemented.

Fig.13 LaO/HfSiO intermixing after thermal annealing, with both Ta$_2$C and TaCNO electrodes. A less-oxygen layer at the bottom of TaCNO electrode is observed.

Fig. 14 EELS, EDS (x1E3) indicate LaO/HfSiO intermixing. Image contrast shows Ta- rich at the bottom of TaCNO.

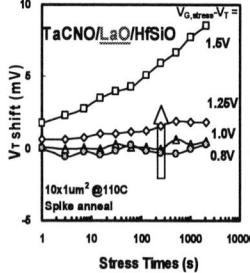

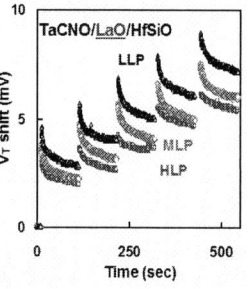

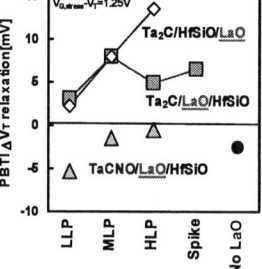

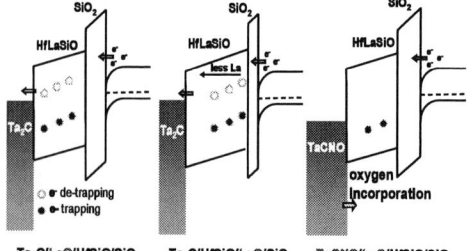

Fig. 15 Normal ΔV_T under PBTI stress for LaO top-capping n-FETs with TaCNO electrodes.

Fig. 16 V_T relaxation vs. time for TaCNO/LaO/HfSiO. Positive ΔV_T and HfSiO-like relaxation behaviors (Figs.9-10) are observed.

Fig. 17 ΔV_T relaxation amplitude vs. laser powers, extracted from the recovery period after 128s stress. TaCNO +HLP (less oxygen vacancies) and Ta$_2$C+LLP (less intermixing) get smaller ΔV_T.

Fig. 18 Schematic diagrams (after thermal anneals) illustrate the negatively charged traps (•) and electron de-trapping (○) during the PBTI stress. Oxygen incorporation from TaCNO results in less trap generation in La-silicate.

50

Impact of the Different Nature of Interface Defect States on the NBTI and $1/f$ noise of High-k / Metal Gate pMOSFETs between (100) and (110) Crystal Orientations

Motoyuki Sato, Yoshihiro Sugita, Takayuki Aoyama, Yasuo Nara, and Yuzuru Ohji

Semiconductor Leading Edge Technologies, Inc. (Selete)

16-1 Onogawa, Tsukuba, Ibaraki, 305-8569, Japan, Phohe:+81-29-849-1336, Fax:+81-29-849-1186, E-mail:sato.motoyuki@selete.co.jp

Abstract

We have clarified the difference in NBTI and $1/f$ noise of high-k/metal gate pMOSFETs between (110) and (100) oriented surfaces. Although the initial interface state density on (110) is higher than that on (100), the NBTI is better on the (110) surface. That is due to the different interface defect nature of interface defect states on (110) surface compared to (100). This difference has a strong impact on $1/f$ noise.

Introduction

The mobility enhancement technique using (110) surfaces is now being intensively studied in order to achieve further drive current improvements [1,2]. This technology is promising in the *hp* 45 nm and beyond CMOSFET with the combination of high-k/metal gate stacks. Thus, more discussions of the reliability of high-k/metal pMOSFEs on (110) are required. However, there have only been a few studies reported on the impact on reliability of using (110) surface. Although the inherent interface state density between Si and dielectrics is higher on (110) [3], its impacts on long-term reliability and analog characteristics has not yet been clarified. In this paper, we clarified the impact of using (110) surface on the NBTI and $1/f$ noise of HfSiON/TiN pMOSFETs.

Experimental

SiO_2(0.7nm)/HfSiON(2.5nm)/TiN(10nm)/W(50nm) pMOSFETs were fabricated on (100) and (110) oriented surface. In both cases T_{inv} was 1.6 nm. To assess the initial characteristics and long-term reliability, charge pumping, ESR, NBTI, $1/f$ noise were compared.

Results and Discussion

1. NBTI

Fig. 1 shows the transconductance (Gm) as a function of gate bias with (100) and (110) pMOSFET. Due to the reduction of effective hole mass, the Gm for devices on (110) substrate is higher than on (100). In addition, there is no impact for the gate leakage current using the (110) surface (**Fig. 2**). There is, however, a strong impact on the Si/SiO_2 interface state density (N_{it}). N_{it} on the (110) surface is 1.5 times higher than that on (100) as shown by the charge pumping current (**Fig. 3**). As a results of the higher N_{it}, the sub-threshold slope for the (110) surface is larger than that for (100) (**Fig. 4**). Even though the (110) surface has higher initial N_{it}, we clarified that this has no impact on the NBTI. Unexpectedly, pMOSFETs on the (100) surface show a higher degradation compared to those on (110) surface (**Fig. 5**). **Fig. 6** show the time evolution of ΔN_{it} on (100) and (110) oriented surface pMOSFETs measured by the stress and charge pumping technique. The figure indicates that degradation in N_{it} for pMOS on (100) is more serious than those on a (110) surface. **Fig. 7** shows the shift of Gm with increasing stress time. This clearly shows that the Gm-Vg curves shifted gradually to negative direction due to the V_{th} instability, and that the peak values are reduced also mainly due to the N_{it} degradation. **Fig. 8** shows that the time evolution of Gm ratio (Gm/Gm(0s)). The rate of degradation on (100) is more serious than that on (110) independent of the initial N_{it}. The NBTI temperature dependence was also different for the two different orientations. **Fig. 9** shows the time evolution of ΔV_{th} at several temperatures (25~125°C). Using a power law analysis ($\Delta Vth = \alpha t^\beta$), the values of β (0.165) were found to be almost independent of the temperature for the (100) surface. On the other hand, the values of β increased with rising temperature for the (110) surface (**Fig. 10**). In order to understand this difference, we observed the property of the dangling bonds between Si and dielectrics using

ESR (**Fig. 11**). The status of the dangling bonds is completely different for (100) and (110). In the case of (100), both Pb_0 and Pb_1 centers were observed. On the other hand, for (110), only the Pb_0 center was observed. Pb_1 is the specific phenomenon on (100) surfaces, which has two bonds at the interface [4]. In NBTI, as mentioned, the values of β were constant of 0.165 (~1/6) and independent of the temperature. This is exactly the same as the H_2^0 diffusion model in the hydrogen R&D model [5]. This means that the combination of two hydrogen atoms is the dominant species for N_{it} degradation. It is thought that it is difficult to generate H_2^0 on a (110) surface with Pb_0 defects, which are caused by one dangling bond compared to a (100) (**Fig. 12**). This Pb center states dependent NBTI behavior on (110) surface was found for the first time.

2. $1/f$ noise

We found that specific physics on (110) introduce the very different $1/f$ noise characteristics compared to (100). **Fig. 13** shows the input-referred spectral noise density (S_{vg}) as a function of frequency. The value of S_{vg} on (110) is higher than that on (100) by about two orders of magnitude. It has been reported that S_{vg} for HfSiON pMOSFETs can be described by the carrier number fluctuation model and given by (1) [6], and should simply be proportional to the interface state density.

$$S_{vg} = \frac{2kT}{LWf} \frac{Nt(E)}{\gamma} \left(\frac{q}{C_{ox}}\right)^2 \cdots (1)$$

(γ : Attenuation coefficient, $N_t(E)$: Trap density)

N_{it}, measured by the charge pumping technique, shows the difference between (100) and (110) to be a factor of 1.5 (**Fig. 3**). It can not be explained only with the trap density. There are three reasons for much S_{vg} degradation. The first is the inversion layer thickness reduction. Since hole wave function on (110) in inversion layer is closer to the interface than that on (100) due to the heavier perpendicular effective hole mass of (110) [2]. Thus, it is more sensitive for the hole transport scattering by the interface defects. The second is higher hole mobility. Based on the mobility fluctuation model, it is thought that it is more sensitive for (110) which have higher hole mobility than that on (100). And the third is the difference of the interface defect state between (110) and (100). Since the energy distribution of density of Pb_0 is much wider than that of Pb_1 [7] (**Fig. 14**), Pb_0 is more sensitive compared to Pb_1 for hole transports. Thus, influence of the Pb center for the (110) mobile hole is much larger than that on (100) surface. Considering the physical mechanism on (110), we can understand the different $1/f$ noise characteristics for (100) surface. (**Fig. 15**) Thus, when using (110) surface for analog devices, we have to fabricate Si/SiO_2 interface with great care.

Conclusion

We have investigated the reliability characteristics of high-k/metal gate pMOSFET on (110) oriented surface and summarized in **Fig. 16**. The NBTI lifetime on (110) is longer than that on (100). On the other hand, $1/f$ noise on (110) is higher than on (100). That difference were caused by the different Pb center condition and different effective hole mass including the perpendicular direction between (110) and (100). When fabricating the analog devices on (110) surface, we need the meticulous attention for Si/SiO_2 interface formation.

References

[1]M.Yang *et al.* :IEDM (2003) 453, [2]M.Saito, :IEDM (2007) 711, [3]S.Chung et al. :IEDM, (2005) 567, [4]A.Stesmans *et al.* :Phys. Rev. B, 58 (1998) 15801, [5]P. E. Nicollian *et al.*, :IRPS, (2007), 197, [6]K.Kojima *et al.* :VISI Symp. (2005) 59, [7] J.Campbell *et al.* :T-ED,6 (2006) 17

978-1-4244-1802-2/08/$25.00 ©2008 IEEE

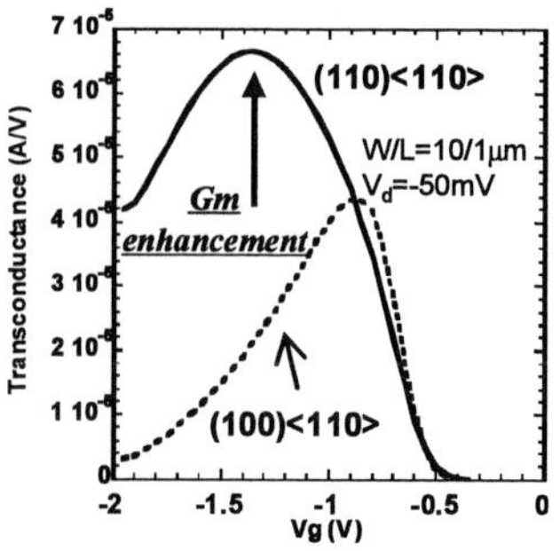

Fig.1 Gm -V_g characterisics of pMOSFETs on (110) and (100) surfaces. PMOS on (110) have larger Gm than that on (100).

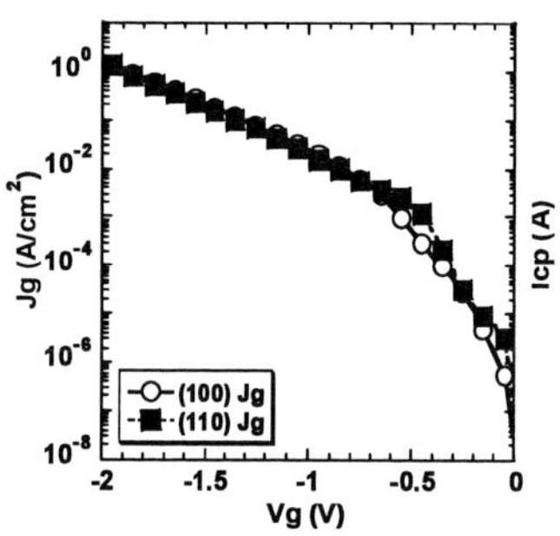

Fig.2 J_g-V_g characteristics of pMOS. This shows that J_g is independent of the wafer orientation.

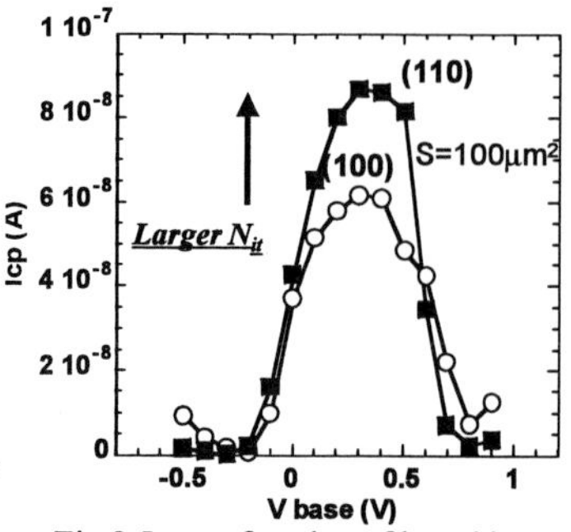

Fig.3 I_{cp} as a function of base bias. N_{it} on (110) is 1.5 times higher than that on (100).

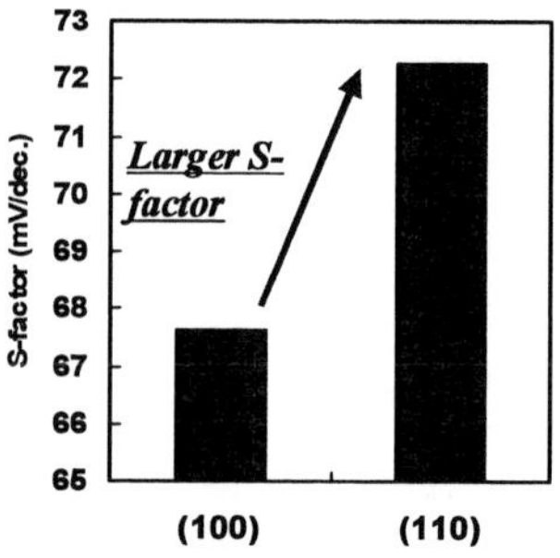

Fig.4 The S-factor of pMOS on (110) and (100). The higher N_{it} on (110) gives a larger S-factor.

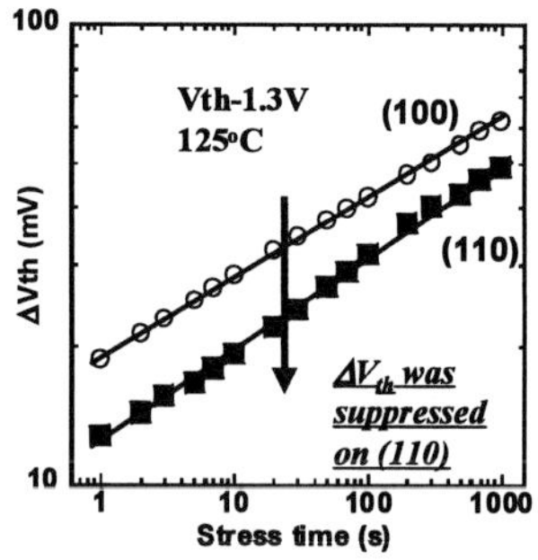

Fig.5 Time evolution of ΔVth in NBTI. ΔVth on (110) was suppressed compared to (100).

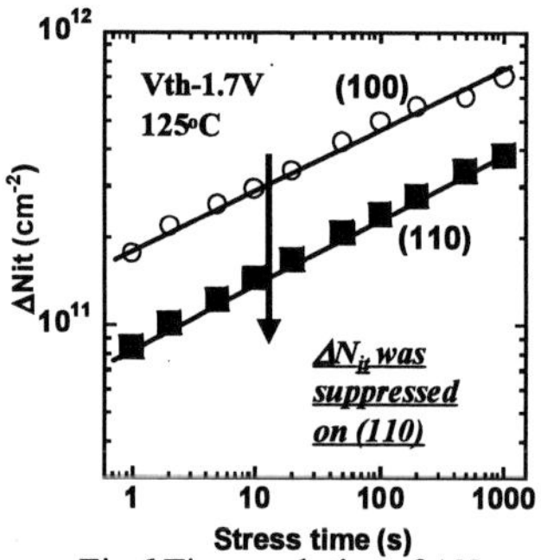

Fig.6 Time evolution of ΔN_{it}. ΔN_{it} on (110) was suppressed compared to (100).

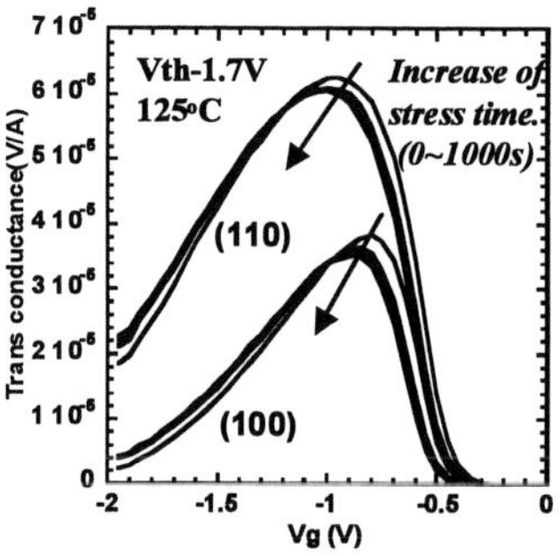

Fig.7 Shift in G_m with increasing stress time. The peak values of G_m reduce with increasing stress time.

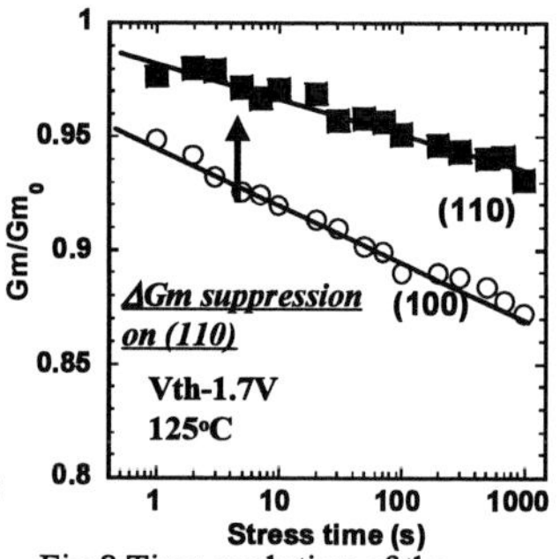

Fig.8 Time evolution of the relative peak values of Gm (G_m/$G_m(0s)$). With the lower rate of increase of N_{it}, the degradation in G_m were suppressed on (110) compared to (100).

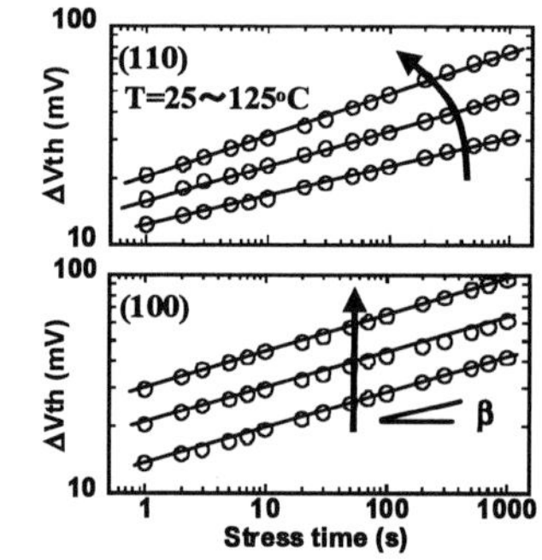

Fig.9 Time evolution of ΔV_{th}. The slope for (110) changes with temperature, whereas the slope for (100) remains constant.

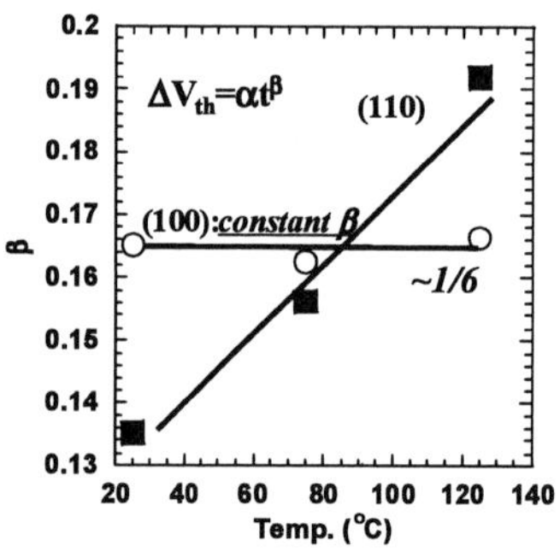

Fig.10 The exponential factors (β) for the power law ($\Delta V_{th}=\alpha t^{\beta}$) as a function of temperature. The values of β for (110) increase with rising temperature.

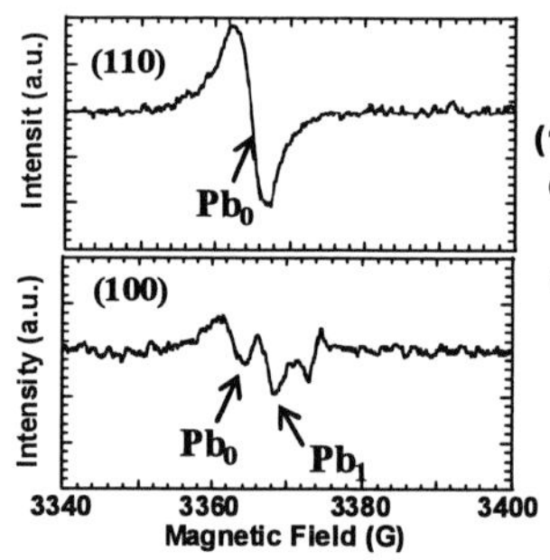

Fig.11 ESR spectra. Only Pb_0 were observed on (110). Both Pb_0 and Pb_1 were observed on (100).

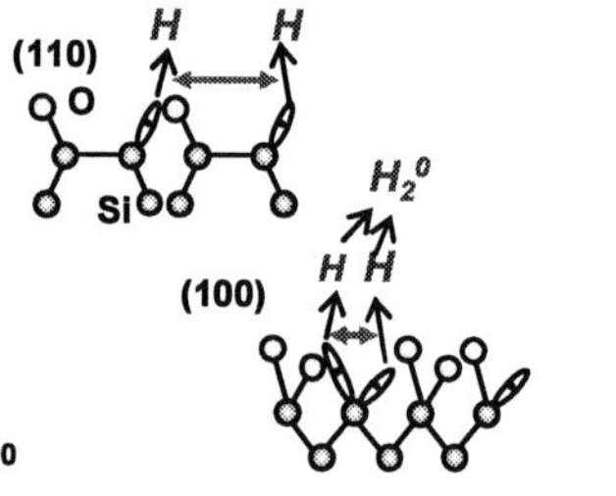

Fig.12 Schematic of Si/SiO$_2$ interface. H_2^0 are the main species of hydrogen release from the Si/SiO$_2$ interface.

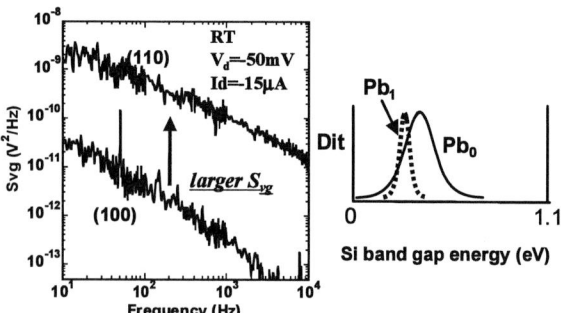

Fig.13 S_{vg} of pMOSFETs. S_{vg} on (110) is larger than that on (100). (W/L=10/1μm)

Fig.14 The energy distribution of Pb_0 and Pb_1 [7].

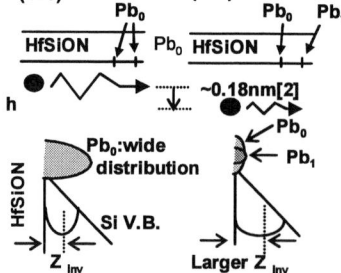

Fig.15 Schematic of the origin of larger S_{vg} on (110) than (100) surface.
1. Z_{inv} reduction
2. Higher hole mobility
3. Stronger Pb_0 defect impact than Pb_1.

	(110)	(100)
Pb center	Pb_0	Pb_0, Pb_1
NBTI	better	worse
Diffusion species	depend on temperature	H_2^0
1/f noise	worse	better

Fig.16 Summary of the effect in interface defect on the NBTI and 1/f noise. Difference of interface state defect strongly affects on the device characteristics and reliability.

Physical Understanding of the Reliability Improvement of Dual High-k CMOSFETs with the Fifth Element Incorporation into HfSiON Gate Dielectrics

M. Sato[1,3], N.Umezawa[2], N. Mise[1], S. Kamiyama[1], M. Kadoshima[1], T. Morooka[1], T.Adachi[2], T.Chikyow[2], K. Yamabe[3],
K. Shiraishi[3], S. Miyazaki[4], A. Uedono[3], K. Yamada[5], T. Aoyama[1], T. Eimori[1], Y. Nara[1], and Y. Ohji[1]
Selete[1],   NIMS[2],   Univ. of Tsukuba[3],   Hiroshima Univ.[4],   Waseda Univ.[5]
16-1 Onogawa, Tsukuba, Ibaraki, 305-8569, Japan, Phohe:+81-29-849-1336, Fax:+81-29-849-1186, E-mail : sato.motoyuki@selete.co.jp

Abstract

We clarified the impact of the fifth material incorporation into HfSiON technology for V_{th} control on the reliability of high-k/metal gate stacks CMOSFETs. HfMgSiON is remarkably effective for suppressing electron traps, giving rise to a dramatic PBTI lifetime improvement for nMOSFETs. With pMOSFETs, Al incorporation is effective for the thermal deactivation of hole traps, resulting in NBTI lifetime improvement. We have established the guidelines of material selection to be incorporated into HfSiON for reliability improvement for nMOS and pMOS individually.

Introduction

A high-k/metal gate stack is required for scaled CMOSFETs, but one of its most serious problems is in V_{th} control. There have been many reports of the incorporation of some material into the Hf-based high-k gate dielectrics to modulate V_{th} [1-4]. However there have been only a few reports on the impact of these incorporations on the reliability [1]. In this paper, we clarified the effect of material incorporation to control V_{th} on the high-k gate dielectrics reliability and its improvement mechanism.

Experimental

Fig. 1 shows the schematic of the dual high-k CMOSFET. Dual high-k were fabricated to control V_{th}. HfSiON was used as the starting material for high-k gate dielectrics. Mg [2] and La [3] incorporations were used for nMOSFETs, whilst Al [4] and Ta were used for pMOSFETs. Each material was incorporated into HfSiON with 1000^oC annealing following a capping deposition. All T_{inv} were between 1.6 and 1.8 nm. For the assessment of the reliability, PBTI and NBTI measurements, charge pumping, and $1/f$ noise measurements were carried out.

Results and Discussion

1. nMOSFETs with Mg or La incorporation

Fig. 2 shows the I_d-V_g characteristics (V_d=50 mV) of nMOSFETs. As an effect of Mg and La incorporation, V_{th} were lowered by 350 mV for La, 500 mV for Mg respectively. Although both materials were effective for V_{th} control, their impact on the reliability was very different. Fig. 3 shows the time evolution of ΔV_{th} in PBTI. This clearly shows that Mg was effective for aggressively suppressing the PBTI degradation. On the other hand, La incorporation accelerated the V_{th} shift. Fig. 4 shows the PBTI lifetime (ΔV_{th}=50 mV) as a function of overdrive stress voltage (V_g-V_{th}). Using the Mg, PBTI lifetime was dramatically improved over four orders of magnitude. Mg was better than La from the point of reliability. Fig. 5 shows the charge pumping current (I_{cp}) as a function of base bias. With La incorporation, N_{it} was reduced, whilst N_{it} was increased with Mg. These results can not explain the PBTI phenomena. Thus, we can say that electron traps causing V_{th} shift in PBTI exist in the bulk high-k not in the interface. Mg incorporation gave rise to the dramatic reduction in electron traps, but not so with La. This electron trap suppressions affect on the $1/f$ noise also. Fig. 6 shows the input-referred spectral noise density (S_{vg}) as a function of frequency for nMOSFETs. Since it has been reported that origin of the S_{vg} of HfSiON nMOSFET is the mobility fluctuation (proportional to T_{inv}^2) [5], S_{vg} were normalized with T_{inv}^2. Fig. 6 clearly shows that using the La or Mg incorporation have changed the frequency exponential factor of γ ($S_{vg} \sim f^{-\gamma}$). Using the La, γ was increased to 1.0 compared with no cap HfSiON (0.85). On the other hand, using the Mg, γ was decreased to 0.79. According to G-R model [6], γ were depend on the trap time constant (τ) and trap center position from the channel. With the trap center apart from the Si/SiO2 interface, τ becomes large, as a results, lower frequency noise is increased and γ is decreased [6] (Fig. 7). The change of the γ values varying indicates that using La increase the bulk trap in high-k, on the other hand, Mg were decrease. The drastic effect of alloying MgO on reduction of the electron traps has been confirmed by first-principles calculations (by VASP [7]). Fig. 8 shows model structures for 3-fold coordinated oxygen vacancy

(V_o) as an electron trap site in the monoclinic HfO2 geometry (model A), and corresponding V_o site neighboring a substitutional Mg atom at Hf site (Mg$_{Hf}$) (model B). It was found that the electron trap level is raised by 0.86 eV in model B compared to that of model A, indicating electron traps are significantly suppressed at the V_o near Mg$_{Hf}$. Elevation of the electron trap level is caused by the low covalency of Mg-O bonding. (no d orbital in Mg) Although there are several reports about the impact of La incorporation on reliability [8, 9], we can not obtain comprehensive understanding. That is why the instability La atoms with highly hygroscopic characteristics and having several valence number.

2. pMOSFETs with Al or Ta incorporation

Fig. 9 shows the I_d-V_g characteristics of pMOSFETs. As an effect of the Al or Ta incorporation into the HfSiON, V_{th} were reduced by about 100 mV. Fig. 10 shows the time evolution of ΔV_{th} in NBTI. This clearly shows that ΔV_{th} were suppressed by both Al and Ta incorporation, results in NBTI lifetimes improvements (Fig. 11). In particular Al is more effective than Ta. To understand the NBTI lifetime improvement mechanism, we have evaluated the $1/f$ noise and NBTI temperature dependence. Differently from nMOSFET, origin of the S_{vg} of HfSiON pMOSFET is thought to be the carrier number fluctuation (proportional to T_{inv}) [5], S_{vg} being normalized to T_{inv}. Fig. 12 clearly shows that S_{vg} were almost same at the room temperature. This indicates the interface and bulk trap densities, which affect on the $1/f$ noise, is not influenced by Al or Ta incorporation at RT. Fig. 13 shows that time evolution of ΔV_{th} in NBTI with several temperatures (25~125°C), with and without Al incorporation. It clearly shows that temperature acceleration of V_{th} shift was suppressed with Al incorporation. Fig. 14 shows the activation energy of NBTI with Al and Ta and without HfSiON. Both Al and Ta were effective for the suppression of NBTI thermal activation. Actually, NBTI involves the increase of N_{it} and hole traps. We have to clarify which is the origin of NBTI improvement. Fig. 15 shows the activation energy of ΔN_{it}. Different from the ΔV_{th} results, the activation energy of ΔN_{it} is almost same, independent of Al and Ta incorporation. It is thought that ΔV_{th} thermal deactivation is due to the hole trap suppression in high-k as an effect of Al and Ta incorporation. One of the plausible cause of the hole trap is the formation of ($\equiv$Hf-OH-Hf$\equiv$)$^+$ bonds in the high-k layer. According to the model proposed in reference [10], hydrogen at SiO2/Si interface are readily diffused into HfSiON layer forming O-H bonds when holes are induced into the high-k film. This effect must be suppressed by alloying AlO or TaO with HfSiON because the Al-O or Ta-O bond has strong covalent character compared to the Hf-O bond, and this interrupts the formation of O-H bonds suppressing hole traps (Fig.16). Since Al have higher electron negativity than Ta (Table I), it is more effective than Ta incorporation

Conclusion

We have clarified the impact of the incorporation of the fifth materials in HfSiON on BTI characteristics for both n(Mg, La) and pMOSFET (Al, Ta). Using HfMgSiON for nMOSFET and HfAlSiON for pMOSFET carrier traps were suppressed resulting in dramatic BTI lifetime improvement. They are explained with the fifth material bonding covalency incorporated into HfSiON. Established guidelines are that low covalency material is suitable for nMOS, high covalency one is suitable for pMOS, respectively (Table I).

References

[1]P.Sivasubramani et al. :VLSI Symp. (2007) 68, [2]N.Mise et al. :IEDM, (2007) 527, [3]S.Kamiyama et al : IEDM, (2007) 539, [4]M.Kadoshima et al : IEDM (2007) 531, [5]K.Kojima et al :VISI Symp. (2005), 59 , [6]R. Brederlow et al.: IEDM (1999) 159, [7] G. Kresse et al., : Phys. Rev. B 47, RC558 (1993), [8]C.Y.Kang et al. : SSDM (2007) 258, [9] B.J.O'Sullivan et al. : SSDM (2007) 376, [10] K. Torii et al. : IEDM (2004) 129

978-1-4244-1802-2/08/$25.00 ©2008 IEEE

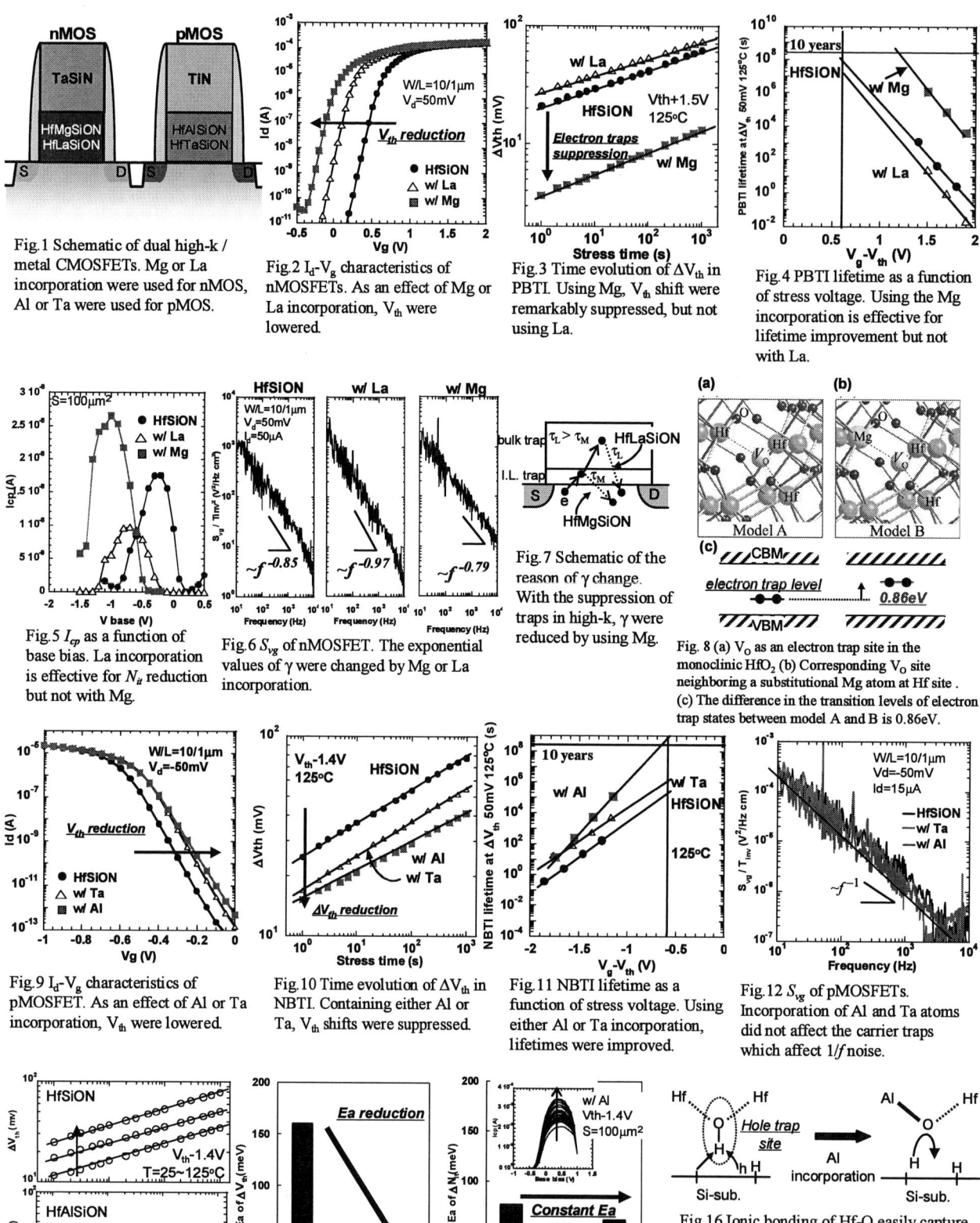

Fig.1 Schematic of dual high-k / metal CMOSFETs. Mg or La incorporation were used for nMOS, Al or Ta were used for pMOS.

Fig.2 I_d-V_g characteristics of nMOSFETs. As an effect of Mg or La incorporation, V_{th} were lowered.

Fig.3 Time evolution of ΔV_{th} in PBTI. Using Mg, V_{th} shift were remarkably suppressed, but not using La.

Fig.4 PBTI lifetime as a function of stress voltage. Using the Mg incorporation is effective for lifetime improvement but not with La.

Fig.5 I_{cp} as a function of base bias. La incorporation is effective for N_{it} reduction but not with Mg.

Fig.6 S_{vg} of nMOSFET. The exponential values of γ were changed by Mg or La incorporation.

Fig.7 Schematic of the reason of γ change. With the suppression of traps in high-k, γ were reduced by using Mg.

Fig.8 (a) V_O as an electron trap site in the monoclinic HfO_2 (b) Corresponding V_O site neighboring a substitutional Mg atom at Hf site. (c) The difference in the transition levels of electron trap states between model A and B is 0.86eV.

Fig.9 I_d-V_g characteristics of pMOSFET. As an effect of Al or Ta incorporation, V_{th} were lowered.

Fig.10 Time evolution of ΔV_{th} in NBTI. Containing either Al or Ta, V_{th} shifts were suppressed.

Fig.11 NBTI lifetime as a function of stress voltage. Using either Al or Ta incorporation, lifetimes were improved.

Fig.12 S_{vg} of pMOSFETs. Incorporation of Al and Ta atoms did not affect the carrier traps which affect $1/f$ noise.

Fig.13 Time evolution of ΔV_{th} in NBTI at several temperatures with Al containing and control HfSiON. With Al, NBTI thermal activation was suppressed.

Fig.14 Ea of ΔV_{th}. Using cap technology, Ea were remarkably suppressed.

Fig.15 Ea of ΔN_{it}. Considering the different behavior of Ea compared to ΔV_{th}, dominant factor of NBTI is bulk hole traps in high-k.

Fig.16 Ionic bonding of Hf-O easily capture proton and form three coordinate O. That becomes hole trap site. Covalent bonding of Al-O suppress the hydrogen capture.

Table I Pauling's electron negativity

Mg	Hf	Ta	Al
1.2	1.3	1.51	1.61

for nMOS   *for pMOS*

Guidelines to improve mobility performances and BTI reliability of advanced High-K/Metal gate stacks

X. Garros[1], M. Cassé[1], G. Reimbold[1], F. Martin[1], C. Leroux[1], A. Fanton[1], O. Renault[1], V. Cosnier[2] and F. Boulanger[1]   xavier.garros@cea.fr

[1]CEA-LETI MINATEC, 17 rue des Martyrs 38054 Grenoble Cedex 9, France
[2]STMicroelectronics, 850, rue J. Monnet, BP. 16, 38921 Crolles, France

Abstract
A systematic study of mobility performances and BTI reliability was done in advanced dielectrics stacks. By reducing the oxide films thicknesses $T_{HK} \leq 2.5nm$, PBTI becomes generally very low and associated lifetimes are always over 10 years. By studying a large variety of dielectric stacks we also clearly demonstrate that *mobility performances, interface defects Nit and NBTI reliability are strongly correlated*. All are affected by *nitrogen species N which is clearly identified as the main mobility killer* when it reaches unintentionally the Si interface during the deposition of nitrided gates or the nitridation steps. However, by optimizing the gate stacks, excellent mobility performances, *up to 100% universal mobility at Eeff=1MV/cm*, and reliability can be achieved.

Introduction
Today advanced dielectric stacks are evaluated for the future 32nm and 22 nodes. However devices integrating these new materials suffer from both mobility μ and reliability degradations. High-K dielectrics are often incriminated to be responsible for these degradations [1-2]. In this study we show that N has actually a detrimental impact on devices performances and reliability rather than the High-K oxide itself. PBTI is first evaluated and then mobility and NBTI are studied.

Experimental
Conventional CMOS transistors were made on 300mm bulk wafers. For this study, several Hf-based oxides were combined with many metal gates TaN, TiN, WN, TaC (see Fig. 1). Generally 1.5-3nm High-K layers were deposited either by ALD or by MOCVD on a 0.7-1nm SiO$_2$ oxide. Then 3-13nm thick metal gates were deposited on the top of the oxide by PVD or ALD. A 10 nm TiN layer is sometimes used as a capping layer before the polySi deposition. To reduce the gate leakage current, Hafnium silicate films can be nitrided by plasma PN. After the gate stack deposition, a spike anneal at 1050°C is used to activate S/D dopants. Carrier mobility is extracted using the CV split technique. To evaluate PBTI and NBTI shift at T=125°C while limiting possible Vt relaxations, we have implemented a similar method than the one proposed by [3]. It consists in monitoring drain currents Id$_{High}$ and Id$_{Low}$ respectively at Vgstress and at Vg~Vt and in evaluating the Vt shift from the Id$_{low}$ variation. Typical Vt shifts vs time obtained with this technique are shown in Fig. 2. "Static" NBTI shift is also measured from Id-Vg curves before and after stress. 3 dies are tested for each stress condition and oxide field is evaluated from CVs. Transistor sizes are W=L=1μm, W=L=10μm and W=10μm L=1μm.

Electrical Results and Discussion
PBTI is first evaluated in many high-k/metal gate stacks (see Fig. 3). It is generally very sensitive to bulk oxide defects [4]. Regardless of the deposition technique and the high-k material, PBTI is low in these thin oxide films $T_{HK} \leq 2.5nm$ integrated with "stable" metal gate like TiN PVD gate (EOTs of oxide/TiN PVD stacks hardly change whatever the thermal budget). Actually PBTI is related to the cristallinity of the high-k layer and can be strongly reduced when the high-k layer changes from a crystalline phase to an amorphous state [5]. With TaN and TiN ALD gates, PBTI is enhanced. Actually defects may be created in the high-k layer since these gate materials can react with the polySi capping during the S/D anneal. A formation of a TaSiN layer after RTP anneal has been already reported by [6]. However even in these most degraded dielectric stacks, PBTI lifetime is largely over 10 years: Vg@10years>Vdd=1V for EOT~1nm (see Fig. 5). Therefore, after optimizing the dielectric stacks (reducing T_{HK}), PBTI is not critical any longer. So we focus now on NBTI issue.
If PBTI is more representative of bulk oxide degradations, NBTI is more sensitive to degradations at the bottom interface Si/SiO$_2$. Like in conventional SiON/PolySi devices, nitrogen in High-K/Metal gate stacks has a detrimental effect on NBTI. To investigate the role played by N on NBTI, we have first compared several Hafnium silicate oxides

where N was voluntarily incorporated by forming a SiON bottom oxide and/or by plasma nitridation PN of a HfSiO layer. As shown in Fig. 6, Nit clearly increases with N content. In the case of SiON/HfSiON, XPS analysis reveals a N concentration ~4% in the SiO$_2$ interlayer (see Fig. 7). The Si–N binding is evidenced on both N1s and Si-2p spectra. Particularly the peak located at 398.2eV corresponds to the contribution of N-(Si)$_3$ bonds in an oxygen rich matrix. Therefore these Si-N bonds could play the role of interface defects and be at the origin of the enhanced NBTI observed in this stack (see Fig. 8). Indeed the SiON/HfSiON stack exhibits the highest NBTI. This confirms that higher is the N content in the bottom oxide higher is the NBTI shift. Furthermore, for the same SiO$_2$ pedestal, NBTI is higher in the HfSiON oxides than in the HfSiO ones. It can be due to N species incorporated during the PN step which has diffused into the pedestal oxide and generate defects inside it. Fig. 9 shows that N also strongly degrades carrier mobility by increasing coulomb scattering. Actually all the results (Nit,NBTI, μ) are very well correlated as proved in Fig. 10. Thus N is both a mobility and reliability killer when it reaches the bottom interface. This fundamental result is again highlighted in Fig. 11. In this graph, many other HK/MG couples are compared and 3 families are easily distinguishable. Excellent results A are obtained with non nitrided oxides and TaN, TaC gates whereas family C corresponds to SiO$_2$/HfSiON where N has reached the bottom interface (equivalent to SiON/HfSiON). Once again these NBTI results are very well correlated to mobility results (see Fig. 12). This confirms that, in most cases, N drives mobility and NBTI.
N can be introduced during the PN step but also *during the metal gate deposition itself*. To study specifically the role of N arising from the metal gate, we have only kept in Fig. 13 data of Fig. 11 for non nitrided oxides HfO$_2$, HfZrO, HfSiO. Both μ$_{eff}$ degradations and enhanced NBTI are observed with nitrided gates like TiN, especially when they are deposited by PVD. As already shown by [7-8], the PVD technique can sometimes sputter N in the oxide which reaches the IL and induces additional defects. This result is further confirmed in Fig. 14. Indeed when a PVD TiN capping is deposited on the top of a TaN gate, Nit is strongly increased due to N incorporation. By comparing HfSiO and HfSiON oxides, we notice that the contribution to NBTI of N coming from the TiN capping adds up to the contribution of N introduced by PN for the HfSiON oxides (see Fig. 15). The electrical signature of N related defects is shown in Fig. 16 by modeling the CV and GV curves [9]. N creates a characteristic single peak in the upper part of the Si bandgap different from peaks associated with Pb centers [10]. It may be related to Si-N bonding. Fig. 17 summarizes the degradation mechanisms of mobility and reliability induced by N.
However, when preventing N from reaching the Si interface, NBTI lifetimes over 10 years and excellent mobility performances *(up to universal mobility@1MV/cm for HfZrO/TaC)* can be achieved in some low EOTs HK/MG stacks (see Fig. 18). All the guidelines to improve mobility and reliability are summarized in Table 1.

Conclusion
Reliability and carrier mobility have been largely investigated. PBTI is not critical any longer in Hf-based technologies contrary to NBTI. By comparing many HK/MG stacks, NBTI is shown to be strongly correlated to mobility results. Most of μ degradations in HK/MG stacks are mainly related to the nitridation processes and not to the high-k oxide itself. N is clearly identified to be responsible for both NBTI and μ degradations when it reaches the Si interface during the deposition of nitrided gates or/and the oxide nitridation PN. Excellent mobility performances and reliability can be achieved with non nitrided stacks.

Acknowledgements
This work was supported by the CEA-Leti and STmicrolectronics and the Pullnano project.

978-1-4244-1802-2/08/$25.00 ©2008 IEEE

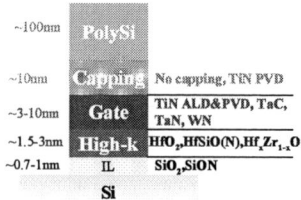

Fig. 1: Schematic view of the various High-K/Metal gate stacks studied here

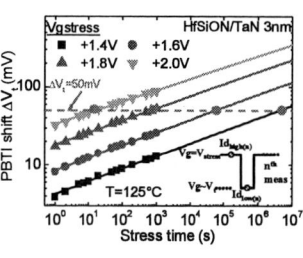

Fig. 2: Typical PBTI shift vs. time at various Vgstress. W=L=1μm. Insert: Vg signal to estimate Id(t)

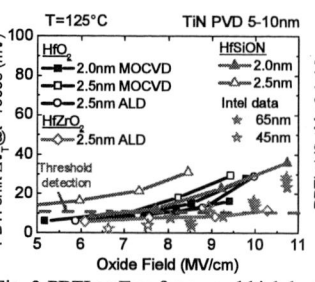

Fig. 3 PBTI vs Eox for several high-k oxides. PBTI is low in thin oxides films T_{HK}<2.5nm. Intel data from [11]

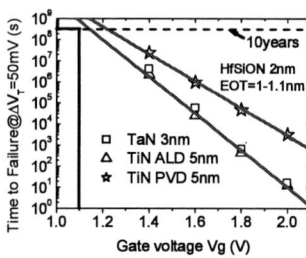

Fig. 4: Impact of metal gate on PBTI on HfSiO(N) oxides. PBTI is increased with TaN or TiN ALD gates

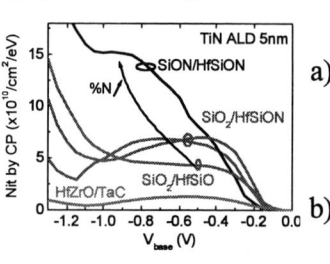

Fig. 5: PBTI lifetimes in HfSiON films (data of Fig. 4)

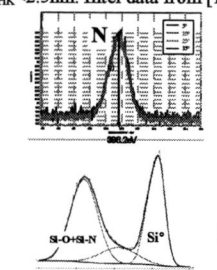

Fig. 6: Nit given by Charge Pumping for HfSiO(N) oxides. Nit increases when N reaches the bottom interface

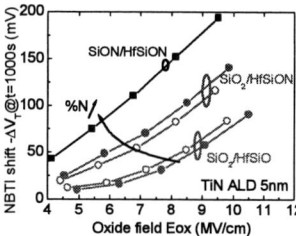

Fig. 7: XPS spectra of ~0.8nm SiON bottom oxide a) N1s b) Si2p

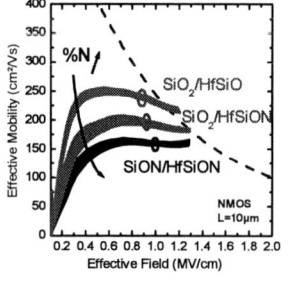

Fig. 8: NBTI shift vs. Eox for several HfSiO(N) oxides.

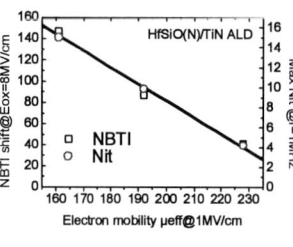

Fig. 9: Electron mobility for the same HfSiO(N) oxides

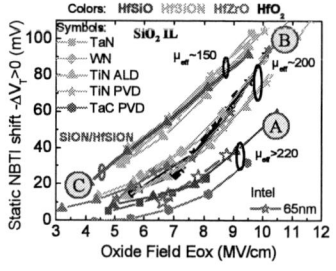

Fig. 10: Correlation between NBTI, Nit and carrier mobility

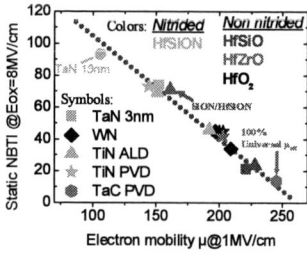

Fig. 11: NBTI vs Eox for many HK/MG stacks. symbols: gate material colors: high-k oxide. Intel data from [11]

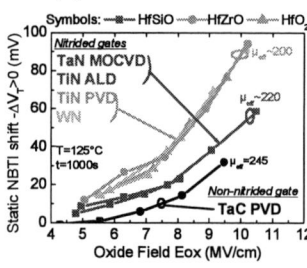

Fig. 12: NBTI is very well correlated to electron mobility on many HK/MG stacks

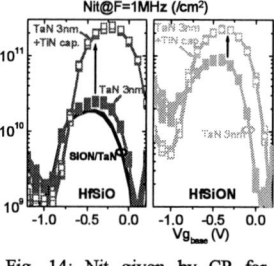

Fig. 13: Impact of metal gate on NBTI and μeff. PVD deposition of TiN induces add. degradations

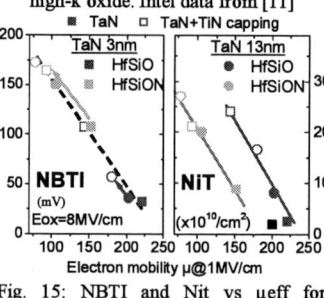

Fig. 14: Nit given by CP for HfSiO(N)/TaN stacks w/wo 10nm PVD TiN capping.

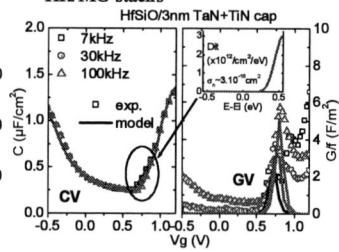

Fig. 15: NBTI and Nit vs μeff for HfSiO/TaN stacks w/wo 10nm PVD TiN capping. Full: TaN Open: TaN+TiN cap.

Fig. 16: Modeling of CVs-GVs of HfSiO/TaN stack with TiN cap. N induces defects in the upper part of bandgap

Fig. 17: Mechanisms of mobility and NBTI degradation

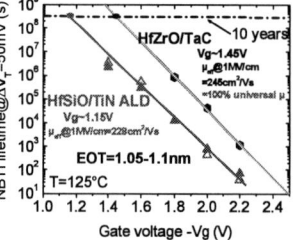

Fig. 18: NBTI lifetimes and mobility on non nitrided oxides. μeff@1MV/cm is ~90-100% of universal mobility

	PBTI	NBTI
Main characteristics	- Low in thin oxides films T_{HK}<2.5nm - Almost independent of the High-k	- Low in non nitrided HK/MG stacks - Strongly enhanced with N content
Main mechanisms which drive ...	Cristallinity of the film → thin "amorphous" oxide film exhibits low PBTI	Nitrogen at the bottom interface Si/SiO₂ incorporated by PN or/and arising from the gate material
Impact of the gate material	Enhanced when using thin "reactive" gates like TaN →reaction at the top interface	Enhanced by nitrided gates especially PVD TiN material
Link with mobility?	No evidence	Strong correlation →Linearly varies with μeff
How to improve...	- Use very thin "amorphous" High-k films T_{HK}<2.5nm - Prefer "stable" thermal gate	- Use non nitrided oxides - Better control of N profile → N must not reach the Si interface - Avoid the PVD techniques able to easily sputter N
Critical or not?	Not critical down to 1nm EOT Lifetimes over 10 years even in "worst cases" conditions	- Critical in some nitrided stacks - Lifetimes over 10 years in non nitrided stacks
Best Trade off	Improve preferentially NBTI which is more critical than PBTI and which affects mobility → HfSiO, HfO₂,HfZrO~2nm + TaC, TaN~5nm	

Table 1: Summary and guidelines to improve devices performances and reliability

References

[1] S.Saito et al., IEDM,2003 [2] M. Fischetti et al., JAP pp. 4587, 2001 [3] B. Kaczer et al., IRPS,2005 [4] J. Mitard et al., IRPS, 2006 [5] X. Garros et al., IRPS, 2008 [6] H-J Cho et al., IEDM, 2004 [7] Rafik et al, IEDM,2007 [8] T. Matsuki, SISC, 2007 [9] P. Batude et al., JAP, pp. 34514, 2007 [10] G. Gerardi et al., APL, pp.348,1986 [11] K. Mistry et al., IEDM, 2007

ELECTRON TRAPPING: AN UNEXPECTED MECHANISM OF NBTI AND ITS IMPLICATIONS

J.P. Campbell[1]*, K.P. Cheung[1], J.S. Suehle[1], A. Oates[2]

[1]Semiconductor Electronics Division, NIST, Gaithersburg, MD 20899, 301-975-8308, *email: jason.campbell@nist.gov
[2]TSMC Ltd. No. 121, Park Ave. 3, Science-Based Industrial Park, Hsin-Chu, Taiwan 300-77, R.O.C

INTRODUCTION

NBTI is the most important reliability problem in advanced CMOS technology.[1] Consequently, the recent CMOS reliability literature is dominated by the characterization and modeling of NBTI. All of the existing literature attributes the NBTI-induced ΔV_{TH} and I_D degradation to interface state generation and/or hole trapping.[2] **In this work, we demonstrate that electron trapping has been overlooked in previous studies and will greatly impact the current understanding of the NBTI phenomenon and consequent lifetime predictions.**

KEY RESULT

The key experimental evidence demonstrating electron trapping during NBTI is shown in figure 1 which illustrates the %peak-G_M degradation as a function of recovery time. 1.6nm SiON pMOSFETs were subjected to an NBTI stress of -2.5V/125C/10secs. Figure 2 illustrates our measurement sequence in which an I_dV_g measurement is taken on the rising edge (pre-stress) and falling edge (post-stress) of the gate pulse as well as after a variable recovery period (post-recovery). Each data point represents an average of 12 measurements and a fresh device is used for each recovery time. The pre-stress/post-stress or "zero–recovery" %G_M degradations are shown in figure 1 for reference. Demonstration of electron trapping is found from a comparison of the pre-stress/post-recovery or "recoverable" data which is clearly dependent on the recovery time. Significant G_M degradation is clearly evident at the shortest recovery time (2 μsecs). However, as the recovery time increases, the G_M degradation clearly reduces and crosses over to G_M values **greater than before stress** (negative values on figure 1). At longer recovery times, the G_M values return to the degraded state. This G_M behavior has never been reported since G_M extraction is usually difficult to obtain from fast-I_dV_g measurements. The full G_M curves are shown in figure 3 for the G_M degradation (fig. 3(a)) and G_M improvement (fig. 3(b)) cases.

The observed G_M behavior can be explained by hole as well as electron trapping and detrapping. Although our stress condition is rather common for NBTI studies of ultra thin gate dielectrics, it represents an electric field that is traditionally categorized as high-field stressing. Under high-field stress, it is known [3] that both electron- and hole-trapping occurs. It is also known that both electrons and holes will detrap once the stress is removed, but with very different detrapping rates (with holes detrapping faster).[4] At the conclusion of stress, both trapped holes and trapped electrons are present, along with positively charged interface states (I-V measurement condition) and we observe a maximum G_M degradation. As the recovery time increases, the much higher rate of hole-detrapping leads to a decrease in net trapped charge (bulk plus interface) as well as Coulombic scattering. Thus, the measured G_M degradation is reduced. This continues until the trapped holes in the bulk are largely depleted and electron-detrapping has a stronger impact on the net charge in the bulk. At this point the net charge in the bulk is negative and it is over compensating the positive interface charges added by the stress, leading to G_M improvement. At longer recovery times, electron detrapping continues and the net negative charge in the dielectric diminishes. This leaves only the positive interface state charge and the G_M returns to degradation.

EXPERIMENTAL

Fully processed 2 x 0.07μm² and 2 x 0.06μm² (physical gate area) SiON pMOSFETs (T_{ox} = 1.6nm) were used in this work. Our fast-I_dV_g measurement (2 digital oscilloscopes, 2 pulse generators, and a fast amplifier circuit)is capable of ≈ 2 μsec measurement time. The **unique capability** that enables this work is the ability to perform stress over a time scale from microseconds to essentially infinite time while maintaining the ability of controlling the recovery time to better than a microsecond. Figures 4(a) and 4(b) illustrate the very good agreement between the raw and filtered data from our fast-I_dV_g measurement and the DC measurement using a parametric analyzer. Figure 4(c) illustrates the extracted G_m characteristic from the fast-I_dV_g measurement which also shows very good agreement with the DC-measurement.

It is important to note that the effect of electron trapping can only be seen with a sufficiently fast I_dV_g measurement. Figure 5 illustrates the measured recoverable (-2.5/125C/10sec) %G_M degradation as a function of measurement time (rise/fall time of the gate pulse) for several different recovery times. It is clear that the measurement time must be less than 10usecs to observe the aforementioned G_M trends. It is also clear, that the recovery time greatly alters the result as the 10μsec recovery time exhibits a G_M degradation while the 100msec and 10sec recovery times exhibit a G_M improvement (negative values in figure 5).

RESULTS AND DISCUSSION

One would expect the hole and electron detrapping phenomenon which is affecting the G_M characteristics to also be reflected in the ΔV_{TH} measurement. However, we find that this trend is much harder to observe in the ΔV_{TH} measurements (figure 6) that were extracted from the same I_dV_g measurements used to produce figure 1 (linear extrapolation at max G_M). This is because this stress conditions results in a relatively small ΔV_{TH} which is overwhelmed by the error of the measurement. We have previously reported [5] zero-recovery and recoverable (5 seconds) ΔV_{TH} and %G_M degradations as a function of stress voltage for various stressing times at 125°C (figure 7). It is clear from these measurements that a large initial ΔV_{TH} is only observable at *exceedingly high* stress voltages and long stress times. Similarly, G_M improvement is easier to observe at higher stress voltages (figure 8).

In an effort to correlate V_{TH} behavior with G_M, we repeated our experiment using a harsher stress condition (-2.7V/125C/1000secs). Figures 9 and 10 illustrate the zero-recovery and recoverable ΔV_{TH} and %G_M degradations as a function of recovery time for this harsher stress case. In this case, the ΔV_{TH} trend (figure 9) mimics the G_M trends as expected (fig 10). Note that both hole-detrapping and electron-detrapping rates are higher in this case which results in a G_M turn around at a much earlier recovery time. Electron-detrapping is assisted by interface states via trap-to-trap tunneling. The detrapping rate increase is a result of the much higher interface state density due to the hasher stress. The higher hole-detrapping rate is the normal result of higher stress voltage [6]. The observed earlier G_M turn around time thus further supports our interpretation.

CONCLUSION

We have clearly demonstrated for the first time that the trapping and detrapping of electrons play a major role during common NBTI stressing and recovery conditions. We also showed that both hole-detrapping and electron-detrapping rates are dependent upon the stress condition and both slow down at less severe stress conditions. Furthermore, it is known that hole-detrapping is insensitive to temperature while electron-detrapping is highly sensitive to temperature. At lower temperatures, electron detrapping can be extremely slow. These factors greatly complicate both the characterization and modeling of NBTI. These results indicate that without accounting for the electron trapping and detrapping, the lifetime projection of NBTI based on the previously suggested physical models may be unreliable.

978-1-4244-1802-2/08/$25.00 ©2008 IEEE

REFERENCES

[1] M.A. Alam, *et al.*, Microelectronics Rel.,**45**, p.71 (2005) [2] V. Huard, *et al.*, Microelectronics Rel., **46**, p. 1 (2006) [3] D.J. DiMaria, *et al.*, J. Appl. Phys., **89**, p. 5015 (2001) [4] K.P. Cheung, *et al.*, *P2ID*, p. 181 (1997) [5] J.P. Campbell, *et al.*, IRPS (2008) (in press) [6] Y. Nissan-Cohen, *et al.*, J. Appl. Phys., p 2024 (1986)

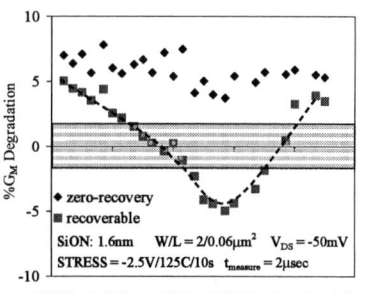

Figure 1: $\%G_M$ degradation as a function of recovery time for pre-stress/post-stress or "zero-recovery" measurements as well as pre-stress/post-recovery or "recoverable" measurements. As the recovery time increases, the recoverable G_M transitions from degradation to improvement and then returns to degradation. This behavior is consistent with both hole and *electron* detrapping. The box centered about 0% represents 1 standard deviation error bar. The lines are only a guide for the eye.

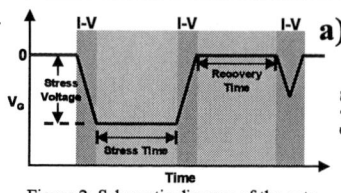

Figure 2: Schematic diagram of the gate voltage pulses during the NBTI stress and measurement sequences. The extracted V_{TH} and G_M values are compared pre-stress/post-stress ("zero-recovery"), and pre-stress/post-recovery ("recoverable").

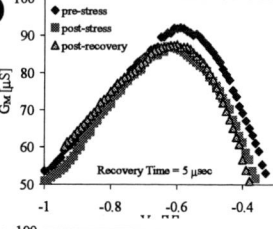

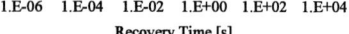

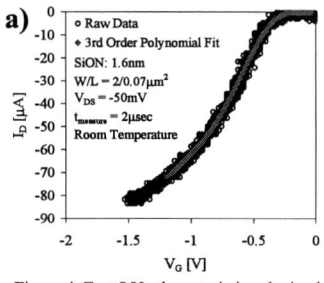

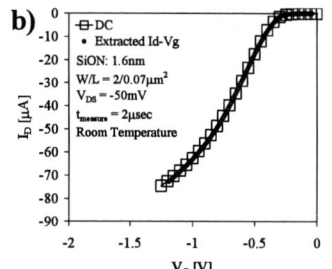

Figure 3: Pre-stress, post-stress, and post-recovery G_M characteristic curves for device subject to -2.5V/125C/10sec. (a) 5μsec recovery time exhibits a post-recovery G_M degradation while (b) 500msec recovery time exhibits a post-recovery G_M improvement.

Figure 4: Fast-I_dV_g characteristics obtained using our measurement approach are subject to significant noise. We utilize a "moving" third-order polynomial fitting procedure to extract the I_dV_g characteristic curves from the raw data (a). Our extracted I_dV_g (b) and G_M (c) characteristics agree very well with DC measurements.

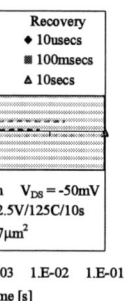

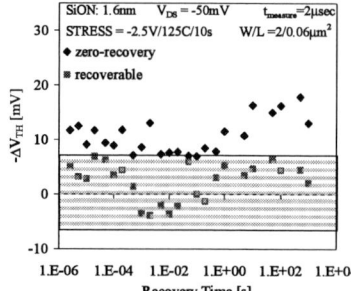

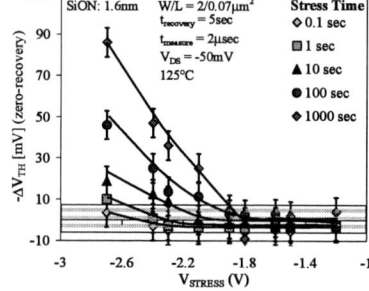

Figure 5: "Recoverable" $\%G_M$ degradation as a function of measurement (rise and fall) time for various recovery times. It is important to note that the G_M degradation/improvement is only observable for measurement times < 10usecs and that the recovery time greatly effects the G_M measurement result. The box centered at 0% represents 1 standard deviation error bar. The lines are only a guide for the eye.

Figure 6: ΔV_{TH} as a function of recovery time for the "zero-recovery" and "recoverable" measurements. The ΔV_{TH} from this stress condition is too small to observe the trend seen in figure 1. The box centered about 0mV represents 1 standard deviation error bar.

Figure 7: "Zero-recovery" ΔV_{TH} as a function of stress voltage for various stress times. It is clear that the fast NBTI degradation only occurs for exceedingly high stress voltages and longer stress times. The box centered about 0mV represents 1 standard deviation error bar. The lines are only a guide for the eye.

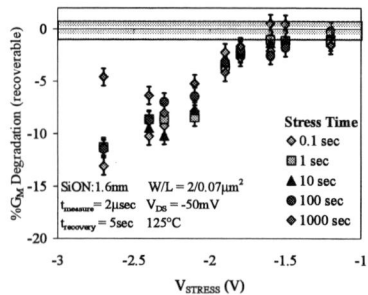

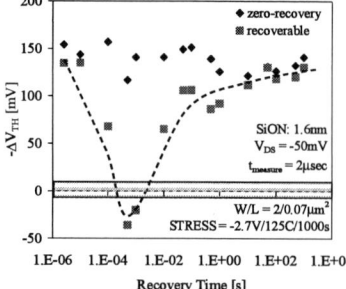

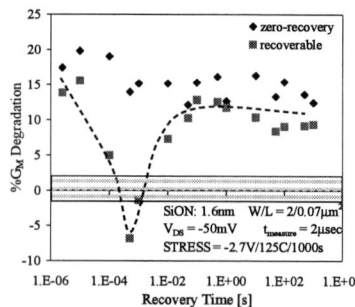

Figure 8: "Recoverable" $\%G_M$ degradation as a function of stress voltage for various stress times. The G_M improvement effect is increases with stress voltage. The box centered about 0% represents 1 standard deviation error bar.

Figure 9: ΔV_{TH} degradation as a function of recovery time for the "zero-recovery" and "recoverable" measurements. At this harsher stress condition (-2.7V/125C/1000sec) the ΔV_{TH} mimics the G_M behavior. The box centered about 0mV represents 1 standard deviation error bar. The lines are only a guide for the eye.

Figure 10: G_M degradation as a function of recovery time for "zero-recovery" and "recoverable" measurements. The GM improvement occurs much faster at this harsher stress condition. The box centered about 0% represents 1 standard deviation error bar. The lines are only a guide for the eye.

I_d Fluctuations by Stochastic Single-Hole Trappings in High-κ Dielectric p-MOSFETs

Shigeki Kobayashi, Masumi Saitoh, and Ken Uchida

Advanced LSI Technology Laboratory, Toshiba Corporation, 8, Shinsugita-cho, Isogo-ku, Yokohama 235-8522, Japan

Phone: +81-45-776-5966, Fax: +81-45-776-4113, E-mail: shigeki8.kobayashi@toshiba.co.jp

Abstract

Random telegraph noise (RTN) in scaled FETs is one of the biggest concerns in the present and future LSIs. However, RTN in high-κ gate dielectric FETs have not been fully studied yet. In this paper, we have studied RTN in high-κ pFETs in comparison with that in SiO_2 pFETs. It is found for the first time that the reduction of the RTN amplitude ($\Delta I_d/I_d$) by the surface holes is smaller in high-κ pFETs, comparing to the SiO_2 pFETs. It is also found that slower traps in the high-κ gate dielectric more severely degrade I_d. It is considered that some key characteristics are understandable in terms of the higher dielectric constant and the smaller barrier height of the high-κ gate dielectric.

Introduction

Random telegraph noise (RTN), which is digital I_d fluctuation induced by the stochastic single-carrier trappings in the gate dielectric, in scaled FETs is one of the biggest concerns in the present and future LSIs, because it unpredictably and greatly fluctuates the dynamic performance of small-size FETs. Therefore, RTNs in small-size FETs are investigated by a number of authors[1-3]. However, these reports are basically on SiO_2 and SiON gate dielectric FETs. In spite of the practical importance of high-κ gate dielectrics, RTNs in high-κ FETs have rarely been studied[4], whereas RTNs in high-κ FETs could be quite different from those in SiO_2 FETs because of the larger dielectric constant and the smaller barrier height of high-κ insulators.

In this paper, I_d fluctuation by the single-hole trappings (RTN) in high-κ (HfSiON) gate dielectric pFETs is experimentally investigated in comparison with that in SiO_2 gate dielectric pFETs.

Experimental Results and Discussion

(a) Experimental Conditions and Two-Level RTN

pFETs with the typical designed gate length (L) of 130nm were fabricated along <110> on Si(001) wafers. The gate dielectric films are SiO_2 or HfSiON with Hf content of 30%. The inversion thicknesses (T_{inv}) are 3.6nm and 3.5nm for SiO_2 and HfSiON, respectively. The drain bias was set to −5 mV in all the measurements.

Fig.1 shows the time dependence of I_d in the HfSiON pFET, clearly demonstrating the two-level random telegraph noise (RTN). It should be noted that RTN is observed in the limited number of devices; we have carefully chosen the samples showing the two-level RTN.

Next, the origin of RTN is investigated. Fig.2 shows the distribution of the duration of low-level I_d in the HfSiON pFET. The exponential distribution clearly indicates that the observed RTN is due to the stochastic charge capture/emission processes[2,5]. In addition, the time constant (τ), which corresponds to the carrier capture/emission time is extracted from the exponential curve.

Fig.3 shows the probability for I_d to be at the low level as a function of the gate overdrive. The probability rises when the gate bias (V_G) increases. The probability increase indicates that the measured trap is acceptor-like[1,2]. When V_G is 0.23V, the trap level is higher than the Fermi level (E_F) of the channel (Fig.4(a)), meaning that a hole is captured by the trap with the small probability of 18%. When V_G is 0.28V, the hole capture/emission frequently occurs, and I_d is at the low level with the 52% probability (Fig.4(b)). When V_G further increases to 0.36V, a hole is captured by the trap with the high probability of 90% (Fig.4(c)). These results reconfirm that the origin of RTN is the single-hole capture/emission by the trap in the gate dielectric.

(b) Smaller Reduction of RTN Amplitude by N_s in High-κ pFETs

Fig.5 shows the RTN amplitude ($\Delta I_d/I_d$) in the SiO_2 pFET as a function of the surface carrier density (N_s). N_s is extracted by multiplying the gate overdrive by the gate capacitance. In the SiO_2 pFET, $\Delta I_d/I_d$ drastically decreases as N_s increases. Figs.6 and 7 show $\Delta I_d/I_d$ in the HfSiON pFETs (devices (A) and (B), respectively) as a function of N_s. Comparing to the SiO_2 pFETs, the reduction of $\Delta I_d/I_d$ by the surface holes is smaller in the high-κ pFETs. The smaller $\Delta I_d/I_d$ reduction in the high-κ pFETs is more clearly seen in Fig.8, where normalized $\Delta I_d/I_d$ in the SiO_2 and the high-κ pFETs is shown as a function of N_s. Here, $\Delta I_d/I_d$ is nor-

malized to that at N_s of $2E12cm^{-2}$.

Next, the physical origin of the smaller reduction of $\Delta I_d/I_d$ in the high-κ pFETs is considered. The large $\Delta I_d/I_d$ observed in this study is not explained by the carrier number fluctuation or the mobility fluctuation, because the total number of holes in the channel is several hundreds. We consider that the large $\Delta I_d/I_d$ is caused by the reduction of the effective channel width (W) by the trapped charge; Coulomb potential by the trapped charge increases the channel potential near the trap, and locally depletes the surface holes (Fig.9)[3,5]. When N_s increases, the surface holes more effectively screen the Coulomb potential. Thus, $\Delta I_d/I_d$ is reduced at higher N_s. Since the screening constant is inversely proportional to the average dielectric constants of Si and the gate dielectric[6], the screening constant in the high-κ pFETs is smaller than that in the SiO_2 pFETs. Thus, the weaker screening of the Coulomb potential causes the smaller $\Delta I_d/I_d$ reduction in the high-κ pFETs.

(c) RTN Amplitude and Trap Position

Fig.10 shows the RTN amplitude ($\Delta I_d/I_d$) versus τ in various pFETs with the almost similar designed W. Here, τ is defined when $<\tau_H> = <\tau_L>$. Since τ generally represents the frequency of the tunneling events, longer τ generally suggests that the trap is farther from the gate dielectric/Si interface. Thus, Fig.10 indicates that $\Delta I_d/I_d$ is larger when the trap is farther from the interface. Note that the trap in the SiO_2 pFET is closer to the interface than that of the high-κ pFET with the same τ, because barrier height of SiO_2 is higher than HfSiON[7]. Although the reason why $\Delta I_d/I_d$ is larger when the trap is farther form the interface is not clear yet, larger $\Delta I_d/I_d$ by the trap farther from the interface has a great impact on the FET operation, particularly in high-κ FETs. In SiO_2 FETs, holes are captured only by the traps near the SiO_2/Si interface due to the high barrier height (Fig.11(a)), whereas in high-κ FETs, holes are captured also by the traps far from the high-κ/Si interface due to the lower barrier height (Fig.11(b)). Thus in high-κ FETs, the traps not only near the interface but also far from the interface induce I_d fluctuation. As a consequence, the huge I_d fluctuation of 20% stochastically occurs, and severely degrades the dynamic characteristics of high-κ pFETs.

(d) Impact of Device Scaling: Strong Vertical Confinement

Impact of the device scaling to the RTN amplitude ($\Delta I_d/I_d$) is also investigated. In short-channel FETs, the high substrate doping concentration (N_{sub}) is inevitable. As a result, the vertical confinement in the inversion layer is enhanced due to the high substrate potential. Fig.12 investigates the effects of enhancement of the vertical confinement to $\Delta I_d/I_d$. The substrate bias (V_{sub}) is used to enhance the vertical confinement. The monotonic $\Delta I_d/I_d$ increase by enhancement of the vertical confinement is clearly confirmed. Thus, high N_{sub} in the scaled FETs further increases the RTN amplitude.

Conclusions

I_d fluctuation by the single-hole trappings (RTN) in the high-κ (HfSiON) pFETs is experimentally investigated in comparison with that in the SiO_2 pFETs. It is found for the first time that the N_s-induced reduction of the RTN amplitude ($\Delta I_d/I_d$) is smaller in the high-κ pFETs, comparing to the SiO_2 pFETs. The smaller $\Delta I_d/I_d$ reduction in the high-κ pFETs is attributed to the weaker screening of the Coulomb potential by surface holes due to the larger dielectric constant in the high-κ gate dielectric. It is also found that $\Delta I_d/I_d$ is larger when the single-hole capture/emission process is longer. In addition, $\Delta I_d/I_d$ enhancement by the stronger vertical confinement is also confirmed. Thus, it is concluded that careful optimization of the high-κ/Si interface in terms of the dielectric constant and the barrier height is essential to realize the stable operation of the short-channel high-κ pFETs.

References

[1]K. S. Ralls et al., PRL, **52** 228 (1984). [2]M. J. Kirton et al., Advanced in Phys., **38** 367 (1989). [3]A. Asenov et al., TED, **50** 839 (2003). [4]A. Lee et al., JCE, 159 (2004). [5]K. Uchida et al., IEDM2002, 177. [6]F. Stern et al., PR, **163** 816 (1967). [7]Y. Kamimuta et al., JJAP, **44** 1301 (2005)

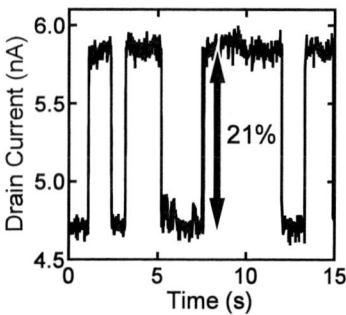

Fig.1 Time dependence of drain current (I_d) in the HfSiON pFET. The two-level random telegraph noise (RTN) is clearly observed.

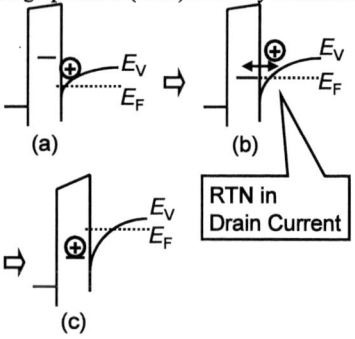

Fig.4 (a) The trap is empty when V_G is low. (b) When the trap level and Fermi level (E_F) in the channel are close, a hole is frequently captured/emitted by the trap. (c) A hole is captured with high probability, when V_G further increases.

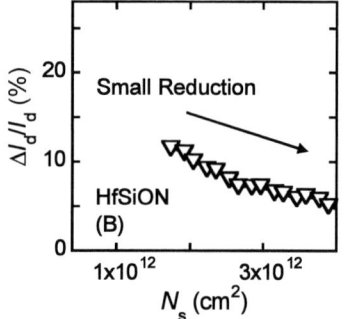

Fig.7 $\Delta I_d/I_d$ in the HfSiON pFET (device B) as a function of N_s. $\Delta I_d/I_d$ reduction at high N_s is small.

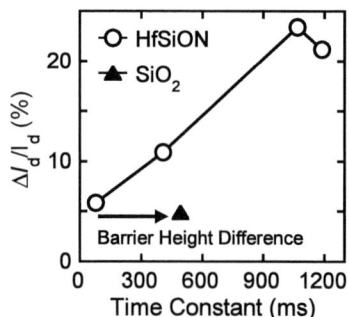

Fig.10 $\Delta I_d/I_d$ versus τ in the SiO$_2$ and the HfSiON pFETs with the almost the same designed W. τ is defined when $\langle\tau_H\rangle = \langle\tau_L\rangle$.

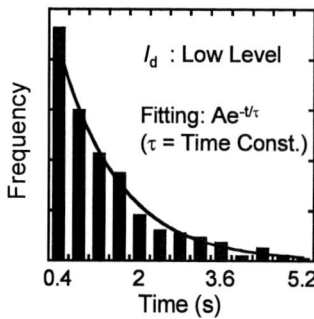

Fig.2 Distribution of duration of low-level I_d (time for which I_d is at the low level). The time constant (τ) is extracted by fitting the distribution with the exponential curve.

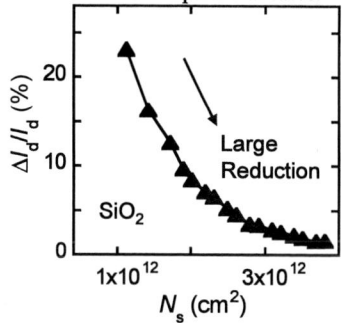

Fig.5 $\Delta I_d/I_d$ in the SiO$_2$ pFET as a function of surface carrier density (N_s). $\Delta I_d/I_d$ drastically decreases when N_s increases.

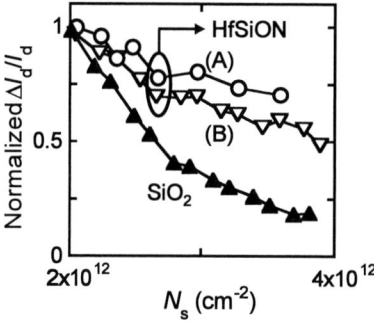

Fig.8 Normalized $\Delta I_d/I_d$ in the SiO$_2$ and HfSiON pFETs (devices A and B) as a function of N_s. $\Delta I_d/I_d$ is normalized to the values at N_s of 2E12cm^{-2}.

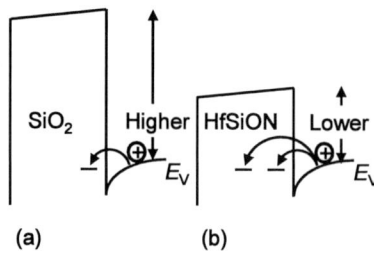

Fig.11 (a) In SiO$_2$ pFETs, holes are captured only by the traps near the SiO$_2$/Si interface because of the high barrier height. (b) In high-κ pFETs, holes are captured also by the traps far from the high-κ/Si interface because of the low barrier height.

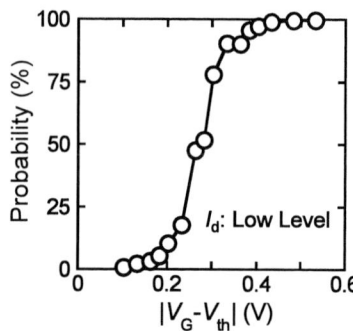

Fig.3 Probability for I_d to be at the low level as a function of the gate overdrive voltage.

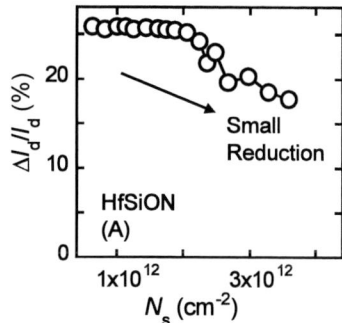

Fig.6 $\Delta I_d/I_d$ in the HfSiON pFET (device A) as a function of N_s. $\Delta I_d/I_d$ reduction at high N_s is small.

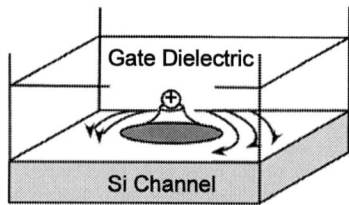

Fig.9 Coulomb potential created by the trapped single-hole locally increases the channel potential. As a result, the effective channel width (W) is greatly reduced.

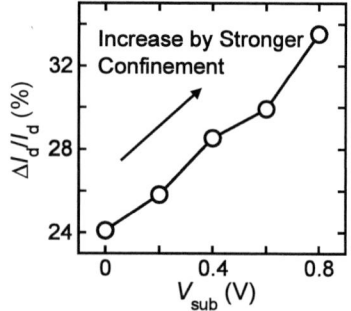

Fig.12 $\Delta I_d/I_d$ v.s. the substrate bias (V_{sub}). $\Delta I_d/I_d$ increases when the vertical confinement is enhanced by V_{sub}.

Roles of oxygen vacancy in HfO_2/ultra-thin SiO_2 gate stacks
- Comprehensive understanding of V_{FB} roll-off -

K. Akiyama[1], W. Wang[2], W. Mizubayashi[2], M. Ikeda[1], H. Ota[2], T. Nabatame[1] and A. Toriumi[2,3]

[1]MIRAI-ASET, AIST Tsukuba West 7, 16-1 Onogawa, Tsukuba 305-8569, Japan,
Tel: +81-298-49-1506, Fax: +81-298-49-1529, email: koji.akiyama@aist.go.jp
[2]MIRAI-ASRC, AIST, Tsukuba 305-8569, Japan, [3]The University of Tokyo, Tokyo 113-8656, Japan

Abstract

This paper discusses a role of the oxygen vacancy in HfO_2/ultra-thin (UT) interfacial layer (IL) SiO_2 gate stacks, focusing on the V_{FB} roll-off. The metal/top-SiO_2/HfO_2/UT IL-SiO_2/Si gate stacks have been studied. It is found for the first time that the V_{FB} roll-off is eliminated by inserting 1~2nm top-SiO_2 between metal gate and HfO_2. This elimination of the V_{FB} roll-off is explained by compensating the bottom dipoles at HfO_2/IL-SiO_2 interface with the counter dipoles at top-SiO_2/HfO_2 interface. Therefore it is concluded that the bottom dipole is assigned to the dominant origin of the V_{FB} roll-off.

Introduction

The V_{FB} roll-off is a very serious issue in terms of EOT scalability and V_{TH} control in high-k p-MOSFETs [1, 2]. We have reported that the V_{FB} roll-off is closely related to the out-diffused oxygen from HfO_2 to Si substrate [1]. Since V_{FB} shift should be understood by an electrical effect, the V_{FB} roll-off has been explained (1) by charged oxygen vacancy (V_o^{2+}) generated in HfO_2 [2, 3], or (2) by dipoles generated at the direct contact HfO_2/Si interface [4]. Our observations, however, cannot be explained by above models.

In this paper, the metal/top-SiO_2/HfO_2/IL-SiO_2/Si gate stack structure has been studied for the first time to clarify a role of UT-SiO_2 for the V_{FB} roll-off. Based on its understanding, a gate stack structure with no V_{FB} roll-off is demonstrated without any EOT increase penalty.

Experimental

Conventional HfO_2/SiO_2 gate stacks with various SiO_2 thicknesses (1 to 17 nm) on a p-Si wafer (hereinafter beveled SiO_2) were fabricated. HfO_2 films were deposited by MOCVD or ALD, followed by PDA at 700 °C in O_2 (100 Pa). W and TiN films were deposited on HfO_2 by dc-sputtering, followed by 50 nm-thick a-Si deposition and annealing in N_2 at 1000°C for 1 sec after gate patterning (MIPS structure). Ni was then deposited and silicided at 500 °C. Finally, all samples were sintered in 100 %H_2 at 400 °C. Specifically, in this work metal/top-SiO_2/HfO_2/IL-SiO_2/Si gate stacks were also prepared with various top-SiO_2 thicknesses (1~8 nm). The top-SiO_2 layer was grown by SiH_4-CVD and then processed as above.

Results and Discussion

Our previous work revealed that the V_{FB} roll-off appeared in metal gates, while it did not in Si-based gates such as poly-Si, FUSI or PASI as shown in **Fig.1** [1]. So, we proposed a model that the oxygen vacancy reaction kinetics in gate stack has a strong relation to the V_{FB} roll-off (**Fig.2**). One can speculate that it might be caused by charged oxygen vacancy (V_o^{2+}) generated when the oxygen in HfO_2 were consumed by interfacial reactions as shown in **Fig. 2**. To extract the V_o^{2+} *charges* quantitatively, HfO_2 thickness dependences of V_{FB} in W/HfO_2/SiO_2 gate stack with thick (the normal region) and thin (the V_{FB} roll-off one) IL-SiO_2 have been measured as shown in **Fig.3**. This result clearly indicates, though unexpected, that no positive V_o^{2+} charges are observed within HfO_2 or HfO_2/IL-SiO_2 interface under the V_{FB} roll-off condition.

Next, we have suspected a *dipole* effect for the V_{FB} roll-off. Metal/top-SiO_2/HfO_2/IL-SiO_2/Si gate stacks, as illustrated in **Fig. 4**, have been intensively studied. If the V_{FB} roll-off is caused by the *dipole* at the bottom HfO_2/IL-SiO_2 interface, the *counter-dipole* formed at the top-SiO_2/HfO_2 interface can cancel (or at least decrease) the V_{FB} roll-off. **Fig. 5 (a)** shows that the 1~2 nm top-SiO_2 insertion can

eliminate the V_{FB} roll-off as a result of the counter balance between the top and the bottom dipole polarities. This is a surprising but clear evidence of the dipole effect for the V_{FB} roll-off. This fact conveys that the V_{FB} roll-off may take place at the top-SiO_2/HfO_2 interface. With further increase of the top-SiO_2 thickness, the metal gate is substantially separated from the HfO_2 and an interaction between metal and HfO_2 should be decreased. **Fig. 5(b)** indicates that by increasing the top-SiO_2 thickness, the V_{FB} roll-off reappears as expected. **Fig. 6** shows the top-SiO_2 thickness dependence of V_{FB} on HfO_2 (3 nm)/IL-SiO_2 (1nm). V_{FB} sharply increases, and then saturates and decreases below 3nm-thick top-SiO_2. This indicates that the top-SiO_2/HfO_2 interface causes the same effect as the bottom, and that the V_{FB} roll-off cannot be understood in terms of the fixed charges. **Fig. 7** also shows the top-SiO_2 thickness dependence of V_{FB} for the HfO_2/ thin IL-SiO_2 (1nm-thick) stack with W gate with various HfO_2 thickness (0~3 nm). We have noticed here that the V_{FB} roll-off is increased by inserting the thick top-SiO_2 (above 2nm) on HfO_2 and saturates above 4 nm regardless of the HfO_2 thicknesses. These results strongly suggest that the oxygen vacancy generation by interfacial reactions has two roles for V_{FB} roll-off; (1) the V_{FB} roll-off dipole formation, and (2) the V_{FB} shift similar to the dipole formation in the reduction ambient annealing [5]. **Fig. 8** schematically illustrates what happen at both interfaces in terms of the oxygen vacancy effects. Furthermore, the V_{FB} roll-off elimination in Si-based gate electrodes [1] is also understandable in terms of the counter dipoles formation with respect to the bottom dipoles.

Finally, based on the dipole model at the HfO_2/IL-SiO_2 interface, two kinds of methods for suppressing the V_{FB} roll-off are demonstrated. The first one is to decrease the V_o-induced dipole density by using $HfSiO_x$. **Fig.9** shows that the V_{FB} roll-off behavior is reduced in $HfSiO_x$ gate stacks and the extent of V_{FB} roll-off decreases gradually with an increase in the Si content of $HfSiO_x$ (**Fig. 10**), though the suppressed V_{FB} roll-off is traded off by the lower dielectric constant. The other one is the metallic Al insertion in the gate electrode (TiN/Al/TiN/HfO_2 structure). It is expected that diffused Al from the gate electrode into HfO_2 layer can scavenge the oxygen in HfO_2/IL-SiO_2 stack, which further causes the positive V_{FB} shift due to the material-induced AlO_x/SiO_2 dipoles [6] with no EOT increase [7]. **Fig. 11** demonstrates that Al-inserted gate stack can effectively suppress the V_{FB} roll-off without any EOT increase.

Conclusion

The V_{FB} roll-off is assigned to the dipoles at the HfO_2/IL-SiO_2 interface, but not to the fixed charges inside the gate stacks. It is found that this dipole is only generated when the oxygen can be released from HfO_2. Furthermore, we have demonstrated the method of Al-inserted gate electrode for suppressing the V_{FB} roll-off without any EOT increase.

Acknowledgement

This work was supported by NEDO.

References

[1] K. Akiyama *et al*., VLSI Tech. Dig., p.72 (2007).
[2] S.C. Song *et al*., IEDM Tech. Dig., p. 337 (2007).
[3] Y. Akasaka *et al*., Jpn. J. Appl. Phys., **45**, p.L1289 (2006).
[4] Y. Abe *et al*., Appl. Phys. Lett., **90**, p.172906 (2007).
[5] Y. Kamimuta *et al*., IEDM Tech. Dig., p. 341 (2007).
[6] K. Iwamoto *et al*., VLSI Tech. Dig., p.70 (2007).
[7] M. Kadoshima *et al*., VLSI Tech. Dig., p.66 (2007).

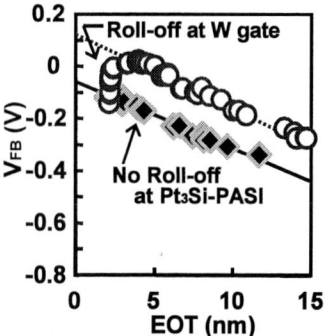

Fig.1 EOT dependence of V_{FB} in W-single layer/HfO$_2$(3nm)/ beveled IL-SiO$_2$ stack. The V_{FB} roll-off appears in metal gate, while it does not in Si-based gate. Note that EOT is changed by changing the IL-SiO$_2$ thickness.

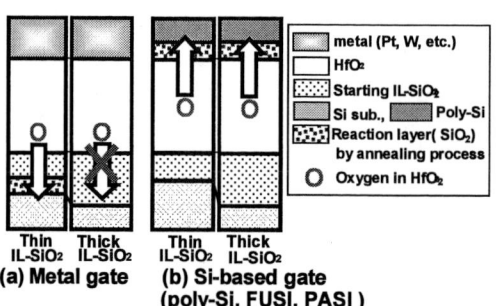

Fig.2 Schematic illustrating the movement of out-diffused oxygen from HfO$_2$ layer for (a) Metal and (b) Si-based gate/HfO$_2$/IL-SiO$_2$ gate stacks with different IL-SiO$_2$ thicknesses cases [1].

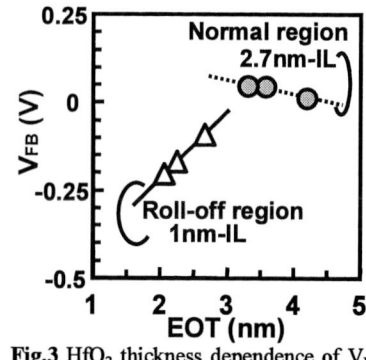

Fig.3 HfO$_2$ thickness dependence of V_{FB} in metal/HfO$_2$/IL-SiO$_2$ gate stacks both with thick IL (2.7nm, normal region) and thin IL (1nm, roll-off region).

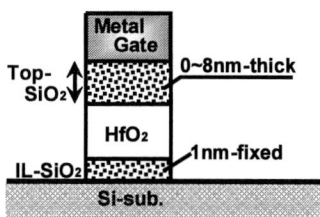

Fig.4 Schematic illustration of top-SiO$_2$/HfO$_2$/IL-SiO$_2$ stack with metal gate. Concerning the HfO$_2$/SiO$_2$ structure, a symmetric gate stack is prepared. The top-SiO$_2$ was grown by SiH$_4$ CVD.

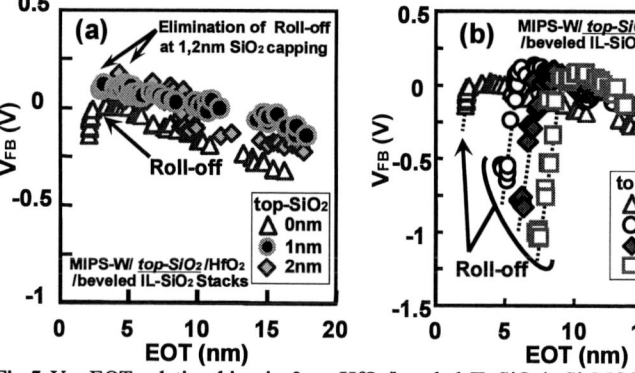

Fig.5 V_{FB}-EOT relationships in 3nm HfO$_2$/beveled IL-SiO$_2$/p-Si MOS capacitors with various thick top-SiO$_2$ layer insertion between MIPS-W and HfO$_2$. **(a)** No V_{FB} roll-off is observed for 1 and 2nm-thick top-SiO$_2$, and **(b)** the V_{FB} roll-off is enhanced with further increase of the top-SiO$_2$ thickness from 3 to 5 nm.

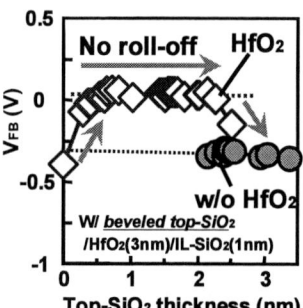

Fig.6 Top-SiO$_2$ thickness dependence of V_{FB} variations for beveled top-SiO$_2$/HfO$_2$(3nm)/IL-SiO$_2$ (1nm) stack with W single layer gate.

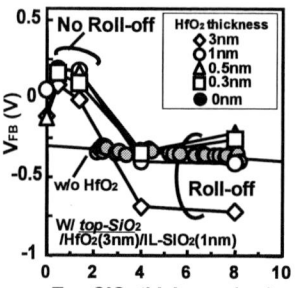

Fig.7 Top-SiO$_2$ thickness dependence of V_{FB} variations for W gate/ top-SiO$_2$/HfO$_2$/IL-SiO$_2$ (1nm) stacks with various HfO$_2$ thicknesses.

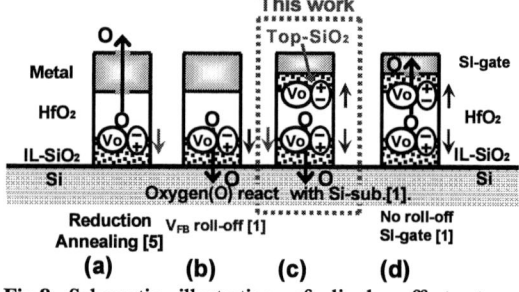

Fig.8 Schematic illustration of dipole effect at each HfO$_2$/SiO$_2$ interfaces. These V_o-induced interfacial dipoles were induced by as follows: (a) reduction annealing [5], (b) & (c) bottom interfacial reaction [1], (c) & (d) top- and bottom-interfacial reactions both at SiO$_2$ inserted metal and Si-based gate/HfO$_2$ interfaces. In (c) & (d) cases, the counter dipoles at top-SiO$_2$/HfO$_2$ interface are compensated with the bottom one, resulting in the suppression of the V_{FB} roll-off.

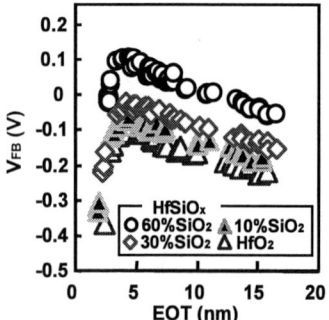

Fig.9 EOT dependence of V_{FB} in MIPS-W/HfSiO$_x$/beveled IL-SiO$_2$ stacks with different Si content in HfSiO$_x$. The roll-off effect is observed in HfSiO$_x$ stack.

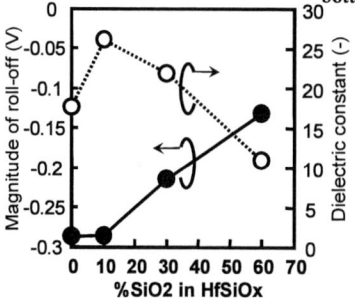

Fig.10 Dependences of V_{FB} roll-off magnitude and dielectric constant on %SiO$_2$ in HfSiO$_x$. The roll-off magnitude is reduced with increasing Si content in HfSiO$_x$.

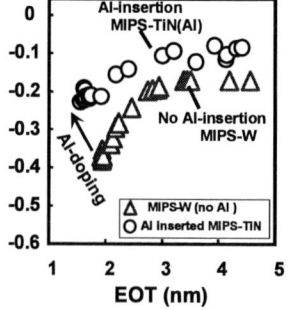

Fig. 11 Improvement of V_{FB} roll-off for MIPS-TiN/HfO$_2$(3nm)/beveled IL-SiO$_2$ stack using Al-insertion on TiN film in MIPS structure. Al-insertion gate can relax the roll-off behavior.

Mechanisms Limiting EOT Scaling and Gate Leakage Currents of High-k/Metal Gate Stacks Directly on SiGe and a Method to Enable sub-1nm EOT

J. Huang, P. D. Kirsch, J. Oh, S.H. Lee, J. Price, [a]P. Majhi, [b]H.R. Harris, D. C. Gilmer, D. Q. Kelly, P. Sivasubramani, G. Bersuker, D. Heh, C. Young, C.S. Park, Y. N. Tan, [a]N. Goel, C. Park, P.Y. Hung, P. Lysaght, [1]K. J. Choi, [2]B. J. Cho, H.-H. Tseng, B .H. Lee, and [c]R. Jammy

SEMATECH 2706 Montopolis Drive, Austin, TX 78741, U.S.A., [a]Intel, [b]AMD, [c]IBM assignee; [1] Jusung Engineering;[2]KAIST, Korea;

e-mail: _paul.kirsch@sematech.org_

Abstract

For the first time, we provide mechanistic understanding of high gate leakage current on surface channel SiGe pFET with high-k/metal gate to enable sub 1nm EOT. The primary mechanism limiting EOT scaling is Ge enhanced Si oxidation resulting in a thick (1.4nm) SiO_x interface layer. A secondary mechanism, Ge doping ($\geq 4\%$) in high-k, possibly by up diffusion, also results in higher leakage. With this understanding, we optimized high-k nitridation reducing O and Ge diffusion to achieve EOT=0.91nm directly on SiGe with leakage equivalent to bulk Si. High I_{on} (1.5× Si), and low subthreshold slope (73mV/dec) are also achieved. This mechanism enables high mobility channel gate dielectric development directly on SiGe without the need for Si cap, simplifying processing and device design.

Introduction

Selective, epitaxial strained SiGe (<40% Ge) is an attractive channel material beyond Si due to its high carrier mobility (~2× Si) and compatibility to the conventional CMOS processes [1]. Besides high mobility, the SiGe valence band offset to Si simplifies achievement of a low pFET threshold voltage (V_t) with high-k/metal gate [2]. However, the SiGe/dielectric interface has several issues: 1) Ge up-diffusion resulting in degradation of device performance [3], 2) interface states [3], 3) Ge condensation [4] and 4) excessive interface oxidation [5]. Consequently, many researchers fabricated SiGe devices with Si cap and low temperature process [6,7]. However, no reports clarify the material and device mechanisms associated with the SiGe/high-k interface for a true surface channel pFET. A known issue with SiGe channel is high gate leakage current (J_g) [2] that makes HfSiON physical thickness scaling unlikely to achieve sub 1 nm EOT. In this work, we explore metal/HfSiON gate stacks on SiGe channel with and without a Si cap including 950°C S/D anneal. Nitrogen dose and placement in HfSiON is proposed to solve the J_g issues. Particular attention is paid to leakage mechanisms such as 1) Ge up-diffusion and 2) accelerated low-k SiO_x bottom interfacial layer (BIL) growth. Sub 1nm EOT surface channel SiGe pFET performance is evaluated and compared to surface channel Si pFETs.

Experiment

After surface clean, selective, epitaxial SiGe was grown. The gate dielectric was grown with atomic layer deposition and nitrided with thermal or plasma nitridation (N*) to form the following stacks: Top nitrided (TN) HfSiON, intensive top nitrided (TN*) HfSiON and Cyclic nitrided (Cyc-N*) HfSiON. Cyc-N* HfSiON consists of multiple HfSiO depositions followed by nitridation. Subsequent processing included a conventional gate first metal gate (MG) process with a 950°C source/drain anneal.

Results

For SiGe pFET gate stacks, EOT does not scale with increasing J_g for physically thin (~2nm) HfSiON [Fig.1(a)]. Compared with MG/TN-HfSiON/**Si** stacks, MG/TN-HfSiON/**SiGe** stacks have higher J_g (>10x) [Fig.1(b)]. Consequently, high-k physical thickness scaling is unlikely to scale HfSiON EOT on SiGe. Increasing the Si cap thickness causes the Si/SiGe to mimic the bulk Si channel and hence

improves the J_g-EOT trend, indicating that interfacial Ge is critical [Fig.1(c)].

HfGeO was intentionally deposited on Si with PVD to simulate Ge up diffusion into the high-k as a possible leakage mechanism. HfGeO (4-15% Ge) increases both J_g and EOT suggesting Hf germanate is undesirable (Fig.2). However, the band gap of Hf germanate is similar to that of HfO_2, suggesting that band gap (barrier height) reduction is unlikely the cause of poor dielectric performance (Fig.3). To understand Hf germanate formation, XPS was done [Fig. 4(a),(b)]. Shifts in Hf4f and O1s chemical states correlate with Hf germanate formation and a similar shift is observed for the HfO_2/SiGe stack, suggesting interfacial germanate does form in the absence of Si cap. Most important, nitridation suppresses these germanate features [Fig.4(c),(d)] suggesting that Ge up diffusion is one key mechanism.

A second key mechanism is Ge catalyzed BIL growth. TEM proves that Ge enhances BIL growth but nitridation reduces BIL thickness (Fig.5). EOT reduction and thin BIL was achieved with plasma N* forming HfSiON (Fig.6). Large EOT reduction (0.4nm) with small J_g increase (<10x) suggests that reduction of low-k BIL is the main factor for EOT scaling (Fig.6). EELS suggests that the BIL is SiO_x but just a small amount of Ge (~3%) up diffuses [Fig.7(a)]. Therefore N content and location is key to achieve thin BIL [Fig.7(b)]. If nitridation is insufficient, O may diffuse down to the SiGe to form SiO_x and Ge segregates below the BIL - the well-known Ge condensation phenomena [4]. Plasma nitridation (TN* and cyc-N*) slows O diffusion and can also locate N close to both SiGe and MG [8]. The Ge up-diffusion and BIL growth was efficiently suppressed and hence can realize EOT=0.91nm and low J_g (Fig.6,8).

Mimimal C-V hysteresis (<10mV) and frequency dispersion was observed on HfSiON/SiGe stacks. However, SiGe Nit (charge pumping) is 10x greater than Si channel (6×10^{11} cm^{-2} vs. 5×10^{10} cm^{-2}) [Fig.9]. Additionally, the Nit of SiGe devices may be under-estimated because the quantum well of SiGe/Si could confine carriers. This high Nit can degrade the mobility. Fortunately, the intrinsic hole mobility of SiGe is high enough to compensate mobility degradation (Fig.10). However, the quantum well of Si/SiGe will increase T_{inv} suggesting formation of a buried channel (Fig.11). Although performance of SiGe pFETs is not as good as Si/SiGe (buried channel) pFETs, SiGe pFETs do not have the issues of high off-state current. Surface channel SiGe pFETs with 0.91nm EOT (TN* HfSiON) can have 1.5x I_{on} improvement vs. surface channel Si pFETs [Fig.12(a)]. The sub-threshold swing (73mV/dec) is small and hence results in low off-state leakage current [Fig.12(b)]. |V_t| is reduced 0.3V making low V_t pFETs more straightforward (Fig.12b) [2].

Conclusion

The mechanisms of high gate leakage current in MG/HfSiON/SiGe stacks are 1) Ge enhanced Si oxidation and 2) Ge up-diffusion, the first being the primary mechanism. Optimized plasma nitridation of HfSiON/SiGe stacks addresses O and Ge diffusion issues enabling EOT=0.91nm. Dielectric deposition directly on SiGe reduces process and device design complexity associated with the common Si cap approach. This scaling and performance demonstration paves the way for surface channel SiGe pFETs at sub 32nm nodes.

978-1-4244-1802-2/08/$25.00 ©2008 IEEE

References

[1] J.L. Hoyt et al., Tech. Dig. IEDM, p. 23, 2002.
[2] H. R. Harris et al., Symp. VLSI Tech., p.154, 2007.
[3] G. K. Dalapati et al., IEEE TED, 53 (5), p.1142 (2006).
[4] T. Tezuka et al., Symp. VLSI Tech., p.198, 2004.

[5] F. K. LeGoues, JAP 65 p.1724 (1989).
[6] S. Suthram et al., Tech. Dig. IEDM, p. 727, 2007.
[7] O. Weber et al., Symp. VLSI Tech., p.42, 2004.
[8] S. Inumiya, et al. Tech. Dig. IEDM, p.23, 2005.

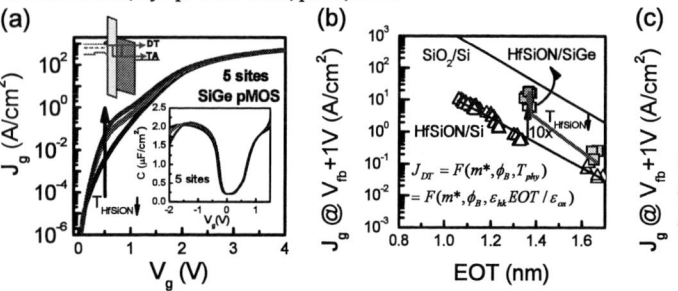

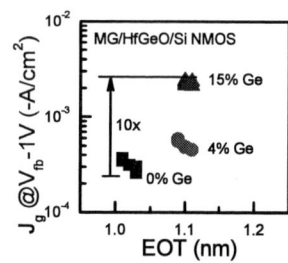

Fig. 1. Scaling issues on SiGe: (a) J_g increases but EOT is unchanged despite HfSiON T_{phys} scaling, (b) J_g is 10× larger on SiGe channel versus equivalent stack on Si channel. (c) J_g and/or EOT is reduced by increasing the Si cap thickness. Therefore, high J_g of SiGe may be due to HfSiON/SiGe interface issues such as 1) small potential barrier, 2) Ge up-diffusion [3], 3) nonstoichiometric GeO_x (k~6) [5], and 4) thick low-k SiO_2 bottom interfacial layer (BIL).

Fig. 2. Ge-doping (4 -15%) in HfO_2 (Si channel) increases both gate leakage current and EOT indicating negative effect of Ge in the high-k.

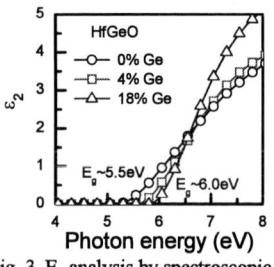

Fig. 3. E_g analysis by spectroscopic ellipsometry. Similar band gap of HfGeO suggests that barrier height reduction is unlikely the cause of poor dielectric performance.

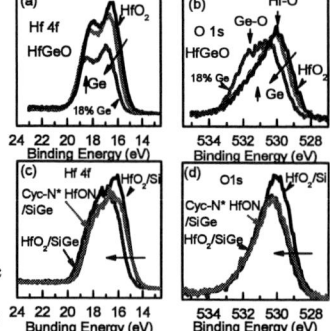

Fig. 4. (a-b) Hf4f and O1s shift with Ge doping forming Hf germanate. (c-d) N* suppresses Hf germanate.

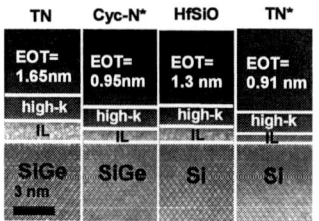

Fig. 5 BIL thickness is 1.4, 1.0, 1.2 and 0.7 nm respectively. HfSiON on SiGe gives thicker BIL (EOT) supporting Ge enhanced oxidation on SiGe(Fig.6). (TN: top nitrided HfSiON; TN* : intensive top nitrided HfSiON; Cyc-N*: cyclic nitrided HfSiON)

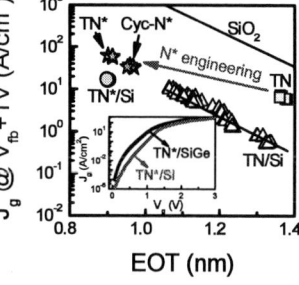

Fig. 6 Large EOT reduction (0.4nm) with small J_g increase (< 10x) suggests that low k material (BIL) is the main factor of EOT scaling instead of Hf germanate (Ge up diffusion).

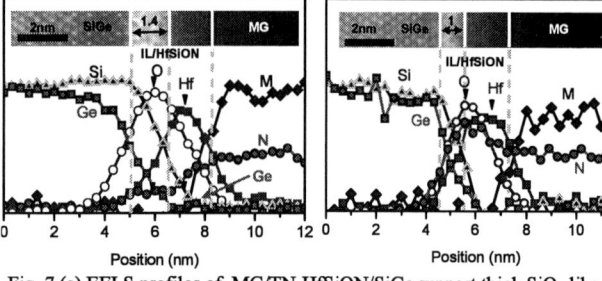

Fig. 7 (a) EELS profiles of MG/TN-HfSiON/SiGe support thick SiO_2-like BIL (1.4 nm) and Ge (~3%) up-diffusion; (b) EELS element profiles of MG/Cyc-N* HfSiON/ SiGe suggest that nitrogen content and location is key to suppress Ge up-diffusion and BIL growth.

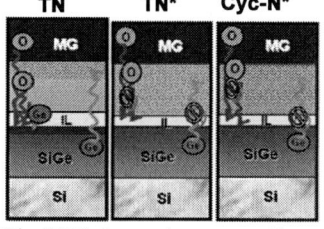

Fig. 8 TN nitrogen dose may not be enough to suppress O and Ge diffusion. Cyc-N* HfSiON and TN* HfSiON were more effective to give thin EOT (Figs.4,5,6,7).

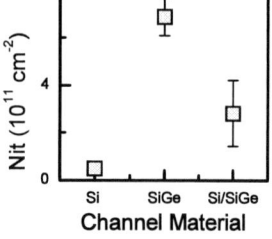

Fig.9 High interface trapped charge density (Nit) of HfSiON/SiGe samples indicates lesser interface quality vs. Si samples. Si cap on SiGe can mimic the Si bulk to reduce Nit.

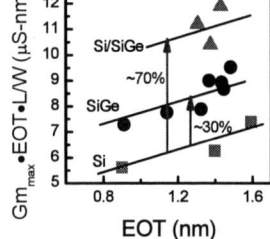

Fig. 10 SiGe channel gives 30% improvement and Si cap can further improve 40% due to the good interface quality (low Nit) and buried channel.

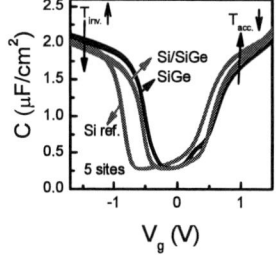

Fig.11 Lower capacitance (thicker T_{inv}) of Si/SiGe samples suggest formation of buried channel in inversion.

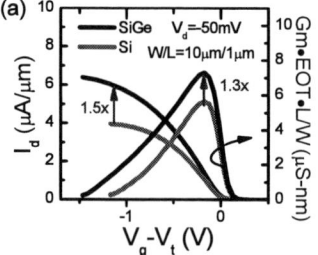

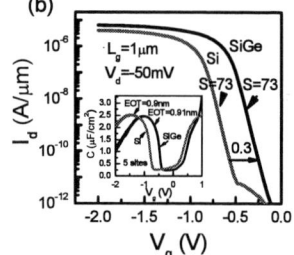

Fig.12 Without Si cap and local strain engineering, I_{on} increases 50 % by SiGe channel with TN* HfSiON dielectric. Sub-threshold swing is the same as Si (73mV/dec.) and I_{on}/I_{off} ratio of 10^6 is achieved.

Fully Integrated and Functioned 44nm DRAM Technology for 1GB DRAM

Hyunjin Lee, Dae-Young Kim, Bong-Ho Choi, Gyu-Seong Cho, Sung-Woong Chung, Wan-Soo Kim, Myoung-Sik Chang, Young-Sik Kim, Junki Kim,
Tae-Kyun Kim, Hyung-Hwan Kim, Hae-Jung Lee, Han-Sang Song, Sung-Kye Park, Jin-Woong Kim, Sung-Joo Hong, and Sung-Wook Park
R&D Division, Hynix Semiconductor Inc., San 136-1 Ami-ri, Bubal-eub, Ichon-si, Kyoungki-do, 467-701, Korea
Email:hyunjin.lee@hynix.com, Phone: +82-31-639-0824, Fax: +82-31-639-0734

Abstract

44nm feature sized $8F^2$ 1Gb DRAM is fully integrated and functioned for the first time with the smallest cell size of $0.015um^2$. A novel cell-transistor structure and new DRAM process technologies are developed. In order to control the threshold voltage uniformity and body-bias sensitivity of saddle-fin cell-transistor, the channel doping profile and saddle-fin geometric dependency were analytically expressed with experimental data. The weak fin height dependency on cell-V_T diminishes the burden of the saddle-fin patterning processes. And the low body-bias sensitivity of the saddle-fin cell-transistor leads wide tWR (write recovery time) margins. Cylinder-like MIM cell capacitor with ZAZ dielectric is exploited on cell capacitor. Copper implemented triple-metal layer and WN barrier-metal techniques were developed to decrease chip size.

Introduction

Three-dimensional cell array transistors, such as RCAT (recess channel array transistor) and ARG (advanced recess gate)-type, have been implemented for sub-70nm DRAM technology to obtain sufficient retention time using enlarged channel length and low channel doping concentration [1,2]. Low resistance tungsten word-line (WL) and cylindrical storage-node (SN) capacitor with ZAZ ($ZrO_2/Al_2O_3/ZrO_2$) have been utilized to increase number of cells per WL and enhance cell capacitance [3].

But to integrate sub-50nm manufacturable 1Gb DRAM with suitable refresh characteristics, gate controllability enhancement with sustaining unscalable gate oxide thickness as well as low channel doping concentration are demanded.

In this paper, the process and the integration technologies of the first 44nm 1Gb DRAM are presented with the smallest $8F^2$ cell size of $0.015um^2$. Detailed saddle-fin cell-transistor schemes with short channel immunity, threshold voltage controllability, and body-bias sensitivity are analyzed [4].

44nm DRAM Technology

Fig. 1 shows the word-line and the bit-line directions from cross-sectional views of fully integrated 44nm DRAM-cell with three-dimensional saddle-fin cell-transistor (cell-TR) and peri-Tr. The key process technologies to integrated 44nm DRAM are summarized and compared with previous technologies. ArF immersion lithography was used for defining critical patterns, such as active, fin, word-line, bit-line, storage node contact, and storage node (Fig. 2). To fabricate saddle-fin cell-TR, active silicon and SOD (spin on dielectric) filled field oxide layer were recessed (Fig. 2b) and wrapped by patterned word line (Fig. 2c).

Radical oxide growth technique was implemented to gate oxide for conformal gate oxide thickness on corner as well as side of the saddle-fin cell-TR. Fig. 3 shows improved Q_{BD} quality compared with dry oxide, and shows no fin height related gate oxide quality degradation. SOD-layer (spin on dielectric) was used for narrow STI region to fulfill the 44nm technology. And partially removed linear nitride layer between the SOD and active silicon substrate induces compressive strain by the SOD-material. The technique was used for PMOS periphery transistors to increase operation current (Fig. 4).

Fig. 5 shows TEM cross-section images of periphery transistors with WN barrier-metal and Tungsten on N+ poly-silicon. As compared to TiN barrier-metal, the ring-oscillator delay was improved to 15%. Line-type storage node contact patterning and cylinder-like storage node with ZAZ dielectric ($ZrO_2/Al_2O_3/ZrO_2$) were implemented to achieve sufficient memory cell capacitance of 25fF/cell for 44nm technology (Fig. 6). Low SN leakage current below 1fA/cell shows negligible effects of floating nitride layer, which sustain the tall and narrow storage node. Copper implemented triple-metal layer was used to decrease chip size as well as DRAM performance improvement with minimization of interconnect delay.

Saddle-Fin Cell-TR

Fig. 7 shows TEM cross-sectional views of the saddle-fin cell-TR on WL and BL direction. The saddle-fin structure was proposed to overcome the non-scalable gate oxide thickness issues with scaled cell size [4]. To fabricate three-dimensional cell-TR with triple-gate on center of the channel and conventional recess channel single-gate on remained channel region, field oxide was over-etched than that of silicon active region as shown in Fig. 7c. Fin width (W_{Fin}) of the saddle-fin is mostly decided by the active patterning process step, when fin height (H_{Fin}) is determined at fin etch process step.

The fin structure combined groove-like long channel length improves low and high margins of 44nm cell-Tr array performances. Due to the effects of partial triple-gate structure, short-channel effects including DIBL (drain induced barrier lowering) and subthreshold swing are improved as 6mV and 85mV/dec. And the superior gate controllability with gate shielded channel region leads improvement of passing-gate [4] and neighbor-gate effects, as shown in Fig. 8.

However, the cell-V_T with fully inversioned channel region is mainly controlled by the recess channel single-gate structure, as shown in Fig. 9 with cell-V_T dependency on $W_{Fin}/(W_{Fin}+2H_{Fin})$. Fig. 10 shows W_{Fin} and H_{Fin} dependency on experimental saddle-fin cell-V_T. Due to the nature of saddle-fin cell-TR, which shows the merits of recess-channel and fin-structure, cell-V_T is more sensitive on W_{Fin} than H_{Fin} and shows contrary dependency. The loose fin height dependency gives large process margins on saddle-fin fabrication using etch-process.

The body-bias of DRAM induces high off-state current on storage node by junction leakage current and GIDL (gate induced drain leakage). And the body-bias sensitivity causes write "V_T" increment and writing current decrement, which induces write "1"-state fail (tWR Fail). The body-bias sensitivity on saddle-fin cell-TR is improved than that of ARG-structure by using triple-gate fin-structure with improved gate controllability. Fig. 11 shows the additional improvement on body-bias sensitivity by fin height increment and channel doping decrement. The fin height dependency on cell-V_T is also loosened by low channel doping concentration. Fig. 12 and Fig. 13 show the tWR characteristics of the 44nm saddle-fin cell-transistor. The tWR time is extracted by using the current driving capability on cell-TR with gate operation voltage decrement, when the storage node is written to "1" state. Compared to previous-technology cell-transistor, the saddle-fin cell-transistor shows superior tWR-performance by low body-bias sensitivity.

Conclusion

$8F^2$ 44nm DRAM with $0.015um^2$ cell size is fully integrated and functioned using novel saddle-fin cell-transistor for the first time. The key processes are ArF immersion lithography, saddle-fin formation, conformal radical gate oxide, WN barrier-metal, cylinder-like MIM cell capacitor with ZAZ dielectric, and triple-metal layer using Copper. These newly developed technologies have good extendibility for the next generation DRAM technology.

978-1-4244-1802-2/08/$25.00 ©2008 IEEE

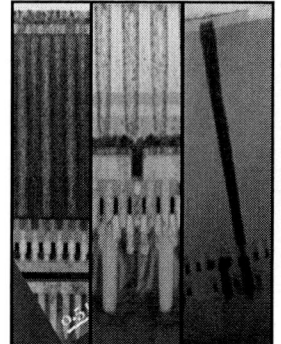

Fig. 1. Fully integrated 44nm DRAM cell on WL/BL direction and peri-Tr.

	66nm	54nm	44nm
Lithography	ArF	ArF	ArF Immersion
Isolation	SOD+ HDP	SOD+ HDP	SOD
Cell Scheme	ARG	ARG	Saddle-Fin
Gox	SiON(57A)	SiON(57A)	SiON(57A)
WL	W-NPG	W-NPG	W-NPG
BL	W	W	W
SN Contact	Hole-Type	Hole-type	Line-type
Storage Node	Cylinder	Cylinder	Cylinder
Cap Dielectric	ZAZ	ZAZ	ZAZ
Metals	Triple-Metal using AL	Triple-Metal using AL	Triple-Metal using Cu

Table 1. Key process technologies of 44nm DRAM integration.

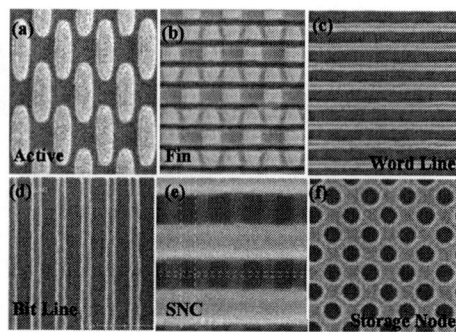

Fig. 2. SEM images of 44nm DRAM cell critical patterns.

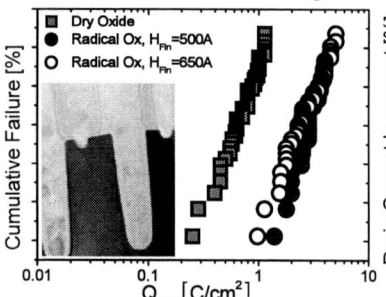

Fig. 3. Q_{BD} characteristics of saddle-fin cell structure with radical oxide.

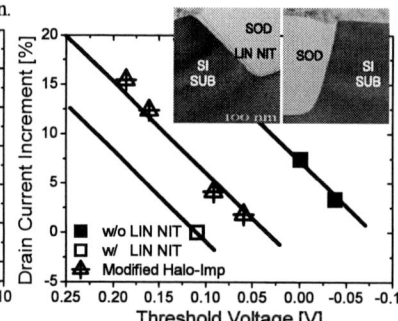

Fig. 4. PMOS operation current increment on liner nitride removes processes.

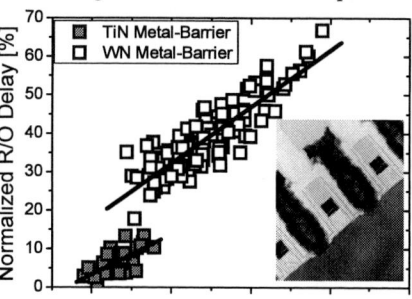

Fig. 5. Normalized RO delay and TEM images of peri-TR with WN metal-barrier.

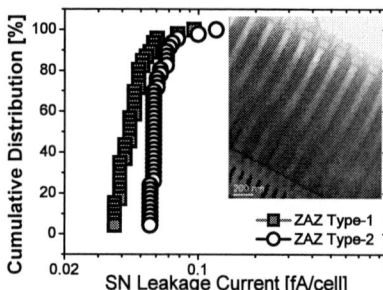

Fig. 6. Cumulative distribution of SN leakage current with different ZAZ-type.

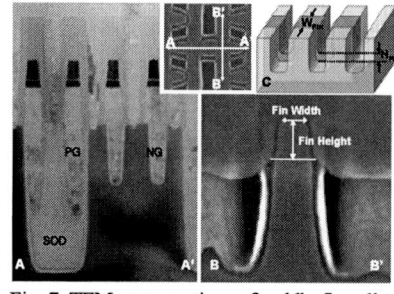

Fig. 7. TEM cross-sections of saddle-fin cell-transistor on BL/WL direction.

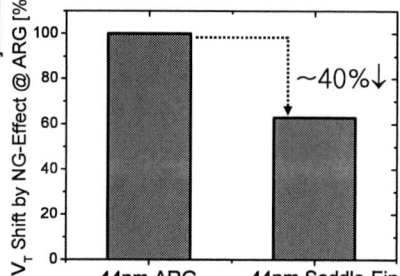

Fig. 8. Suppression of Neighbor gate effects on V_T shift using saddle-fin cell-Tr.

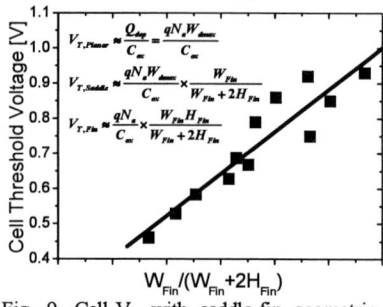

Fig. 9. Cell-V_T with saddle-fin geometric parameter from TEM and SEM images.

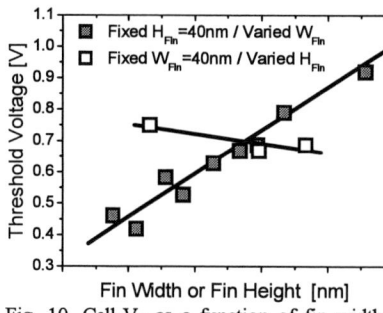

Fig. 10. Cell-V_T as a function of fin width and fin height with fixed $W_{Fin}=H_{Fin}=40nm$.

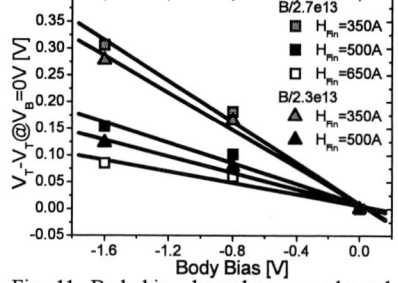

Fig. 11. Body-bias dependency on channel doping concentration with fin height.

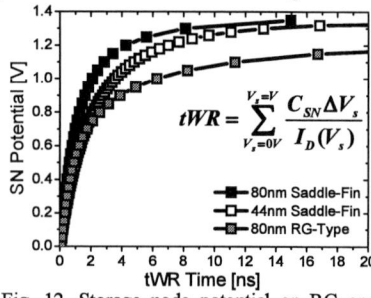

Fig. 12. Storage node potential on RG and 44nm and 80nm saddle-fin with tWR time.

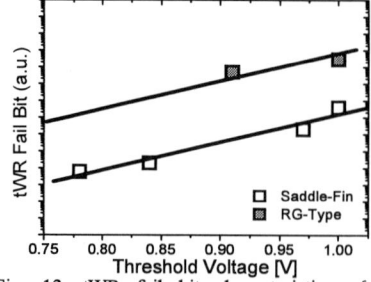

Fig. 13. tWR fail bit characteristics of saddle-fin and RG-type cell-transistor.

References

[1] J.Y. Kim et al., *VLSI.*, p.11, 2003.

[2] J.Y. Kim et al., *VLSI*, p.34, 2005.

[3] D.-S. Kil et al., *VLSI*, p.38, 2006.

[4] S.-W. Chung et al., *VLSI*, p.32, 2006.

A Cost Effective 32nm High-K/ Metal Gate CMOS Technology for Low Power Applications with Single-Metal/Gate-First Process

X. Chen, S. Samavedam[1], V. Narayanan, K. Stein, C. Hobbs[1], C. Baiocco, W. Li, D. Jaeger, M. Zaleski[1], H. S. Yang, N. Kim[2], Y. Lee, D. Zhang[1], L. Kang[1], J. Chen[2] H. Zhuang[3], A. Sheikh, J. Wallner, M. Aquilino, J. Han[3], Z. Jin, J. Li, G. Massey, S. Kalpat, R. Jha, N. Moumen, R. Mo, S. Kirshnan, X. Wang, M. Chudzik, M. Chowdhury[1], D. Nair, C. Reddy[1], Y. W. Teh[2], C. Kothandaraman, D. Coolbaugh, S. Pandey[2], D. Tekleab[2], A. Thean[1], M. Sherony, C. Lage[1], J. Sudijono[2], R. Lindsay[3], J. H. Ku[4], M. Khare, A. Steegen

IBM Semiconductor Research and Development Center (SRDC), Hopewell Junction, NY 12533
[1]Freescale Semiconductor, [2]Chartered Semiconductor Manufacturing, [3]Infineon Technologies AG, [4]Samsung Electronics Co., Ltd
e-mail: chenxian@us.ibm.com

Abstract

For the first time, we have demonstrated a 32nm high-k/metal gate (HK-MG) low power CMOS platform technology with low standby leakage transistors and functional high-density SRAM with a cell size of 0.157 μm². Record NMOS/PMOS drive currents of 1000/575 μA/μm, respectively, have been achieved at 1 nA/μm off-current and 1.1V V_{dd} with a low cost process. With this high performance transistor, V_{dd} can be further scaled to 1.0V for active power reduction. Through aggressive EOT scaling and band-edge work-function metal gate stacks, appropriate Vts and superior short channel control has been achieved for both NMOS and PMOS at L_{gate}=30nm. Compared to SiON-Poly, 30% RO delay reduction has been demonstrated with HK-MG devices. 40% Vt mismatch reduction has been shown with the Tinv scaling. Furthermore, it has been shown that the 1/f noise and transistor reliability exceed the technology requirements.

Introduction

Fast growing mobile applications require low power CMOS technology that can offer enhanced performance, low standby power and active power without increasing the cost [1]. High-k metal gate is considered to be one of the most promising technology enabler for low power application to meet the above requirements [2]. One concern for HK-MG in low power application is the cost from dual metal gates [3, 4]. In this paper, we present a cost-effective low power 32nm technology with single-metal/gate-first process[5]. The use of Hf based high-k gate dielectrics allows us to maintain a low-gate leakage of <0.1A/cm² while continuing to provide substantial room for EOT scaling. The dramatic short-channel improvements with EOT-scaling enable us to scale L_{gate} down to 30nm. Moreover, the strong performance offered by HK-MG allow us to trade off some performance for a cost-balanced low-power technology, by foregoing embedded SiGe S/D stressor. V_{dd} of the core LSTP (low stand-by power) transistors has been scaled down to 1.0V for active power reduction. Furthermore, a low operating power (LOP) transistor is offered with the same gate stack as the LSTP transistor at reduced 0.8V V_{dd} (Table1). A HK-MG compatible programmable element, eFUSE, is offered for redundancy and chip-id applications.

Process Integration

LSTP and LOP core transistors are fabricated on the same chip with thick oxide I/O transistors using a dual-oxide process. As shown from TEM (Fig.1), 30nm transistors were fabricated with gate first high-k/metal gate process employing Hf-based high-k gate dielectrics and a thermally stable single metal with conventional Si capping layer. After gate stack patterning and etch, conventional spacer process and self-aligned implants were used. Less than 3% total process cost is added with HK-MG compared with Poly-SiON gate stack. Advanced millisecond-based RTA annealing has been implemented for shallow junction engineering to scale gate length to 30nm without any negative impact to the gate stack. Dopant optimization enabled eFUSE elements with high post-programmed resistance. A 1.2 NA/193 nm wet immersion lithography was used for critical levels to enable a 70% scaling of key ground rules from 45nm. Table 2 shows the key FEOL and BEOL ground rules and scaling factors from 45nm. BEOL is composed of 6 to 11 layers of copper metallization in ultra low-K (k=2.4) dielectric (Fig.2). Due to the increased effects of scaling on 1X (100nm pitch) wires, additional use of hierarchy is employed by the addition of 1.3X pitch intermediate wiring levels for lower resistance.

Device Characteristics and SRAM Cell

With high-k/metal gate, Tinv is reduced by more than 1 nm with same gate leakage as the SiON-poly gate stack (Fig.3). Leading edge NFET and PFET performance has been achieved in this 32nm technology due to Tinv scaling and optimized high-k/Si interface. High drive currents of 1000/575 μA/μm (N/P) have been achieved at I_{off} =1nA/μm and V_{dd}=1.1V without additional stress engineering complexity and cost, such as e-SiGe process (Figs. 4, 5). With high drive capability, V_{dd} can be scaled down to 1V for the 32nm technology. One concern for 45nm to 32nm node scaling is the performance degradation from series resistance and stress loss due to gate pitch scaling [6]. With an optimized process, <3% (N) and 0% (P) performance degradation was observed with gate pitch scaling from 252 nm to 126 nm (Fig.6). Good subthreshold characteristics have been shown for both NFET and PFET with HK-MG stack (Fig.7). With effective band-edge work-function gate stacks, appropriate Vt and short channel control was achieved (Figs. 8, 9). Most importantly, EOT scaling has enabled L_{gate} scaling down to 30nm. Compared with SiON-poly, HK-MG devices shows less DIBL at same L_{gate}. The transistor performance at 0.8V operation (LOP) is also shown in Fig.10. Without an additional gate oxide [7] to maintain process simplicity, good LOP transistor performance (Ion = 700/340 μA/μm @ 100nA/μm off-current) was achieved at V_{dd}=0.8V. HK-MG ac performance is shown in Fig.11. Compared with SiON-poly, 30% delay reduction has been demonstrated with HK-MG due to improved transistor DC performance and the L_{gate} scaling from 40nm (SiON-poly control) to 30nm (HK-MG).

Top-down SEM of a functional 0.157 μm² SRAM cell is shown in Fig.12. Device widths and lengths have been scaled from the 45nm technology to achieve a high density SRAM cell. With a conventional SiON-gate stack, Vt mismatch would increase and degrade SRAM stability. However, due to aggressive Tinv reduction in this 32nm HK-MG low power technology, the mismatch is reduced 40% compared with SiON-poly gate stack (Fig.13), significantly improving SRAM operating voltage range. The butterfly curve of the 0.157μm² dense SRAM cell is shown in Fig. 14 and a static noise margin (SNM) of 250 mV is achieved.

Potential 1/f noise and reliability degradation have been critical concerns for the HK-MG devices. Impacts of HK-MG gate stack on 1/f noise have been examined and no degradation was observed compared with poly-SiON gate stack from previous node (Fig. 15). NBTI and PBTI were assessed for LTSP transistors. Both NFETs and PFETs meet the 10 year lifetime at V_{dd}=1.2V after stress at 125 °C (Fig. 16).

Conclusion

We have successfully demonstrated a cost-effective 32nm LSTP and LOP technology platform which uses a gate-first and single-metal HK-MG process with an effective band-edge work-function solution for both NFETs and PFETs. EOT scaling with HK-MG enables superior transistor performance with aggressive gate length scaling and a dense SRAM cell with good SNM. Good noise performance and reliability of HK-MG devices was also confirmed.

This work was performed at the IBM Microelectronics Div., Semiconductor Research & Development Center, Hopewell Junction, NY 12533. The authors would like to thank J. Jagannathan, P. Gilbert, G. Patton and T-C Chen for their support.

978-1-4244-1802-2/08/$25.00 ©2008 IEEE

References:
[1] T. Hook, IEDM 2006, p.701
[2] ITRS 2005.
[3] K. Mistry, IEDM2007, P. 247
[4] H. Huang, IEDM2007, P. 285
[5] M. Chudzik, VLSI2007, P.194
[6] S. Wu, IEDM 2007, P. 263
[7] E. Josse, IEDM 2006, p.693

Trans	LSTP	LOP	I/O
Vdd (V)	1.0	0.8	1.5/1.8
Lgate (nm)	30	30	150

Table 1: Key 32nm Transistors.

Fig.1: TEM of the Low power transistor with Lgate=30nm.

Rule	Pitch (nm)	Scaling factor from 45nm
Contacted Gate Pitch	126	70%
CA pitch	100	63%
N+/P+	56 (space)	70%
1X metal	100	71%
1.3X metal	130	New level
2X metal	200	71%

Table 2: Key Ground rule of this 32nm CMOS foundry technology.

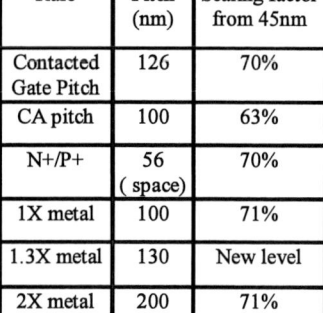

Fig.2: SEM cross-section through BEOL wiring of SRAM. First 9 layers of copper metal are shown where 7 are in ULK dielectric.

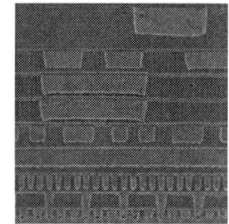

Fig.3: Gate leakage versus Tinv which shows >10nm Tinv reduction with HK-MG at same gate leakage.

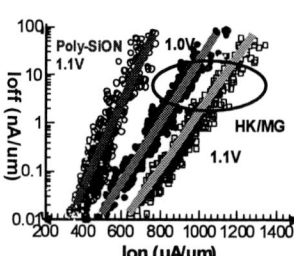

Fig.4: LSTP NFET I_{off}-I_{on} with 1000uA/um I_{on} at 1nA/um I_{off} and 1.1V V_{dd}.

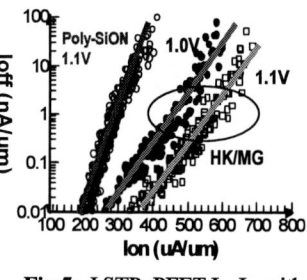

Fig.5: LSTP PFET I_{off}-I_{on} with 575uA/um I_{on} at 1nA/um I_{off} and 1.1V V_{dd}.

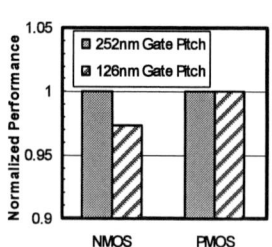

Fig.6: Transistor performance at different gate pitch.

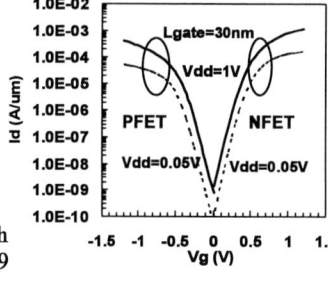

Fig.7: NFET and PFET I_d-V_g curve. SS is 90mV/dec for NFET and PFET.

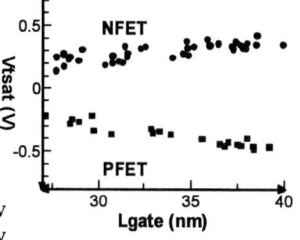

Fig.8: V_t roll-off characteristics of NFET and PFET.

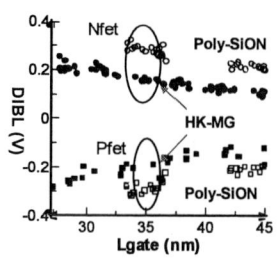

Fig.9: Comparison of DIBL between HK-MG and Poly-SiON.

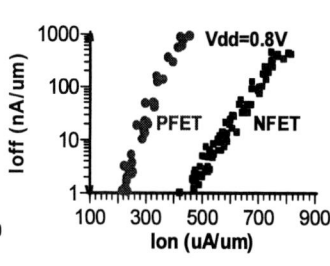

Fig.10: I_{off}-I_{on} of the LOP transistors at 0.8V.

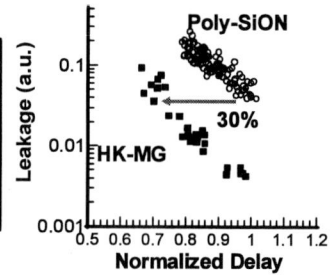

Fig.11: 30% RO delay reduction with HK-MG at same Vdd and leakage.

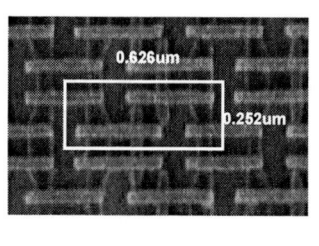

Fig.12: Top-down SEM of a 0.157 um² SRAM bit-cell.

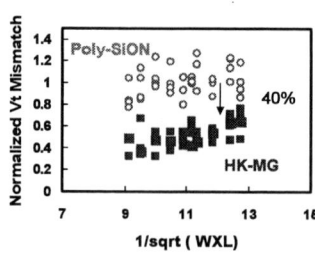

Fig.13: Vt mismatch is reduced 40% with HK-MG due to Tinv scaling.

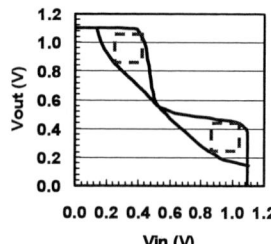

Fig.14: Measured output for the 0.157um² SRAM bit-cell with SNM 250mV.

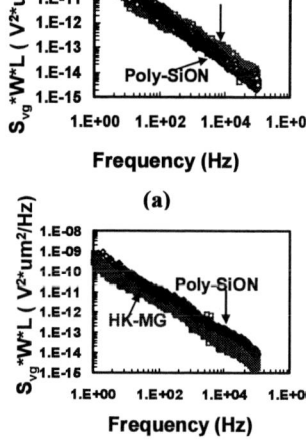

Fig.15: Normalized gate input refered voltage noise for 32nm HK-MG technology compared with Poly-SiON gate stack for NFET (a) and PFET (b).

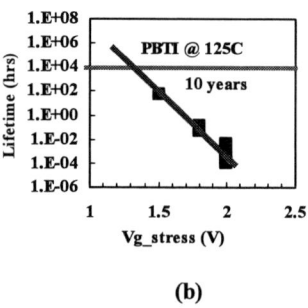

Fig.16: NBTI (a) and PBTI (b) measurements of HK-MG wafer meet 10 year lifetime requirement.

68

Variability Aware Modeling and Characterization in Standard Cell in 45 nm CMOS with Stress Enhancement Technique

H. Aikawa, E. Morifuji, T. Sanuki, T. Sawada, S. Kyoh*, A. Sakata, M. Ohta, H.Yoshimura, T. Nakayama, M. Iwai, F. Matsuoka

Advanced Logic Technology Dept. System LSI Division, Toshiba Corporation Semiconductor Company

8, Shinsugita-cho, Isogo-ku, Yokohama, 235-8522, Japan

Phone: +81-45-776-5756, FAX: +81-45-776-4111, E-mail: h.aikawa@semicon.toshiba.co.jp

*Process & Manufacturing Engineering Center, Toshiba Corporation Semiconductor Company

Abstract

Gate density is ultimately increased to 2100 kGates/mm^2 by pushing the critical design rules without increasing the circuit margin in 45 nm technology. Layout dependences for stress enhanced MOSFET including contact positioning, 2nd neighboring poly effect, and bent diffusion are accurately modeled for the first time. With the constructed design flow, gate length change of −2.8% to +3.6% and Idsat change of −10% to +14% are removed from uncertain margin in 45 nm corner libraries.

Introduction

The performance of MOSEFT has been improved with using intensive stress enhancement techniques. Fig.1 shows the obtained Idsat-Ioff plots and TEM pictures for NMOS and PMOS in 45 nm technology. High on-current is achieved thanks to dual stress liner (DSL) and eSiGe techniques[1]. On the other hand, variability caused by layout effects such as stress and rounding effects with lithography has become the critical obstacle against shrinking the design rule and corner margins. This variability has been handled both by adding this uncertainty into the corner margin and by setting the limiting design rules as shown in fig.2 (case1). If the design rules are scaled, the corner margin should be increased because of increase of uncertainty on layout effects (case2). By realizing the SPICE simulation including the compact model of layout effects as listed in table 1, this margin is removed from corner margin (case3), therefore chip size can be minimized. In this work, gate density is ultimately increased by pushing the critical design rules to limit of operation fail without increasing the circuit margin in 45 nm technology for the first time. The achieved gate density in 45 nm technology is 2100 kGates/mm^2 which is 2.6 times denser than in 65 nm technology as shown in fig.3. The scaling factor is accelerated as 2.6X from 2.0X and this strategy can be applied also in 32 nm technology.

Methodology

Table 1 indicates the layout dependent variations handled in this work. Each factor is represented in fig.2(b). The origins of these layout dependences are composed of stress, dopant, and lithography. Variability originated from stress is reflected by modulating the mobility parameter defined in SPICE for each transistor[3]. Variability with dopant effect in various layouts is reflected by shifting Vth. Rounding shape caused by lithography is fed into SPICE with modified gate length (L) and width (W) by extracting the effective L and W from contours obtained from simulation of lithography and etching(Fig.4)[2]. Fig. 5 represents the design flow with device extraction from actual layout. In this work, modeling of layout dependence on contact positioning, gate space effect adding 2nd neighboring poly, and bent shape diffusion are carried out for the first time.

Modeling of stress effect

DSL makes layout dependence complicated since not only boundary effect but also the other neighbors such as poly gate and contact modulate the stress in MOSFET as shown in Fig.6. Neighboring poly gates may relax the actual strain in channel[4]. Contact hole also relaxes the strain locally. Fig.7 shows the dependence of contact to gate distance on Idsat change in PMOSFET. It can be clearly seen that the Idsat degrades as contacts are placed closer to the gate. The impact of contact position in diffusion is studied in 3-D TCAD stress simulation. In fig.8, Idsat change is plotted as a function of contact location in y direction. It should be noted that the influence of contact is negligible when the contact is located outside of transistor width. Stress change in x direction is large, but the change in y direction is very small as can be seen in stress distribution in fig.6. Hence, contact effect only in x-direction is considered. Fig.9 indicates the Idsat change caused by 1st neighboring gates. In addition, Idsat change by 2nd neighboring gates is evaluated by comparing the transi1ors with two dummy gates and with four dummy gates shown in Fig.10. It should be noted that Idsat is influenced not only by 1st neighboring gate but also by 2nd neighboring gate. To model this higher order effect, Idsat changes of transistors with two dummy gates located in both sides are simulated by TCAD. In fig.11, Idsat changes for transistors with various gate spaces (s1 and s2) are plotted both with the distance of 1st neighboring gate and with the effective gate space defined as s1+a(s1-s2). It should be noted that all the Idsat change with various s1 and s2 can be accurately modeled by defining the effective gate space including 2nd neighboring gate effect. Finally, transistor with bent shape diffusion is also studied as it has measurable influence especially on PMOS with eSiGe. Fig.12 shows the measured impact on bent diffusion. Idsat increase caused by bent shape is clearly observed. By changing the layout parameters randomly by TCAD, it is revealed that the uniaxial stress Sxx makes the major contribution to the total variation compared to Syy as shown in Fig. 13. The model is then constructed by analyzing the influence of each layout parameters on the Sxx (Fig. 14).

Application to Cell Libraries

To evaluate variability in actual 45 nm cell libraries, effective gate lengths and Idsat changes are extracted through constructed *Calibre* design flow for hundreds of actual cells. Figs. 15 and 16 show the distribution of extracted gate length with litho effect and Idsat change with stress effect extraction, respectively. Peaks in the distribution are related to patterns used frequently in cells. The gate length change of −2.8% to +3.6% and Idsat change of −10% to +14% are observed in aggressively scaled cells. It should be noted that this variability can be completely removed from corner margin thanks to the models constructed in this work.

Conclusions

Gate density is ultimately increased to 2100 kGates/mm^2 by pushing the critical design rules without increasing the circuit margin in 45 nm technology for the first time. Layout dependences for stress enhanced 45 nm MOSFET including contact positioning, gate space effect adding 2nd neighboring poly, and bent diffusion are accurately modeled. With the constructed models and design flow, gate length change of −2.8% to +3.6% and Idsat change of −10% to +14% are removed from uncertain margin in 45 nm corner libraries.

Acknowledgment

The authors would like to acknowledge A. Shiraishi and other staffs in Mentor Graphics Japan for helpful discussions.

References

[1] H. Nii et.al., p.685, IEDM2006 [2] R. S. Fathy et.al., [6521-24], SPIE Lithography 2007 [3] H. Tsuno et.al., p.204, VLSI Symp2007 [4] A. Oishi et.al.,p.239, IEDM2005

978-1-4244-1802-2/08/$25.00 ©2008 IEEE

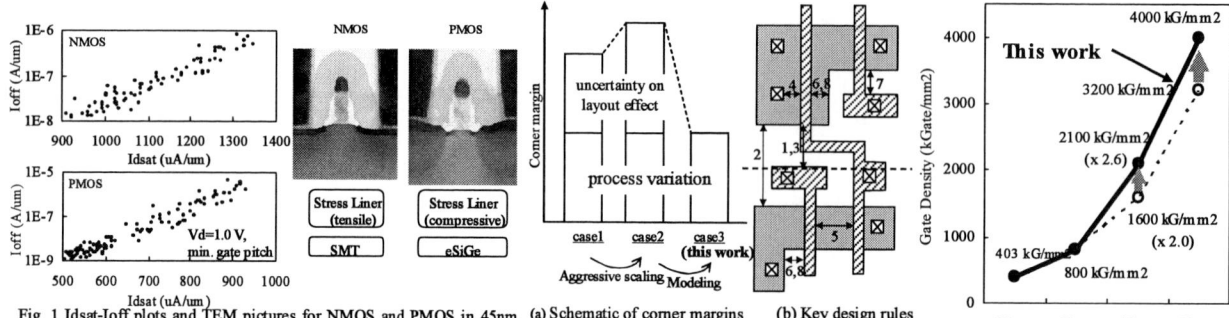

Fig. 1 Idsat-Ioff plots and TEM pictures for NMOS and PMOS in 45nm technology.

(a) Schematic of corner margins   (b) Key design rules

Fig. 2 Schematic of corner margins and key design rules on cell layout. The numbers correspond to those defined in Table 1.

Fig. 3 Gate density trend achieved in this work.

Table 1 List of layout dependences handled in this work.

	item	remarks	litho	stress	dopant
			physical origin		
1	Well proximity effect	The same method with BSIM4 model.			O
2	STI width	The same method with Ref. [3].		O	O
3	Dual stress liner (DSL)	Mobility change near the boundary of DSL is considered in both L and W directions.		O	
4	Contact position	Relaxation of DSL stress by contact is considered. Contacts located outside of gate width can be neglected. (Figs. 7, 8)		O	
5	Gate space	Mobility modulation due to neighboring gates is considered. Effective gate space is defined. (Figs. 9-11)		O	
6	Bent shape diffusion	Stress change is modeled with three parameters. (Figs. 12-14)	O	O	O
7	Gate rounding	Effective gate length and gate width are extracted from contours of lithography simulation. (Figs. 4 and 15)	O		
8	Diffusion rouding		O		

Fig. 4 Modification of transistor length (L) and width (W) in SPICE with taking lithography effect into account.

Fig. 5 Design flow for device extraction applied in this work.

(a) Stress change near contact

(b) The effect of neighboring gates

Fig. 6 DSL Stress modulation caused by contact(a) and neighboring gate(b).

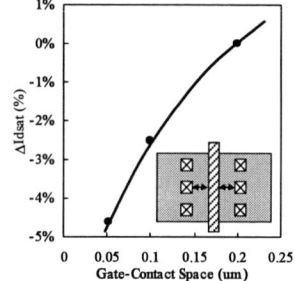

Fig. 7 Dependence of contact to gate distance on Idsat change in PMOS.

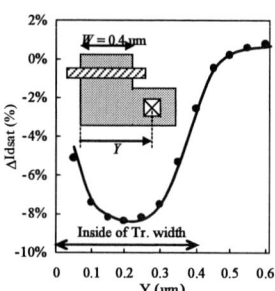

Fig. 8 Idsat change as a function of contact location in y direction.

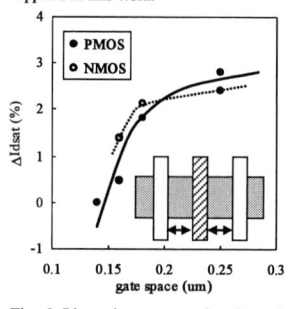

Fig. 9 Idsat change as a function of space between neighboring gates.

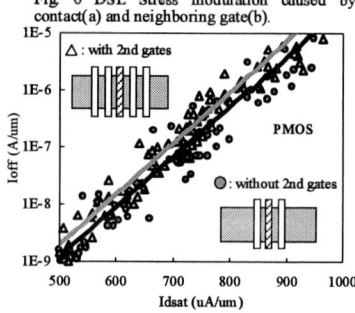

Fig. 10 Idsat-Ioff plots comparing the transistor with and without 2nd neighboring gates.

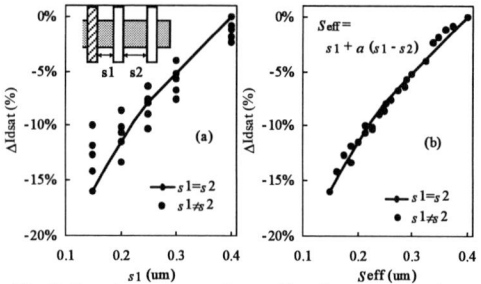

Fig. 11 Idsat changes for transistors with various gate spaces (s1 and s2) are plotted both with the distance of 1st neighboring gate and with the effective gate space defined as Seff = (s1+a(s1-s2)).

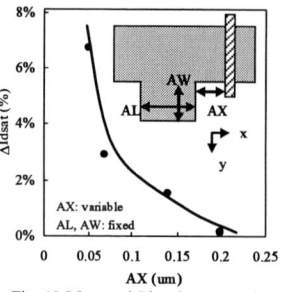

Fig. 12 Measured Idsat impact on bent diffusion. AW and AL are fixed. AX is a variable.

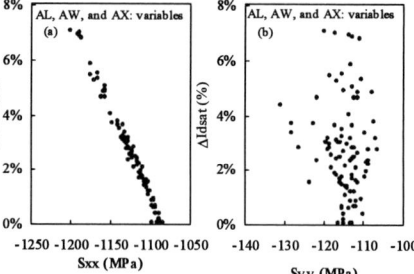

Fig. 13 Stress simulation results for layout with bent shape diffusion. AL, AW, and AX are varied as implicit parameters.

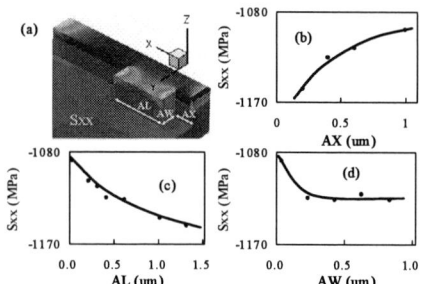

Fig. 14 Influence of each layout parameters such as AX, AL and AW on Sxx.

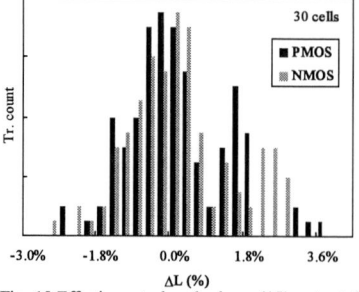

Fig. 15 Effective gate length change(ΔL) extracted from lithography contours(fig.4) for 30 standard cells.

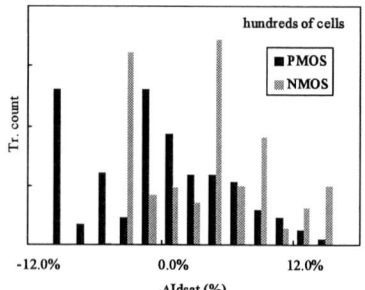

Fig. 16 Distribution of Idsat change extracted from hundreds of 45 nm cell libraries. Idsat change ranges from −10% to +14%.

A Scaled Floating Body Cell (FBC) Memory with High-k+Metal Gate on Thin-Silicon and Thin-BOX for 16-nm Technology Node and Beyond

Ibrahim Ban, Uygar E. Avci, David L. Kencke[*], and Peter L.D. Chang

Component Research, [*]Process Technology Modeling, Technology Manufacturing Group (TMG)

Intel Corporation, RA3-252, Hillsboro, OR, 97124, USA

Phone: 503-613-8442, Fax: 503-613-7997, E-mail: ibrahim.ban@intel.com

Abstract

A scaled, undoped, thin-BOX, planar FBC technology is demonstrated for the first time, featuring 10-nm BOX, 25-nm SOI, high-k, metal gate, separate back-gate (BG) doping, and raised source-drain epitaxy. Retention of a minimum 3-μA sensing window for 100 ms, in devices with 60-nm gate-length (L_g) and 70-nm diffusion width (W), represents the best retention time of all sub-100-nm FBC devices. FBC scaling is predicted to be feasible at least to 40-nm L_g, enabling memory cell sizes much smaller than 6T-SRAM at 16-nm technology node. Functional 32-nm L_g devices suggest the feasibility at the 11-nm technology node.

Introduction

FBC promises high-density memory for embedded applications. While recent studies continue to highlight improvements, demonstration of scaled FBC devices has been elusive in the sub-100-nm L_g regime [1-4]. An Independently-Controlled Double Gate (IDG) architecture on thick-BOX SOI [5], however, has enabled us to study fully-depleted FBC design and operation in the thin BG oxide regime ($t_{BGox} \leq 40$ Å.) With this capability, simulation models were calibrated in the short-L_g and thin-t_{BGox} regime, and used to develop a planar thin-BOX architecture with asymmetric front-gate (FG) and BG oxide thicknesses and dopings. Fabricated 60-nm L_g devices demonstrated low intrinsic threshold voltage (V_t) variations and good device and memory characteristics in agreement with model predictions that forecast FBC scalability to 16-nm technology node and beyond.

Experimental

To capture the scaling benefits of high-k+metal gate, 45-nm logic technology [6] was adapted for thin-BOX SOI. Certain key processing steps of the 45-nm logic technology were modified (Fig. 1) for 10-nm-BOX SOI substrates. A SiO$_2$ underlayer was deposited to target the FG oxide thickness. Well doping was eliminated in the body to minimize the random dopant fluctuation (RDF) effect, and tip and source/drain (S/D) implants were adjusted to produce low-leakage junctions. A TEM cross-section of a typical device, nominally targeted at 32-nm L_g and 25-nm T_{si}, is shown in Fig. 2.

Results and Discussion

Double-gate operation of thin-BOX devices is verified since BG bias modulates FG characteristic curves (Fig. 3). Read currents for both states "1" and "0" from a 60-nm L_g device held at zero drain bias (Fig. 4) show that the signal window remains open at 1s. After a worst-case drain-disturb operation, a dI_d (dI_d $=I_{d@state\ "1"} - I_{d@state\ "0"}$) of ~ 4 μA is maintained at 10 ms. Although this device has better retention than typical devices, their disturb characteristics are similar. Simulations suggest that

dI_d can be increased by reducing S/D and tip resistance.

Undisturbed retention longer than 100 ms was measured in a wide, short 32-nm L_g device, and demonstrates the possibility of scaling the technology (Fig. 5). To our knowledge, this is the shortest FBC device reported to date.

Physical mechanisms leading to loss of states have been investigated through transient simulations (Fig. 6). Shockley-Read-Hall (SRH) recombination at the source end is the cause of hole loss for state "1" during disturb. During state "0" disturb, band-to-band tunneling (BTBT) at the drain end supplies excess holes into the body.

Standard SRH, BTBT, and impact ionization (II) models have been calibrated using experimental IDG-FBC data [5] and verified on thick-BOX and thin-BOX processes. Typical simulated retention curves show good agreement between model predictions and silicon data (Fig. 7).

The calibrated models were used to forecast FBC scaling. A 55-nm L_g device was first designed to meet target performances. The 40-nm L_g device was then defined by scaling all dimensions by ~0.7X to maintain a constant maximum electric field in the channel. Bias conditions have been optimized for each L_g to balance retention and programming time. Short programming time (~1-ns) and long ~100-ms worst-case disturbed retention is achieved with a 3-μA dI_d sense window for ~1-ns read (Fig. 8). Assuming that the tail bit retention is ~100X worse than the median and is degraded by ~10X compared to that at room temperature (Fig. 9), the 40-nm L_g device can support a 100 μs refresh time for embedded applications.

RDF also needs to be addressed for scalability. The intrinsic V_t variations measured on paired FBC devices (Fig. 10) show that the 1-σ value is less than 15 mV, a sufficient intrinsic margin for large array design.

Published FBC data to date are benchmarked (Fig. 11). Given the limited data in the sub-100-nm L_g regime, this work represents the smallest FBC devices with the highest dI_d at the lowest operating voltages. Moreover, simulation suggests that FBC is scalable at least down to 40-nm L_g while achieving target performance.

At 16-nm technology node, assuming 40-nm L_g, 60-nm space between gates and 60-nm metal pitch, the cell size can be 0.012 μm^2 or 0.006 μm^2 for folded bit line and open bit line architectures, respectively. The cell size is much smaller than projected 6T-SRAM cell at the same technology node. Even if a 60-nm L_g is used, cell size is only 0.0072 μm^2, still significantly smaller than 6T-SRAM. If L_g scales to 32 nm, the FBC cell size advantage will remain down to 11-nm technology node.

References

[1] F. Matsuoka et al., *IEDM, 2007*, [2] S. Okhonin et.al., *IEDM, 2007*, [3] C. W. Oh, et al., VLSI Tech., 2007, [4] R. Ranica et al., IEDM, 2004, [5] I. Ban et al, *IEDM, 2006*, [6] K. Mistry et al., *IEDM, 2007*.

978-1-4244-1802-2/08/$25.00 ©2008 IEEE

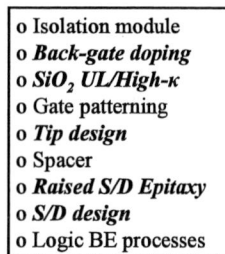

- o Isolation module
- o ***Back-gate doping***
- o ***SiO_2 UL/High-κ***
- o Gate patterning
- o ***Tip design***
- o Spacer
- o ***Raised S/D Epitaxy***
- o ***S/D design***
- o Logic BE processes

Fig. 1 Thin-BOX FBC process flow chart. The process modules developed for thin-BOX FBC are highlighted in *italic*. Back End (BE) processes are identical to Intel's 45-nm Logic Technology.

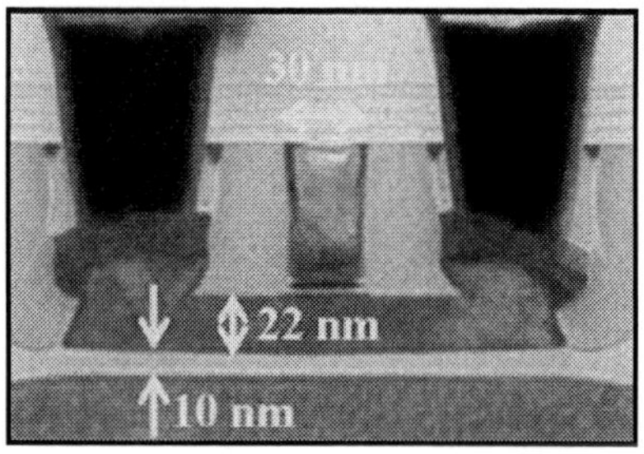

Fig. 2 A typical TEM micrograph of a 10-nm BOX FBC device, nominally targeted at L_g=32 nm and T_{si}=25 nm.

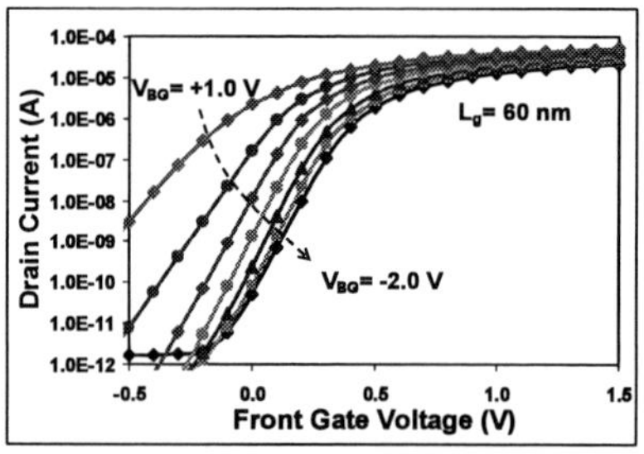

Fig. 3 BG bias effect at 0.5V increments. The increased threshold voltages (Vt) and the improved subthreshold slopes are the verifications of double-gate operation. Low BG bias operation becomes possible with thin-BOX and BG doping.

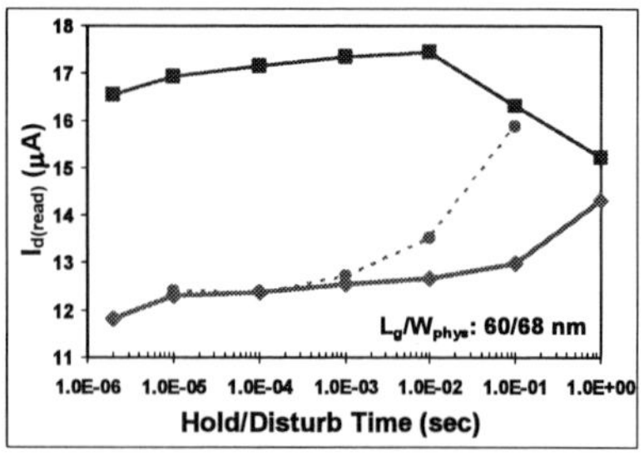

Fig. 4 Retention characteristics of a device with L_g=60 nm. State "0" is the only state impacted by the disturb (dashed line).

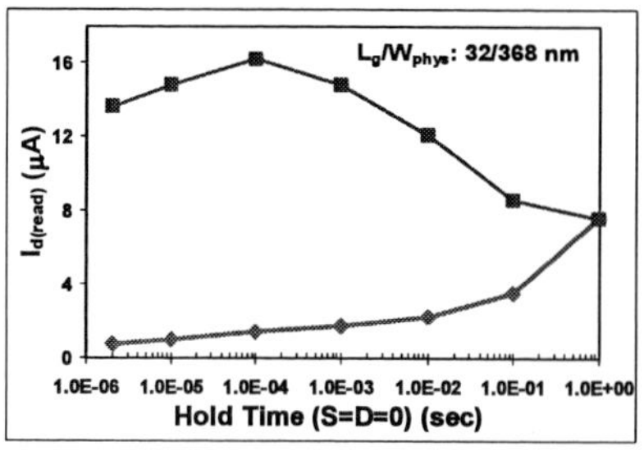

Fig. 5 Retention characteristics of a device with L_g=32 nm.

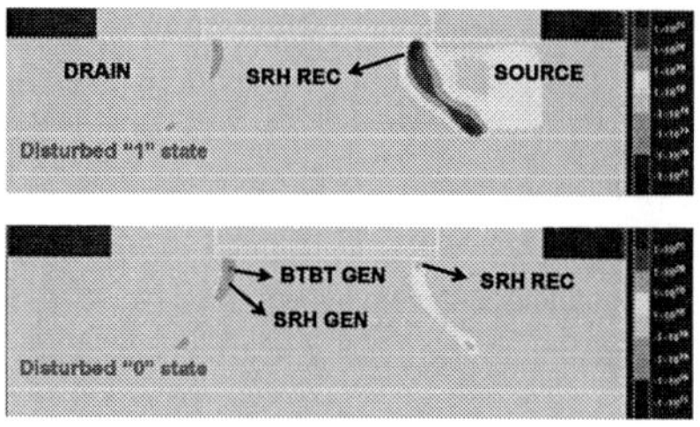

Fig. 6 Simulated disturb mechanisms in state "1" (top) and state "0" (bottom).

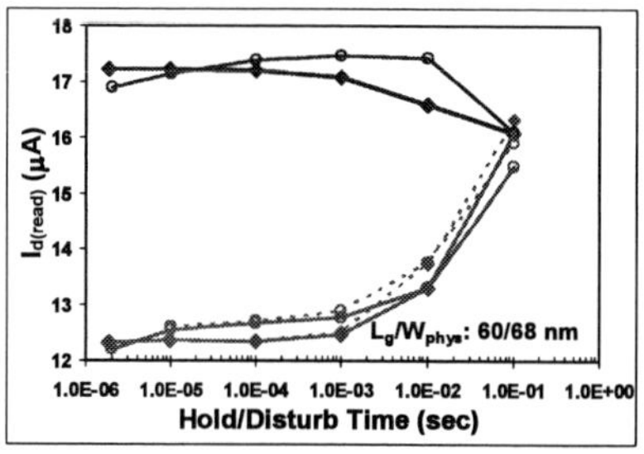

Fig. 7 Simulation curves calibrated to an average of eight representative devices across the wafer. Filled symbols are experimental, empty symbols are simulation data. Dashed lines are the disturbed data points from experiments and simulations.

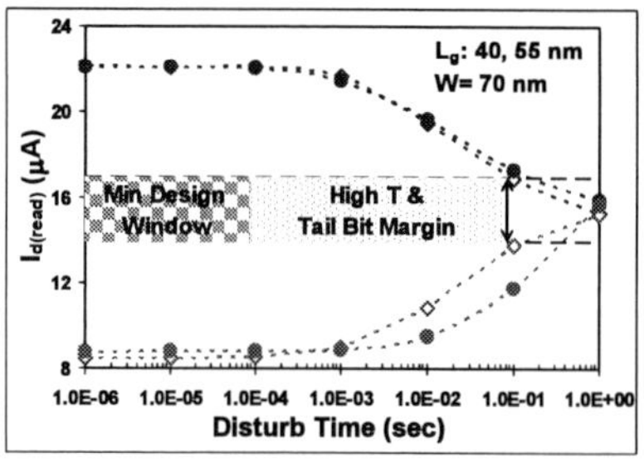

Fig. 8 Simulated target devices with disturb conditions at L_g=55 nm (●), and L_g=40 nm (◊). ~100X margin for tail bits and ~10X margin for high temperature operation are assumed to target ~100 µsec refresh time.

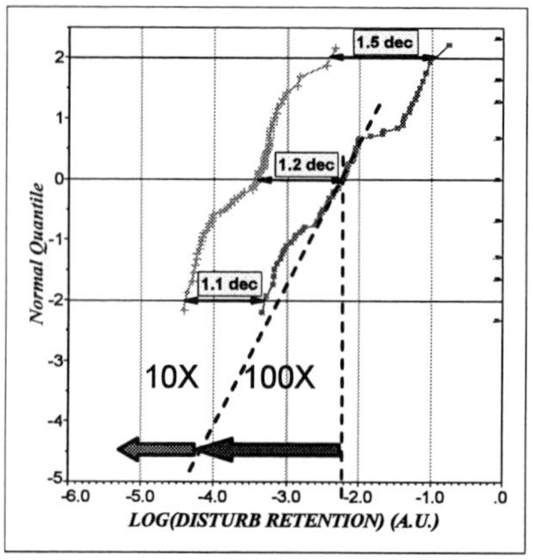

Fig. 9 Disturbed-retention distributions measured at 23 °C and 95 °C on 48-nm Lg devices. 4.5σ tail bits are ~100X off the median and their high temperature retention degraded by ~10X compared to room temperature.

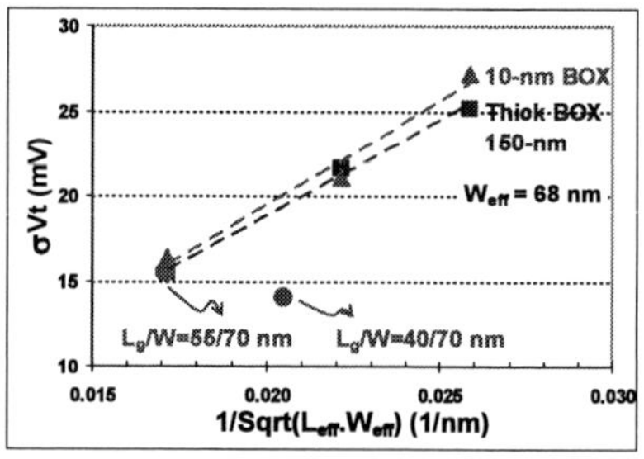

Fig. 10 σVt (mV) measured from pairs of matched FBC devices are shown for both 150-nm (■) and 10-nm BOX (▲) processes. σVt values for L_g=55 nm and L_g=40 nm target devices were calculated based on their respective FG oxide thicknesses and are shown with filled symbols (●).

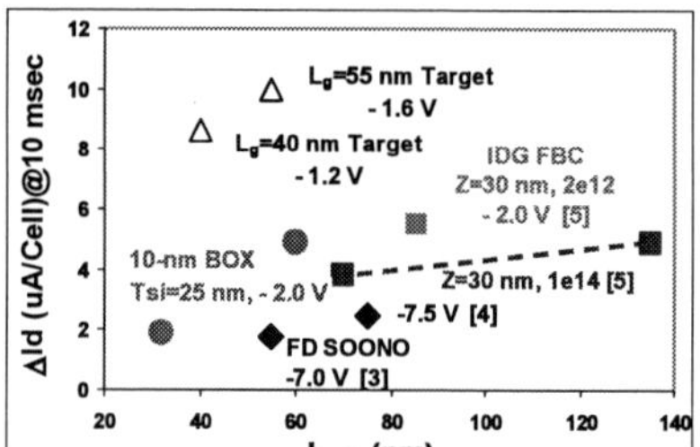

Fig. 11 Benchmarked experimental data to date on FBC devices. Y-axis represents read current difference between state "1" and state "0" at 10 msec normalized per cell (W=70 nm). Back-gate biases in published retention measurements are also given. Simulation data points for target devices are shown with empty triangles. Note that dimensions used in target device simulations are the scaled values.

On the Dynamic Resistance and Reliability of Phase Change Memory

B. Rajendran[†], M-H. Lee[°], M. Breitwisch[†], G. W. Burr[⊕], Y-H. Shih[°], R. Cheek[†], A. Schrott[†], C-F. Chen[°],
M. Lamorey[‡], E. Joseph[†], Y. Zhu[†], R. Dasaka[†], P. L. Flaitz[□], F. H. Baumann[□], H-L. Lung[°] and C. Lam[†]

IBM Macronix PCRAM Joint Project

[†]IBM T.J. Watson Research Center, 1101 Kitchawan Road, Yorktown Heights, NY, 10598, USA
([°]Macronix International Co., Ltd., [‡]IBM Essex Junction, [⊕]IBM Almaden Research Center, [□]IBM Hopewell Junction)
Tel: +1-914-945-1809, email: brajend@us.ibm.com

Abstract

A novel characterization metric for phase change memory based on the measured cell resistance <u>during</u> RESET programming is introduced. We show that this 'dynamic resistance' (R_d) is inversely related to the programming current (I), as $R_d = [A/I] + B$. While the slope parameter A depends only on the intrinsic properties of the phase change material, the intercept B also depends on the effective physical dimensions of the memory element. We demonstrate that these two parameters provide characterization and insight into the degradation mechanisms of memory cells during operation.

Keywords: PCRAM, NV memory and chalcogenide

Introduction

As phase change memory (PCM) moves from technology feasibility into technology qualification, one of the significant remaining challenges to commercialization is the lack of accurate methods for characterizing reliability-degradation mechanisms. In this paper, we introduce a novel non-destructive characterization methodology for PCM, allowing a deeper physical understanding of such degradation mechanisms.

Measurement Methodology

Each memory cell characterized in this study consisted of a mushroom phase change element (PCE) using $Ge_2Sb_2Te_5$ (GST) with TiN electrodes, in series with an nMOSFET (Fig. 1) based on 180nm CMOS technology. The 'dynamic resistance' is measured (Fig. 2) while programming the memory element using a current larger than the nominal RESET current of the cell ($\sim$750μA). Relatively large access transistors (W/L=4.28μm/0.18μm), biased in the active region to keep transistor resistance constant, were used (Fig. 3). To extract R_d for different programming currents, the bit-line amplitude V_{BL} was stepped from 0-1.5 V while pulsing the word-line (V_{WL}=3.5V, 20ns rise-time, 75ns pulse length). The current flowing through the cell remained constant during the measurement, indicating that steady-state conditions were achieved within the cell (Fig. 4).

Dynamic Resistance Measurement

When such a steady-state is achieved, a hemispherical molten region of radius x plugs the BEC of the memory cell (Fig. 5). An approximate expression for the resistance of the region between the electrodes at that instant is given by:

$$R = \frac{\rho_s - \rho_m}{2\pi}\left[\frac{1}{x}\right] + \frac{1}{2\pi}\left[\left(1 + \pi/2\right)\frac{\rho_m}{r_0} - \frac{\rho_s}{H}\right] \quad (1)$$

where ρ_s and ρ_m are the resistivity of the crystalline and molten GST (at that temperature), r_0 denotes the effective contact radius at the GST/Bottom Electrode Contact (BEC) interface, and H denotes the effective thickness of the GST above the BEC interface.

Electro-thermal simulations (Fig. 6) indicate that this radius x increases linearly with current as $x = cI$, with a constant of proportionality, c, independent of the cell dimensions. This implies that (1) can be written in the form:

$$R_d = \frac{\rho_s - \rho_m}{2\pi c}\left[\frac{1}{I}\right] + \frac{1}{2\pi}\left[\left(1 + \pi/2\right)\frac{\rho_m}{r_0} - \frac{\rho_s}{H}\right] \equiv A\left[\frac{1}{I}\right] + B \quad (2)$$

While the slope A depends only on the intrinsic material properties of the cell, the intercept B also depends on the physical structure of the cell. Despite the simplistic assumptions inherent in this model, the measured R_d obeys the trend predicted by (2) quite closely (Fig. 7). Given the linear dependence of R_d over a wide range of $1/I$, the A and B parameters can be uniquely and reliably determined, from even a handful of measurements near the nominal RESET condition.

Reliability characterization

It is well-known that the endurance of PCM depends strongly on the specific programming conditions [1]. In order to ensure that our cells were strongly RESET, the cycling conditions were chosen so that the applied current ($I_{applied}$=2 mA) for the RESET operation was significantly larger than the nominal RESET current. We find that the minimum RESET current (Fig. 8) and the programmed resistance levels (Fig. 9) drift with cycling, suggesting a change in either the material or physical properties of the critical volume of the memory cell. We can use the dynamic resistance measurement to understand the physical mechanisms behind these observations. Based on R_d measurement, A and B parameters were extracted for a collection of cells that were cycled between 1 and $\sim$10 million cycles (Figs. 10, 11). The parameter A is stable up to 3E7 cycles, suggesting that the resistivities of the molten and crystalline GST are not changing. Thus the observed decrease in B with cycling must be due to either an effective increase in r_0 or a decrease in H.

Energy Dispersive X-ray analysis of the active volume of memory cells indicates that Sb atoms begin to agglomerate near the BEC/GST interface with cycling (Fig. 12), resulting in a dome-like metallic region at the interface. This effectively increases the area of contact between the BEC and GST (as if the BEC were 'protruding' into the GST), explaining the steady decrease in SET resistance with cycling. To first order, this does not affect the resistivity of the critical volume at melting temperatures, keeping A invariant. Simulations confirm that the presence of a metallic dome at the GST/BEC interface would result in an increase in the RESET current (Fig. 13). A lower R_d also implies a faster quench time, leading to higher RESET resistance levels with cycling.

Conclusions

A novel non-destructive characterization method based on the dynamic resistance measurement for PCM has been introduced. The extracted parameters from this measurement provide a method to characterize the intrinsic properties of the memory cell and help to develop insight into the physical mechanisms for cell degradation, which is crucial for the commercialization of PCM.

Acknowledgments

Expert processing support from the Microelectronic Research Line at IBM Watson Research Center and valuable discussions with W. Gallagher and R. Liu are gratefully acknowledged.

References

[1] S. Lai, *IEDM Tech. Digest*, 2003.

978-1-4244-1802-2/08/$25.00 ©2008 IEEE

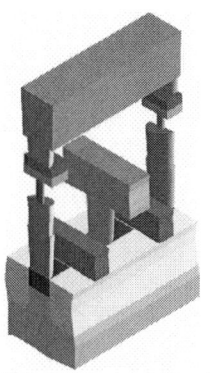

Fig. 1: Two neighboring mushroom-type PCM devices.

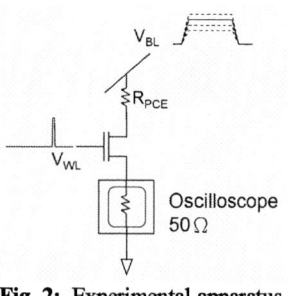

Fig. 2: Experimental apparatus for R_d measurements. The bit-line voltage controls the pulse amplitude while the word-line controls the pulse timing.

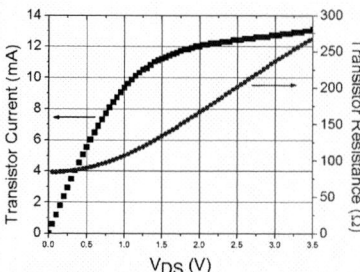

Fig. 3: FET device characteristics. By using a large transistor (W/L=4.28/0.18), we ensure that the transistor resistance is relatively constant across the range of measurement currents (0-3mA).

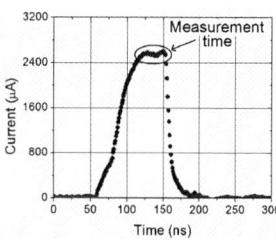

Fig. 4: Variation of measured current with time, implying that the memory cell reaches and holds a steady-state temperature distribution before the cell is quenched.

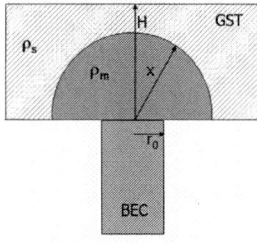

Fig. 5: During an R_d measurement, the molten portion of the GST plugging the Bottom Electrode Contact (BEC) is a hemisphere of radius x.

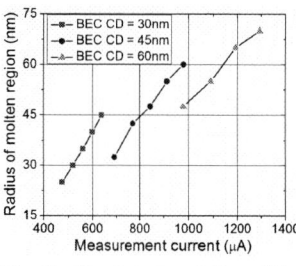

Fig. 6: Simulated variation of the radius x of the molten region, as a function of the programming current. The slope of the lines is independent of the diameter of the BEC.

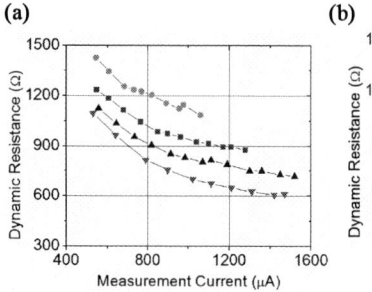

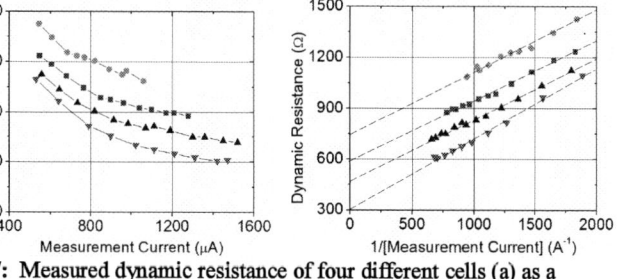

Fig. 7: Measured dynamic resistance of four different cells (a) as a function of the measurement current, and (b) as a function of the inverse of the current. The linear relationship between R_d and $1/I$ allows a simple extraction of the slope and intercept ($R_d = [A/I] + B$).

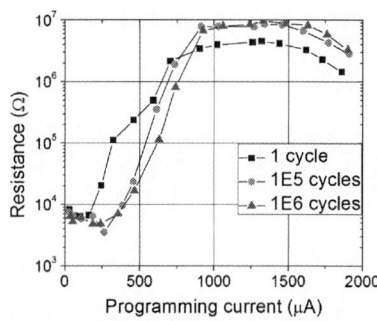

Fig. 8: Shift in the minimum RESET current with cycling.

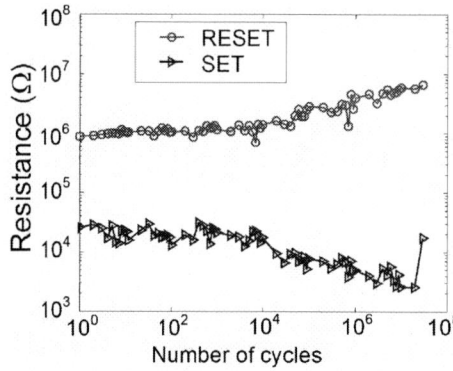

Fig. 9: Shift in resistance levels with cycling.

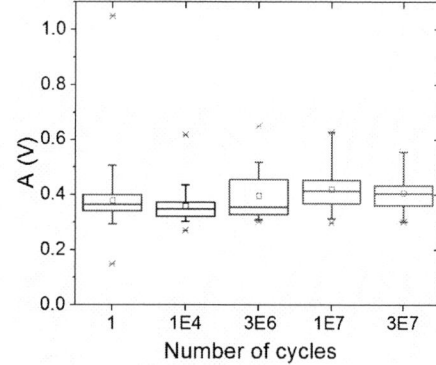

Fig. 10: Variation in parameter A (which has units of volts) with cycling.

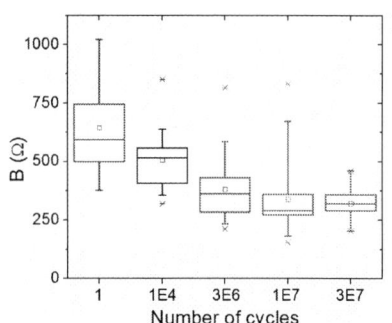

Fig. 11: Variation in parameter B (in units of Ω) with cycling. Constancy in A with cycling implies that the decrease in B is due to an effective increase in r_0 or a decrease in H.

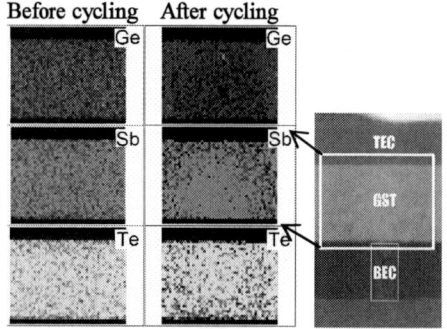

Fig. 12: Results of EDX elemental analysis showing an agglomeration of Sb atoms at the GST/BEC interface after cycling.

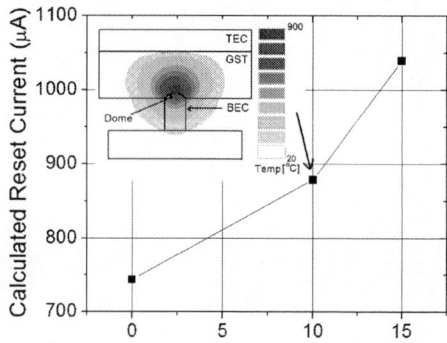

Fig. 13: Simulated RESET current as a function of the thickness of a dome-like metallic region at the BEC-GST interface.

Two-bit Cell Operation in Diode-Switch Phase Change Memory Cells with 90nm Technology

D.-H. Kang, J.-H. Lee, J.H. Kong, D. Ha, J. Yu, C.Y. Um, J.H. Park, F. Yeung, J.H. Kim, W.I. Park, Y.J. Jeon, M.K. Lee, J.H. Park, Y.J. Song, J.H. Oh, G.T. Jeong, and H.S. Jeong

Advanced Technology Development Team 2, Memory R&D Div., Samsung Electronics Co., Ltd.
San #24, Nongseo-Dong, Giheung-Gu, Yongin-City, Gyunggi-Do, 449-711, South Korea
Phone: +82-31-209-9140, Fax: +82-31-209-3274, E-mail: daehwan.kang@samsung.com

Abstract

This paper firstly reports key factors which are to be necessarily considered for the successful two-bit (four-level) cell operation in a phase-change random access memory (PRAM). They are 1) the write-and-verify (WAV) writing of four-level resistance states and 2) the moderate-quenched (MQ) writing of intermediate resistance levels, 3) the optimization of temporal resistance increase (so-called resistance drift) and 4) of resistance increase after thermal annealing. With taking into account of them, we realized a two-bit cell operation in diode-switch phase change memory cells with 90nm technology. All of four resistance levels are highly write endurable and immune to write disturbance above 10^8 cycles, respectively. In addition, they are non-destructively readable above 10^7 read pulses at 100ns and 1uA.

Introduction

Phase-change random access memory (PRAM) is most promising to realize a multi-level cell (MLC) operation because it has very wide range of resistance across two orders of magnitude or the higher, with respect to writing current. According to the PRAM road map [1], it is expected that highest memory densities of PRAM become comparable to conventional memories such as NOR Flash and DRAM in coming years when MLC operation is fully accomplished. In this paper, we systematically investigated a four-level (two-bit) cell operation in diode-switch phase change memory cells with 90nm technology and discussed its possibilities and issues as well.

Write-and-Verify and Moderate-Quenched Writings of Four Resistance Levels

Two-bit cell operation needs a delicate control of resistance distribution from the initial writing step. For this, we introduced two writing techniques, that is 1) the write-and-verify (WAV) writing and 2) the moderate-quenched (MQ) writing.

The WAV writing is to program data repeatedly to have a target value which has been widely used in Flash memory even though it has a disadvantage of time-consuming. Figure 1 clearly shows the well-defined initial distributions of four resistance levels [(00), (01), (10), (11)] by using this WAV writing technique. The poor resistance distributions by normal writing are also shown, for comparison.

The MQ writing is necessary to obtain stable intermediate resistances of (01) and (10) states. A conventional current-modulated writing method is inadequate since the resistance change with respect to writing current is too abrupt to have the intermediate resistances as shown in R-I curve of Fig. 2(a). However, the MQ writing, which is modified from a known slow-quenched set writing [2], is to melt and moderately quench a Ge-Sb-Te material to have a targeted intermediate resistance value by modulating the falling width of fixed writing current, as seen in Fig. 2(b).

Optimization of Temporal and Thermal-Annealed Resistance Increases

There are two intrinsic resistance-increment phenomena in a Ge-Sb-Te memory cell with respect to an elapsed time and/or thermal history after writing, which should be optimized or minimized for two-bit cell operation because these undesirable increase esp. of intermediate (01) and (10) resistances may overlap neighboring levels not to differentiate four resistance levels from one another. One is the temporal resistance increase known as a resistance drift and the other is the resistance increase after thermal annealing or history [3].

Figure 3 shows the temporal increases of several resistance levels at room temperature during an elapse of 10^3 sec, where it is known that all of them follow an empirical equation of $R \sim R_i t^d$ (R_i is the initial resistance right after writing). Here, the drift coefficient d is widely varied from 0.004 to 0.108 depending on the resistance level and

device temperature [Fig. 4]. Due to this, the resistances of four levels are significantly increased and the distributions become broader than initial one, as an example of Fig. 5(a) up to an elapse of 7.2×10^5 sec. This temporal resistance increase becomes considerably slowdown after 10^5 sec, as replotted by the elapsed time in Fig. 5(b).

If the memory cell undergoes a thermal annealing or history after writing, an additional resistance increment is observed with the change of E_a (the activation energy for electrical conduction) and it may make worse the resistance distribution. Figure 6 represents four resistance levels experienced by thermal annealing at 130°C for 12 hr, vividly showing that the resistances of (10) and (11) levels are approximately two-times higher than those of 1.4×10^6 sec elapse only. Figure 7 plots resistance change of four levels with respect to thermal history. The resistance R_0 at T_0 before thermal annealing is changed to $R_0^{'}$ after thermal history at T_1, which can be given by

$$ R_0^{'} = R_0 \exp\left[\left(\frac{1}{kT_0} - \frac{1}{kT_1}\right)(E_a^{'} - E_a) \right] $$

Here, E_a and $E_a^{'}$ are activation energies for electrical conduction before and after thermal history. Table I summaries activation energy values of four levels before and after 85°C thermal history. These resistance-increments phenomena can be further suppressed by the structural and compositional modification of phase change memory cell or by a new reading scheme using the reference cell with same thermal history to cells of each level.

Reliability Characteristics of Two-Bit Cells

Cell reliability characteristics of four resistance levels are also investigated in terms of data retention, write endurance, write disturbance, and read endurance. Firstly, as already known in Fig. 6, the crystallization-induced resistance decrease is not observed in (11) and (10) levels under 130°C and 12 hr bake, which guarantees data retention of 85°C and 10 yr with the crystallization activation energy of 2.5eV. However, the (01) resistance is somewhat decreased, which may be attributed to the partial crystallization of amorphous phase in (01) and is to be optimized. Secondly, cells in four levels are highly write endurable and immune to write disturbance of neighboring cell above 10^8 cycles at 85°C [Figs. 8 and 9]. It should be commented in Fig. 8 that the resistance of (01) or (10) level is somewhat fluctuated with writing cycles, which is probably due to the variation in the relative fraction of amorphous and crystalline phases in intermediate levels. Finally, they are non-destructively readable up to the cumulated time of 1 sec (equivalent to 10^7 read pulses of 100ns) at 1uA and 85°C [Fig. 10].

Conclusion

Two-bit (four-resistance-level) cell operation was reported in diode-switch phase change memory cells with 90nm technology. Deliberate and time-consuming writing techniques such as write-and-verify and moderate-quenched writings are very effective to have discrete and stable four resistance levels at an initial writing step. In addition, the optimization of intrinsic resistance-increment phenomena gives distinct four levels even after 1.4×10^6 sec elapse and 130°C bake.

References

[1] J. H. Oh et al., IEDM Tech. Dig., pp.2.6.1-2.6.4 (2006).
[2] K.-J. Lee et al. ISSCC Dig. of Tech. Papers, p. 472 (2007).
[3] A. Pirovano et al., IEEE Trans. on Elect. Dev., 51(5), 714 (2004).

978-1-4244-1802-2/08/$25.00 ©2008 IEEE

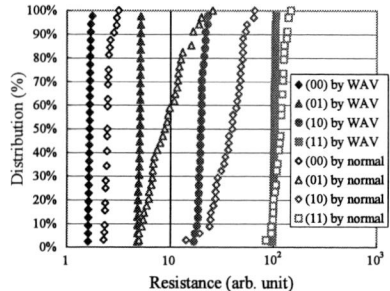

Figure 1. Initial distributions of four resistance levels by write-and-verify (WAV) writing (solid symbols) and normal writing (open symbols) at room temperature.

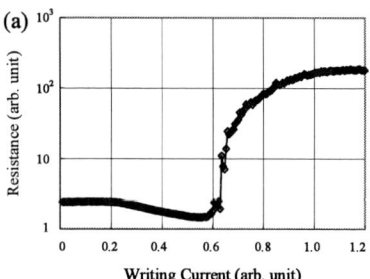

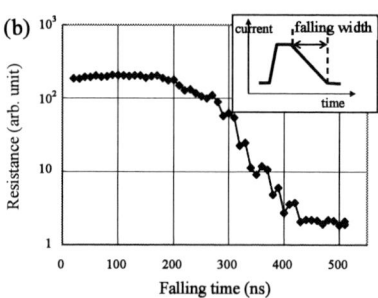

Figure 2. (a) Resistance-current curve by a conventional current-modulated writing and (b) Resistance-falling width curve by a moderate-quenched (MQ) writing.

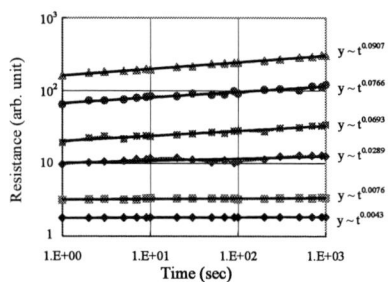

Figure 3. Temporal increases of several resistance levels at room temperature right after writing up to 10^3 sec.

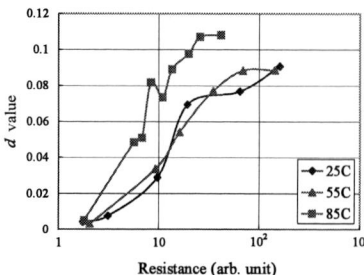

Figure 4. Drift coefficients as functions of resistance and temperature.

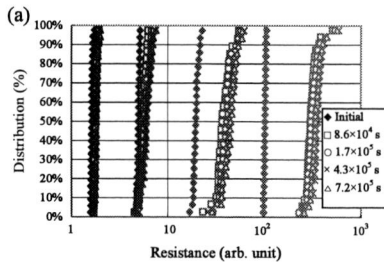

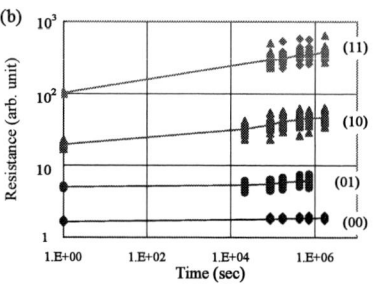

Figure 5. (a) Resistance distributions four resistance levels of (00), (01), (10), and (11) up to an elapse of 7.2×10^5 sec at room temperature and (b) Resistance change of four levels with respect to the elapsed time.

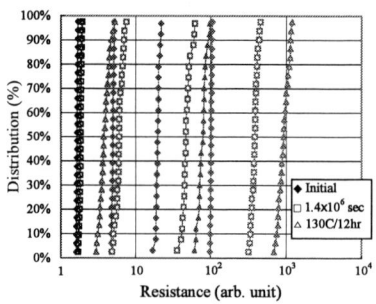

Figure 6. Resistance distributions four resistance levels of (00), (01), (10), and (11) after an elapse of 400 hr (1.4×10^6 sec) and an additional thermal annealing at 130°C for 12 hr bake.

Table I. Activation energies for electrical conduction before and after thermal history at 85°C in four levels.

	(00)	(01)	(10)	(11)
E_a (eV)	0.018	0.080	0.132	0.165
E_a' (eV)	0.023	0.041	0.227	0.290

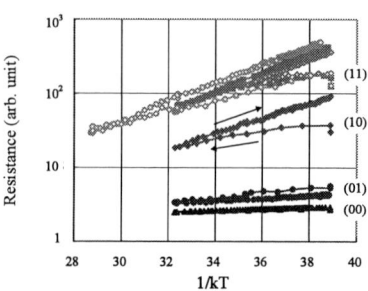

Figure 7. Resistance change of four levels level according to thermal history (i.e., the temperature is increased from room temperature to 55°C, 85°C, or 130°C and then is decreased to room temperature).

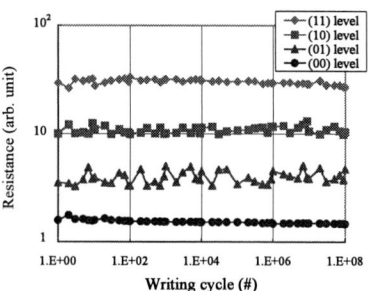

Figure 8. Write endurance characteristics of four resistance levels at 85°C.

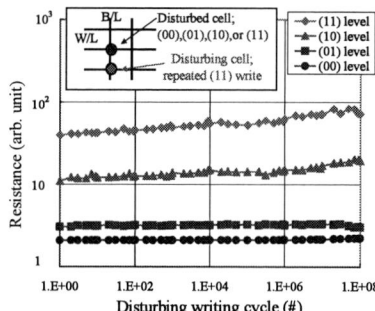

Figure 9. Resistance change of four levels with respect to the disturbing write cycles of neighboring cell at 85°C.

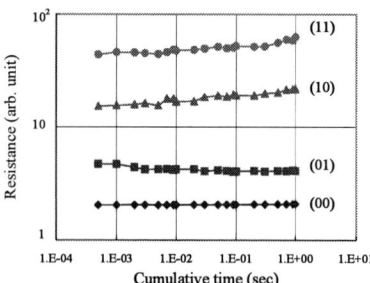

Figure 10. Resistance change of four levels as a function of the cumulative time of read current of 1uA.

A Unified Physical Model of Switching Behavior in Oxide-Based RRAM

N. Xu, B. Gao, L.F. Liu*, Bing Sun, X.Y. Liu, R.Q. Han, J.F. Kang**, and B. Yu†

Institute of Microelectronics, Peking University, Beijing 100871, P.R. China;
*E-mail: lfliu@ime.pku.edu.cn; **E-mail: kangjf@pku.edu.cn
†NASA Ames Research Center, Moffett Field, CA 94035, USA

Abstract

Excellent bipolar resistive switching (RS) behavior was achieved in TiN/ZnO/Pt resistive random access memory (RRAM) devices. A unified physical model based on electrons hopping transport among oxygen vacancies along the conductive filaments (CFs) is proposed to elucidate the RS behavior in the RRAM devices. In the unified physical model, a new reset mechanism due to the depletion of electrons in oxygen vacancies and the recovery of electron-depleted oxygen vacancies (V_O^+) with non-lattice oxygen ions (O^{2-}) is proposed and identified.

Introduction

Resistive switching in transition-metal oxides such as NiO_x[1], TiO_2[2], WO_x[3] has attracted extensive attention due to its potential application in future universal memory technologies. Recently, significantly improved resistive switching characteristics have been achieved by introducing new materials and cell structures [1-3]. However, understanding transport and switching mechanisms is still a great challenge. The formation and rupture of conductive filaments (CFs) in oxide layer has been adopted to elucidate resistive switching behavior by many researches but the switching mechanism is still unclear [4,5]. In this paper, TiN/ZnO/Pt RRAM devices were fabricated and excellent bipolar RS characteristics is demonstrated. A unified physical model is proposed to elucidate the RS behavior in the RRAM devices, where new understanding on the set, reset and switching failure behaviors are discussed.

Experiments

ZnO based RRAM devices were fabricated. About 30nm ZnO films were deposited on $Pt/Ti/SiO_2/Si$ substrates by reactive sputtering followed by a $450^{\circ}C$ furnace annealing in O_2/N_2 mixture ambient for 20min. After top electrodes (TiN, Ti, W) were deposited at room temperature, devices were patterned by the traditional lithography technique to form isolated square-shape memory cells, with the size varied from $10 \times 10 \mu m^2$ to $200 \times 200 \mu m^2$. Electrical measurements were performed using Agilent4156C and Agilent4284 at different substrate temperatures to evaluate the switching characteristics of the RRAM devices.

Resistive Switching Characteristics

Devices with different top electrodes were fabricated. Only the devices with TiN top electrode (TiN devices) show high device yield and reproducible RS behavior. The bipolar RS characteristics of the TiN devices are shown in Figs.1-3. The measurement of the Electrical Pulse Induced Resistance Switching (EPIRS) behavior indicated that the $50 \times 50 \mu m^2$ devices can be set and reset using a 4V/<20ns pulse and a -4V/60ns pulse, respectively. These characteristics illustrate that TiN/ZnO/Pt stack is a promising candidate for emerging high-performance nonvolatile data storage applications.

Fig. 4 shows the cell-area dependence of the resistance values in HRS (R_{HRS}) and in LRS (R_{LRS}). Fig. 5 shows the relation of R_{LRS} to the current-compliance for the last set process (I_{COMP}). The weak cell-area dependence of R_{LRS} compared to R_{HRS} and the I_{COMP}-dependent R_{LRS} values match the CF mechanism [5]. A reversible multi-level resistive switching behavior from HRS to LRS (shown in Fig.6 and the insert) was observed in the TiN devices when a 1V durable voltage stress was applied on the TiN-TE. This multi-level resistive switching is very close to a dielectric soft-breakdown phenomenon, suggesting the set process is equivalent to a soft-breakdown under a low durable voltage stress. Fig. 7 shows the relaxation current as a function of stress time under a stress voltage lower than set voltage, which can be fitted well by the Pillai model [6], indicating the polarization effect caused by ions migration in dielectrics occurs.

The conduction mechanism in LRS was investigated based on the temperature dependence of the DC conductance (Fig.8) and the frequency dependence of the AC conductance (Fig.9). The decreased DC resistance in LRS with increased temperature and the well-fitted frequency response of AC conductance by the Mott's formula suggest that the conduction transport in LRS is electron hopping through localized oxygen vacancies [7]. The XPS O1s core levels spectra of ZnO film is shown in Fig.10, indicating that non-lattice oxygen ions (O^{2-}) exist in the ZnO film surface [8]. These O^{2-} can be movable when the electrical field at interface is sufficient high (10^7 V/cm) [9].

Physical Model and Prediction

Based on the above observations, a unified physical model is proposed to explain the conduction in LRS/HRS and the switching between LRS/HRS as shown in Fig. 11. In the unified physical model, 1) the conduction of LRS and HRS is due to electron hopping transport among localized oxygen vacancies (V_O^+) in the CFs; 2) the switching between LRS and HRS is due to the formation and rupture of the CFs; 3) SET process is similar as dielectric soft breakdown which generates and move oxygen vacancies to form CFs, like a percolation effect; 4) RESET is due to the depletion of electrons in some V_O^+ along CFs @ V_{RESET} and the recovery of the electron-depleted V_O^+ with O^{2-}; 5) Switching Failure between LRS/HRS is due to insufficient non-lattice oxygen ions (O^{2-}) to recover the electron-depleted V_O^+ after multi-cycles reset process. The transport and switching characteristics can be simulated by using the flow as shown in Fig. 12.

Fig. 13 shows the calculated electrons occupation rates in V_O^+ along CFs under various applied voltage based on the electron hopping transport among V_O^+. Electrons depletion in V_O^+ (defined as the occupation rate reaches a low level) near the cathode can be observed when the applied voltage reaches a critical value (V_{RESET}). Since the electron-depleted V_O^+ (positive charged) can significantly increase its capture section to O^{2-} (negative charged), the recover probability of V_O^+ with O^{2-} significantly increases. This supports the new proposed reset mechanism.

Fig.14 shows the simulated reset behaviors of the RRAM devices with a single CF and multiple (10) CFs. The sharp reset process is observed in the single CF device. The simulated reset behavior was confirmed by the measured I-V curves as shown in Fig.15. Since the scaled cell size can cause the reduction of CF numbers, the scaled RRAM devices can achieve better reset characteristics.

Fig.16 shows the measured I-V curves of the device when the resistive switching failure occurs. The simulated I-V curves are shown in the insert, assuming the switching failure is due to insufficient non-lattice oxygen ions to recover the electron-depleted oxygen vacancies under a reset voltage. The agreement between measured and simulated I-V curves supports the proposed switching failure mechanism.

Based on the switching failure mechanism, we can deduce that increasing the non-lattice oxygen ions density or the storing ability of non-lattice oxygen ions will improve the switching cycle endurance. This deduction was partly identified by excellent switching characteristics of the TiN devices compared to other devices since TiN is regarded as oxygen reservoir [2]

Conclusion

A unified physical model including new understandings on the reset and resistive switching failure mechanisms is proposed based on the hopping transport among oxygen vacancies along the conductive filaments (CFs). Based on the model it can be deduced that 1) increasing the O^{2-} density and storing O^{2-} ability will benefit to improve the switching cycle endurance; 2) the scaled cell size of RRAM device will help to increase the resistance window and set/reset speed.

Acknowledgment: This work is partly supported by 973 Program (2006CB302700) and NSFC (90407023)

978-1-4244-1802-2/08/$25.00 ©2008 IEEE

Reference

[1]K. Tsunoda et al. Tech. Dig. Int. Electron Device Meet. 2007, p.767-770. [2]Masayuki Fujimoto et al. APL 89,223509(2006) [3]ChiaHua Ho et al. IEEE Symp. on VLSI Technol.2007 p228-229 [4]Rainer Waser et al. Nature materials,6. p834-840(2007) [5] D. Lee et al. Tech. Dig. Int. Electron Device Meet. 2006, p.796 [6]P. Pillai et al. European Polymer Journal,17,p.611 (1981) [7]N.F.Mott Electronic Processes in Non-crystalline Materials, Clarendon Press, Oxford 1979 [8]P.-T. Hsieh et al. Appl.Phys.A 90,317-321(2007) [9]R. Dong et al. APL, 90,182118(2007)

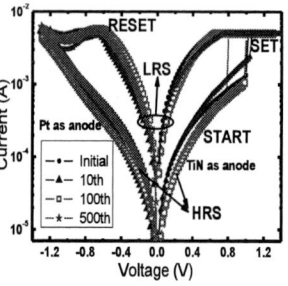

Fig.1 I-V curves of TiN/ZnO/Pt device for initial and 10th, 100th and 500th DC cycles using double voltage sweeping mode with I_{COMP}=5mA.

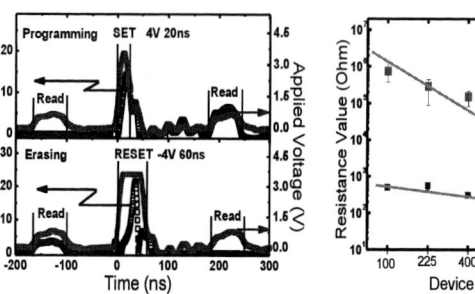

Fig. 2 Memory data retention in HRS and LRS, the current values were tested under a high durable stress (500mV) by using sampling mode

Fig. 3 EPIRS' respondent current under applied switching voltage on TE (TiN). a) 4V/20ns Set pulse, and b) -4V/60ns Reset pulse.

Fig. 4 Area dependences of the resistance values in HRS and LRS. The weak area dependence in LRS supports the conductive filament (CF) mechanism.

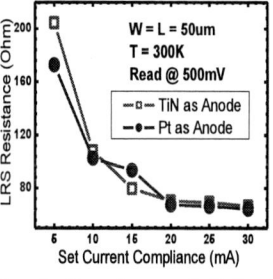

Fig. 5 Dependence of LRS resistance values on the previous set current compliance, supporting the CF mechanism [5].

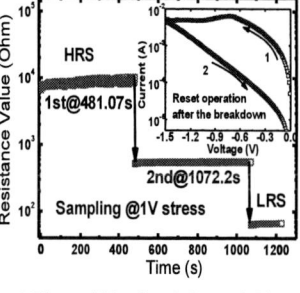

Fig. 6 The multi-level resistive switching from HRS to LRS @1V stress is similar to a dielectric soft breakdown (SB). The insert shows the device can be reset from LRS to HRS after SB.

Fig. 7 Relaxation current as a function of stress time under a 500mV stress voltage applied on TiN top electrode.

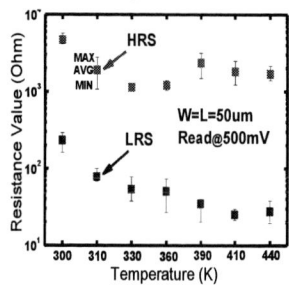

Fig. 8 Temperature dependence of R_{HRS} and R_{LRS}. The reduced R_{LRS} with increased temperature supports the electrons hopping transport.

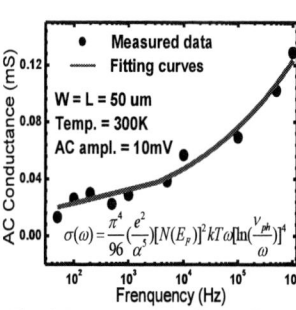

Fig. 9 Frequency dependence of the AC conductance in LRS, which can be fitted by the Mott's electrons hopping theory.

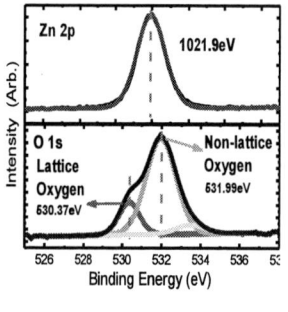

Fig. 10 The XPS spectra of Zn 2p and O 1s core levels in ZnO film. Non-lattice oxygen ions were observed in the ZnO film.

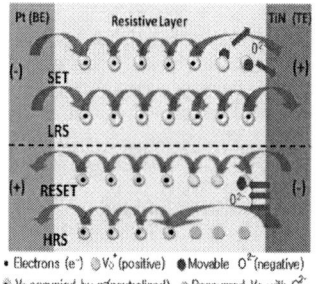

Fig. 11 Schematic views of the unified physical model for the conduction transport in and the switching processes between LRS and HRS.

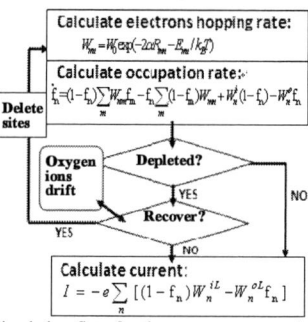

Fig. 12 Simulation flow for the current transport in LRS/HRS and switching process based on the unified model, where f, W, R are electron occupation rate, hopping rate, and hopping distance.

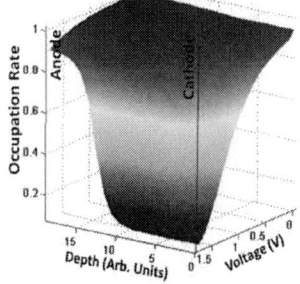

Fig. 13 Calculated electrons occupation rate in the oxygen vacancies of CF as a function of depth and applied voltage. The electron depletion under a critical voltage near cathode occurs in LRS

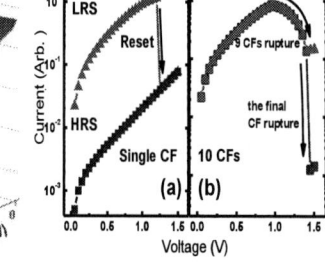

Fig. 14 Simulated I-V curves of a) single CF and b) multiple CFs (10) in the RRAM device. The slower I-V slope of reset process similar to the measured curves was simulated compared to single CF. This means that reduced CFs with the scaled cell size can cause fast reset process

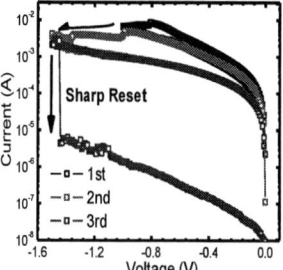

Fig. 15 Measured I-V curves of Multi-step Reset phenomena. The last voltage sweeping cycle during Reset caused an abrupt current decrease, which is seldom observed in stable Reset process.

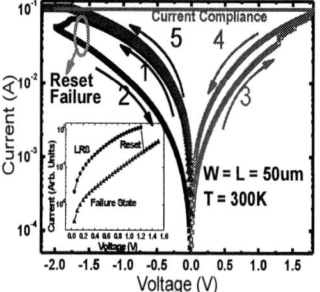

Fig. 16. Measured I-V curves when the switching failure occurs after continuous cycles. The insert shows simulated I-V curves by assuming insufficient O^{2-} to recover the electron-depleted V_O^+

An Endurance-Free Ferroelectric Random Access Memory as a non-volatile RAM

D. J. Jung*, W. S. Ahn, Y. K. Hong, H. H. Kim, Y. M. Kang, J. Y. Kang, E. S. Lee, H. K. Ko, S. Y. Kim, W. W. Jung, J. H. Kim, S. K. Kang, J. Y. Jung, H. S. Kim, D. Y. Choi, S. Y. Lee, K. H. A, and C. Wei, and H. S. Jeong

Technology-Development Team 2, Semiconductor R&D Center, Memory Division, Samsung Electronics Co. Ltd.,
*San #24, Nongseo-Dong, Giheung-Gu, Yongin-City, Kyungki-Do, S. Korea, Tel: +82-31-209-0369, *Email: djjung@samsung.com*

Abstract

We demonstrate endurance characteristics of a 1T1C, 64 Mb FRAM in a real-time operational situation. To explore endurance properties in address access time t_{AA} of 100 ns, we establish a measurement set-up that covers asymmetric pulse-chains corresponding to D1- and D0-READ/RESTORE/WRITE over a frequency range from 1.0 to 7.7 MHz. What has been achieved is that endurance cycles approximate 5.9×10^{24} of cycle times in an operational condition of $V_{DD} = 2.0$ V and 85 °C in the developed 64 Mb FRAM. Donor concentration due to build-up of oxygen vacancy in a ferroelectric film has also been evaluated to 2.3×10^{20} cm^{-3} from I-V-t measurements.

Introduction

None of existing non-volatile memories including many emerging candidates provide ever-increasing demands for NVRAM-use in applications with adequate solutions. FRAM is regarded as one of the non-volatile and byte-addressable random access memories. Fig.1 shows not only how many non-volatile RAM (NVRAM) could be taken into account in sever and laptop markets in the not-too-distant future, but also how FRAMs can be competitive in throughput performance when adopted for NVRAM-use.[1] Among many emerging memories such as phase-change memory and magneto-resistive memory, FRAM can be a perfect candidate as a NVRAM due to two: one is performance superiority as shown in Fig.1(b). The other reason is excellent immunity against temperature. Fig.2 shows temperature dependency of 64 Mb FRAM in holding stored information: (a) changes in D1-D0 distributions of bit-line potential at high temperatures up to 125 °C and (b) charge-to-voltage characteristics of memory-cell capacitors in a temperature range from 25 to 200 °C. This superiority against temperature is due to a fundamental nature of ferroelectric materials having a high Curie temperature T_C, more than 430 °C in PZT films, for example. In spite of those excellences, one of the reasons why ferroelectric memories have not been adopted in today's market place, in particular, as a NVRAM, is lack of endurance, which has never been explored in a real-time READ/WRITE operational situation, which is much different from a conventional way, in which fatigue characteristics of a ferroelectric capacitor are commonly evaluated with symmetric bi-polar pulse chains of 50 or 100% duty-cycle at 1 MHz. In this report, we explore endurance characteristics of a 1T1C 64 Mb FRAM in a real-time operational situation having address access time t_{AA} of 100 ns.

Device and Process Features

180-nm technology has been adopted to integrate an 1T1C FRAM in 64 Mb density, organization of which has 16 IOs. Fig.3 shows micrographic views of cross-sectional images after full integration of the 64 Mb FRAM both (a) in a peripheral circuitry region and (b) in a cell array region, containing 15F^2 cells. In process features, 80-nm MOCVD PZT serves as a ferroelectric film; SrRuO$_3$ as a top electrode (TE); and Ir as a bottom electrode (BE). The 64 Mb FRAM operates at $V_{DD} = 2.0 \pm 0.2$ V; cycle/read-access time are 120 ns/100 ns; and CMOS stand-by current is less than 20 μA[2].

Results and Discussion

Based on an 1T1C architecture having a WL parallel PL scheme, it should be noted that in real-time operations, voltage applied to cell capacitors differs from that of a TEG-level case that cell capacitors are directly connected to a pulse source. This difference stems from contribution of parasitic bit-line capacitance C_{BL} in real-time operations. In an early stage of READ, plate-line voltage V_{PL} is divided into two components: ferroelectric capacitance C_F and C_{BL}, both of which are connected in series. In the subsequent restoration of READ, so-called RESTORE, configuration of C_F and C_{BL} is now so parallel that cell capacitors experience a full V_{DD}, and then D1 capacitors are going back to where it came from but D0 capacitors stay the original state. Fig.4 represents typical D1-D0 distributions of bit-line potential V_{BL}, obtained from all the possible locations in 8"-wafers at 85 °C. As shown in Fig.4, while the developed V_{BL} for D1, V_{BLD1}, is about 800 mV (equivalently, voltage applied to ferroelectric cell capacitors $V_F = 1.20$ V at $V_{DD} = 2.00$ V), the V_{BL} for D0, V_{BLD0},

approximates 250 mV (likewise, $V_F = 1.75$ V). Fig.5 is schematic drawings of how pulse forms to apply cell capacitors vary in each operation of D1 and D0 when real-time access occurs. For example, as shown in Fig.5(b), in READ and RESTORE, asymmetric pulse chains at 7.7 MHz are applied to D1 capacitors when address access time t_{AA} is 100 ns.

In FRAM, it is not readily achieved to assure whether a memory device can suffer virtually infinite endurance cycles of 10^{16} due to memory size and thus a long time span to test. For instance, Fig.6(a) shows changes in V_{BLD1}-V_{BLD0} distributions before and after a HTOL (High Temperature Operational Life) test during 336 hours at 125 °C. Despite such a hash condition, endurance cycles of 64 Mb FRAM is merely a few millions. Even taking into account minimum number of cells (in this case 128 bits because of 16 IOs) as shown in Fig.6(b), time to take for 10^{13} cycles is at least more than 20 days.

Not only to overcome these limits but also to consider endurance cycles of 64 Mb FRAM as a realistic situation, we have explored endurance characteristics over a frequency range between 1.0 and 7.7 MHz at 85 °C. The real-time operational pulse chains differ from those for probing fatigue immunity in a conventional way. First is that pulse chains in shape are asymmetric bi-polar for D1 and uni-polar for D0. Second is that duty cycles are 39% for D1 and 42% for D0 as denoted in Fig.5. Fig.7 is a schematic diagram to establish these evaluations. Fig.8 shows pulse forms generated by this measurement set-up illustrated in Fig.7. Fig.9 shows frequency-dependent endurance characteristics as a function of V_{DD} acceleration. Cycle-to-failure is defined as the cycle at which remanent polarization Pr has yet to collapse. As seen in Fig.9, the cycle-to-failure at 7.7 MHz is almost a billion times of 10^{16} cycles (5.9×10^{24}). This suggests that the developed 64 Mb FRAM is free from any concerns of READ/WRITE endurance. The reason that measured data deviate from each mean value, is because PZT films has a rough surface so that thickness variation is $\pm 15\%$ approximately, as indicated in the insets of Fig.9. Normalized Pr are plotted against pulse cycles in Fig.10. Almost no changes in 2.0-V measured hysteresis loops at V_{DD} acceleration of 5.0 V have appeared as endurance cycles increase in both D1 and D0 pulse-forcing, as plotted in the inset of Fig.10(a) and (c). It is expected to have a linear relationship between log(cycle-to-failure) and pulse-forcing frequency as shown in Fig.11. From I-V-t measurements, we have also investigated donor concentrations due to build-up of oxygen vacancy. The increase in oxygen vacancy in a ferroelectric film is supposed to be responsible for losing polarization.[3] Provided the ferroelectric film is partially depleted obtained I-V-t data has been fitted linearly in the entire range of measurement voltage.[4] The donor concentrations (N_D) has been calculated at specific fatigue cycles in 1.0 MHz bi-polar pulse. We can estimate the donor concentration after the 10^{16} cycles would be 2.3×10^{20}/cm^3 by assuming this linear tendency as illustrated in Fig. 12.

Conclusions

We present endurance-free 64 Mb FRAM devices, which can be applicable to a NVRAM. To probe this, we develop a method to evaluate endurance characteristics in a real-time operational situation such as address access time t_{AA} of 100 ns/2 byte. The evaluated 64 Mb FRAM turns out to be endurance-free because in real-time READ/RESTORE/WRITE (7.7 MHz), cycle-to-failure extrapolated to $V_{DD} = 2.0$ V at 85 °C, is about 6.0×10^{24}. Donor concentration due to build-up of oxygen vacancy in a ferroelectric film has also been investigated about 2.3×10^{20} cm^{-3} from I-V-t measurements and a partially depleted Schottky model. These results could strongly support that the developed 64 Mb FRAM is endurance-free.

References

[1] S. L. Min and E. H. Nam, *Design Automation 2006. Asia and South Pacific Conference.*

[2] D. J. Jung. et. als., *The 16th International Symp. on the Application and Ferroelectrics* (ISAF), Nara, Japan, May 27-31, 2007; Y. M. Kang, et. als, *Symp. on VLSI Tech., Dig.* , 2006, p.124.

[3] A. M. Bratkovsky and A. P. Levanyuk, *Phys. Rev. Lett.* **84** 3177 (2000).

[4] E. S. Lee, et. als, *Jap. J. Appl. Phys.*, will appear in 2008

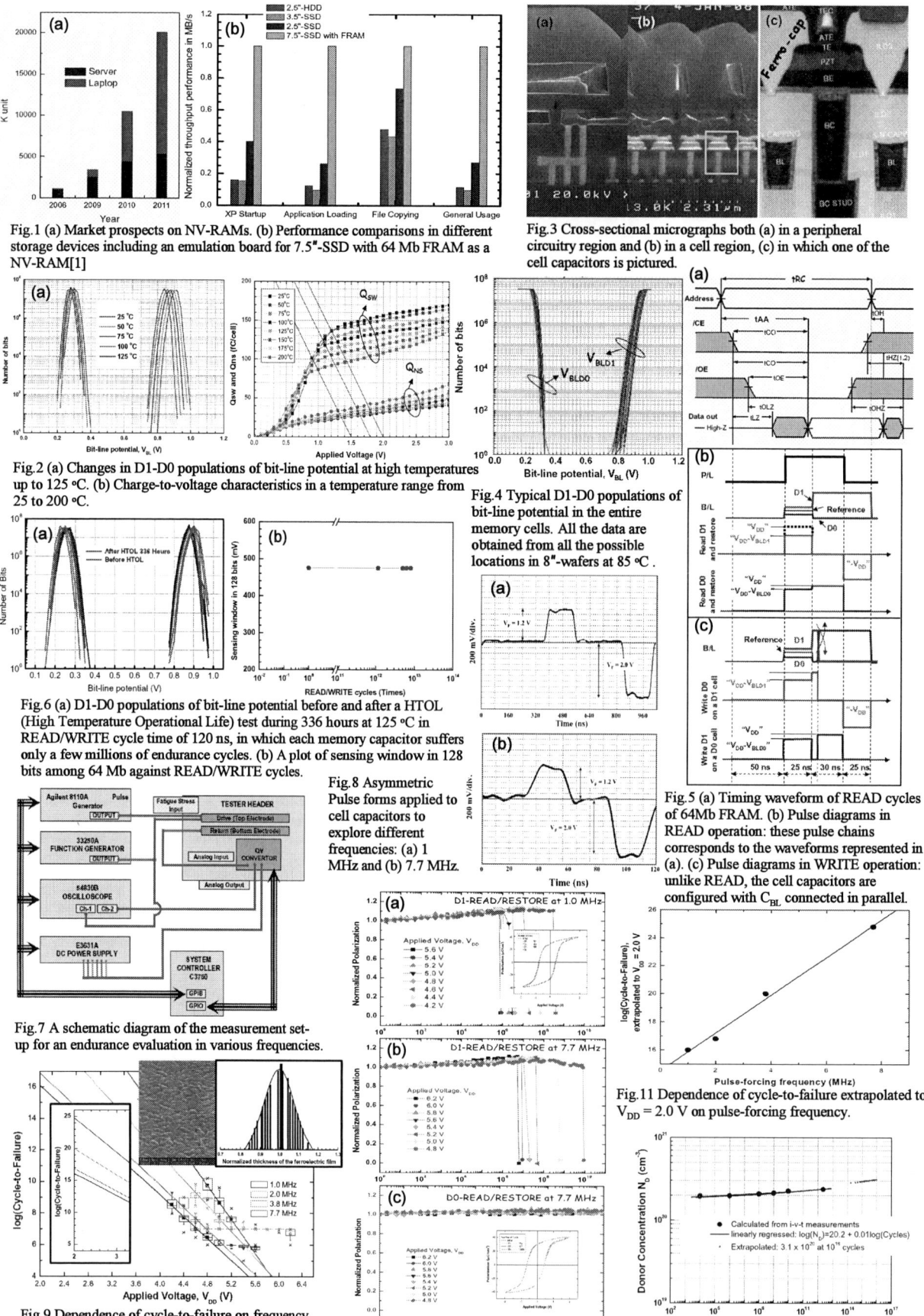

Fig.1 (a) Market prospects on NV-RAMs. (b) Performance comparisons in different storage devices including an emulation board for 7.5"-SSD with 64 Mb FRAM as a NV-RAM[1]

Fig.3 Cross-sectional micrographs both (a) in a peripheral circuitry region and (b) in a cell region, (c) in which one of the cell capacitors is pictured.

Fig.2 (a) Changes in D1-D0 populations of bit-line potential at high temperatures up to 125 °C. (b) Charge-to-voltage characteristics in a temperature range from 25 to 200 °C.

Fig.4 Typical D1-D0 populations of bit-line potential in the entire memory cells. All the data are obtained from all the possible locations in 8"-wafers at 85 °C.

Fig.6 (a) D1-D0 populations of bit-line potential before and after a HTOL (High Temperature Operational Life) test during 336 hours at 125 °C in READ/WRITE cycle time of 120 ns, in which each memory capacitor suffers only a few millions of endurance cycles. (b) A plot of sensing window in 128 bits among 64 Mb against READ/WRITE cycles.

Fig.8 Asymmetric Pulse forms applied to cell capacitors to explore different frequencies: (a) 1 MHz and (b) 7.7 MHz.

Fig.5 (a) Timing waveform of READ cycles of 64Mb FRAM. (b) Pulse diagrams in READ operation: these pulse chains corresponds to the waveforms represented in (a). (c) Pulse diagrams in WRITE operation: unlike READ, the cell capacitors are configured with C_{BL} connected in parallel.

Fig.7 A schematic diagram of the measurement set-up for an endurance evaluation in various frequencies.

Fig.11 Dependence of cycle-to-failure extrapolated to $V_{DD} = 2.0$ V on pulse-forcing frequency.

Fig.9 Dependence of cycle-to-failure on frequency of asymmetric pulse chains, ranging from 1.0 to 7.7 MHz. A SEM image in the insets represents surface morphology of the ferroelectric film, thickness variation of which roughly approximates ±15%.

Fig.10 Typical fatigue characteristics of cell capacitors in the case of D1-READ and RESTORE at (a) 1.0 MHz and (b) 7.7 MHz. (c) D0-READ/RESTORE at 7.7 MHz.

Fig.12 Changes in donor concentration N_D due to oxygen vacancy (V_O") as a function of fatigue cycles at 1.0 MHz.

A New Direct Low-k/Cu Dual Damascene (DD) Contact Lines for Low-loss (LL) CMOS Device Platforms

J. Kawahara, M. Ueki, M. Tagami, K. Yako, H. Yamamoto, F. Ito, H. Nagase, S. Saito, N. Furutake, T. Onodera, T. Takeuchi, H. Nakamura*, K. Arita*, K. Motoyama*, E. Nakazawa*, K. Fujii*, M. Sekine*, N. Okada* and Y. Hayashi

Device Platforms Research Labs., NEC Corporation, *NEC Electronics Corporation

1120 Shimokuzawa, Sagamihara, Kanagawa 229-1198, JAPAN

Phone: +81-42-771-4267 Fax: +81-42-771-0886 e-mail: j-kawahara@bq.jp.nec.com

Abstract A new direct low-k/Cu dual damascene (DD) contact line has been developed for low loss (low parasitic capacitance and low resistance) CMOS device platforms by on-current BEOL technologies. The excellent low contact resistance is realized in the low-k pre-metal-dielectrics (PMD) with a reduced aspect ratio, achieving 5.4Ω for 75nmϕ contact which is only 1/4 relative to a conventional W-plug. The CMOS active performance was improved with no reliability degradation, featuring in cost-effective RF/ubiquitous applications.

Keywords: Cu contact, Low-k, CMOS, Low parasitic elements.

Introduction

Scaled-down CMOS devices are needed not only with high current drivability but also low parasitic losses especially for RF/ubiquitous applications. High performance CMOS with high-k gate dielectric [4-5] and full low-k/Cu interconnects [1] tried to be integrated together with miniature RF/analog passive components [2-3, 6] (Fig. 1). Recently, the parasitic elements inside of the scaled-down CMOS itself such as the large contact (CT) resistances limit the performance, and Cu contacts instead of conventional W ones attempted to be implemented [7-11].

Here, we have proposed a new direct low-k/Cu dual damascene (DD) contact lines in the scaled-down CMOS for the first time, achieving a very low contact resistance.

Features of the process and structure

Fig. 2 shows the cross-sectional STEM image of the direct low-k/Cu DD contact lines demonstrated in a high-k CMOS device. The pre-metal dielectric (PMD) consisted of a rigid SiOCH film (k=3.1) and a silica-amorphous-carbon composite film (SACC, k=3.1), which had high adhesion strength and the Cu diffusion barrier ability as well as the better step coverage on the narrow-pitched gate patterns than the SiOCH (Fig. 3). This is due to the surface reaction dominant deposition by the plasma polymerization [12]. After CMP planarization, a molecular-pore-stack SiOCH film (MPS, k~2.5) with the SACC hard-mask was deposited [1]. The DD contact lines were patterned by multi-resist process in common with leading-edge BEOL process, revealing the self-aligned Cu DD contacts with the aspect ratio of 2.8 to the 250nm-pitched gate lines covered with the SiN-stress liner (Fig. 4). The DD contact lines were filled with a ECD-Cu film on PVD Ta/TaN barrier-metals, which was contacted to the NiSi source/drain in the CMOS directly. The low-k PMD enables us to reduce the contact aspect ratio (AR) from 3.8 to 2.8, facilitating the Cu metallization without increasing the parasitic capacitance.

Electrical Properties

Direct Cu DD Contact (CT) lines: The Cu DD CTs (Fig. 2) had the excellent low contact resistance such as 5.4Ω for 75nmϕ contact, which was only 1/4 (-75% reduction) referred to a conventional W-plug (Fig.5). The reduction ratio was enlarged by the contact size scaling. This is not only because the Cu DD

CT was made of lower resistivity of Cu than W-plug, but also the DD structure eliminated the high-resistance interface between the plug and the M1 line.

The junction leakage current of the Cu DD CT indicated as the same level as the W plug (Fig.6). No dependency of the miss alignment on the junction leakage was observed (Fig.7). The PVD-Ta/TaN barrier metal at the bottom of contact with the reduced aspect ratio (AR~2.8) prevented Cu from the silicidation reaction to NiSi/Si perfectly.

CMOS Performances: No degradation was observed in the I_d-V_g characteristics (Fig.8) and the I_{on}-I_{off} characteristics (Fig. 9) with 50-nm gate MOSFETs with high-k dielectrics [4-5]. The negative bias temperature instability (NBTI) of the threshold voltage V_{th} and the time-dependent dielectric breakdown (TDDB) of n/pMOSFETs were unchanged by the Cu DD contacts, ensuring the reliability of the gate dielectrics (Figs.10 and 11). The propagating delay in the 199-stages CMOS ring-oscillator with the direct low-k/Cu DD contacts decreased by 7% relative to that with the conventional SiO_2-PMD/W-plug (Fig. 12) due to decreasing the parasitic effects just by applying the cost-effective leading-edge BEOL technology (Table. 1).

Conclusion

A direct low-k/Cu DD contact line was developed for the low loss (LL) CMOS device just by applying the leading-edge BEOL technologies. The excellent low contact resistance of 5.4Ω for 75nmϕ contact was achieved in the low-k PMD with the reduced aspect ratio. The CMOS active performance was improved by the low-k/Cu contact structure without any degradation of CMOS reliability. The direct low-k/Cu DD contact is a promising technology for cost-effective RF/ubiquitous applications.

Acknowledgment

The authors would like to thank to Mr. M. Yasuda, T. Fukai, T. Nakayama, Y. Kasama, A. Mitsuiki, K. Taniguchi, NEC Electronics and Drs. Y. Mochizuki and H. Watanabe, NEC for their technical discussion and the research support.

References

[1] M.Ueki, et al., IEEE IEDM 973-976(2007). [2] N.Inoue, et al., IEEE IEDM 989-992(2007). [3] K.Hijioka, et al., SSDM 910-911(2007). [4] H.Nakamura, et al., Symp. VLSI Tech. 158-159(2006). [5] G.Tsutsui, et al., Symp. VLSI Tech. 176-177(2007). [6] Y. Hayashi, IEEE IEDM 363-366(2006). [7] K.Kawamura, et al., ADMETA 128-129(2007). [8] G. Van den bosch, et al., IEEE IEDM(2006). [9] S. Demuynck, et al., IITC 178-180(2006). [10] A. Topol, et al., Symp. VLSI Tech. 142-143(2006). [11] M. Inohara, et al., IEEE IEDM 931-933(2001). [12] J.Kawahara, et al., Plasma Source Sci. Technol., **12**(2003) S80-S88G.

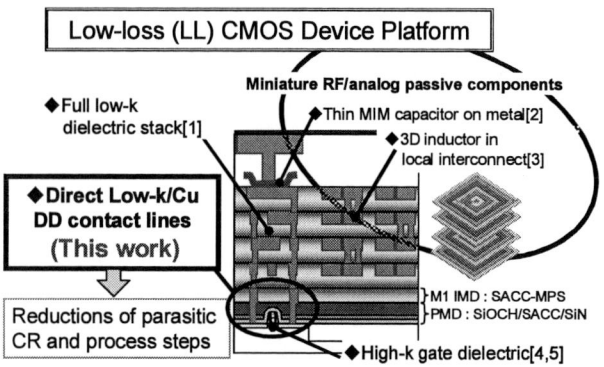

Low-loss (LL) CMOS Device Platform

- ◆ Full low-k dielectric stack[1]
- ◆ Thin MIM capacitor on metal[2]
- ◆ 3D inductor in local interconnect[3]

Miniature RF/analog passive components

◆ **Direct Low-k/Cu DD contact lines (This work)**

Reductions of parasitic CR and process steps

M1 IMD : SACC-MPS
PMD : SiOCH/SACC/SiN

◆ High-k gate dielectric[4,5]

Fig.1 Schematic illustration of a low-loss CMOS device platform for next generation RF/ubiquitous applications [6].

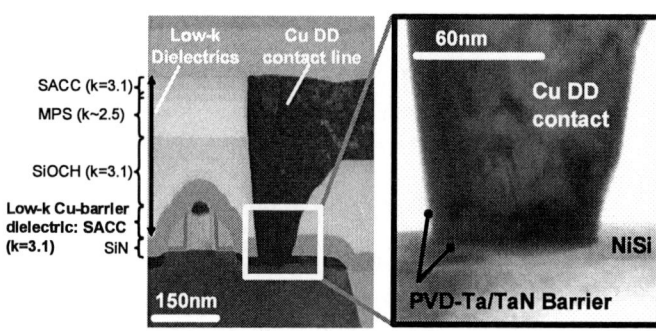

Fig.2 Cross-sectional STEM image of the direct low-k/Cu DD contact lines integrated in a high-k CMOS device. Magnified image shows clear interface of the PVD-Ta/TaN barrier metal and the NiSi at the 80nmφ-contact of AR=2.8.

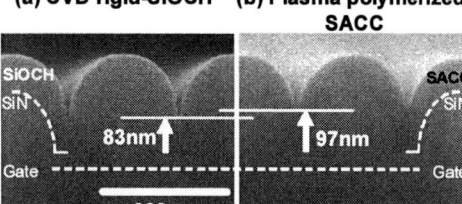

(a) CVD rigid-SiOCH **(b) Plasma polymerized SACC**

Fig.3 Cross sectional SEM images of 100nm-thick (a) the CVD SiOCH and (b) the plasma polymerized SACC on the 270nm-pitched gate covered with the SiN-stress liner. The narrow gap is filled better with SACC than SiOCH, and no slit void was detected.

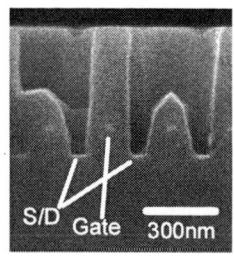

Fig.4 A cross-sectional SEM image after the DD contact etching to the 250nm pitched gate. The self-align DD contact with AR of 2.8 had a well-controlled feature through the SiN-stress liner.

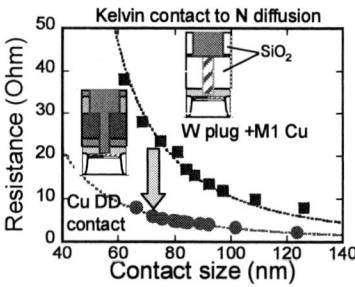

Fig.5 Kelvin contact resistance as a function of the contact size. The low-k/Cu DD contact decreases the resistance extremely in the scaled-down contacts.

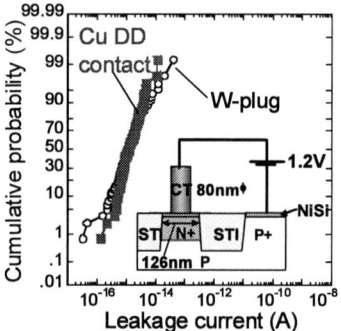

Fig.6 Junction leakage current with the W plug and the Cu DD contact, indicating that no Cu penetration into NiSi.

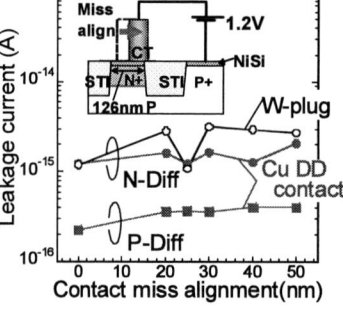

Fig.7 The junction leakage as a function of the miss alignment. The step of the STI and NiSi did not affect the junction leakage in the Cu DD contacts.

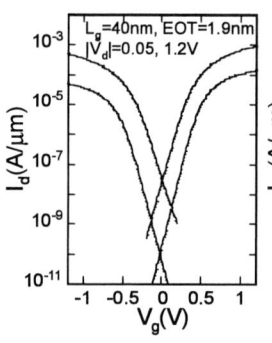

Fig.8 I_d-V_g of 40nm gate-length, high-k n/p MOSFETs with the Cu DD contacts.

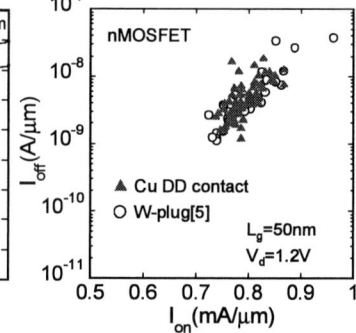

Fig.9 I_{on}-I_{off} characteristics of nMOSFETs. The variation of the Cu DD contacts was comparable to the W plugs.

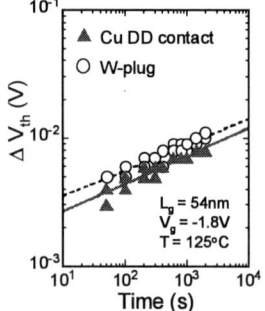

Fig.10 NBTI of V_{th} at 125°C, -1.8V for pMOSFETs. The Cu DD contact did not affect the gate dielectric reliability.

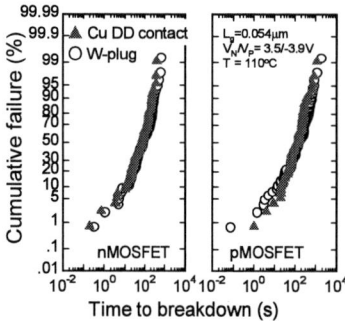

Fig.11 TDDB lifetimes of n/pMOSFETs at 110°C, 3.5/-3.9V. No degradation of failure time was observed in the Cu DD contacts.

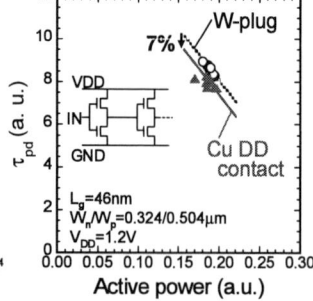

Fig.12 Power-delay plots of ring oscillators with the low-k/Cu DD contact and SiO$_2$-PMD/W-plug.

Table 1. Summary of the characteristics and process of Cu contacts.

	Ref.[10]	Ref.[7]	This work
Structure	Cu plug + M1 Cu	Cu plug + M1 Cu	Cu DD
PMD	SiO$_2$	SiO$_2$	Low-k
M1 IMD	SiO$_2$	SiO$_2$	Low-k
Contact resistance (Referred to W-plug (%))	No data (-25%)	20Ω @75nmφ (-50%)	5.4Ω @75nmφ (-75%)
Barrier metal	Ta based	Ta/WSi$_x$/ CVD-TiN/Ti	PVD-Ta/TaN
Contact A/R	No data	5	2.8

82

A Novel CVD-SiBCN Low-K Spacer Technology for High-Speed Applications

C.H. Ko, T.M. Kuan, *Kangzhan Zhang, *Gino Tsai, *Sean M. Seutter, C.H. Wu, T.J. Wang, C.N. Ye,
H.W. Chen, C.H Ge, K.H. Wu, and W.C. Lee

Taiwan Semiconductor Manufacturing Company, Ltd., Hsinchu, Taiwan, R.O.C.
*Applied Materials, Inc., Santa Clara, California, U.S.A.
Email: chko@tsmc.com   Tel: 886-3-505-5126   Fax: 886-3-563-7525

Abstract

State-of-the-art low-K spacer technology featuring novel CVD-SiBCN material is demonstrated for the first time. A significant 20% CMOS ring speed enhancement is demonstrated with SiBCN (K=5.2) spacer, compared to Si_3N_4 (K=7.5) spacer, due to reduced fringing capacitance and enhanced strain effects by spacer-PSS and CESL techniques. Electron mobility is improved by 6% for long channel NMOS transistor and $g_{m,max}$ is increased by 11% for short 35nm physical gate length NMOS using a preferable spacer structure that is comprised of a low stress SiBCN spacer on thin SiO_2 liner and a final 600°C rapid thermal post-anneal. Superior GIDL and better gate leakage is obtained because low permittivity SiBCN alleviates gate-fringing field effects (GF effects), and device reliability is not adversely impacted by this new process.

Introduction

As channel length scales further into in the nanometer regime, the parasitic capacitances and series resistance are going to seriously impact the transistor performance. Low-K spacer is a good candidate to minimize parasitic capacitance for high-speed applications. Total overlap capacitance (Cov) consists of three components as shown in Fig. 1: direct overlap capacitance (Cdo), outer fringing capacitance (Cof), and inner fringing capacitance (Cif). As the gate geometry shrinks for the next few technology nodes, the gate electrode to source/drain and contact fringe fields are growing and the effect of Cof becomes more prominent. It has been reported that adopting air spacer can reduce Cof and improve RC delay [1]. The effect of spacer-induced strain used to increase Idsat has also been demonstrated [2]. However, no literature has reported the combination of low-K spacer and process-strained Si (PSS) techniques to achieve higher CMOS performance.

In this work, we used a novel thermal CVD process to deposit low permittivity SiBCN films and have successfully demonstrated CMOS ring oscillator with channel length down to 35nm. The integration process is compatible with conventional CMOS processes as well as cost effective. Material and device characteristics and device integrity are carefully evaluated.

Material Characteristics and Experimental

Integration sequence of CMOS transistors is illustrated in Fig. 2. Low dielectric constant SiBCN film was deposited at 550°C by thermal CVD and conventional-Si_3N_4 was used for a control sample. Fig. 3 shows SiBCN films have step coverage >95% and pattern loading effect <10% confirmed by SEM. Fourier Transform Infrared Spectroscopy (FTIR) in Fig. 4(a) shows boron is bonded to nitrogen in the SiBCN film, and Fig. 4(b) shows the SiBCN process enables low temperature film deposition even without plasma assistance. Table 1 depicts wet etching sensitivity of SiBCN spacer in ozone (Std. Clean1), ammonia-based (Std. Clean2), and HF strippers which are usually used for removing surface organics and oxide after spacer formation. Wet etch rate (WER) of SiBCN spacer is higher than Si_3N_4, and rapid thermal process (RTP) post-anneal for 30 seconds at 600°C and 700°C in a nitrogen atmosphere reduces WER significantly. Fig. 5 shows the SiBCN film stress increases as annealing temperature increases. Table. 2 summarizes the properties of the SiBCN film deposited at 550°C and its composition measured by Rutherford backscattering (RBS) and hydrogen forward scattering (HFS).

Device Characteristics and CMOS Performance

Device simulation using Sentaurus software has been conducted to verify the AC and DC performance affected by spacer structures. Fig. 6 shows the electric field distribution in two different permittivity spacers (K=7.5 and K=3.1) on top of two different SiO_2 (K=3.9)

thicknesses. We observe that either increasing SiO_2 thickness or adopting low permittivity spacer can lower the gate-fringing field effects (GF effects) [3,4]. Fig. 7 illustrates Cof and Cov decrease significantly due to reduced GF effects with low dielectric spacer or thicker SiO_2 liner, and the extreme case is spacer with K-value less than 3.9 with ultra thin SiO_2 liner. Spacer-PSS technique and contact-etch-stop layer (CESL) [2] have been reported to enhance DC performance of transistors. From NMOS device simulation shown in Fig. 8, thin SiO_2 liner helps to transfer stress from spacer and CESL to channel region and low stress spacer is preferred to enhance tensile channel strain. It is therefore proposed that SiBCN spacer as presented in this paper with dielectric constant of 5.2 and low tensile film stress of 430MPa when combined with thin SiO_2 spacer can achieve much better AC and DC CMOS performance compared to Si_3N_4 (K=7.5/ 1.5GPa) spacer.

Fig. 9 shows I_D-V_D CMOS data characteristics with SiBCN spacer with and without RTP post-anneal. With 600°C post-anneal PMOS performance increases by 4% with negligible impact on NMOS. With 700°C post-anneal, PMOS performance shows 6% enhancement but NMOS performance shows 6% degradation. Therefore 600°C post-anneal is the proposed optimum condition for CMOS performance. This different NMOS versus PMOS dependence on annealing temperature can be attributed to the change of stress of the SiBCN spacer and that NMOS and PMOS favor opposite longitudinal strain. As shown in Fig. 10, the spacer-PSS benefit for NMOS using the SiBCN spacer with 600°C post-anneal approach results in 6% electron mobility improvement for long channel devices and 11% $g_{m,max}$ gain for 35nm short channel devices compared to standard Si_3N_4 spacer. Fig. 11 shows ~20% improvement in ring oscillator speed at Wn/Wp=0.6μm/0.3μm, and this is attributed to a combination of fringing capacitance reduction and enhanced strain effects by using the SiBCN spacer technique. Fig. 12 is a plot of ring speed versus leakage of the SiBCN structure and demonstrates AC performance enhancement with acceptable leakage level.

Device Integrity and Reliability

Fig. 13 shows SiBCN spacer device exhibits superior gate-induced drain leakage (GIDL) ascribed to low permittivity that reduces GF effects resulting in lower vertical electric field and suppressed depletion region in the S/D extension at V_G=-1V and V_D=1V. Fig. 14 shows SiBCN spacer device has lower gate oxide leakage at V_G=1V and V_D=0V due to the weaker vertical electric field underneath the gate sidewall compared to Si_3N_4 spacer, as shown in Fig. 15. Fig.16 (a) and (b) show no negative impact of this new process on either substrate leakage or hot carrier lifetime (HCL) of NMOS devices.

Conclusion

For the first time, we report a novel low-K spacer technology featuring CVD-SiBCN material with low dielectric constant of 5.2 and film stress of 430MPa to boost AC and DC CMOS performance. The SiBCN spacer approach in NMOS achieves 6% electron mobility improvement for long channel transistors and 11% $g_{m,max}$ gain for 35nm physical gate length. This novel approach can deliver up to 20% increase in CMOS ring speed. In addition, GIDL and gate leakage are improved by reduction of GF effects and device reliability is not adversely impacted. The SiBCN spacer technology has been demonstrated to be an effective solution for future high-speed CMOS applications.

Reference

[1] M. Togo et al., VLSI Tech. Dig. pp. 38, June. 1996.
[2] C.H. Ko et al., VLSI-TSA-Tech pp. 25, April. 2005
[3] T. Mizuno et al., IEEE TED., vol.39, pp.982, 1992.
[4] J.C. Guo et al., IEEE TED., vol.41, pp.1239, 1994.

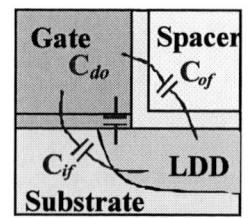

$Cov = Cdo + Cof + Cif$
Cdo : direct overlap capacitance
Cof : outer fringing capacitance
Cif : inner fringing capacitance

Fig. 1. Schematic cross-section of parasitic capacitance around the gate electrode.

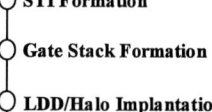

- STI Formation
- Gate Stack Formation
- LDD/Halo Implantation
- SiBCN Spacer
- S/D formation
- Silicide Formation & CESL Deposition

Fig. 2. Integration sequence with SiBCN low-K spacer.

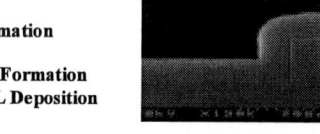

Fig. 3. SEM pictures show that SiBCN film has step coverage >95% and pattern loading effect <10%.

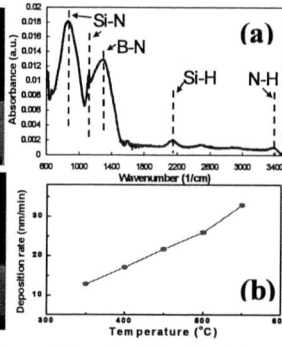

Fig. 4 (a). Bonding structure of SiBCN. (b) SiBCN deposition rate vs. wafer temperature - 550°C for device test.

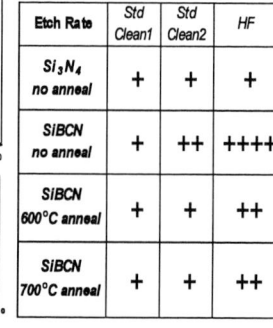

Etch Rate	Std Clean1	Std Clean2	HF
Si_3N_4 no anneal	+	+	+
SiBCN no anneal	+	++	++++
SiBCN 600°C anneal	+	+	++
SiBCN 700°C anneal	+	+	++

Table 1. RTP post-anneal after SiBCN spacer structure formation is adopted to avoid high wet etching rate.

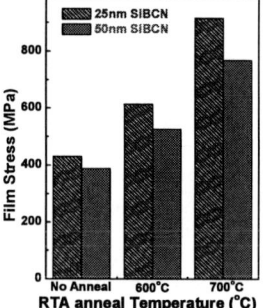

Fig. 5. Stress measurements of a SiBCN film deposited on a bare Si blanket wafer and then annealed in an RTP chamber.

Summary of CVD SiBCN film	
wafer temperature	550°C
deposition rate	360Ang/min
step coverage	>95%
pattern loading	<10%
k-value	5.2
film stress (250Ang thickness)	430MPa (tensile)
bonding structure (by FTIR)	boron bonded to nitrogen
composition (by RBS/HFS):	
silicon	18.5%
boron	26.5%
carbon	5.0%
nitrogen	31.0%
hydrogen	19.0%
refractive index	2.05

Table 2. Summary of film properties for SiBCN film deposited at 550°C (no anneal).

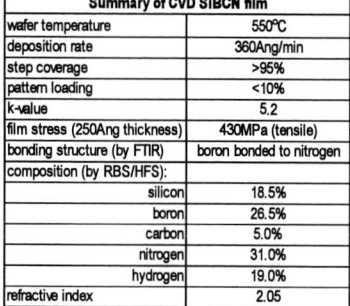

Fig. 6. Simulated electric field distributions in spacer regions.

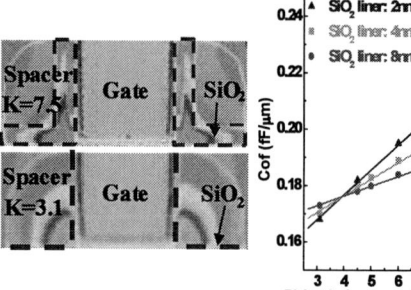

Fig. 7. Simulated outer fringe capacitance vs spacer permittivity with different SiO_2 liner thickness.

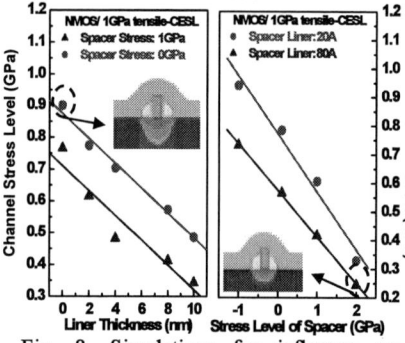

Fig. 8. Simulations for influence on channel stress from spacer stress and SiO_2 liner. Low spacer stress or thin SiO_2 liner significantly enhances tensile strain in the channel region.

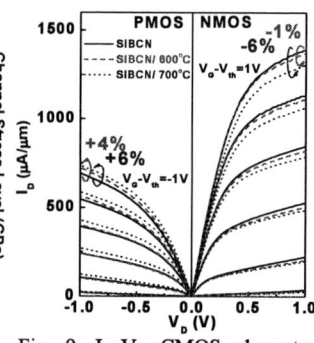

Fig. 9. I_D-V_D CMOS characteristics at 35nm gate length with SiBCN spacer without anneal and with 600°C or 700°C anneal.

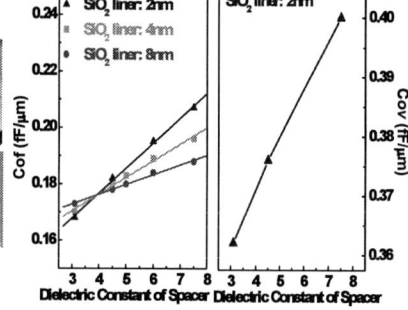

Fig. 10. Higher mobility for long channel NMOS and higher g_m for short channel NMOS with SiBCN spacer and 600°C anneal.

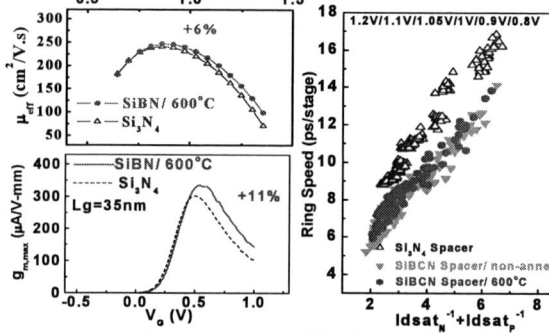

Fig. 11. ~20% improvement in ring oscillator speed is achieved with SiBCN spacer technique at Wp/Wn=0.6 μm/0.3μm.

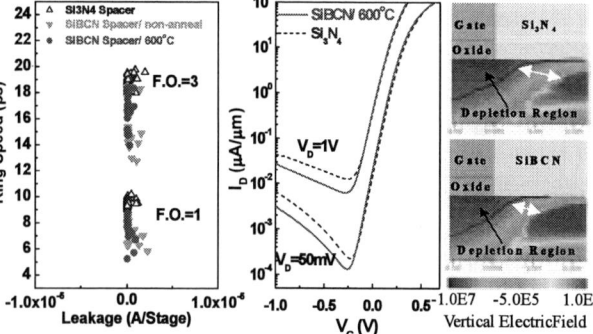

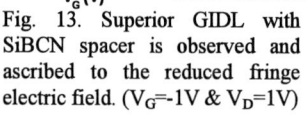

Fig. 12. Ring speed versus leakage of SiBCN spacer structure demonstrates good AC performance without leakage.

Fig. 13. Superior GIDL with SiBCN spacer is observed and ascribed to the reduced fringe electric field. (V_G=-1V & V_D=1V)

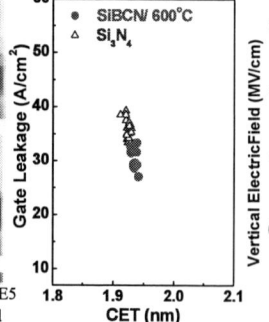

Fig. 14. Lower gate oxide leakage is observed for SiBCN devices.

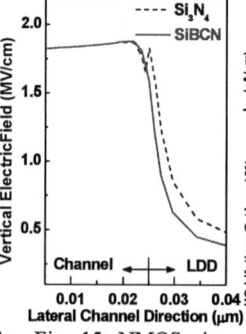

Fig. 15. NMOS simulation shows lower vertical electric field with SiBCN spacer at V_G=1V and V_D=0V.

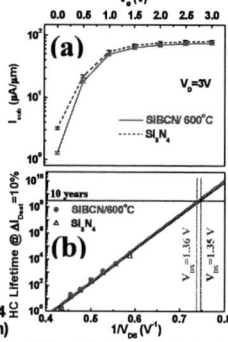

Fig. 16. Low substrate leakage and better hot carrier lifetime with SiBCN spacer in NMOS.

A Proposal of New Concept Milli-second Annealing: Flexibly-Shaped-Pulse Flash Lamp Annealing (FSP-FLA) for Fabrication of Ultra Shallow Junction with Improvement of Metal gate High-k CMOS Performance.

Takashi Onizawa, Shinich Kato, Takayuki Aoyama, Yasuo Nara and Yuzuru Ohji.

Semiconductor Leading Edge Technologies, 16-1 Onogawa, Tsukuba-shi, Ibaraki-ken, 305-8569, Japan.

Phone: +81-29-849-1219, Fax: +81-29-849-1186, E-mail:onizawa.takashi@selete.co.jp

Abstract

We propose the suitable milli-second annealing (MSA) for metal/high-k device performance and ultra-shallow-junction (USJ) fabrication: flexibly-shaped-pulse flash lamp annealing (FSP-FLA). The conventional FLA treatment on metal/high-k device degrades its effective electron mobility (μ_{eff}) and bias temperature instability (BTI) characteristics. A recovery annealing (RA) treatment after FLA is most effective to recover those degradations. However, the annealing after dopant activation causes deactivation and diffusion. The FSP-FLA allowed us sub-10-milli-second annealing after activation FLA; it realizes high BTI reliability and high μ_{eff} without deactivation and diffusion.

Introduction

The fabrication of USJ by using MSA is a topic for 32-nm-node CMOS device and beyond. Recently, the impacts of MSA such as flash lamp or laser on CMOS device performance have been reported [1-3]. To construct the gate-first metal/high-k devices, we have to investigate the MSA compatibility with metal/high-k device. In this study, we investigated the impact of FLA on metal/high-k devices, demonstrated the new concept FLA, and discovered the suitable MSA for metal/high-k device fabrication.

Sample preparations

Figure 1 shows the concept of this study and the process flow for the sample preparations. Metal/high-k devices (EOT = 1.0 nm) were fabricated with the gate-first process flow for NMOS (TaSiN/ HfSiON) and PMOS (TiN/HfSiON) [4]. The activation annealing was performed by FLA or sRTA of 850 to 1050 °C. For the FLA sample, RA by 850 °C spike was performed in various timing of the sample preparation (before ion implant (I/I), before and after FLA). Additionally, we demonstrated the FSP-FLA for the activation annealing by using the FLA equipment of DAINIPPON SCREEN MFG. CO., LTD.

Results and discussions

I. Comparison of sRTA and FLA

Figure 2 shows the gate leakage current density (J_g) for NMOS and PMOS fabricated with FLA or 1000 °C sRTA. Both of electron and hole currents are in the same level for FLA and sRTA. It indicates the use of FLA don't degrade the J_g characteristics. Figure 3 shows flatband voltage (V_{FB}) for the samples. In both NMOS and PMOS, the V_{FB} are shifted toward midgap with sRTA temperature increase, and the V_{FB} of FLA sample is the nearest for the silicon-band-edge than other > 850 °C sRTA samples. The V_{FB} shift seems to be caused by the Fermi-level pinning due to high temperature annealing [5], and it is found that the V_{FB} shift occurs gradually. The thermal budget of FLA for metal/high-k gate stack seems to be lower than that of 850 °C sRTA. In the view point of J_g and V_{FB}, FLA is preferable treatment for activation annealing of metal/high-k device. However, it had been reported that the metal/high-k device fabricated with FLA shows lower μ_{eff} and BTI reliability [3]. Figure 4 shows the (a) PBTI for NMOS and (b) NBTI for PMOS, respectively. Similarly, the FLA sample used in this study shows lower BTI reliabilities than that of sRTA samples.

II. Timing of Recovery annealing for BTI & μ_{eff} improvement

For evaluate the BTI improvement, we prepared the three kinds of the samples with various RA timing in sample preparation.

Figure 5 shows the impact of RA timing on NBTI and PBTI. Figure 6 shows the impact of the RA on μ_{eff} of NMOS. As shown in here, the improvements of BTI and μ_{eff} [3] are observed only on the sample with RA after FLA, and it particularly effective in PBTI (NMOS). We consider the reason for PBTI and μ_{eff} degradation to increase of the electron trapping sites during FLA treatment (e.g. wafer warping), and it can be recovered by RA.

Figure 7 shows the SIMS profiles for NMOS extension (As 2 keV 10^{15} cm^{-3}) with various temperatures of RA after FLA treatment. The RA after FLA causes over 4nm dopant diffusion even when RA is treated under 850°C. From the view point of higher BTI and μ_{eff}, RA after FLA is desirable. However, the thermal treatment after FLA activation may cause a dopant deactivation and thermal diffusion. To resolve this problem, we used the new concept FLA for the SD activation by using FSP-FLA.

III. Flexibly-shaped-pulse FLA

Figure 8 shows (a) the concept of FSP-FLA pulse used in this study and (b) measured pulse form. A low power sub-10-milli-second lamp annealing was added after high power milli-second FLA (for dopant activation) to improve the BTI and μ_{eff}. Figure 9 shows the SIMS profiles for NMOS extensions after conventional FLA or FSP-FLA. Figure 10 shows the sheet resistance (Rs) vs. junction depth at 10^{18} cm^{-3} (X_j) for FSP-FLA, FLA, and RA after FLA. With RA temperature increase from 850 to 950 °C, Rs increased by dopant deactivation. With RA temperature increase from 950 to 1050 °C, Rs is reduced by dopant diffusion and activation. As shown in these figures, FSP-FLA realizes the high activation with less-diffusion similar to conventional FLA. This indicates the sub-10-milli-second annealing did not deactivate and diffuse the dopant. Additionally, further recovery of implantation defects is observed in FSP-FLA sample. Figure 11 shows Therma Wave signal (T.W.) for the samples. The increase of T.W. means more defect and/or deeper X_j. The lower T.W. value of FSP-FLA is caused by recovery of the implant defect with less-diffusion.

Fig. 12 and Fig. 13 show PBTI and μ_{eff} of the samples, respectively. FSP-FLA sample shows improved PBTI and μ_{eff} as same as or more than 850 °C RA. We consider those recoveries are gained by increased thermal budget by FSP-FLA.

Conclusions

The suitable MSA for metal/high-k devices to fabricate USJ and to obtain high BTI reliability and mobility is performed by using FSP-FLA. To improve the BTI and μ_{eff}, thermal treatment after FLA annealing is useful. FSP-FLA provides these improvements without dopant deactivation and diffusion. More adjustment of FSP-FLA pulse shape surely achieves the more improvement of BTI and μ_{eff}.

References

[1] T. Hoffmann et al., IWJT, p. 137 (2007).
[2] A. Shima, et al., VLSI Tech. Dig., p. 174 (2004).
[3] P. Kalra et al., IEDM, p. 353 (2007).
[4] F. Ootsuka et al., SSDM, p. 1116 (2006).
[5] K. Akiyama et al., IWDTF, p. 63 (2006).

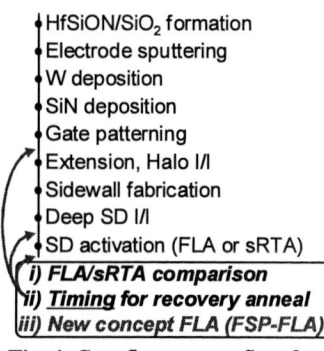

Fig. 1. Gate first process flow for the preparation of metal/high-k gate stacks with various Annealing.

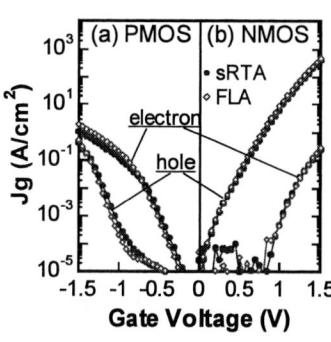

Fig. 2. Gate leakage current density of (a) TiN/HfSiON and (b) TaSiN/ HfSiON fabricated with FLA or 1000°C sRTA.

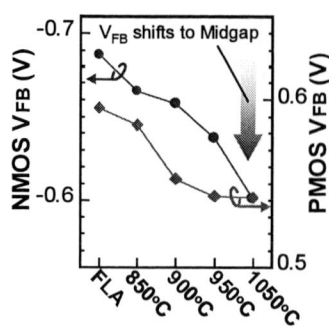

Fig. 3. Flat band voltages of NMOS and PMOS fabricated with FLA or 850 ~ 1050°C sRTA.

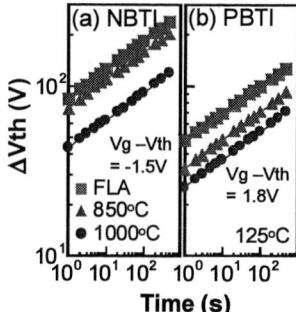

Fig. 4. V_{th} shift for PMOS(NBTI) and NMOS(PBTI) devices fabricated with FLA, 850°C or 1000°C sRTA.

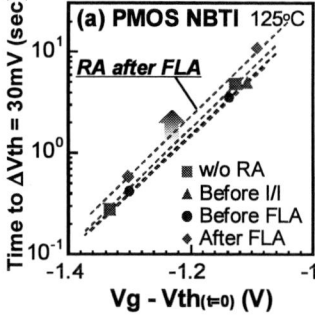

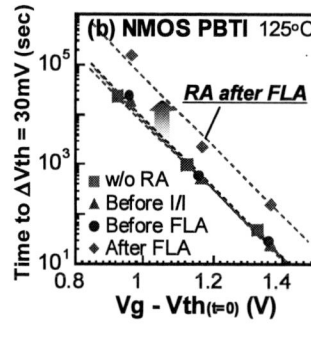

Fig. 5. Comparison of time to ΔV_{th} = 30mV for (a) PMOS and (b) NMOS under BT stresses with various timing of 850°C recovery annealing in the sample preparation.

Fig. 6. Comparison of effective electron mobility of NFET samples for various timing of 850°C recovery annealing.

Fig. 7. SIMS pofiles of NMOS extensions. Spike after FLA causes dopant diffusion even RA = 850 °C.

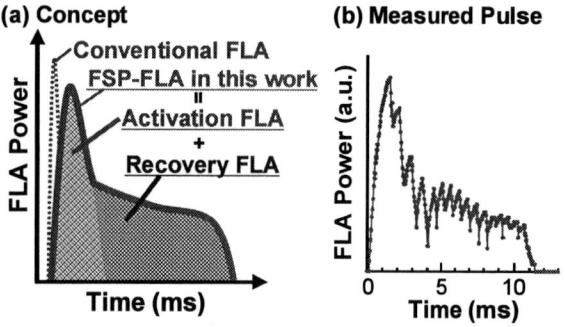

Fig. 8. (a) The concept of FSP-FLA pulse shape used in this study and (b) measured pulse shape. Additional low power sub-10-milli-sec pulse was added follow high power pulse FLA for recovery annealing.

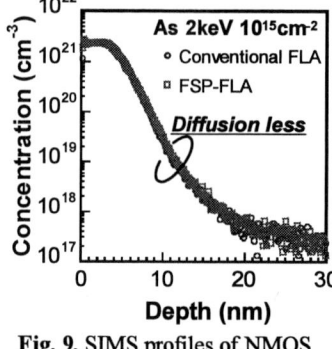

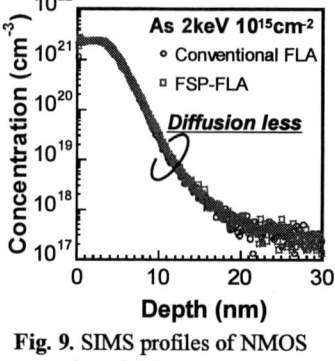

Fig. 9. SIMS profiles of NMOS extensions fabricated with conventional FLA or FSP-FLA.

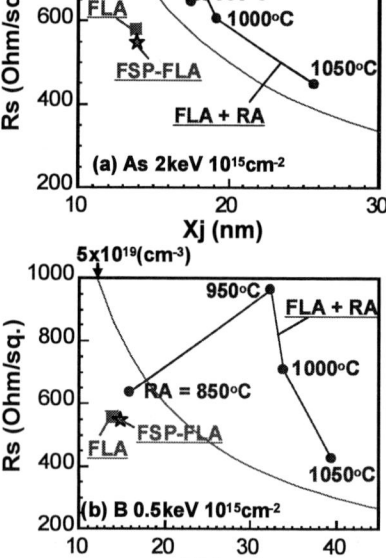

Fig. 10. Rs-Xj characteristics of (a) NMOS and (b) PMOS extensions, respectively. The doses are (a) As 2 keV 10^{15} cm^{-2}, (b) Ge PAI + B 0.5 keV 10^{15} cm^{-2}. FSP-FLA realize diffusion less activation the same as conventional FLA.

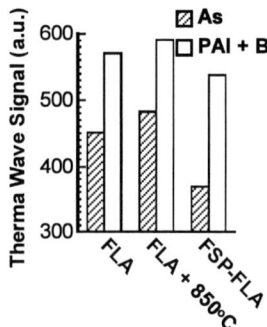

Fig. 11. Therma wave signal for various FLA. Higher T.W. means higher defect and/or deeper X_j.

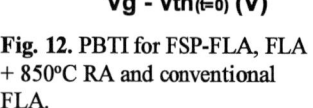

Fig. 12. PBTI for FSP-FLA, FLA + 850°C RA and conventional FLA.

Fig. 13. Effective electron mobility of NFET fabricated with FSP-FLA, FLA + 850°C RA and conventional FLA.

Steep Channel & Halo Profiles utilizing Boron-Diffusion-Barrier Layers (Si:C) for 32 nm Node and Beyond

A. Hokazono [1], H. Itokawa [2], N. Kusunoki [3], I. Mizushima [3], S. Inaba [1], S. Kawanaka [1], and Y. Toyoshima [1]

[1] Center for Semiconductor Research & Development, [2] Process & Manufacturing Engineering Center, [3] System LSI Division,
Toshiba Corporation, Semiconductor Company 8 Shinsugita-cho, Isogo-ku, Yokohama, Kanagawa 235-8522, Japan
Phone: +81-45-776-5663, Fax: +81-45-776-4104, Email:akira.hokazono@toshiba.co.jp

Abstract

Si:C layers under non-doped-Si epitaxial channel (Epi-channel) produces steep channel profile for 25 nm-L_G nMOSFET. Si:C layers work as the dopant-diffusion-barriers from the boron doped regions. Moreover, retrograde Halo profiles are also realized in this structure. Steep channel profiles at scaled device are confirmed, and the benefits of its profile at L_G of 25 nm are discussed.

1. Introduction

In order to enhance device performance while suppressing short channel effect (SCE), steeper channel profiles is indispensable [1]; however, adequate benefit has not been achieved at scaled devices due to the collapse of steep channel profile. In terms of steep channel profile formation, there are several papers proposing Si epitaxial growth after V_{TH} adjustment ion implantation (II) [2]. However, the final channel profile broadens because of the thermal budget throughout the following process steps (Fig. 1 (a)). Therefore, dopant-diffusion-barriers are required to suppress unintentional dopant diffusion. Since carbon has been used as the boron (B)-diffusion retarding materials, Si:C layers has been inserted under non-doped-Si epitaxial layers for nMOSFET [3-5] (Fig. 1 (c)). There are few reports for very short channel region down to 25 nm and 30 nm gate length (L_G). Advantage of steep channel profiles at scaled device has not been confirmed so far.

In this paper, the benefits of steep channel profiles at scaled devices are experimentally studied for the first time. Compatibility of Si:C-Si Epi-channel structure with advanced CMOS technology is evaluated, where higher mobility was achieved by the tensile channel stress and the elimination of the Coulomb scattering. Remarkable improvement in V_{TH} mismatch for Si:C-Si Epi-channel is also achieved. Finally, the benefits of steep channel profile inherent in scaled devices are pointed out.

Fabrication of Si:C-Si Epi-Channel MOSFET

After V_{TH} adjustment boron implantation, Si:C layers were grown by reduced pressure CVD (RPCVD) with SiH_2Cl_2-SiH_3CH_3-HCl-H_2 gas system followed by Si epitaxial growth. Substitutional carbons of 0.09% in Si:C layers were observed by X-Ray Diffraction (XRD). nMOSFET with SiON gate dielectrics (1.1-nm EOT) is fabricated with Si:C (5 - 10 nm) and non-doped Si epitaxial growth (5 - 10 nm) (Fig 1 (b)).

Device Design Optimization (Si:C-Si Thickness, Bottom Doped Regions)

B-diffusion from the bottom doped regions into the channel surface is suppressed by Si:C layers (Fig. 2), where the device design optimization of bottom doped regions is significant for further device scaling. Bottom doped regions need additional doping in order to decrease the depletion layer width (X_{DEP}) (③ in Fig. 3). If the bottom doped regions are formed without additional implantation (② in Fig. 3), X_{DEP} becomes wider, which cause the degradation of SCE and current drive at fixed off-current. Despite the higher doping concentration in bottom doped regions in Si:C-Si Epi-channel devices, V_{TH} remains the low value due to the suppression of B-diffusion by Si:C layers (Fig. 4); whereas, V_{TH} becomes high for Si Epi-channel only structure owing to the B-diffusion. 19% improvement of peak linear transconductance (G_M) at L_G of 20 nm was obtained (Fig. 5) in spite of the increased effective electric field (E_{eff}) due to decreased X_{DEP} (See Fig. 3). This result is explained by the elimination of Coulomb scattering. Higher G_M at thicker Si-thickness is due to the lower B-concentration at channel surface. Suppression of drain-induced barrier lowering (DIBL) in Si:C-Si Epi-channel device is attributed to X_{DEP} scaling by additional doping to bottom doped regions (Fig. 6 (a)). If the thickness ratio of Si:C to epitaxial Si becomes large, additional implantation for bottom doped regions should be required (Fig. 6 (b)). This is because thicker Si:C reduces the effect of SCE control by Halo doping, which is explained in later (See Fig. 13). Si:C-Si Epi-channel transistor showing higher I_{DSAT} indicates 13% improvement of current drive than that of conventional channel transistor (Fig. 7). Appropriate V_{TH} adjustment is also indispensable; in fact, V_{TH} can be controlled by additional implantation for bottom doped regions under Si:C layers (Fig. 8). One of the concern of using Si:C layers is the current drive degradation by the interface states (D_{it}) generation reported in Ref [4]. However, D_{it} induced modulation in C-V curves and frequency dependence of C-V curve was not observed in this experimental Si:C–Si Epi-channel transistor (Fig. 9).

Impact of Channel Structures on V_{TH} Mismatch

V_{TH} variation is the pressing problem for high-density SRAM; therefore, V_{TH} mismatch in pair MOSFETs with various channel structures was evaluated. V_{TH} sensitivity on channel implantation dosage was investigated by the simulation calibrated from the experimental results. V_{TH} in Si:C-Si Epi-channel transistor is insensitive to channel dosage due to the B-diffusion barriers of Si:C layers (Fig. 10). In typical Pelgrom plots, smaller A_{VT} (slope of $\Delta\sigma V_{TH}/\Delta(1/(LW)^{0.5})$) was observed in Si:C-Si Epi-channel devices due to the lower B-concentration on channel surface (Fig. 11). Furthermore, even with the higher doping for bottom doped regions in Si:C-Si Epi-channel, A_{VT} is still kept small (Fig. 12).

Compatibility of Si:C-Si Epi-channel with Advanced CMOS Technology

(a) Channel & Halo Profiles at Scaled device Another concern for Si:C-Si Epi-channel structure is whether a steep channel profile is remained after the thermal budget of activation, especially at the scaled device. Simulation analysis confirmed that steep channel profile is maintained even at L_G of 25 nm (Fig. 13 (a)); therefore, this technique is promising for device scaling. Also, much steeper B-profile is realized by only Halo implantation, compared to the conventional process (Fig. 13 (b)). Retrograde Halo profile can be designed thanks to the less B-diffusion in Si:C (Fig. 13 (c)). Body effect factor (γ) is the indicator of the channel profile steepness [6], and higher γ was observed in Si:C-Si Epi-channel device at $L_G = 30$ nm as well as at long L_G (Fig. 14), which indicates that the very steep channel is maintained at scaled device.

(b) Mobility Enhancement Technique Compatibility of Si:C-Si Epi-channel structure with mobility enhancement technique is indispensable. Since embedded Si:C layers under Si channel yields compressive stress for channel regions [7], the degradation of electron mobility might be the third concern. Both the reduction of Coulomb scattering (①→② in Fig. 15) and the mobility enhancement by 1.8 GPa contact etching stress liner process (①→③ in Fig. 15) work well in Si:C-Si Epi-channel transistor with tensile stress (①→④ in Fig. 15). This result suggests that the compressive stress by Si:C layers could be negligible because of the following reasons; one is the small stress due to the small amount of substitutional carbon, and the other is thin Si:C layers thickness not to relax tensile stress.

Device Characteristics (Scaled Transistor with Steep Channel)

I_{DS}-V_{DS} characteristics of nMOSFET with Si:C-Si Epi-channel and conventional channel were compared, where the devices with almost the same V_{TH} were evaluated. Despite 7 times higher dosage in bottom doped regions under Si:C layers, 5.3% I_{DSAT} improvement was realized (Fig. 16). Ref [8] reported that large V_{GS}, or high E_{eff} reduces the mobility advantage in steep channel profile, because the mobility is determined by surface roughness scattering instead of Coulomb scattering. Therefore, it has been considered that the advantage of the steep channel profile is limited to the low E_{eff} regions. This is consistent with the long channel transistor in this study (Fig. 17); however, as the gate length is scaled down, the benefit of steep channel profile comes again because of the following two reasons; one is explained by E_{eff}-mobility curves. Since channel impurity concentration is very high due to Halo implantation in short channel transistor, Coulomb scattering degrades the carrier mobility even at high E_{eff} regions. Therefore, the elimination of Coulomb scattering improves the transistor performance. The other is explained by the carrier injection velocity (V_{inj}). I_{DSAT} in scaled devices is mainly determined by V_{inj} at source edge, where virtual V_{inj} at source regions is strongly correlated with the remote Coulomb scattering by channel & Halo doping; hence, steep channel profile at scaled device increases V_{inj}. The higher V_{inj} in Si:C-Si Epi-channel transistor was confirmed by simulation (Fig. 18). In conclusion, steep channel & Halo profiles using Si:C layers is preferable for 32 nm generation and beyond.

Conclusion

Applicability of Si:C layers to the channel engineering is studied for the first time in scaled device down to 25 nm gate length. This proposed structure has other benefits, such as the elimination of the Coulomb scattering or the acceleration of the carrier injection velocity at the source edge. Therefore, the importance of the steep channel profile has been confirmed to keep the advantage of the MOSFET scaling.

References [1] R-H. Yan *et al.*, IEEE-ED39, 1704 (1992). [2] K. Noda *et al.*, IEEE-ED45, 809 (1998). [3] T. Ernst *et al.*, *Symp. VLSI Tech.*, 51 (2003). [4] O. Weber *et al.*, *ESSDERC'03*, 271 (2003). [5] F. Ducroquet et al., *IEDM Tech. Dig.*, 437(2004). [6] H. Koura et al., JJAP, Vol. 39, 2312(2000). [7] K-W. Ang *et al.*, *Symp. VLSI Tech.*, 42 (2007). [8] J. B. Jacobs *et al.*, IEEE-ED42, 870 (1995).

978-1-4244-1802-2/08/$25.00 ©2008 IEEE

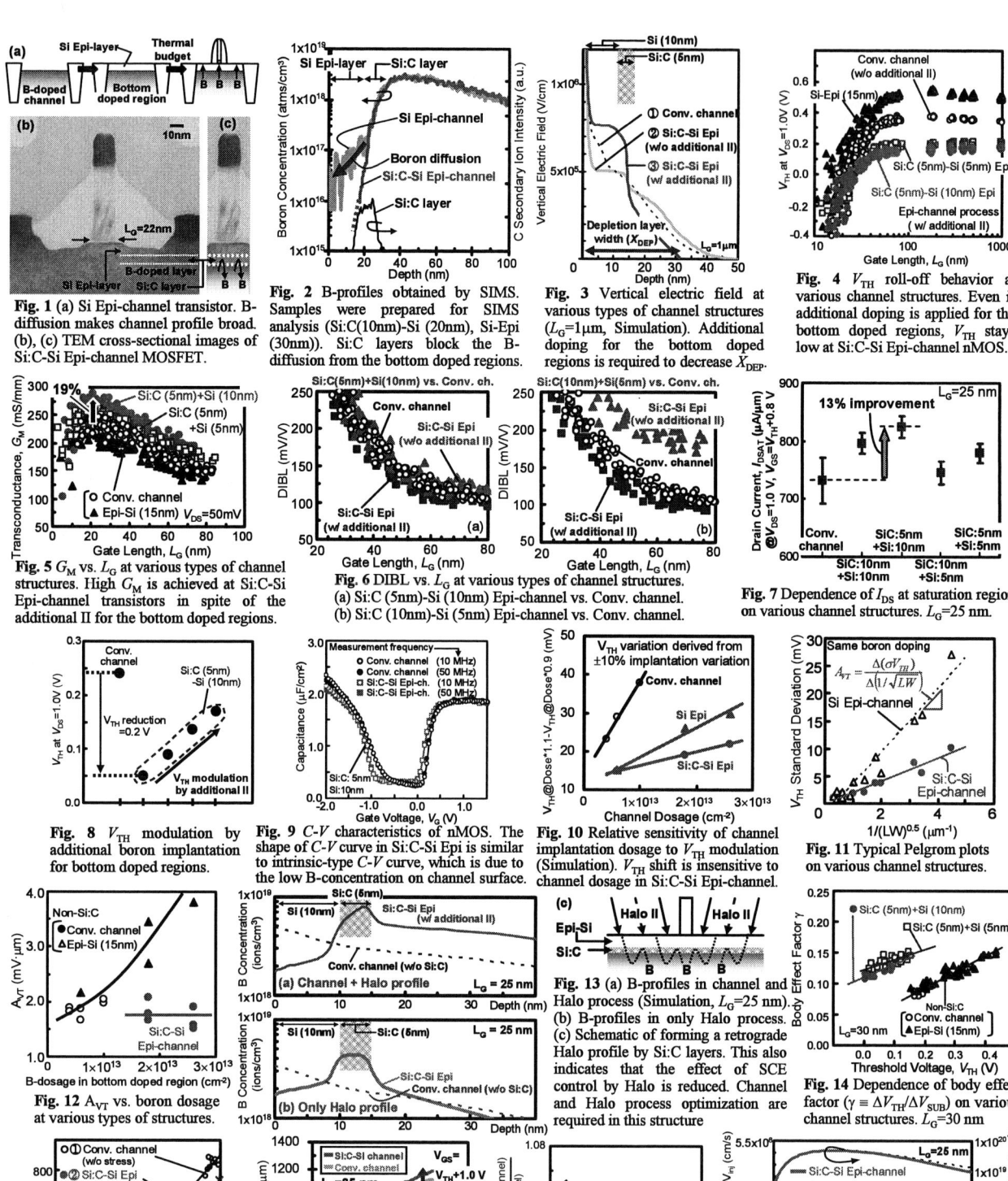

Fig. 1 (a) Si Epi-channel transistor. B-diffusion makes channel profile broad. (b), (c) TEM cross-sectional images of Si:C-Si Epi-channel MOSFET.

Fig. 2 B-profiles obtained by SIMS. Samples were prepared for SIMS analysis (Si:C(10nm)-Si (20nm), Si-Epi (30nm)). Si:C layers block the B-diffusion from the bottom doped regions.

Fig. 3 Vertical electric field at various types of channel structures (L_G=1μm, Simulation). Additional doping for the bottom doped regions is required to decrease X_{DEP}.

Fig. 4 V_{TH} roll-off behavior at various channel structures. Even if additional doping is applied for the bottom doped regions, V_{TH} stays low at Si:C-Si Epi-channel nMOS.

Fig. 5 G_M vs. L_G at various types of channel structures. High G_M is achieved at Si:C-Si Epi-channel transistors in spite of the additional II for the bottom doped regions.

Fig. 6 DIBL vs. L_G at various types of channel structures. (a) Si:C (5nm)-Si (10nm) Epi-channel vs. Conv. channel. (b) Si:C (10nm)-Si (5nm) Epi-channel vs. Conv. channel.

Fig. 7 Dependence of I_{DS} at saturation region on various channel structures. L_G=25 nm.

Fig. 8 V_{TH} modulation by additional boron implantation for bottom doped regions.

Fig. 9 C-V characteristics of nMOS. The shape of C-V curve in Si:C-Si Epi is similar to intrinsic-type C-V curve, which is due to the low B-concentration on channel surface.

Fig. 10 Relative sensitivity of channel implantation dosage to V_{TH} modulation (Simulation). V_{TH} shift is insensitive to channel dosage in Si:C-Si Epi-channel.

Fig. 11 Typical Pelgrom plots on various channel structures.

Fig. 12 A_{VT} vs. boron dosage at various types of structures.

Fig. 13 (a) B-profiles in channel and Halo process (Simulation, L_G=25 nm). (b) B-profiles in only Halo process. (c) Schematic of forming a retrograde Halo profile by Si:C layers. This also indicates that the effect of SCE control by Halo is reduced. Channel and Halo process optimization are required in this structure

Fig. 14 Dependence of body effect factor ($\gamma \equiv \Delta V_{TH}/\Delta V_{SUB}$) on various channel structures. L_G=30 nm

Fig. 15 R_{ON} vs. L_G at various types of channel structures. Smaller slope means the higher electron-mobility.

Fig. 16 I_{DS}-V_{DS} characteristics of Conv. and Si:C-Si Epi-channel device. Mobility enhancement technique is not applied.

Fig. 17 L_G vs. transistor performance enhancement by utilizing Si:C-Si Epi-channel. I_{DSAT} improvement is prominent at scaled device.

Fig. 18 Carrier injection velocity, V_{inj} at source potential peak (Simulation, L_G=25 nm). V_{DS} = 0.9 V. V_{GS} is applied (~ V_{TH}+0.9 V) to set the same carrier concentration on the channel surface. Though inversion layer in Si:C-Si Epi-channel is push onto the channel surface, V_{inj} in Si:C-Si Epi-channel is higher due to the elimination of Coulomb scattering.

Scaling Evaluation of BE-SONOS NAND Flash Beyond 20 nm

Hang-Ting Lue, Tzu-Hsuan Hsu, S. C. Lai, Y. H. Hsiao, W. C. Peng, C. W. Liao, Y. F. Huang, S. P. Hong*, M. T. Wu*, F. H. Hsu*, N. Z. Lien*,
S. Y. Wang*, L.W. Yang*, T. Yang*, K.C. Chen*, K.Y. Hsieh, Rich Liu, and Chih-Yuan. Lu
Emerging Central Lab, *Technology Development Center,
Macronix International Co., Ltd., 16 Li-Hsin Road, Hsinchu Science Park, Hsinchu, Taiwan.   E-mail: htlue@mxic.com.tw

Abstract

We have successfully fabricated and characterized sub-30 nm and sub-20 nm BE-SONOS NAND Flash. Good device characteristics are achieved through two innovative processes: (1) a low-energy tilt-angle STI pocket implantation to suppress the STI corner edge effect, and (2) a drain offset using an additional oxide liner to improve the short-channel effect. The conventional self-boosting program-inhibit and ISPP (incremental step pulse programming) for MLC storage are demonstrated for 20nm BE-SONOS NAND operation. Read current stability and read disturb life time are also evaluated. The estimated number of storage electrons is only 50-100, and for the first time we have demonstrated successful data retention after 150℃ baking in the "few-electron" regime. Our results strongly suggest that BE-SONOS is a promising charge-trapping (CT) technology for NAND Flash scaling.

I. Introduction

Recently, we proposed BE-SONOS [1] to solve NAND scaling problems below 30nm nodes. Compared with the conventional SONOS or MANOS, BE-SONOS uses a thin ONO tunneling barrier that allows hole tunneling during erase, while eliminates the direct tunneling leakage at retention. Therefore, fast erase speed and good data retention can be simultaneously achieved. Because the device has a planar structure and has no floating gate interference, BE-SONOS is a promising CT device at 30nm and below nodes.

In this work, we address issues for further scaling NAND flash below 20nm. Especially, self-boosting and ISSP characteristics, and data retention characteristics when the storage electrons fall under 100 are critically examined.

II. Process

Device cross-sectional views for sub-30 and sub-20 nm BE-SONOS devices are shown in Fig. 1. The key processes to improve the sub-20 nm characteristics are illustrated in Fig .2. In Fig. 2(a), we introduce a low-energy tilt-angle STI pocket implantation into the sidewall at the STI corners, just after the nitride hard mask etching, and followed by the STI trench etching. A higher p-well doping concentration suppresses the sidewall parasitic transistor and thus reduces the STI edge effect [2]. In order to provide better short-channel effect, we use an additional oxide liner before junction implantation to enlarge the effective channel length. In Fig. 1(a), a near-planar structure is fabricated, which facilitates the pitch scaling. Many identical devices (~1000) are tested within a whole wafer to provide a clear statistical evaluation. The whole-wafer measurement provides a worst-case estimation for the device variations.

NAND operation is evaluated using a 32-WL NAND array. Typical thickness for the O1/N1/O2/N2/O3 layers are 13/20/25/60/60 Å, respectively.

III. Device Characteristics

The typical program/erase characteristics of the sub-20 and sub-30 nm BE-SONOS devices are shown in Fig. 3. ISPP method is used for programming by applying a constant voltage increment (e.g. 0.2 V) at each successive programming step. Figure 3(a) shows that ISPP programming is linear (with ISPP slope = 0.7 for sub-30 nm device), and converges together for various starting V_{PGM}. The ISPP slope for the sub-20 nm devices is also linear but lower (~0.5) than sub-30 devices.

Figure 3(b) compares the erase speed. Sub-20 nm device also shows lower erase speed than sub-30 nm devices. At the same channel width, longer channel length shows faster speed and lower erase saturation.

Figure 4 explains the above geometric effects. Since the ONO stack height (~17 nm) is already comparable with the device dimension, significant fringing field exists at the edge. The simulation results show that the bottom oxide electric field is reduced when channel length is scaled down, leading to degraded program/erase efficiency. On the other hand, the electric field across the top oxide is slightly increased, leading to enhanced gate injection and larger erase saturation.

Figure 5 compares the ISPP of BE-SONOS with various EOT and O1. All BE-SONOS capacitors (Fig 5(a)) show ideal linear programming (ISPP slope~1). However, Fig. 5(b) shows that the NAND devices generally has degraded ISPP slope than capacitors. Our previous analysis [2] indicates that the STI edge effect degrades the ISPP slope. Through our novel STI pocket implantation, the ISPP slope is well maintained even for the sub-20 nm devices. This fact is of crucial importance, since ISPP programming self-corrects any ONO thickness variation, and provides a tight Vt distribution control.

The Vt distribution during dumb-mode ISPP programming (without program-verify) is shown in Fig. 6(a). Although the distribution is wide (measured for the whole wafer), the programmed-state has a Gaussian distribution, and is uniformly shifted during ISPP. This behavior is consistent with Fig. 5, since ISPP slope are similar for all the devices. The well-behaved ISPP is a critical attribute for BE-SONOS that enables tight Vt distribution.

A checkerboard pattern with adjacent cells defined at different levels (A, B, C, D) are designed to study the MLC window (Fig. 6(b)). After cell A is programmed using ISPP, self-boosting method is applied and cell B is continuously programmed. Likewise for cell C and D. We apply the same testing procedure within the whole wafers. The results in Fig. 6(c) prove that ISPP together with self-boosting method provides a tight Vt distribution for the sub-30 nm MLC BE-SONOS NAND.

Erase distribution is shown in Fig. 7. In general the erase distribution is wide. However, we found that recessed-STI (or FinFET-like) shows lower erased Vt distribution than near-planar structure. This is due to the larger field enhancement effect [3]. This offers a promising way to solve the erase saturation.

Typical endurance is shown Fig. 8. Excellent endurance is maintained after 1K cycling stress. The corresponding IV (inset) curves are well-behaved.

The stored electron number for the sub-30 nm BE-SONOS device are estimated to be 50 ~ 100 (trap density (~10^{13}/cm^2) times the channel area). Figure 9 shows that the sub-30 nm BE-SONOS devices possess excellent data retention after P/E cycling stress and very long-term (>300 hour) 150 ℃ baking. This proves that BE-SONOS has excellent few-electron (<100) storage capability. For comparison, the retention of sub-50 nm BE-SONOS is shown in the inset. It shows negligible charge loss after baking.

IV. Read Current Stability and Read Disturb:

The read current stability under continuous reading is shown in Fig. 10(a). The current fluctuation (noise) is very small even though the device is very small. The inset shows the evaluation of BE-SONOS device using pulse-IV technique. Pulse-IV measurements show no transient response from μsec to msec ranges. This shows that there is no transient charge trapping or de-trapping in the thin N1.

Read disturb life time is evaluated using a larger pass gate voltage for acceleration. Figure 10(b) shows that the sub-30 nm BE-SONOS can sustain more than 1M read cycles at Vread<7 V.

VI. Summary

Sub-20 nm BE-SONOS NAND is demonstrated. Excellent few electron storage and tight Vt distribution control capabilities are demonstrated.

References

[1] H. T. Lue et al, in *IEDM Tech. Dig.*, 2005, pp. 547-550.
[2] H. T. Lue et al, in *IEDM Tech. Dig.*, 2007, pp. 161-164.
[3] T. H. Hsu et al, in *IEDM Tech. Dig.*, 2007, pp. 913-916.

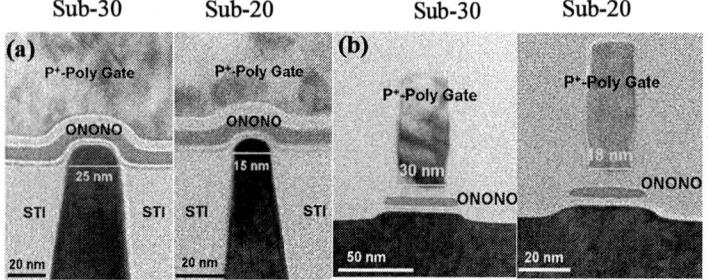

Fig. 1 TEM cross-sectional views of the near-planar BE-SONOS devices. (a) channel-width direction. Small STI recess (<10 nm) is obtained. (b) channel-length direction.

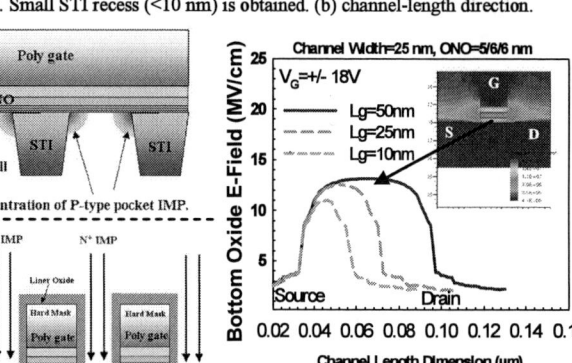

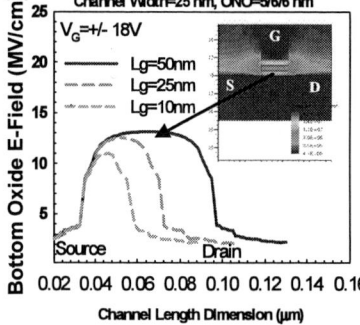

Fig. 2 (a) STI pocket implantation. A higher well doping is introduced to the sidewalls of STI to suppress the edge parasitic transistor. (b) Drain offset using oxide liner.

Fig. 4 Bottom oxide E field simulation along the channel length (Lg) direction for various Lg. Smaller channel length shows degraded bottom tunnel oxide field, while increased top oxide field.

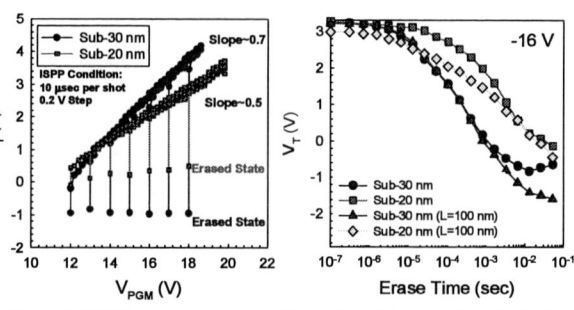

Fig. 3 (a) ISPP programming and (b) erasing characteristics of sub-20 and sub-30 nm BE-SONOS NAND devices. Sub-20 device shows degraded program/erase efficiency than sub-30 devices.

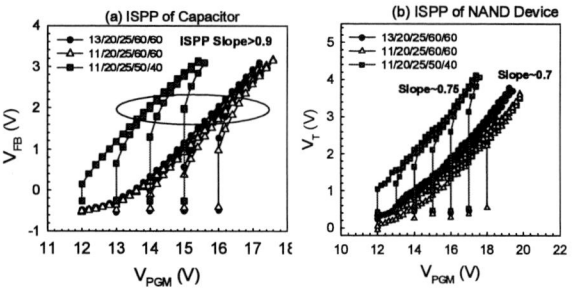

Fig. 5 ISPP comparison of sub-30 nm BE-SONOS with various EOT. (a) Capacitor. (b) NAND devices. All capacitors show ideal ISPP slope~1, even for very thin BE-SONOS. However, the NAND devices shows lower ISPP slopes.

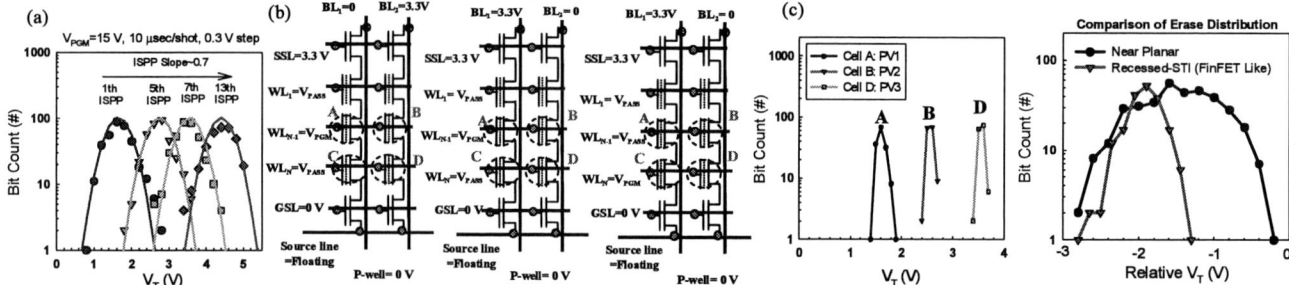

Fig. 6 (a) Vt distribution (collected from the whole wafer) after dumb-mode ISPP programming (without verify) for sub-30 nm BE-SONOS devices. Program state distribution is uniformly shift with ISPP slope ~0.7. (b) Procedure to define the checkerboard pattern of MLC. When the selected cell is program-verified, self-boosting method (V_{CC}=3.3 V, V_{PASS}=12 V) is applied for program-inhibit. (c) Vt distribution using ISPP and self-boosting methods. Many identical NAND arrays are tested. The final distribution of cell A (PV1) is only slightly broadened because of the program disturb. Tight distribution is obtained within the whole wafer.

Fig. 7 Comparison of erased state distribution for the near-planar and recessed-STI (FinFET-like) structures. Recessed-STI shows lower erased Vt and smaller distribution.

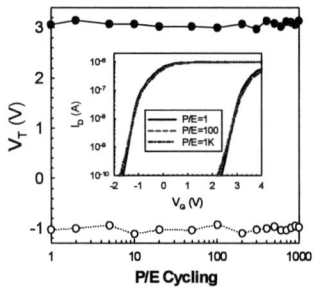

Fig. 8 (a) Typical P/E cycling endurance of sub-30 nm NAND cell. The inset shows the corresponding IV curve for the 32-WL NAND cell.

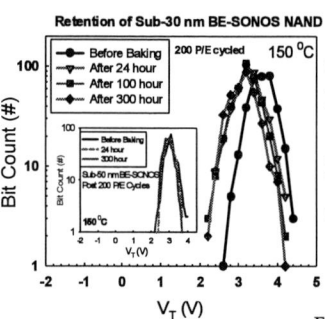

Fig. 9 150℃ retention of the sub-30 nm BE-SONOS NAND devices after 200 P/E cycling. Excellent few-electron storage (<50) is demonstrated after long-term (>300 hr) baking. The inset shows the retention of the sub-50 nm BE-SONOS for comparison.

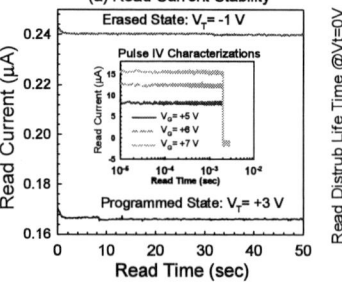

Fig. 10 (a) Read current stability measurement for the sub-20 nm BE-SONOS NAND device. A constant voltage (6V) is applied at pass gates and select gates to continuously read the 32-WL NAND current. The inset shows the pulse-*IV* measurement result for a single BE-SONOS MOSFET using Keithley 4200 system. Read current is stable from usec to msec, indicating no transient charge-trapping/de-trapping issues. (b) Read disturb life time evaluation. Various large gate voltage is applied to accelerate the read disturb. Read disturb can exceed 1M read cycles (assuming 1 msec read time for each read).

Highly Scalable NAND Flash Memory with Robust Immunity to Program Disturbance Using Symmetric Inversion-Type Source and Drain Structure

Chang-Hyun Lee, Jungdal Choi, Youngwoo Park, Changseok Kang, Byeong-In Choi,
Hyunjae Kim[1], Hyunsil Oh[1], and Won-Seong Lee

Advanced Technology Development Team 2, CAE Team[1], Semiconductor R&D Center, Samsung Electronics Co., LTD.
San #16, Banwol-Dong, Hwasung-City, Gyeonggi-Do, Korea, 445-701
Phone:+82-31-208-2598 Fax:+82-31-208-4799 E-mail:changhyun@samsung.com

Abstract

The symmetric inversion-type S/D structure has been employed for achieving available program disturbance for scaled NAND flash memory beyond sub-40nm node. The inversion S/D structure enables the channel doping to be reduced due to non-existence of n-lateral diffusion and it suppresses charge sharing between program-inhibit channels, resulting in superior program disturbance. Moreover, the cells show better current drivability in the technology node less than 50nm by more successful working of gate fringing field with smaller word-line gap, compared to those with the n-diffused S/D junction.

Keywords: NAND flash, Inversion S/D, Fringing field, Program disturbance, TANOS

Introduction

One of the scaling issues of the NAND flash memory is to guarantee robust characteristics to program disturbance without losing the on-off characteristics of cells. As the NAND cells shrink, the tolerance of charge leakage responsible for dropping the boosting potential of program-inhibit channel is reduced as shown in Fig. 1. From the reason, the program-inhibit channel should hold boosting-induced charge more tightly not to be lost for keeping proper boosting potential of program-inhibit channel with scaling down. The overall leakage path including of the leakages via selector transistors and the band-to-band tunneling in channel/junction should be blocked more tightly for scaled NAND flash memory. On the contrary, the increase in channel doping is inevitable for suppressing the short channel effects and keeping the available on-off characteristics with scaling of gate length. However, the increase of channel doping causes to degrade program disturbance owing to the band-to-band tunneling at junction and channel and the hot carrier problem [1].

In this work, the novel NAND flash memory with symmetric inversion S/D structure has been proposed to solve the program disturbance for scaled NAND flash memory beyond sub-40nm node.

Results and Discussions

The NAND flash memory with the inversion S/D structure was fabricated by using a 40nm TANOS-NAND standard process. The TANOS cells have $SiO_2/SiN/Al_2O_3$/metal gate and its string consists of 32 cells and SSL/GSL transistor. The key features were described in the previous papers [2,3].

Figures 2 shows the schematic view of NAND string with the inversion S/D structure. For the inversion S/D cells, the current flow is governed by the amount of the inversion charge induced by fringing field of neighboring gate controls instead of majority carrier in case of n-diffused S/D cells. The virtual gate potential onto the S/D region during a read operation was simulated with the NAND technology node in Fig. 3. In this simulation, the unselected cells are biased at 6V of Vread with -3V of erase state and the "on" cell is biased at 0V of Vsense. The virtual potential is revealed not to be enough to induce sufficient inversion charge at 90nm of WL gap, but beyond 40nm node, the minimum virtual potential is induced up to 4V enough to bring about electron carrier density of high $10^{19} cm^{-3}$. In terms of cell current drivability, the n-diffused cells have higher electron concentration at S/D region than that of the inversion S/D cells in regime less than 60nm

of WL gap. However, beyond 50nm of WL gap, the inversion S/D cell is revealed to have better current drivability than n-diffused cells because the inversion S/D cells have the same electron density onto the S/D region as the n-diffused cells but lower channel resistance originated from reducing channel doping. Such a superior current drivability for the inversion S/D cells coincides with the cell current comparison of the 40nm-node TANOS-NAND flash memeory, as shown in Fig. 5 and Fig. 6.

In terms of the short channel effects, the inversion S/D cells are proven to have stronger immunity to the short channel effects than the n-diffused cells as shown in Fig. 7. We can maintain lower off-cell current for the inversion S/D cells as the cell threshold voltage decreases because the n-diffused S/D layer enhances off-leakage path by lateral diffusion of n-type dopants.

The characteristics of program and pass disturbances are compared in Fig. 8. The inversion S/D cells show superior characteristics of program disturbance than the n-diffused S/D cells. The erased state of the inversion S/D cells is not disturbed by program-inhibit stress in the regime of lower pass voltage, even though the cell is evaluated by the conventional self-boosting scheme. The simulation result reveals that the channel self-localization phenomena happen in program-inhibit channel for the inversion S/D cells as shown in Fig. 9.

The inversion S/D layer suppresses the sharing of the boosting-induced charge among the channels of the cells biased by pass voltage, effectively [4] and so the boosting efficiency is enhanced. Moreover, the n-p junction in the string dose not exist for the inversion S/D cells and then the junction leakage is suppressed as shown in Fig. 10. Such two causes bring about superior program disturbance for the inversion S/D cells than the n-diffused cells.

The multi-level cell (MLC) distribution was achieved successfully by using two phase MLC program [3]. The cells in the "11" state still keep erase state without fail bit originated from program disturbance when the other cells are programmed from the erase one to the "01" state of highest program threshold voltage.

From the result, it is proven that the program disturbance-free window can be obtained for the NAND flash memory with the inversion S/D structure.

Conclusion

A new NAND flash memory with the symmetric inversion S/D structure has been successfully developed. For 40nm-node TANOS-NAND flash cell, the inversion S/D cells is proved to have better current drivability and superior program disturbance because of the successful working of fringing field and better short channel effects, compared to the conventional n-diffused S/D cells. The symmetric inversion S/D structure is expected to be a promising solution to overcome program-disturbance problem for the scaled NAND flash memory beyond sub-40nm node.

References

[1] J.D. Lee, et al, IEEE 21st NVSMW, pp. 31-33. 2006.
[2] Y. Park, et al, IEDM Tech. Dig., pp. 29-31. 2006.
[3] C.H. Lee, et al, VLSI Tech. Dig., pp. 21-22. 2006
[4] D.Y. Oh, et al, IEEE 22nd NVSMW, pp. 39-41. 2007.

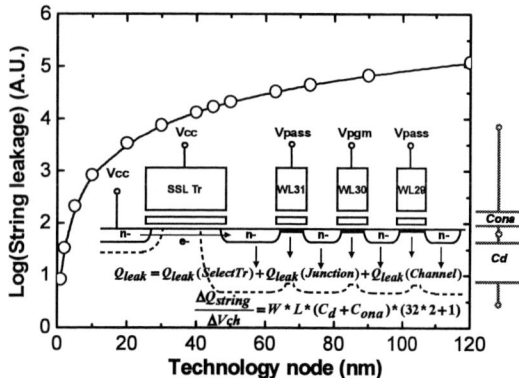

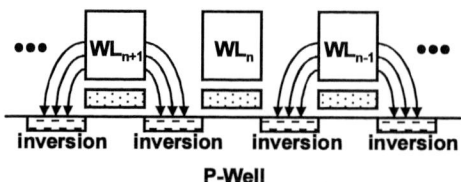

Fig 2(a). Schematic view of proposed NAND flash memory with symmetric inversion S/D structure.

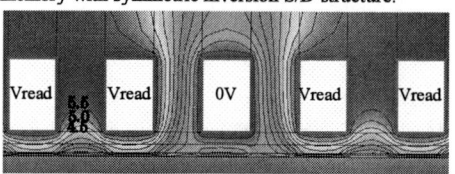

Fig 2(c). Simulated potential contour induced by neighboring gates for 40nm-node TANOS-NAND flash memory.

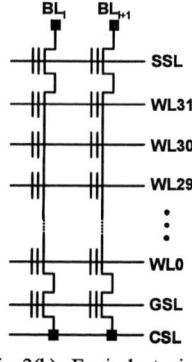

Fig 2(b). Equivalent circuit diagram of NAND flash memory with symmetric inversion S/D structure.

Fig. 1. Leakage requirement of boosted channel for maintaining a drop of less than 1V during programming time as a function of technology node.

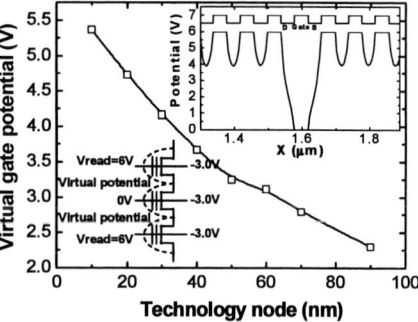

Fig. 3. Simulated virtual gate potential onto S/D region induced by fringing field of neighboring gates as a function of technology node.

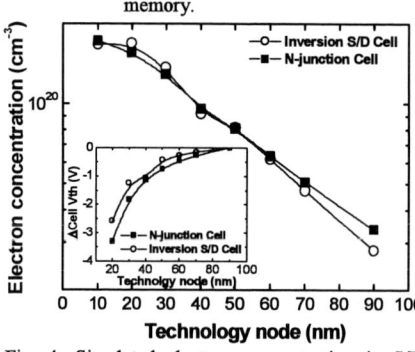

Fig. 4. Simulated electron concentration in S/D region for N-junction cells and inversion S/D cells as a function of technology node (Inset graph is V_{TH} roll-off characteristics).

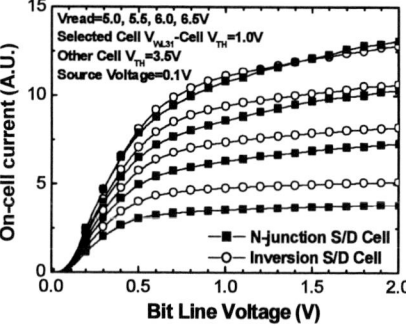

Fig. 5. "On" cell currents of fabricated 40nm-node TANOS-NAND flash memories with N-junction cells and inversion S/D cells.

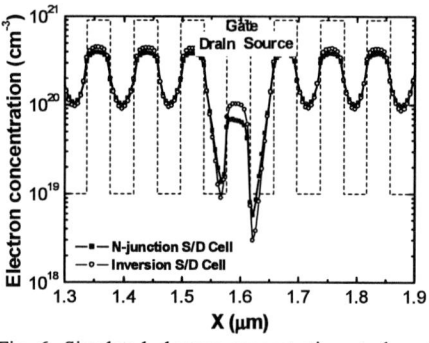

Fig. 6. Simulated electron concentration at channel surface of 40nm-node TANOS-NAND flash memories with N-junction cells and inversion S/D cells.

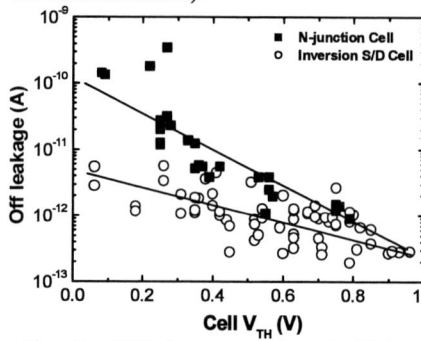

Fig. 7. Off-leakage currents of fabricated 40nm-node TANOS-NAND flash memories with N-junction cells and inversion S/D cells as a function of cell threshold voltage.

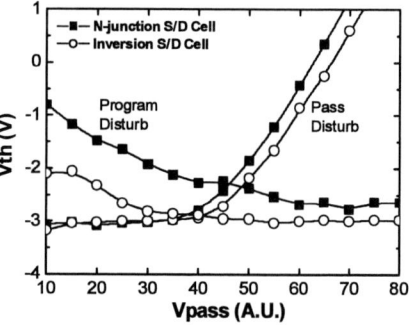

Fig. 8. Program and pass disturbances of fabricated 40nm-node TANOS-NAND flash memories with N-junction cells and inversion S/D cells.

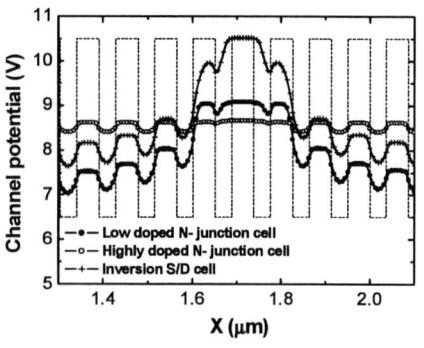

Fig. 9. Simulated potential of boosted channel under program-inhibit situation for 40nm-node TANOS-NAND flash memories with N-junction cells and inversion S/D cells.

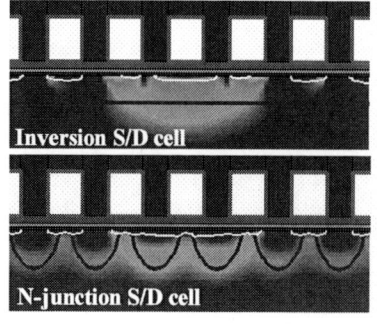

Fig. 10. Simulated leakage current of boosted channel under program-inhibit situation for 40nm-node TANOS-NAND flash memories with N-junction cells and inversion S/D cells.

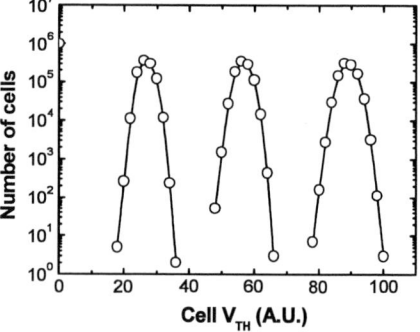

Fig. 11. Multi-level cell V_{TH} distribution with 4M-bit density for fabricated 40nm-node TANOS-NAND flash memory with inversion S/D cells.

92

Vertical Structure NAND Flash Array Integration with Paired FinFET Multi-bit Scheme for High-density NAND Flash Memory Application

June-Mo Koo, Tae-Eung Yoon, Taehee Lee, Sungjae Byun, Young-Gu Jin, Wonjoo Kim, Sukpil Kim, Jongbong Park, Junseok Cho, [2]Jeong-Dong Choe, [2]Choong-Ho Lee, [3]Jong Jin Lee, [4]Je-Woo Han, [5]Yunseung Kang, [4]Sangjun Park, [4]Byoungho Kwon, [4]Yong-Ju Jung, Inkyoung Yoo, and Yoondong Park

Samsung Advanced Institute of Technology, San 14-1, Giheung-Gu, Yongin-City, Gyeonggi-Do, 449-712, South Korea
[2]Advanced Technology Development Team 2, [3]DRAM Process Architecture Team, [4]Process Development Team, [5]FAB Process Technology Development Group 1, Semiconductor R&D Center, Samsung Electronics Co., LTD., South Korea
Tel) 82-31-280-9293, Fax) 82-31-280-9308, E-mail) junemo.koo@samsung.com

Abstract

Multi-bit Vertical Structure NAND (VsNAND) Flash memories with 32-paired FinFET cell string have been successfully integrated for the first time. Its array integration issues regarding the sub-10nm vertical structure fin could be solved by proper choices of isolation material, ion implantation, and word line patterning. VsNAND Flash array cells with TANOS (TaN/Al$_2$O$_3$/SiN/SiO$_x$/Si) charge trap structure show possibilities of acceptable program/erase properties and cell Vth distribution characteristics for multi-level NAND Flash application.

Introduction

Demands of low cost and high-density NAND Flash memory have rapidly increased due to its small cell size compared to other types of non-volatile memories. Although scaling NAND Flash memory has been aggressively achieved compared to other memory devices [1], NAND Flash memory will face severe problems of short channel effect and charge coupling interference in an array [2].

For the scaling down of the device, FinFET structure is regarded as a good candidate to solve the reliability problem with a larger storage area compared to planar short channel structure for a given design rule [3]. However, FinFET Flash memory also has its limitation with scaling-down. At this point, TANOS is now the most possible candidate due to its charge coupling free structure [4].

In this work, we adapt a combination of TANOS and VsNAND Paired FinFET [5] for high-density NAND Flash memory. TANOS VsNAND Flash Array, using two sidewall channels on a single bit line, is integrated first on the test vehicle based on 63nm design rule and its electrical properties are discussed.

Device Fabrication

Fig. 1 shows the VsNAND array structure and two sidewall channels of a single bit line. The process flow, sequence and key parameters for the device integration are shown in Fig. 2 and Table I. Silicon fin was formed using conventional STI (shallow trench isolation) process. To make the multi-bit cell on a single fin, the silicon fin was simultaneously etched with SiN spacer hard mask while DC and CSL areas were protected by SiN mask. After active fin separation, boron ion implantation is implemented into fin-to-fin space (in Fig. 3) and the trench is filled with SiN material with small void (in Fig. 4(a)) in order to reduce the interference of cell-to-cell from the simulation results of Fig. 3(d). Since the channel of the VsNAND is formed at the vertical sidewall of the split-fin, channel area can be achieved by adjusting field oxide recess. Increasing the amount of field recess, larger on-cell current is possible due to wide channel area, but there are trade-offs between the on-cell current and the difficulty of gate patterning process. In this work, field oxide is recessed at about 100nm.

Results and Discussions

As shown in Fig. 4 (a), the integrated VsNAND string has a TANOS cell structure with 4nm SiO$_2$ / 6nm SiN /13nm Al$_2$O$_3$ / 7nm TaN / 50nm Poly-Si /35nm WN/W. When the word lines of gate stack were patterned, etch process of TaN sidewall layer was a big concern for the TANOS VsNAND structure. We adopted a two-step etch process with 0 bias to remove the TaN layers between word line to word line, as shown in Fig. 4 (b). SEM and TEM images the final integrated VsNAND array structure are shown in Fig. 5 including a 32-cell string (a), a gate word line (b) and DC contacts.

Initial threshold voltages were measured on 32-cell string as shown in Fig. 6. After erasing at 18V, Fig. 7 indicates the Id-Vg characteristics of memory cell transistors for various program voltages. Fig. 8 show program and erase characteristics of the integrated TANOS VsNAND array cell. The program voltage and time for reaching threshold voltage of 3V is about 18.5V and 100us in Fig. 8 (a), while erase voltage of -3V and time were 19V and 10ms in Fig. 8 (b), which are enough for multi-bit cell operation in Fig. 9. The conventional checkerboard pattern to observe program inhibit characteristics is shown in Fig. 10 and Fig. 11. The Id-Vg results of the checkerboard pattern were apparently distinguished programmed cells from un-programmed cells. Fig. 12 and Fig. 13 show endurance and HTS (high temperature storage) characteristics. In these test, the string cells are stressed by 1.2k cycling and then baked at 200℃ for 2hr. There was a charge loss of 0.5V in program Vth.

Conclusion

TANOS VsNAND Flash of 32-cell string array was successfully integrated on sub-10nm paired fin body with 100nm sidewall channel using conventional 63nm NAND Flash vehicle. The integrated VsNAND array not only has good Id-Vg characteristics for MLC but also no program disturbance property of adjacent cells from checkerboard pattern.

References

[1] D. Kwak, et al., VLSI Tech. Dig., pp. 12, 2007.
[2] K. Kim, IEDM Tech. Dig., pp. 333, 2005.
[3] P. Xuan, et al., IEDM Tech. Dig., pp. 609, 2006.
[4] Y. Shin, et al., IEDM Tech. Dig., pp. 337, 2005.
[5] S. Kim, et al., VLSI Tech. Dig., pp. 104, 2006.

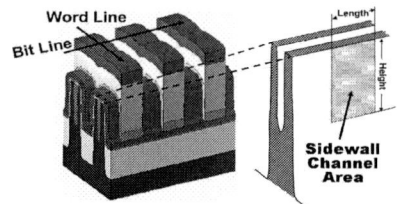

Fig. 1 Schematic diagrams of the VsNAND Flash array structure with two sidewall channels.

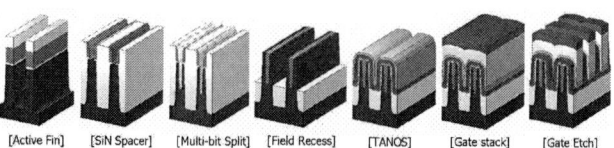

[Active Fin] [SiN Spacer] [Multi-bit Split] [Field Recess] [TANOS] [Gate stack] [Gate Etch]

Fig. 2 Process flow of TANOS VsNAND Flash Array Integration.

Table I. Process Sequence and Key Parameter of VsNAND Array Integration

Process Sequence		Key Parameter	
1. Well Form	9. Bit to Bit Isolation IIP	Parameter	nm
2. Cell Active Fin Form	10. SiN Fill and Air Gap	Fin Width	<10
3. STI and STI CMP	11. SiN CMP	Channel Height	>100
4. SiN Removal	12. Field Oxide Recess	Gate Length	63
5. SiN Spacer Depo	13. Gate Stack	Tunnel SiO_2	4
6. DC and CSL Masking	14. W/L Pattern (TaN)	SiN	7
7. Spacer Form	15. Junction Form	Al_2O_3	15
8. Cell Multi-bit Form	16. BEOL	TaN	10

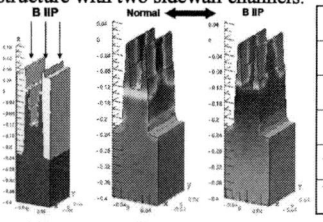

(a) (b) (c) (d)

Fig. 3 TCAD simulation results of implantation for Fin to Fin interference.

MLC Spec < 0.30 ~ 0.35 V (Cell-to-cell Interference sub 3x nm)	
Structure (weak S/D doping)	Interference ΔVTh
TANOS Planar	0.45 (Tox 3nm) 0.56 (Tox 8nm)
High-k IPD Planar	0.57
TANOS VsNAND	0.45 (SiNx) 0.37 (Void)
TANOS VsNAND BF2 doping	0.21 (SiNx) 0.09 (Void)

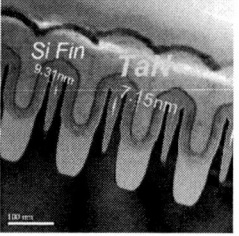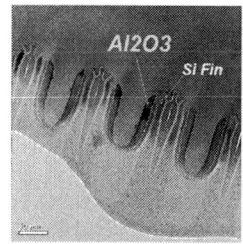

(a) (b)

Fig. 4 Cross-sectional TEM images (a) before and (b) after TaN dry etch results. (TANOS sidewall cell with 7nm TaN / 13nm Al_2O_3 / 6nm SiN / 4nm SiO_2/ 10nm Si)

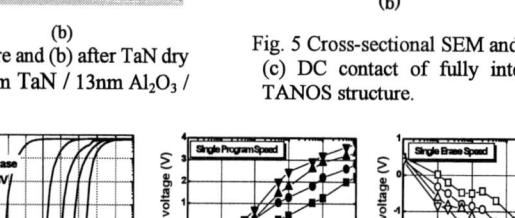

(a)

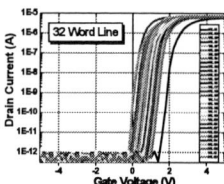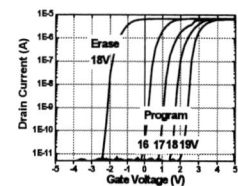

(b) (c)

Fig. 5 Cross-sectional SEM and TEM images (a) cell string (b) gate and (c) DC contact of fully integrated VsNAND Flash memory with TANOS structure.

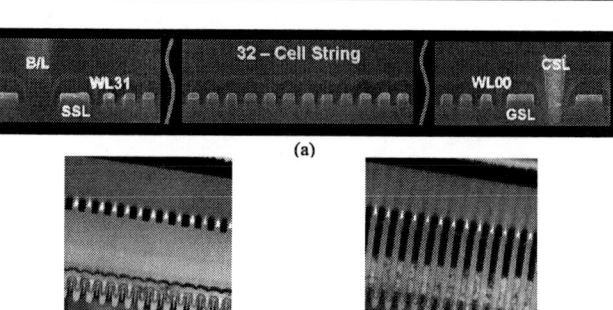

(a) (b)

Fig. 6 Initial Id-Vg curves of 32-word lines in VsNAND string.

Fig. 7 Id-Vd characteristics of memory cell transistor for various program voltages after erase 18V.

Fig. 8 (a) Program and (b) erase characteristics of VsNAND Flash cell.

Fig. 9 Id-Vg characteristic of VsNAND cell after program and erase at 100us and 10nm respectively.

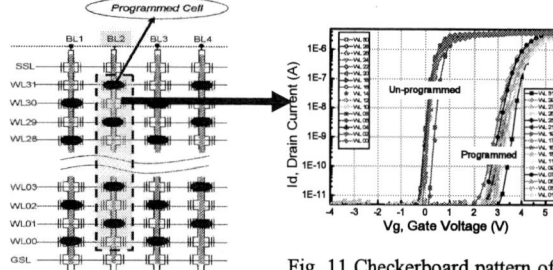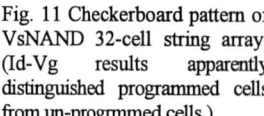

Fig. 10 Checkerboard pattern of programmed VsNAND array.

Fig. 11 Checkerboard pattern of VsNAND 32-cell string array. (Id-Vg results apparently distinguished programmed cells from un-progrmmed cells.)

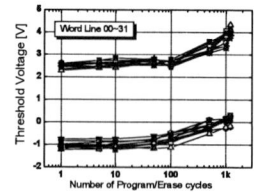

Fig. 12 Endurance characteristics of VsNAND Flash with 32-cell string.

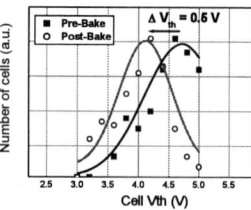

Fig. 13 Cell Vth distribution after baking at 200℃ for 2hr. The string cells are stressed by 1.2k cycling.

Novel 3-D Structure for Ultra High Density Flash Memory with VRAT (Vertical-Recess-Array-Transistor) and PIPE (Planarized Integration on the same PlanE)

Jiyoung Kim, Augustin J. Hong, Masaaki Ogawa, Siguang Ma, Emil B. Song, You-Sheng Lin*, Jeonghee Han**, U-In Chung** and Kang L. Wang

Electrical Engineering Department, University of California, Los Angeles, CA 90095 USA

UCLA Nanoelectronics Research Facility*, Process Development Team, Samsung Electronics Co., Korea **

E-mail) hbt100@ee.ucla.edu

Abstract

A 3-D Flash memory cell of VRAT (Vertical-Recess-Array-Transistor) has been fabricated using a unique and simple 3-D integration method of PIPE (Planarized Integration on the same PlanE), which allows for the successful implementation of ultra high density Flash memory. In addition, procedures to increase the memory density further using another advanced structure, Zigzag-VRAT (Z-VRAT), are developed.

Introduction

Since the late 1990s, NAND Flash has been driving the semiconductor industry, surpassing the DRAM business with a faster development speed due to its simple cell structure and the rapidly growing market of mobile applications. Nowadays, NAND Flash technology has reached the 30nm node and 64Gb storage capacity has been obtained by using a remarkable technology of the CTL (Charge Trap Layer) rather than a traditional floating gate technology [1]. The information amount generated world-wide has been doubled every year and the memory industry has been able to meet this level of demand in a cost-effective manner, primarily by simply scaling the device laterally [Table 1]. However, the technological improvement by scaling is getting difficult, since it requires high cost lithography processes, suppression of short channel effects, and ever increasing difficulty in isolation among the adjacent cells. Recently, in order to overcome the problems of the scaling, several 3-D structures have been proposed but they are often complex technologically and thus have high bit cost [2-3]. In this paper, a new 3-D VRAT using a simple integration PIPE scheme is developed for the future Flash memory architecture with the potential to achieve tera-bit level memory densities. A crucial benefit for the VRAT structure is to increase the memory density through stacking multi-layers, rather than by simply scaling down the size of each device lithographically. The use of the vertical structure and fully depleted channels allows for keeping the dimension of the gate length the same to suppress subthreshold leakage current. Furthermore, in the simple PIPE scheme, vertical interconnects can be simultaneously and conveniently formed along with the main array of VRAT, providing a new paradigm for 3-D integration in a simple and cost-effective architecture. Finally, an advanced structure, Z-VRAT, is introduced for achieving even higher memory densities.

Experiments and Results

1. Basic concept of VRAT with PIPE

The VRAT consists of vertically chained transistors and interconnects, which are processed by the unique 3-D integration scheme, PIPE. The vertical stacking of the array enables a much higher memory density while the PIPE provides a novel convenient method to interconnect the stacked transistors and WLs (Word Lines). Multi-stacks of the nitride and undercut oxide layers provide a mold to be covered with a thin and fully depleted polysilicon channel. Moreover, the elongated channel length due to the recessed shape circumvents the short channel effects while the double gate structure increases the driving current. Fig. 2a shows a conventional stair-like method to connect each WL to the corresponding contact, which usually requires as many lithography and etch steps as the number of the stacked layers while the PIPE scheme does not need any additional lithography step [Fig. 2b]. During the deposition of the oxide and nitride layers on to the Si substrate, the stacked layers run along the lateral directions and are bent vertically when they meet the Si mesa [Fig. 2b]. After filling the undercut space with the WL electrode and then planarization by CMP process, all the WLs are exposed on the same plane allowing for forming the contact at the same depth from the top.

2. Fabrication Processes

Fig. 3 shows the process sequence for building the VRAT structure. The multi layers of the oxide and nitride films are deposited on the Si mesa. After patterning the multi-stack layers by lithography, followed by etching, the active region is formed [Fig. 3a]. The oxide layers are selectively etched by wet process, which makes undercut space [Fig. 3b]. The next process is to cover the mold with a polysilicon which becomes a channel material of active transistors [Fig. 3c]. For the storage material, a nitride CTL is introduced between the tunnel oxide and control oxide, followed by deposition of the WL electrode. A subsequent etch-back process removes the WL electrodes on the sidewall, leaving the WL electrodes in the undercut space and isolating them from one another [Fig. 3d]. After planarization by CMP process, multi-stack layers over the Si mesa are removed and all the WL electrodes are exposed [Fig. 3e]. The next step is to isolate each vertical array by cutting through the polysilicon to separate the mesa into several stacked arrays [Fig. 3f]. Finally, BL and WL contacts are made and connected to periphery circuits [Fig. 3g]. In this experiment, the VRAT with two stacked layers was fabricated. The undercut space was formed by using BOE [Fig. 4a]. The polysilicon of 200nm thick was deposited by using LPCVD at $600^\circ C$ to improve a step coverage [Fig. 4b]. As a WL electrode, a highly doped N-type polysilicon was introduced after deposition of a tunnel oxide of 7nm thick by RTO at $600^\circ C$, a nitride CTL of 15nm thick using LPCVD at $800^\circ C$ and a control Al_2O_3 layer of 17nm thick using ALD at $300^\circ C$ [Fig. 4c]. Fig. 5 shows SEM images of the process of the PIPE scheme starting from formation of the Si mesa using KOH etching [Fig. 5a], to deposition of the four stacked layers [Fig. 5b] and then to planarization by CMP process [Fig. 5c].

3. Electrical Characteristics

Fig. 6 shows a layout of VRAT, which has three terminals and all the gates are chained to the same gate terminal and biased together. The basic FET function was confirmed from V_{GS}-I_{DS} and V_{DS}-I_{DS} characteristics [Fig. 7, 8]. The high V_{th} of 2.5V came from Fermi level pinning due to the polysilicon gate and the thick gate oxide; these and parasitic resistance degrade the V_{DS}-I_{DS} performance. The device performance may be improved by using a metal gate with a thin gate oxide as well as optimized S/D implants. The VRAT shows a stable and repeatable intrinsic hysteresis characteristic [Fig. 9]. The programming and erasing operations show excellent performance, of which a V_{th} margin was more than 4V on the programming bias of 16V [Fig. 10].

4. Advanced structure: Z-VRAT

Another novel structure, Z-VRAT, is introduced to further increase memory density while maintaining the merits of VRAT and PIPE. As shown in Fig. 11, Z-VRAT can be fabricated by splitting a VRAT into two narrow mesas through the middle and hence doubling the memory density. It can be fabricated using the same process sequence of VRAT up to the deposition process of the polysilicon channel, beyond which the polysilicon on top is removed to expose the nitride layer, and a wet process is used to remove all the mold layers leaving just the skeleton of the polysilicon, on which another set of WL electrodes can be formed inside [Fig. 12a]. The processes of filling the gate electrode, planarization and formation of contacts are similar with those of VRAT. The Z-VRAT with eight transistors in a vertical stack has been simulated using MEDICI and the result shows a zigzag current path along the polysilicon channel [Fig. 13]. Parasitic resistance effects are also evaluated by comparing a driving current on the best and worst array conditions and the results are comparable to the planar type array. To investigate the interactions among cells in a vertical stack, the memory programming function has also been simulated by using PD-AAA from MEDICI, assuming the CTL as a thin conductor. The potential of the CTL is changed only in the programmed cell as intended, but not in the adjacent cells [Fig. 15], demonstrating negligible interactions to the neighboring cells. The memory density is anticipated to reach as high as 128Gb using VRAT and as high as 256Gb with Z-VRAT, both on the 50nm node technology with 16 multi-stacks [Table 2].

Conclusion

The VRAT has been successfully developed, demonstrating 3-D integration with a unique, convenient and simple process, PIPE. The technology developed affords vertically stacked multi-layer Flash memory. Our simulation also shows that the new Z-VRAT structure will push the Flash memory density to 256Gb using the 50nm technology.

References

[1] Kinam Kim *et al.*, IEDM Tech. Dig., pp 27-30, 2007

[2] Soon-Moon Jung *et al.*, IEDM Tech. Dig., pp 37-40, 2006

[3] H. Tanaka *et al.*, Symp. on VLSI Tech. Dig., pp 14-15, 2007

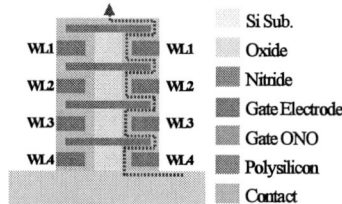

	2004	2005	2006	2007
DR[nm]	60s	50s	40s	30s
Density	8G	16G	32G	64G
Litho. (NA)	ArF (0.85)	ArF (0.93)	ArF (1.20)	ArF DPT

Table 1 Development Roadmap of the NAND Flash memory.

Fig.1 Schematic cross sectional view of VRAT. The broken line with red color shows the zigzag current path.

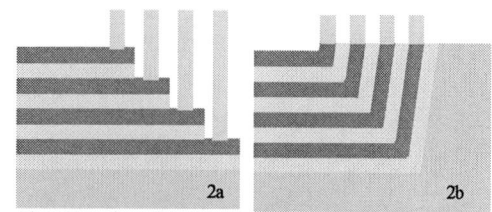

Fig.2 PIPE scheme [Fig. 2b] showing that no additional lithography step is needed comparing with the conventional stair-like methods, which need many extra steps [Fig. 2a].

Si Sub.
Oxide
Nitride
Gate Electrode
Gate ONO
Polysilicon
Contact

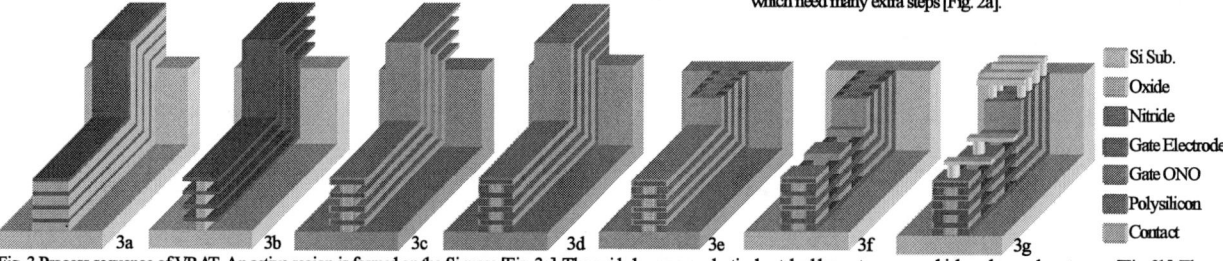

Fig. 3 Process sequence of VRAT. An active region is formed on the Si mesa [Fig. 3a]. The oxide layers are selectively etched by wet process, which makes undercut space [Fig. 3b]. The polysilicon is deposited on the multi-staking mold, which becomes the channel material [Fig. 3c]. The undercut space is filled with gate oxide and electrode followed by etch-back process to isolate the gates from each other [Fig. 3d]. The gate electrodes are exposed on the same plane after planarization by CMP and N-type implantation is added with tilt angle to form the S/D [Fig. 3e]. Each string is isolated [Fig. 3f]. The contacts of WLs and BLs are completed [Fig. 3g].

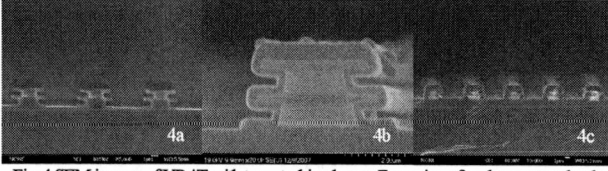

Fig. 4 SEM images of VRAT with two-stacking layers. Formation of undercut space by the BOE process [Fig. 4a]; the polysilicon deposition on the mold by LPCVD [Fig. 4b]; filling the undercut space by gate oxide and polysilicon [Fig. 4c]

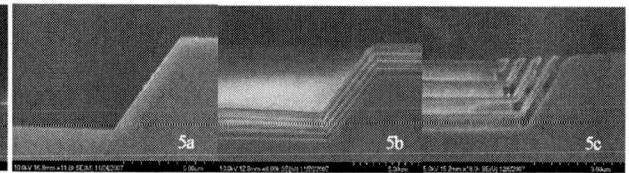

Fig. 5 SEM images of PIPE with four-stacking layers. Formation of the Si mesa using KOH wet process [Fig. 5a]; Deposition of four layers of oxide and nitride films [Fig. 5b]; planarization by CMP and exposure of electrodes on the same plane [Fig. 5c]

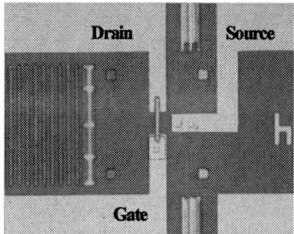

Fig. 6 Top view layout of VRAT which has three terminals.

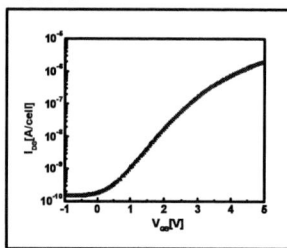

Fig. 7 V_{GS} vs. I_{DS} plot of VRAT.

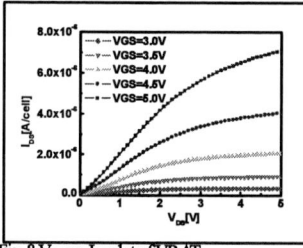

Fig. 8 V_{DS} vs. I_{DS} plot of VRAT

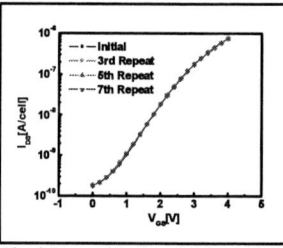

Fig. 9 Repeatable characteristic. Intrinsic hysteresis is nearly zero.

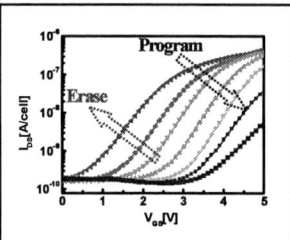

Fig. 10 Memory performance. V_{th} margin is more than 4V through programming and erasing operations.

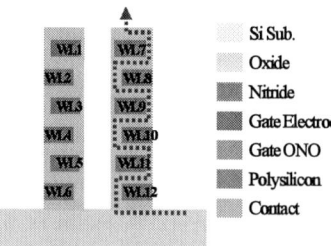

Fig. 11 Schematic cross sectional view of Z-VRAT with six transistors in a vertical string. The dashed line in red color shows the current path.

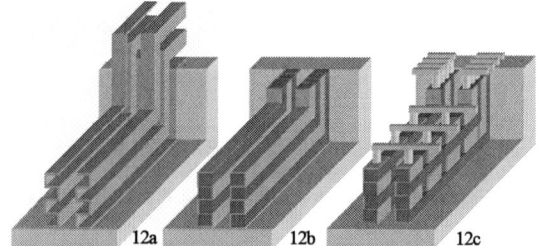

Fig. 12 Z-VRAT process. The polysilicon on the top is removed and oxide/nitride layers are etched by wet process [Fig. 12a]; filling with a gate electrode and planarization [Fig. 12b]; forming contacts and connections with metal lines [Fig.

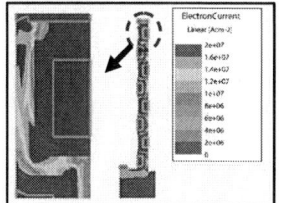

Fig. 13 MEDICI simulation for Z-VRAT with eight transistors in a vertical array. Electron current flows in a zigzag path along the polysilicon channel.

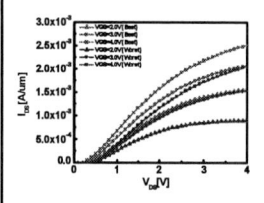

Fig. 14 MEDICI simulation for the driving current on the best and worst cell conditions, which are comparable to the planar type device.

Fig. 15 Interactions between cells in a vertical string, being negligibly small from transient simulation of the voltage difference of the CTL.

	Planar	VRAT		Z-RAT	
DR	50nm	50nm		50nm	
Density	16G	64G	128G	128G	256G
Note	MLC	8 Lys. SLC	16Lys. SLC	8 Lys. SLC	16Lys. SLC

Table 2 Comparison of possible memory density on the 50nm node. The Z-RAT with 16-staking layers is expected to reach 256Gb.

Channel-Stress Study on Gate-Size Effects for Damascene-Gate pMOSFETs with Top-Cut Compressive Stress Liner and eSiGe

S. Mayuzumi, S. Yamakawa, *D. Kosemura, *M. Takei, J. Wang, T. Ando, Y. Tateshita, M. Tsukamoto,
H. Wakabayashi, T. Ohno, *A. Ogura and N. Nagashima

Semiconductor Business Group, SONY Corporation, Atsugi Tec., 4-14-1 Asahi-cho, Atsugi-shi, Kanagawa, 243-0014, Japan
Phone: +81-46-230-5508, Fax: +81-46-230-6556, E-mail: Satoru.Mayuzumi@jp.sony.com
*School of Science and Technology, Meiji University

Abstract

Damascene gate process enhances the drivability in shorter gate length region, as compared to conventional gate 1st process for pFETs with compressive stress SiN liner and embedded SiGe. The origin of the gate length effect is investigated for the first time by using the UV-Raman spectroscopy. Moreover, the relationship between channel strain and gate width for damascene gate pFETs is analyzed and the effect is also demonstrated. It is found that channel strain is considerably enhanced in shorter gate length and narrower gate width by the combination of damascene gate process and stress enhancement techniques.

Introduction

High-performance metal/high-k gate MOSFETs with top-cut dual stress SiN liners and embedded SiGe (eSiGe) using damascene gate process have been recently reported [1, 2]. This process considerably enhances channel strain due to dummy gate removal [1, 3]. Id gain using damascene gate process is greater than that using gate 1st process, especially for shorter gate length (Lgate) (Fig. 1) [1, 4]. It seems that high channel strains for shorter Lgates are achieved by the damascene gate process because of using c-SL and eSiGe. It is necessary to make an analysis of the channel strain by physical analysis technique in order to understand the mechanisms in detail. In recent years, strain measurement techniques for Si are being developed such as UV-Raman spectroscopy system, which is useful to allow nondestructive test for evaluating both local and global strains [5, 6]. In this paper, the origin of the drivability enhancement in shorter Lgate region for damascene gate pFETs is investigated by the UV-Raman spectroscopy for the first time. Moreover, the behaviors of channel strain for gate width (Wgate) (ref. [7] for gate 1st pFETs) are also characterized, and the drivability enhancement in narrower Wgate is demonstrated.

Device Fabrication

The process flow used in this work is similar to that described in ref. [1]. Fig. 2 schematically shows a cross sectional structure of damascene metal/high-k gate stack for pFET with top-cut c-SL and eSiGe. Si recess and 80-nm eSiGe epitaxical deposition were carried out by using dummy gates and those spacers. 40-nm c-SLs having the stress of 2.0-GPa were deposited after nickel silicidation. After ILD deposition, c-SLs were cut only on the top of dummy gates by using CMP. The devices after dummy gate removal were used for the UV-Raman spectroscopy ($\lambda = 364$ nm). Metal/high-k gate stacks were formed with TiN/HfO$_2$. For reference, the devices without c-SL and eSiGe were also fabricated.

Results and discussion

A. Dependence on gate length

Simulated lateral stress (Sxx) distributions around the channel for gate 1st and damascene gate pFETs with three different Lgates are shown in Fig. 3. Fig. 4 shows the comparison of Sxx in the combination of c-SL and eSiGe for damascene gate pFETs with 40-nm and 0.2-um Lgates. Fig. 5 shows Sxx and vertical stress (Szz) dependence on Lgate for the same devices as shown in Fig. 3. Sxx and Szz are the stresses around the channel center and edge at 1-nm depth below the Si surface. Fig. 6 shows relative hole mobility dependence on Lgate, where hole mobilities were calculated using the calculated average channel stress and piezo-resistance coefficients [1, 8]. Hole mobility increases with the increases in compressive Sxx and tensile Szz. Compressive Sxx in shorter Lgate for damascene gate pFETs becomes much larger than those for gate 1st ones (Figs. 3 and 5). Moreover, compressive Sxx becomes significantly large, due to the combination of damascene gate process and c-SL even if eSiGe isn't applied (Fig. 4). On the other hand, the differences in tensile Szz between gate 1st and damascene gate pFETs are comparatively small in shorter Lgate as compared to compressive Sxx (Fig. 5). Therefore, hole mobilities in shorter Lgate for damascene gate pFETs become much larger than those for gate 1st (Fig. 6). Figs. 7 and 8 show Raman peak shift dependence on Lgate for damascene gate pFETs. The Raman peak shifts around the channel center are plotted in Fig. 8. High strains are observed at the edge of the channel for 10-um Lgate

devices as supported by simulations (Figs. 3, 5 and ref. [6]). Moreover, as Lgate shortens, Raman peak shifts become large. Although the devices with gate removal are used for the UV-Raman spectroscopy, stress differences between prior to and after metal/high-k gate process would be negligible [1]. Fig. 9 shows the comparison of Sxx at the channel center between simulations and UV-Raman analyses for pFETs with c-SL and eSiGe, where the simulated Sxx are the stresses at 5-nm depth below of the Si surface because of coordinating with the penetration depth of the UV-Raman spectroscopy. The increases in the Raman peak shift are transformed into Sxx by using ref. [9], assuming that channel stresses are induced by uniaxial stresses. These results of UV-Raman analyses significantly agree with those of simulations. Therefore, it is confirmed that damascene gate process with top-cut c-SL and eSiGe remarkably enhance the channel stresses in shorter Lgates.

B. Dependence on gate width

Raman peak shift dependence on Wgate position for damascene gate pFETs with c-SL and eSiGe is shown in Fig. 10, where Lgate is 1 um, and Wgates are 1 um and 10 um. Fig. 11 shows Raman peak mapping around the edge of the channel scanning in Wgate direction for the same device as shown in Fig. 10. High compressive strain is applied around the edge of the channel beside the STI. Moreover, compressive strain around the center of Wgate position is enhanced in narrower Wgate. As mentioned above, the differences of Raman peak in Figs. 10 and 11 practically correspond to Sxx. These results indicate that high uniaxial compressive Sxx induced by c-SL and eSiGe in Lgate direction expand the lattice constant especially at the edge of channel beside the STI in Wgate direction. Therefore, Sxx around the center of Wgate position increases in decreases with Wgate. Fig. 12 shows the dependence of Raman peak shift at Wgate center on Wgate for damascene gate pFETs with and without eSiGe. As Wgate shortens, Raman peak shifts become large regardless of eSiGe. The high compressive Sxx induced by the combination of damascene gate process and c-SL without eSiGe in Lgate direction (Fig. 4) must be accelerated by lattice-constant expansion at the channel edge in Wgate direction. Therefore, it is considered that the Wgate dependence (Fig. 12) is obtained regardless of eSiGe. These results suggest that damascene gate pFETs achieve the high drivability in narrow Wgate region. Figs. 13 and 14 show channel resistance dependence on Lgate and Ion-Ioff characteristics for damascene gate pFETs with c-SL, eSiGe and various Wgates, respectively. These results demonstrate the higher drivability in both linear and saturation regions for narrower Wgate pFETs. Fig. 15 shows normalized Ion dependence on Wgate. Ion enhancements in narrower Wgates are demonstrated by using c-SL or eSiGe for damascene gate pFETs, as supported by the results of UV-Raman spectroscopy. Eventually, high drive current of 1090 uA/um at Ioff = 100 nA/um, Vdd = 1.0 V and Tinv = 1.4 nm is obtained for 0.3-um Wgate pFETs.

Conclusion

The channel strain dependences on gate length and gate width for pFETs with top-cut compressive stress SiN liner and eSiGe using damascene gate process are precisely investigated by the UV-Raman spectroscopy for the first time. Damascene gate process enhances channel strain and the drivability in both shorter gate length and narrower gate width. These technologies have advantages in device scaling.

Acknowledgment

The authors would like to thank the STARC for the supports of UV-Raman spectroscopy.

References

[1] S. Mayuzumi et al., IEDM, p.293 (2007) [2] K. Mistry et al., IEDM, p.247 (2007) [3] J. Wang et al., VLSI, p.46 (2007) [4] S. Pidin et al., IEDM, p.213 (2004) [5] A. Ogura et al., Jpn. J. Appl. Phys., Vol.45, p.3007 (2006) [6] D. Kosemura et al., SSDM, p.390 (2007) [7] S.K.H. Fung et al., IEDM, p.1037 (2007) [8] C. S. Smith, Phys. Rev., Vol.94, p.42 (1954) [9] I.D. Wolf, Semicond. Sci. Technol. Vol.11, p.139 (1996)

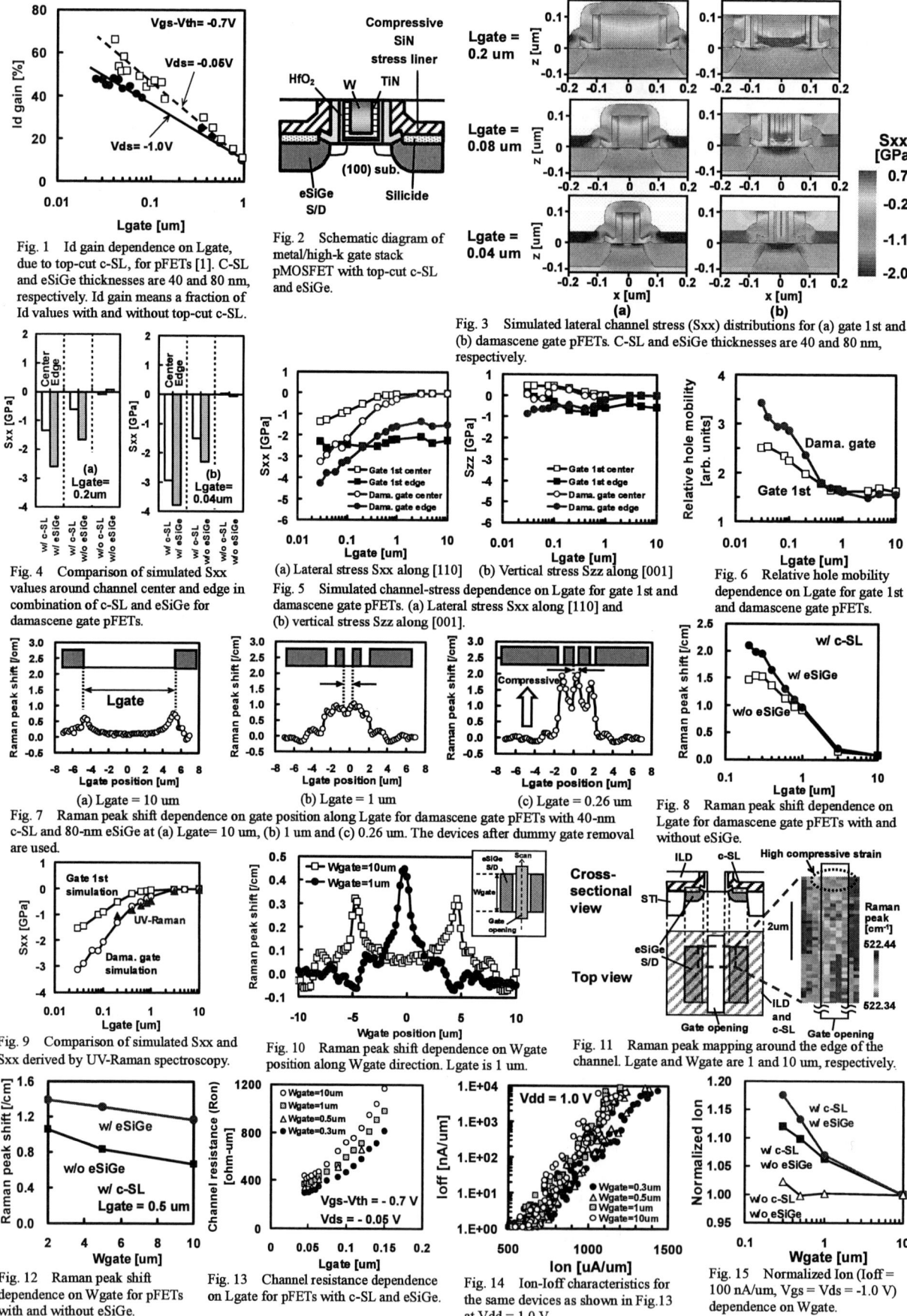

Fig. 1 Id gain dependence on Lgate, due to top-cut c-SL, for pFETs [1]. C-SL and eSiGe thicknesses are 40 and 80 nm, respectively. Id gain means a fraction of Id values with and without top-cut c-SL.

Fig. 2 Schematic diagram of metal/high-k gate stack pMOSFET with top-cut c-SL and eSiGe.

Fig. 3 Simulated lateral channel stress (Sxx) distributions for (a) gate 1st and (b) damascene gate pFETs. C-SL and eSiGe thicknesses are 40 and 80 nm, respectively.

Fig. 4 Comparison of simulated Sxx values around channel center and edge in combination of c-SL and eSiGe for damascene gate pFETs.

(a) Lateral stress Sxx along [110] (b) Vertical stress Szz along [001]
Fig. 5 Simulated channel-stress dependence on Lgate for gate 1st and damascene gate pFETs. (a) Lateral stress Sxx along [110] and (b) vertical stress Szz along [001].

Fig. 6 Relative hole mobility dependence on Lgate for gate 1st and damascene gate pFETs.

(a) Lgate = 10 um (b) Lgate = 1 um (c) Lgate = 0.26 um
Fig. 7 Raman peak shift dependence on gate position along Lgate for damascene gate pFETs with 40-nm c-SL and 80-nm eSiGe at (a) Lgate= 10 um, (b) 1 um and (c) 0.26 um. The devices after dummy gate removal are used.

Fig. 8 Raman peak shift dependence on Lgate for damascene gate pFETs with and without eSiGe.

Fig. 9 Comparison of simulated Sxx and Sxx derived by UV-Raman spectroscopy.

Fig. 10 Raman peak shift dependence on Wgate position along Wgate direction. Lgate is 1 um.

Fig. 11 Raman peak mapping around the edge of the channel. Lgate and Wgate are 1 and 10 um, respectively.

Fig. 12 Raman peak shift dependence on Wgate for pFETs with and without eSiGe.

Fig. 13 Channel resistance dependence on Lgate for pFETs with c-SL and eSiGe.

Fig. 14 Ion-Ioff characteristics for the same devices as shown in Fig.13 at Vdd = 1.0 V.

Fig. 15 Normalized Ion (Ioff = 100 nA/um, Vgs = Vds = -1.0 V) dependence on Wgate.

45nm High-k + Metal Gate Strain-Enhanced Transistors

C. Auth, A. Cappellani, J.-S. Chun, A. Dalis, A. Davis, T. Ghani, G. Glass, T. Glassman, M. Harper, M. Hattendorf,
P. Hentges, S. Jaloviar, S. Joshi, J. Klaus, K. Kuhn, D. Lavric, M. Lu, H. Mariappan, K. Mistry, B. Norris,
N. Rahhal-orabi, P. Ranade, J. Sandford, L. Shifren%, V. Souw, K. Tone, F. Tambwe, A. Thompson, D. Towner,
T. Troeger, P. Vandervoorn, C. Wallace, J. Wiedemer, C. Wiegand

Logic Technology Development, %PTM, Intel Corp., Hillsboro, OR, U.S.A.

Contact: email chris.auth@intel.com

Abstract

Two key process features that are used to make 45nm generation metal gate + high-k gate dielectric CMOS transistors are highlighted in this paper. The first feature is the integration of stress-enhancement techniques with the dual metal-gate + high-k transistors. The second feature is the extension of 193nm dry lithography to the 45nm technology node pitches. Use of these features has enabled industry-leading transistor performance and the first high volume 45nm high-k + metal gate technology.

Introduction

High-k + metal gate transistors have been incorporated into our 45nm logic technology to provide improved performance and significantly reduced gate leakage [1]. Hi-k + Metal gates have also been shown to have improved variability at the 45nm node [2]. The transistors in this work feature 1.0nm EOT high-k gate dielectrics with dual workfunction metal gate electrodes and 35nm gate lengths. The addition of new gate materials is complicated by the need to mesh the process requirements of the metal gate process with the uniaxial strain-inducing components that have become central to the transistor architecture. The resultant process flow needs to ensure that the performance benefits of both elements are fully realized.

The standard scaling requirements for the strained silicon components and for the gate and contact pitches also needs to be addressed at the 45nm node. Using 193nm dry lithography for critical layers at the 45nm technology node is preferred over moving to 193nm immersion lithography due to lower cost and greater maturity of the toolset. In order to achieve the 160nm gate and contact pitch requirements, unique gate and contact patterning process flows have been implemented.

Strain + Metal Gate: Key process considerations/results

The most commonly used techniques for implementing strain in the transistors include embedded SiGe in the PMOS S/D, stress memorization for the NMOS and a nitride stress capping layer for NMOS and PMOS devices. The two common methods for introducing a metal gate to the standard CMOS flow include, either "gate-first" or "gate-last" process. Most comparisons of these two process flows focus on the ability to select the appropriate workfunction metals, the ease of integration or the ability to scale but typically fail to comprehend the interaction with the strain-inducing techniques.

In the gate-first flow (Fig. 1), the dual-metal processing is completed prior to the polysilicon gate deposition. The metal-gates are then subtractively etched along with the poly gates prior to S/D formation. In contrast, for the gate-last flow, a standard polysilicon gate is deposited after the high-k gate dielectric deposition, which is followed by standard polysilicon processing through the salicide and the 1st ILD deposition. The wafer is then planarized and the dummy poly gate removed. The dual-metal gates are then deposited along with a low-resistance gate fill material. The excess metal is then polished off and followed by contact processing (Fig. 2).

By removing the poly gate from transistor after the stress-enhancement techniques are in place, it has been shown that the stress benefit from the embedded S/D SiGe process can be enhanced [3]. This is a key benefit for the gate-last process and can be illustrated in simulation with an estimated 50% increase in lateral compressive stress by removal of the polysilicon gate (Fig. 3). The Ge concentration of the SiGe stressors was increased from 22% in our 65nm technology [4] to 30% in 45nm. The combined impact of the increased Ge fraction and the strain enhancement from the gate last process allow for 1.5x higher hole mobility compared to 65nm despite the scaling of the transistor pitch from 220nm to 160nm.

Two methods of stress enhancement have been employed on the NMOS in this technology. First, the loss of the nitride stress layer benefit due to scaling the pitch from 65nm has been overcome by the introduction of trench contacts and tailoring the contact fill material to induce a tensile stress in the channel. The NMOS response to tensile vs. compressive contact fill materials is shown in figure 4. The trench contact fill material impact on the PMOS device is mitigated by use of the raised S/D inherent to the embedded SiGe S/D process.

The S/D component of stress memorization is compatible with the gate-last flow while the poly gate component would be compromised [5]. The poly gate component is replaced by Metal Gate Stress (MGS): modifying the metal-gate fill material to directly induce stress in the channel [6]. By introducing a compressive stress gate fill material the performance of the NMOS device is enhanced and additive to the contact fill technique [Fig. 5]. By use of a dual-metal process with PMOS 1st, the stress of the NMOS gate is decoupled from the PMOS gate through optimization of the PMOS gate stack to buffer the stress.

Through the strain enhancement and elimination of poly depletion both the saturation and linear drive currents improved (Fig. 6,7). Subthreshold characteristics are well-behaved (Fig. 8). Ring oscillator data for a fanout of 2 gate delay shows an improvement of 23% is demonstrated (Fig. 9). The table in figure 10 breaks out the RO gains between Idsat, Idlin and the gate and junction capacitances.

193nm Dry Patterning @ 45nm

The gate patterning process uses a double patterning scheme. Initially the gate stack is deposited including the polysilicon and hardmask deposition. The first lithography step patterns a series of parallel, continuous lines. Only discrete pitches are allowed, with the smallest at 160nm. A second masking step is then used to define the cuts in the lines. The 2-step process enables abrupt poly endcap regions allowing tight CTG design rules (Fig. 11).

The contact patterning process also uses a similar restriction to facilitate lithography. Trench diffusion contacts run parallel to the gates with discrete pitches, while trench gate contacts run orthogonal to the gates. Use of trench contacts has the added benefits of lowering the contact resistance by >50% and allowing use as a local interconnect which improves SRAM/logic density by up to 10%.

Conclusion

High-k + metal gate transistors have been integrated into a manufacturable 45nm CMOS process using 193nm dry lithography. The significant strain enhancement benefits of the gate-last process flow have been highlighted. The process has demonstrated record drive current at low leakage.

References

1. K. Mistry, et al, IEDM Tech. Dig. pp 247-250, 2007
2. K. Kuhn, IEDM Tech. Dig. pp. 471-474, 2007
3. J. Wang, et al, VLSI Sym. Tech. Dig. pp 46-47, 2007
4. S. Tyagi, et al, IEDM Tech. Dig. pp 1070-1072, 2005
5. A. Wei, et al, VLSI Sym. Tech. Dig. pp 216-217, 2007
6. C. Kang, et al, IEDM Tech. Dig. pp885-888, 2006

Gate-First	Gate-Last
-Isolation	-Isolation
-Hi-k gate deposition	-Hi-k gate deposition
-Dual Metal-Gate Dep	-Poly-Si gate dep/patterning
-Poly-Silicon deposition	-S/D formation
-Poly-Si/metal etch	-Salicide/Contact etch stop
-S/D formation	-1st ILD dep/polish
-Salicide/Contact etch stop	**-Poly Si gate removal**
-1st ILD dep/polish	**-Dual-Metal Gate dep**
-Contact formation	-Contact formation

Fig.1 Comparison of unique steps in gate-first and gate-last process flows. Key differences are highlighted in bold.

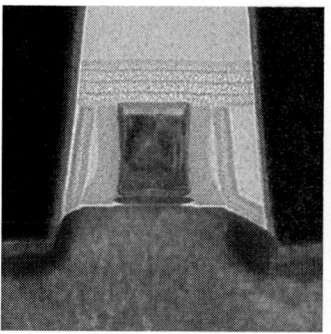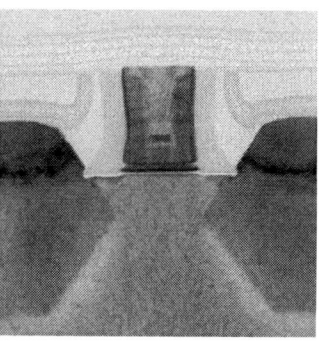

Fig.2 TEMs of High-k + Metal Gate NMOS and PMOS transistors

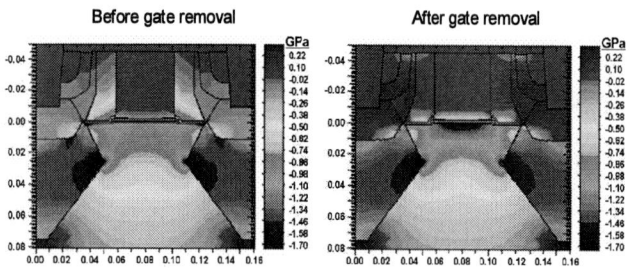

Fig.3 Stress contours in the PMOS transistor before and after the removal of the polysilicon dummy gate. Stress in the channel is shown to increase 50% from ~0.8GPa to >1.2 GPa.

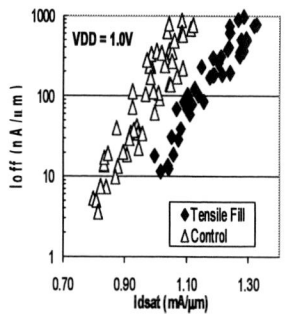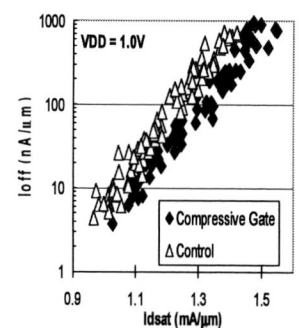

Fig.4 Ion-Ioff benefit of tensile Contact Fill showing a 10% NMOS Idsat benefit. Contact resistance is matched for the two fill materials.

Fig.5 Ion-Ioff benefit of compressive gate stress showing a 6% NMOS Idsat gain. Tensile Contact Fill is used on both sets of data.

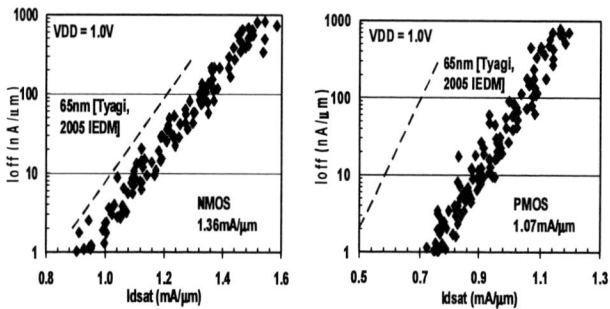

Fig.6 Overall Ion-Ioff for NMOS & PMOS relative to previous generation. 45nm transistors are benchmarked at 160nm pitch.

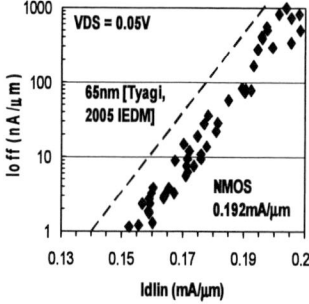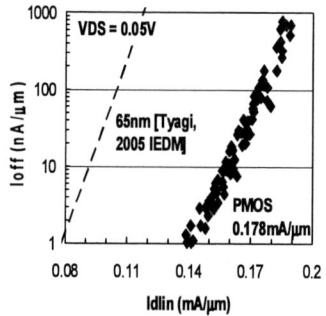

Fig. 7 NMOS/PMOS Idlin vs. Ioff relative to previous generation. 45nm transistors are benchmarked at 160nm pitch

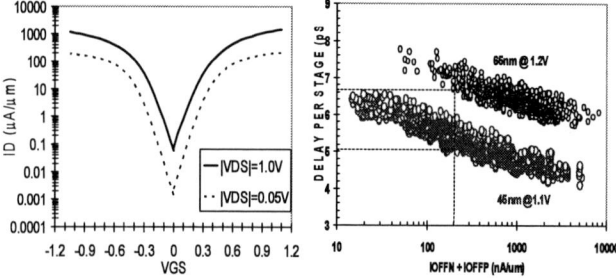

Component	Benefit
PMOS Idsat	+13
PMOS Idlin	+18
NMOS Idsat	+3
NMOS Idlin	+2
Cjunction	+2
Cgate/Cov	-8%
Voltage Scaling	-7%
Total	+23%

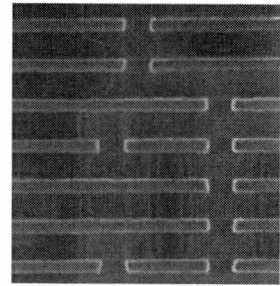

Fig.8 Subthreshold Id-Vgs for both NMOS and PMOS transistors

Fig.9 RO data for a fanout of 2 showing 23% improvement in gate delay vs. 65nm, despite voltage scaling from 1.2V to 1.1V.

Fig. 10 Breakdown of RO gains vs. 65nm results. The voltage scaling term accounts for the reduction in VDD from 1.2V (65nm) to 1.1V (45nm).

Fig.11 Top-down SEM post poly patterning process showing 160nm poly pitch and square poly ends, devoid of rounding.

Strain enhanced Low-V_T CMOS featuring La/Al-doped HfSiO/TaC and 10ps Invertor Delay

S. Kubicek, T.Schram, E.Rohr, V.Paraschiv, R.Vos, M.Demand, C.Adelmann, T.Witters, L.Nyns, A.Delabie , L.-Å.Ragnarsson, T.Chiarella, C.Kerner, A.Mercha, B.Parvais, M.Aoulaiche[†], C.Ortolland, H.Yu, A.Veloso, L.Witters, R.Singanamalla[†], T. Kauerauf[†], S.Brus, C.Vrancken, V.S.Chang[1], S-Z.Chang[1], R.Mitsuhashi[2], Y.Okuno[2], A.Akheyar[3], H.-J.Cho[4], J.Hooker[5], B. J. O'Sullivan, S.Van Elshocht, K.De Meyer[†], M.Jurczak, P.Absil, S.Biesemans and T.Hoffmann

IMEC, assignee at IMEC from [1]TSMC, [2]Matsushita, [3]Infineon, [4]Samsung, [5]NXP and [†]IMEC and KULeuven

Kapeldreef 75, B-3001, Leuven, Belgium

Abstract

We discuss several advancements over our previous report [1]:

- Introduction of conventional stress boosters resulting in 16% and 11% for nMOS and pMOS respectively. For the first time the compatibility of SMT (Stress memorization technique) with High-κ/Metal Gate is demonstrated. In addition, we developed a blanket SMT process that does not require a photo to protect the pMOS by selecting a hydrogen-rich SiN film.

- A comprehensive study of HfSiO and HfO2 as function of La/Al doping and spike/laser annealing. Parameters studied include Vt tuning, reliability and process control.

- Demonstration of fast invertor delay of 10ps including high frequency response analysis revealing the negative impact of high metal sheet resistance and parasitic metal-poly interface oxide.

Introduction

Low V_T High-κ /Metal Gate CMOS transistors were reported recently [1, 2]. In [1], low V_T's with High- κ and Metal Gates were achieved in a gate first process flow using dielectric capping layers [4-6]. In this work, we have focused on the compatibility of the gate first High-κ/Metal Gate CMOS transistors with the conventional performance enhancement techniques and on the thorough characterization of the gate stack integrity and high frequency performance.

High-κ doping with cappings

LaO and AlO cappings have been integrated into a gate-first CMOS flow in order to tune the effective work function of nMOS and pMOS devices, respectively. The details of the process integration were presented in [1]. A simplified process flow is reported in Fig.1, along with the previously demonstrated V_T-L characteristics in Fig.2. Now we review in more details the manufacturability of such process.

A. Effective workfunction modulation process window – A systematic study of the cappings has been performed. As illustrated in Fig.3, the capping location, "below" or "above" the host high-κ film has been studied. Different high-κ films (HfSiO, HfO2) and annealing conditions (Spike or Laser at different power) have been investigated as well. Fig.4 summarizes the observed V_T shift obtained for various conditions. Contrarily to placing the capping "above", a large shift is observed for all laser powers if the capping is "below", indicating that a much lower thermal budget is required to fully intermix La or Al into the HiK/IL layers. Note that the kinetic seems slower for La than for Al, which has been also confirmed experimentally. At sufficiently high Laser power or with a conventional Spike-RTA, similar V_T shifts are observed, whether the cap is "below" or "above" HfSiO or HfO2.

B. Gate dielectric integrity – Fig.5 shows the impact of capping (AlO here) on J_G-EOT. The use of capping does not intrinsically change the J_G-EOT trade-off characteristics. Using Laser anneal instead of Spike appears to help in achieving aggressively scaled EOTs. This is believed to come from reduced interfacial oxide re-growth. The impact on mobility is illustrated in Fig.6. Even though no e-mobility loss is seen with capping ("below" or "above", with Spike or Laser), a slight h-mobility loss is seen with AlO. Most of it can be recovered by using Laser instead of Spike (EOT reduction without mobility loss). But just like for LaO (nMOS), no apparent intrinsic differences are observed if the cap is placed "below" or "above" the high- κ film. Similar observations are derived from NBTI characterization, where a slight degradation is seen with AlO, independently of its location (Fig.7a). Noise properties are not much degraded either (Fig.7b).

C. Process control and manufacturability – We also verified electrically that the use of capping to dope high-k films can be well controlled. Excellent wafer uniformity is shown in Fig. 8, for both 9Å AlO (pMOS), and 7Å LaO (nMOS). Also, no degradation of the matching characteristics is observed (Fig. 9).

In summary, we have verified that the capping scheme has a large process window, and does not degrade the gate dielectric integrity.

Stress boosters compatibility

We have first verified in unichannel transistors that the gain from traditional stress boosters (CESL, embedded-SiGe, channel orientation) was maintained on High-κ/Metal gate. In particular, we demonstrate for the first time that this remains true for SMT, as shown in Fig. 10. This implies that the polysilicon strain induced by re-crystallization is effectively transferred into the channel, through the thin metal gate. Also, by using an optimized hydrogen-rich nitride layer, pMOS degradation can be recovered while maintaining nMOS gain (insert Fig 10). By combining Si <100>/(100), tensile-CESL and SMT, we achieve significant improvement of 11%/16% for pMOS/nMOS, over our previous report [1] (Figs. 11-13).

High-Frequency

The AC performance of our gate stacks has been evaluated with ring oscillator (top-down view in Fig. 14). During the processing, an intentional gap (of ~100nm) is used to separate n- and p- metals. We verified electrically that no issue is seen down to 110nm n-p active spacing (Fig.15). The ring oscillator performance of 2 types of DMDD CMOS circuits is shown in Fig16. We call one pseudo-SMDD (pSMDD), since it uses the same metal (Ta2C) for nMOS and pMOS. We observe 100% more degradation in power-delay for DMDD compared to pSMDD. x-TEM revealed that both metal shows a thin parasitic oxide between metal and poly, but TaCNO has also a very rough interface (Fig. 17). From RF characterization, a large drop in gate capacitance is seen for TaCNO, as early as 300MHz, which can be modeled by an additional parasitic RC component in series (Fig 18). Note that TaCNO has also ~100x higher resistivity than Ta2C (Fig 19a). Path to a simple single metal (Ta2C) CMOS is shown in Fig 19b. The EOT benefit of Ta2C versus TaCNO, comes with a ~200mV V_T penalty. But, almost half of it can be readily recovered by simple well optimization. After optimization of this single metal process, we achieve similar RO performance as our previously reported FUSI devices [10] (Fig.20)

Conclusion

Applying conventional stress boosters to the gate first DMDD High-κ/Metal Gate nMOS and pMOS transistors results in 16% and 11% performance improvement, respectively. These transistors are compatible with SMT. Adding the dielectric cappings into the gate stack in order to tune V_T does not degrade the gate dielectric integrity.

References

[1] S.Kubicek, IEDM Tech. Dig., p.49, 2007, [2] K.Misry, IEDM Tech. Dig., p.247, 2007, [3] H-S. Jung, Symp. VLSI Technology, p.204, 2006, [4] M.M.Frank, VLSI-TSA Tech. Dig. P.97, 2005, [5] H. Alshareef, APL, Vol.89, 2006, [6] P. F. Hsu, Symp. VLSI Technology, p.14, 2006, [7] Y. Nishida, Symp. VLSI Technology, p.214, 2007, [8] M. Chudzik, Symp. VLSI Technology, p.194, 2007, [9] W. J. Taylor et.al., IEDM Tech. Dig., p.625, 2006, [10] A. Rothschild et al, VLSI Technology, p.198, 2007.

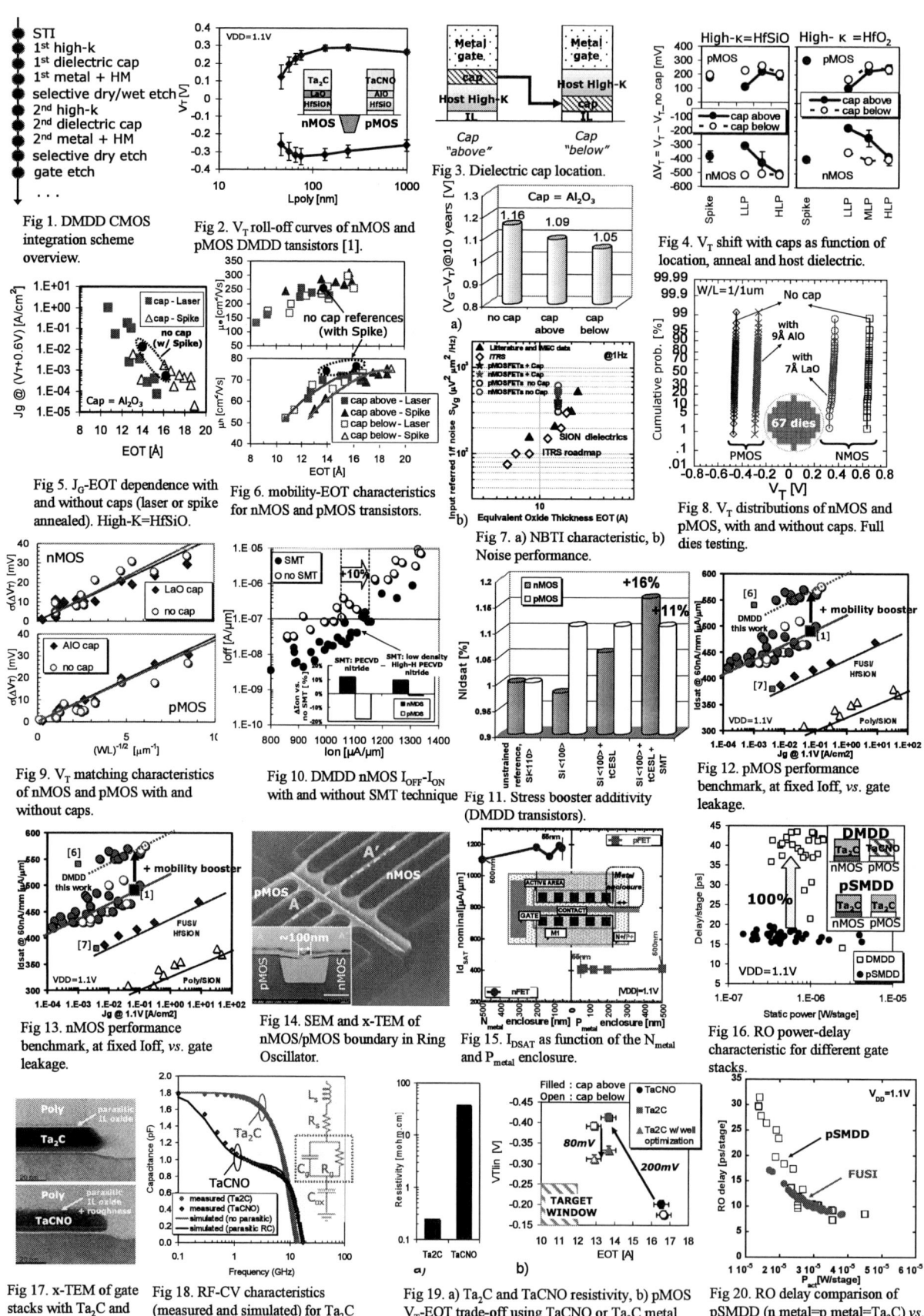

- STI
- 1st high-k
- 1st dielectric cap
- 1st metal + HM
- selective dry/wet etch
- 2nd high-k
- 2nd dielectric cap
- 2nd metal + HM
- selective dry etch
- gate etch
- ...

Fig 1. DMDD CMOS integration scheme overview.

Fig 2. V_T roll-off curves of nMOS and pMOS DMDD tansistors [1].

Fig 3. Dielectric cap location.

Fig 4. V_T shift with caps as function of location, anneal and host dielectric.

Fig 5. J_G-EOT dependence with and without caps (laser or spike annealed). High-K=HfSiO.

Fig 6. mobility-EOT characteristics for nMOS and pMOS transistors.

Fig 7. a) NBTI characteristic, b) Noise performance.

Fig 8. V_T distributions of nMOS and pMOS, with and without caps. Full dies testing.

Fig 9. V_T matching characteristics of nMOS and pMOS with and without caps.

Fig 10. DMDD nMOS I_{OFF}-I_{ON} with and without SMT technique

Fig 11. Stress booster additivity (DMDD transistors).

Fig 12. pMOS performance benchmark, at fixed Ioff, vs. gate leakage.

Fig 13. nMOS performance benchmark, at fixed Ioff, vs. gate leakage.

Fig 14. SEM and x-TEM of nMOS/pMOS boundary in Ring Oscillator.

Fig 15. I_{DSAT} as function of the N_{metal} and P_{metal} enclosure.

Fig 16. RO power-delay characteristic for different gate stacks.

Fig 17. x-TEM of gate stacks with Ta$_2$C and TaCNO metal gate.

Fig 18. RF-CV characteristics (measured and simulated) for Ta$_2$C and TaCNO gate stacks (L=0.6μm).

Fig 19. a) Ta$_2$C and TaCNO resistivity, b) pMOS V_T-EOT trade-off using TaCNO or Ta$_2$C metal (W/L=1/1um).

Fig 20. RO delay comparison of pSMDD (n metal=p metal=Ta$_2$C) vs. FUSI/High-k reference [10].

102

Impact of Tantalum Composition in TaC/HfSiON Gate Stack on Device Performance of Aggressively Scaled CMOS Devices with SMT and Strained CESL

M. Goto[1], K. Tatsumura[2], S. Kawanaka[1], K. Nakajima[3], R. Ichihara[2], Y. Yoshimizu[3], H. Onoda[4], K. Nagatomo[1],
T. Sasaki[3], T. Fukushima[3], A. Nomachi[3], S. Inumiya[3], H. Oguma[3], K. Miyashita[4],
H. Harakawa[3], S. Inaba[1], T. Ishida[1], A. Azuma[1], T. Aoyama[3], M. Koyama[2], K. Eguchi[3], Y. Toyoshima[1]

[1]Center for Semiconductor Research and Development, [2]Advanced LSI Technology Laboratory, Corporate R&D Center
[3]Process and Manufacturing Engineering Center, [4]System LSI Division, Semiconductor Company
Toshiba Corporation
8 Shinsugita-cho, Isogo-ku, Yokohama, Kanagawa 235-8522, Japan
Phone: +81-45-776-5680, FAX: +81-45-776-4104, E-mail: ma.goto@toshiba.co.jp

Abstract

We report TaC_x/HfSiON gate stack CMOS device with simplified gate 1st process from the viewpoints of fixed charge generation and its impact on the device performance. Moderate Metal Gate / High-K dielectric (MG/HK) interface reaction is found to be a dominant factor to improve device performance. By optimizing TaC_x composition, fixed charge free TaC_x/HfSiON device is successfully fabricated. Also, we have demonstrated that the strain effect in deeply scaled devices can be enhanced by eliminating the fixed charges in HfSiON, for the first time. Utilizing Stress Memorization Technique (SMT) and strained Contact Etch Stop Layer (CESL), L_g=35nm high performance TaC_x/HfSiON devices is achieved.

Introduction

MG/HK stack has many advantages for 45nm node technology and beyond [1, 2]. HfSiON dielectric [3, 4] and TaC_x gate [5-10] has been focused as a candidate to satisfy the ITRS requirement. Especially, Ta-based gate material has been investigated from the viewpoint of Effective Work Function (EWF) control by changing material composition [9-10]. However, the effect of metal gate material composition on short channel device performance has not been extensively studied yet. For example, Ta-rich TaC_x metal gate shows higher current drive in long channel device, but degradation in short channel device compared with Carbon-rich TaC_x metal gate (Fig. 1). In this paper, fixed charge generation mechanism is clarified based on the extensive physical analyses. TaC_x composition impact on deeply scaled device performance is also investigated especially from the viewpoint of strain effect, which is reduced by fixed charges in gate dielectric [11].

Device Fabrication

TaC_x/HfSiON devices with poly-Si capping layer (MIPS) were fabricated. Following HfSiON ([Hf] = 50%) formation, three types of TaC_x gates were processed by PVD method controlling composition of TaC_x. (a) Ta-rich, flat profile (TaC_x-1), (b) Carbon-rich, flat profile (TaC_x-2), and (c) Carbon-rich (top) to Ta-rich (bottom), gradual profile with conclusively optimization (TaC_x-3), as shown in SIMS profiles (Fig. 2). The EWF of all TaC_x metals is approximately 4.4-4.5 eV. Here, PVD TiN capping layer was inserted between poly-Si and TaC_x as a barrier material to avoid Si diffusion into TaC_x layer. After the formation of MG/HK stack, conventional CMOS process such as spike annealing and NiSi process were applied. To examine the strain effect, SMT and tensile CESL process were additionally utilized.

Effect of TaC_x Composition Control on Device Performance

The relationship of J_g vs T_{inv} is shown in fig. 3, where T_{inv} in any TaC_x / HfSiON stack devices are successfully reduced from poly-Si gate without J_g increase. Among MG/HK devices, apparent J_g difference is observed between TaC_x-1 and the others. J_g-T_{inv} line of TaC_x-1 is shifted toward thinner T_{inv}. This implies that permittivity of HfSiON in TaC_x-1 becomes higher than those of TaC_x-2 and TaC_x-3 [8]. Both electron and hole mobility are clearly degraded, especially at low E_{eff} region in TaC_x-2 (Fig.4). This is consistent with long channel current drive degradation of TaC_x-2 device in fig. 1. In this case, mobility degradation is attributed to increase of remote coulomb scattering due to fixed charges in HfSiON. On the other hand, as shown in fig. 1, TaC_x-1 has a problem in short channel devices. As seen in V_{th} roll-off curve (fig.5), TaC_x-1 device shows weak SCE in nMOSFET but strong SCE in pMOSFET comparing to others. T-CAD simulation verified gate-edge negative fixed charge modulation of SCE [3] as shown in fig. 5. Here, it is also confirmed that gate-edge fixed charge effect is suppressed in TaC_x-2 and TaC_x-3 structures. From device characteristics and T-CAD simulation, it is summarized that flat composition profile in TaC_x gates (TaC_x-1 and TaC_x-2) possess additional fixed charge in HfSiON dielectrics, while

optimized TaC_x-3 gate shows no fixed charge issues. Consequently, both highest current drive and lowest V_{th} fluctuation [12] are achieved with TaC_x-3, as shown in fig.6 and fig.7.

Fixed-Charge Suppression Model Based on Physical Analyses

Figure 8 shows HRTEM of all TaC_x devices after spike RTA. The thinning of HfSiON film is observed in TaC_x-1 device. Back side SIMS profiles shows significant Si diffusion from HfSiON into TaC_x in TaC_x-1 device (fig. 9). From back side XPS, HfO_x peak is observed in TaC_x-1 from O1s spectra (Fig. 10 (a)). From Hf4f spectra, higher intensity of metallic-Hf peaks (Hf or HfC_x,) are observed in TaC_x-1 and TaC_x-3 than TaC_x-2 (Fig. 10 (b)). Figure 11 shows schematic summary of HRTEM, SIMS and XPS analysis. Here, flat Ta-rich profile (TaC_x-1) leads to significant MG/HK interface reaction. In this case, HfSiON dielectric becomes Hf-rich composition due to Si diffusion, while flat C-rich profile (TaC_x-2) has no interface reaction. MG/HK interface reaction also takes place in TaC_x-3 which has Ta-rich composition at MG/HK interface. However, thanks to C-rich layer, interface reaction is less than TaC_x-1 device. This is consistent with J_g-T_{inv} results (fig. 3). In Figure 12, we would like to propose a model which explains fixed charge behavior combining with process impact. After the MG/HK stack formation, all samples could have sheet fixed charge at near MG/HK interface due to PVD damage [13]. In the case of TaC_x-2 device, sheet fixed charge remains through all processes because of no MG/HK interface reaction. In contrast, the devices which have Ta-rich layer at MG/HK interface (TaC_x-1 and TaC_x-3) eliminate sheet fixed charge by MG/HK interface reaction. In this case, flat Ta-rich profile (TaC_x-1) leads to significant Si diffusion due to excessive Ta. This results in Hf-rich HfSiON film which tends to have gate-edge fixed charge. On the contrary, owing to precisely controlled TaC_x composition, sheet fixed charge has been removed by moderate MG/HK reaction without inducing gate-edge fixed charges in TaC_x-3 device.

Performance Booster effect on TaC_x/HfSiON gate stack

The relationship between TaC_x composition and effects of mobility enhancement techniques is investigated utilizing SMT and strained CESL. The reported mechanisms of SMT are the stress memorization into gate poly-Si [14, 15]. In this work, the effect of metal gate insertion in SMT-boosted devices is firstly examined. As shown in fig. 13, same performance improvement was observed between metal gate and poly gate devices. The result suggests that the MIPS structure can still enjoy the benefit of SMT, though, the strain effects are found to be affected by TaC_x composition. Figure 14 shows the comparison of the performance enhancements by each stress techniques in TaC_x-1 and TaC_x-3. The transconductance enhancement in TaC_x-3 devices is larger than that of TaC_x-1 devices. Therefore, the elimination of the fixed charges is highly effective for enhancing strain effect. This result is also consistent with the previous report which says fixed charge degrades the mobility enhancement by strain [11]. By utilizing fixed-charge free TaC_x-3 gate stack, not only the carrier mobility without strain, but also the impact of stress technique on the performance of short-channel devices are significantly improved. As a result, we have achieved additive performance enhancement from both SMT and tensile CESL (tSL) with significant (23 %) total performance enhancement (Fig. 15).

Conclusion

By careful optimization of TaC_x composition, we have succeeded in fabricating fixed-charge free TaC_x / HfSiON devices with high carrier mobility. We demonstrated that strain effects in short channel devices are enhanced by eliminating the fixed charges inside HfSiON, for the first time. As a result of high carrier mobility by enhanced strain effect, L_g=35nm high-performance TaC_x/HfSiON device with SMT and tensile CESL have been fabricated.

978-1-4244-1802-2/08/$25.00 ©2008 IEEE

Reference

[1] M. Chudzik, VLSI Tech., p.194 (2007)
[2] K. Mistry, IEDM, p.247 (2007)
[3] T. Watanabe, IEDM, p.507 (2004)
[4] S. Inumiya, IEDM, p.25 (2005)
[5] Y. T. Hou, IEDM, p.31 (2005)
[6] L. A. Ragnarsson, EDL, p.486 (2007)
[7] S. Kubicek, IEDM, p.49 (2007)
[8] K. Tatsumura, IEDM, p.349 (2007)
[9] J. K. Schaeffer, JAP, 014503 (2007)
[10] W. S. Hwang, VLSI Tech., p.156 (2007)
[11] M. Saitoh, VLSI Tech., p.132 (2007)
[12] K. Takeuchi, IEDM, p.467 (2007)
[13] W. P. Bai, EDL, p.231 (2007)
[14] K. Ota, IEDM, p.27 (2002)
[15] A. Eiho, VLSI Tech., p.218 (2007)

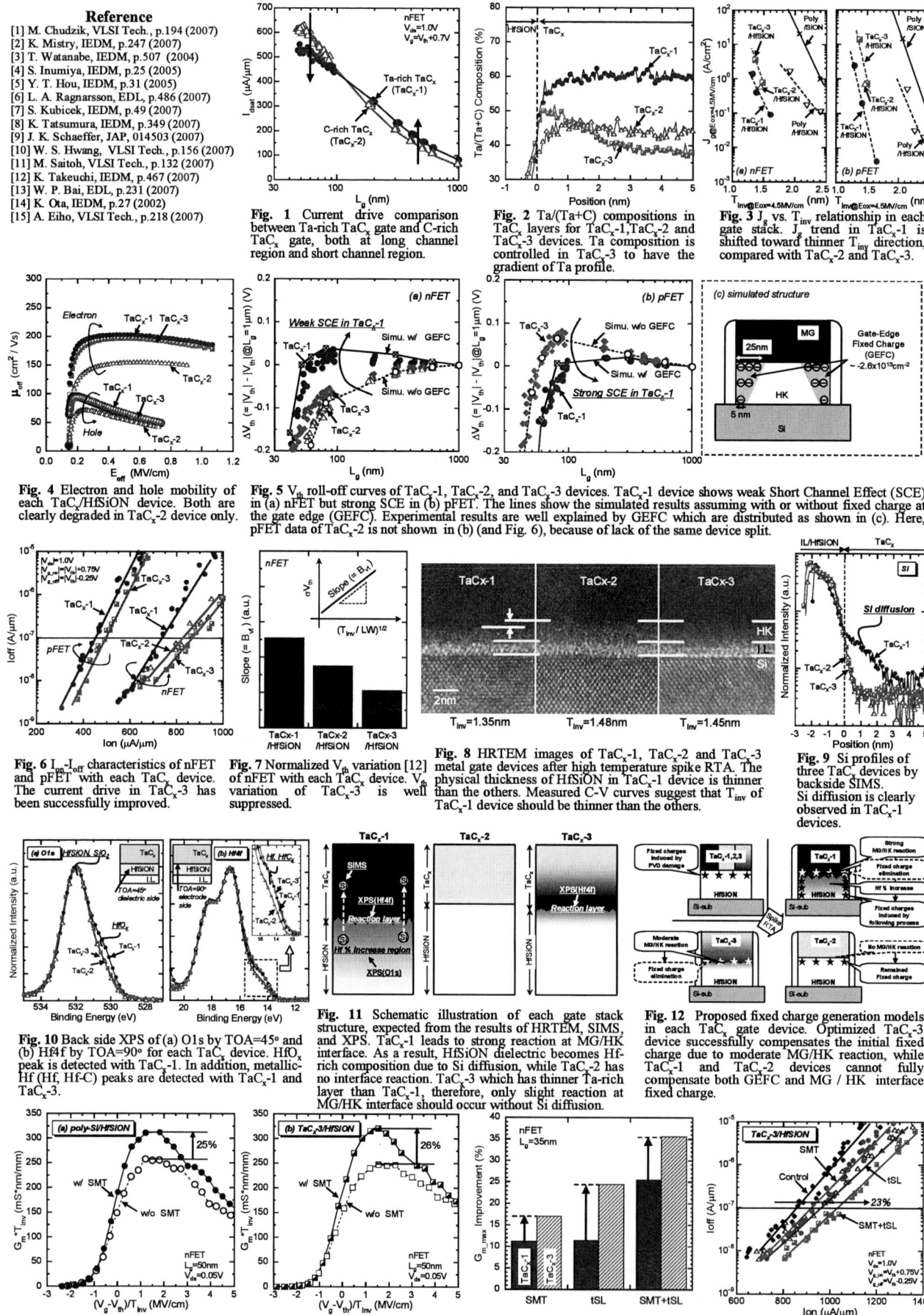

Fig. 1 Current drive comparison between Ta-rich TaC_x gate and C-rich TaC_x gate, both at long channel region and short channel region.

Fig. 2 $Ta/(Ta+C)$ compositions in TaC_x layers for TaC_x-1, TaC_x-2 and TaC_x-3 devices. Ta composition is controlled in TaC_x-3 to have the gradient of Ta profile.

Fig. 3 J_g vs. T_{inv} relationship in each gate stack. J_g trend in TaC_x-1 is shifted toward thinner T_{inv} direction, compared with TaC_x-2 and TaC_x-3.

Fig. 4 Electron and hole mobility of each TaC_x/HfSiON device. Both are clearly degraded in TaC_x-2 device only.

Fig. 5 V_{th} roll-off curves of TaC_x-1, TaC_x-2, and TaC_x-3 devices. TaC_x-1 device shows weak Short Channel Effect (SCE) in (a) nFET but strong SCE in (b) pFET. The lines show the simulated results assuming with or without fixed charge at the gate edge (GEFC). Experimental results are well explained by GEFC which are distributed as shown in (c). Here, pFET data of TaC_x-2 is not shown in (b) (and Fig. 6), because of lack of the same device split.

Fig. 6 I_{on}-I_{off} characteristics of nFET and pFET with each TaC_x device. The current drive in TaC_x-3 has been successfully improved.

Fig. 7 Normalized V_{th} variation [12] of nFET with each TaC_x device. V_{th} variation of TaC_x-3 is well suppressed.

Fig. 8 HRTEM images of TaC_x-1, TaC_x-2 and TaC_x-3 metal gate devices after high temperature spike RTA. The physical thickness of HfSiON in TaC_x-1 device is thinner than the others. Measured C-V curves suggest that T_{inv} of TaC_x-1 device should be thinner than the others.

Fig. 9 Si profiles of three TaC_x devices by backside SIMS. Si diffusion is clearly observed in TaC_x-1 devices.

Fig. 10 Back side XPS of (a) O1s by TOA=45° and (b) Hf4f by TOA=90° for each TaC_x device. HfO_x peak is detected with TaC_x-1. In addition, metallic-Hf (Hf, Hf-C) peaks are detected with TaC_x-1 and TaC_x-3.

Fig. 11 Schematic illustration of each gate stack structure, expected from the results of HRTEM, SIMS, and XPS. TaC_x-1 leads to strong reaction at MG/HK interface. As a result, HfSiON dielectric becomes Hf-rich composition due to Si diffusion, while TaC_x-2 has no interface reaction. TaC_x-3 which has thinner Ta-rich layer than TaC_x-1, therefore, only slight reaction at MG/HK interface should occur without Si diffusion.

Fig. 12 Proposed fixed charge generation models in each TaC_x gate device. Optimized TaC_x-3 device successfully compensates the initial fixed charge due to moderate MG/HK reaction, while TaC_x-1 and TaC_x-2 devices cannot fully compensate both GEFC and MG / HK interface fixed charge.

Fig. 13 Transconductance (Gm) comparison for (a) poly-Si/HfSiON devices and (b) TaC_x-3/HfSiON devices with or without SMT. Similar performance improvement was observed both in metal gate and poly gate devices.

Fig. 14 Comparison of Gm enhancement for SMT only, tensile CESL (tSL) only and the combination of SMT & tSL in TaC_x-1 and TaC_x-3 devices.

Fig. 15 I_{on}-I_{off} characteristics in TaC_x-3 devices for each booster techniques. Significant performance enhancement up to 23 % is achieved for SMT+tSL.

Embedded Split-Gate Flash Memory with Silicon Nanocrystals for 90nm and Beyond

Gowrishankar Chindalore, Jane Yater, Horacio Gasquet, Mohammed Suhail, Sung-Taeg Kang, Cheong Min Hong, Nicole Ellis, Glenn Rinkenberger, James Shen, Matthew Herrick, Wendy Malloch, Ronald Syzdek, Kelly Baker, Ko-Min Chang

Technology Solutions Organization, Freescale Semiconductor, Inc.
6501 William Cannon Drive, Austin, Texas, USA
Tel: (512) 895-8381; Fax: (512) 895-8605; g.chindalore@freescale.com

Abstract

We present a split-gate based NOR flash memory array with silicon nanocrystals as the storage medium. 128KB memory arrays have been evaluated with this technology and the results presented here show a nanocrystal memory that has been demonstrated to achieve a minimum 1.5V operating window that is maintained through 10K program/erase cycles; well controlled array threshold distributions; fast source-side injection programming (10-20us); fast tunnel erase into the gate; and robust high temperature data retention for both uncycled and cycled arrays. Results presented here with focus on the array operation demonstrate the maturity of this technology for implementation into consumer, industrial, and automotive microcontrollers. (Keywords: Microcontrollers, Nanocrystals, Flash Memories, Memory Array, Source-side Injection, Tunnel Erase, Endurance, Data Retention)

Introduction

The increasing intelligence in the products that affect our lives such as consumer (appliances, electronics, etc); industrial (robotics, power tools, etc); and automotive (engine control, safety features, etc) products require microcontrollers with high density of flash memory that are not only reliable but also cost-effective to integrate. The traditional floating gate based embedded flash memories are either implemented as one transistor cells (1-T) mostly in high density applications, or as split-gate cells (1.5T) in medium-low density applications [1]. However, combining nanocrystal storage and split-gate architectures, the scaling limitations imposed by floating gate memories are overcome while availing the benefit of smaller module size achievable with split-gate architecture. In this paper, we report the progress made in developing such a broadly applicable nanocrystal memory supported with an extensive set of data from memory arrays.

Description

A. Memory Array Design and Implementation

The nanocrystal memory has been integrated into the 90nm technology node along with high voltage (HV) devices necessary for the array operation. The HV devices are used to drive the control gate (CG) and source, while core logic transistors (low voltage and I/O devices) control the select gate (SG) and the bit lines. The array test vehicle is a 16Mb part partitioned into sixteen 1Mb modules (Fig. 1). Each of the 1Mb modules is soft sectored into 2KB sectors, creating a selectable array from 2KB-2MB.

The memory stack consists of thermally grown bottom oxide, a layer of deposited nanocrystals and high quality deposited top oxide. The dielectric layers are thinner than those used in the floating gate memories. The control gate portion of the channel is counter-doped in order to eliminate the read disturb that is inherent to memories with thin dielectrics [2,3]. The array threshold (Vth) distributions in the natural state (fully discharged nanocrystals) have a width of 1.3-1.5V (Fig. 2), close to the expectation based on channel dopant fluctuations. The natural Vth distributions have been found to be repeatable in over 10,000 1Mb modules covering multiple silicon lots.

B. Program and Erase Operations

The source-side injection programming is fast (<20us) and requires low power (2-15uA/bit). The program efficiency is a strong function of the select gate bias (Fig. 3) due to its effect on the field at the gap. The tunnel erase operation into control gate using positive bias is also very fast at the beginning of life (Fig. 4), but slows with cycling due to oxide trap up [4]. Neither operation requires negative bias, thus increasing array efficiency.

C. Endurance and Reliability

Over one thousand 1Mb modules have been cycled with a minimum operating window of at least 1.5V. The window is maintained after 10K write/erase cycles (Fig. 5) and is large enough to generate the read currents necessary for high performance products (read speeds < 20ns), while also providing sufficient margin for high temperature data retention (DR) from the program state. The memory stack is inherently vulnerable to trap-up during programming due to the discontinuous nature of nanocrystals [3]. However, we have minimized trap-up with process optimization so that sufficient operating window can be maintained throughout the life of the product.

Figure 6 shows long-term DR for both uncycled and 10K cycled parts as a function of temperature. The data shows well controlled Vth distributions with an intrinsic shift that is consistent with expectation based on single bitcell evaluations [4], and in addition, no extrinsic bits have been observed in the arrays as seen from the Vth distributions collected during DR bake (Fig. 6) despite the use of thinner dielectrics.

Summary

Nanocrystal memory technology combined with split-gate architecture allows for efficient design solutions that span a wide range of applications from low to high density due to minimal analog circuit requirements and relatively high array efficiency. Array results from split-gate nanocrystal flash memories show controllable Vth distributions, fast program (<20us) and erase operations (<250us at the beginning of life), at least 1.5V operating window that is maintained through 10K cycles, and robust high temperature DR without extrinsic bits.

978-1-4244-1802-2/08/$25.00 ©2008 IEEE

References

[1] S. Saha, IEEE TED, vol.54, pp. 3049-3055, 2007
[2] C.T. Swift, 2006 21st IEEE NVSMW, pp. 56-57.

[3] J.A. Yater, 2007 22nd NVSMW, pp.77-78.
[4] C.M. Hong, 2007 22nd NVSMW, pp.75-76.

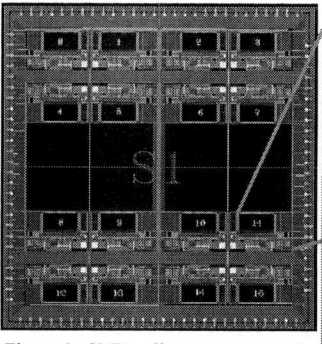
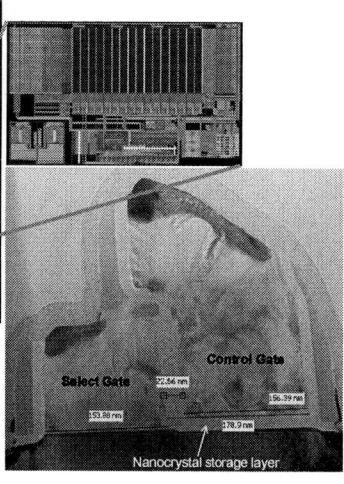

Figure 1: 2MB split-gate nanocrystal memory test vehicle partitioned into sixteen 1Mb modules, with sector size as small as 2KB creating selectable arrays from 2KB-2MB. TEM cross-section of a typical array bitcell is shown on the right.

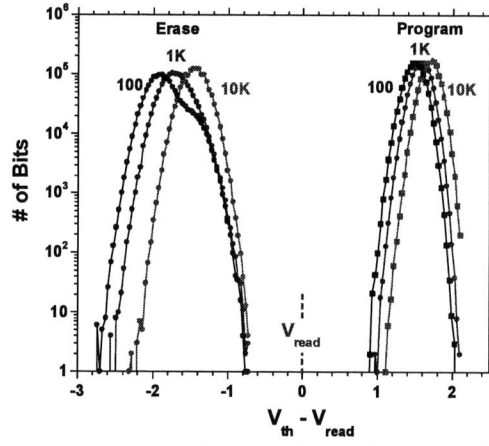

Figure 4: Erase rate distribution with 13V control gate. The erase speed is fast with all bits are erased under 200μs with 13V. Both program and erase operations are positive bias operations, thus eliminating the need for negative charge pump entirely.

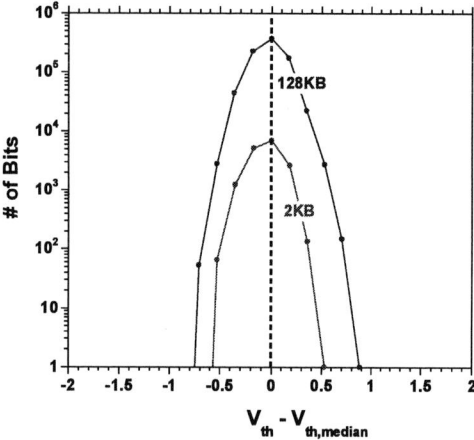

Figure 2: Natural Vth distributions (fully discharged nanocrystals) for 2KB and 128KB memory sizes. The widths are consistent with expectation based on channel dopant fluctuations under the control gate.

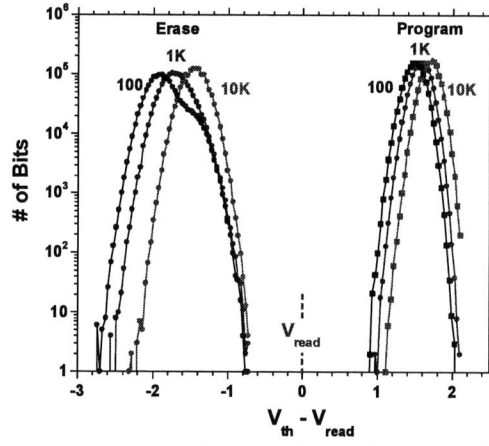

Figure 5: Program and erase Vth distributions as a function of cycling for 1Mb module. The minimum window is about 1.5V which is sufficient to balance the read performance and the data retention requirements.

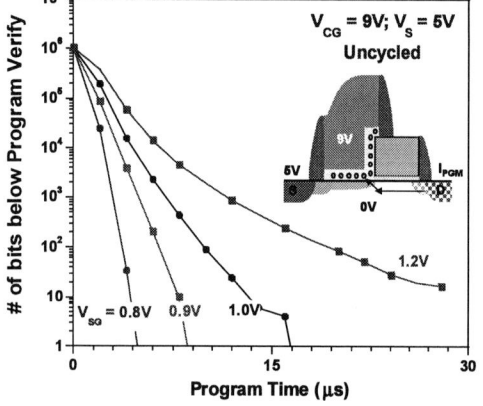

Figure 3: Program rate distribution as a function of select gate bias. With increasing SG bias, the inversion field at the gap region increases resulting in less efficient heating of the carriers.

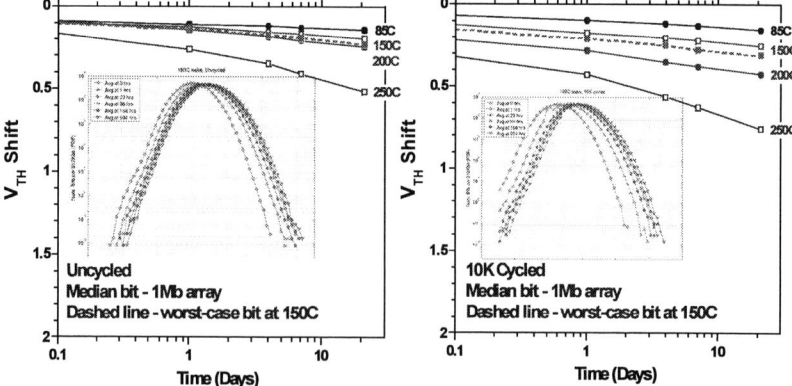

Figure 6: Threshold shift for program state as a function of temperature for uncycled and 10K-cycled 1Mb arrays. The Vth shift is higher after cycling due to oxide detrapping. The Vth distributions are also well controlled without evidence of extrinsic bits as seen by the Vth distributions shown in the inset (only 150C distributions shown here for clarity).

Gate-all-around Single Silicon Nanowire MOSFET with 7 nm width for SONOS NAND Flash Memory

Kyoung Hwan Yeo, Keun Hwi Cho, Ming Li, Sung Dae Suk, Yun-young Yeoh, Min-Sang Kim, Hyunjun Bae,
Ji-Myoung Lee, Suk-Kang Sung[1], Jun Seo[2], Bokkyoung Park[3], Dong-Won Kim, Donggun Park, and Won-Seoung Lee

ATDT1, [1]ATDT2, [2]PTDG1, [3]MTT2, R&D Center, Samsung Electronics Co.

San 24, Nongseo-Dong, Kiheung-Ku, Yongin-City, Kyoungi-Do, 449-711, KOREA

Phone : +82-31-209-6668, Fax : +82-31-209-3274, E-mail : mbc39010.yeo@samsung.com

Abstract

Gate-all-around (GAA) MOSFET with single silicon nanowire is fabricated and applied to SONOS memory as a cell transistor for NAND flash string. Driving current over 1 uA, which is sufficient to NAND string, is obtained with single nanowire of ~7 nm width. Using FN tunneling conditions, V_{TH} window of 4.5V and fast program/erase (P/E) speed of ~10 us are obtained, respectively. The smaller nanowire width is, the faster program speed and the larger V_{TH} shift are achieved. P/E operations in NAND string with GAA SONOS nanowire are demonstrated for the first time.

Introduction

Recently, nanowire channels have been proposed and studied using conventional processes [1~5]. They showed high immunity to short channel effect and high performance with strong gate control even in the ultimately scaled regime. And, several studies showing good electrical characteristics for memory device have been published with strong advantages of nanowire [6~9]. We already developed GAA SONOS device with ultra thin twin silicon nanowires and showed its electrical characteristics with channel hot electron injection, hot hole injection, and FN tunneling mechanism [7].

In this paper, GAA single nanowire MOSFET (NWFET) is fabricated and applied to SONOS flash memory. Memory characteristics with different nanowire width and tunnel oxide thickness (T_{TO}) are investigated. In addition, P/E characteristics of NAND string with nanowire of 7 nm width are presented.

Fabrication of single nanowire SONOS memory

Fabrication of single NWFET is similar to that of twin NWFET except for several differences [1~3, 7]. The major difference is trimming step of active SiN. Single nanowire can be formed if active SiN is trimmed before trench etching. Fig. 1 shows the process flow and schematic diagram of SONOS memory with single silicon nanowire. The single NWFET has narrow source/drain (S/D) regions as well as the nanowire channel. Fig. 2 shows SEM and TEM images of single nanowires. Nanowires of 10 to 25 nm width are formed after removing SiGe layer and surrounded by ONO of 3.0/6.8/6.7 nm.

Electrical characteristics

Single NWFET shows similar driving current per nanowire with twin NWFET due to gate all around structure and well optimized processes as shown in Fig. 3.

Figs. 4 ~ 11 show electrical characteristics of GAA single nanowire SONOS memory. In Fig. 4, I_D-V_G curves show sufficient current over 1 uA for NAND application. Large V_{TH} shift of 4.5 V is achieved with nanowire of 7 nm. Fig. 5 shows P/E speed using FN tunneling mechanism. Large V_{TH} shifts over 3 V and fast P/E speed of ~10 us are obtained at all gate bias although nanowire region is not fully workable at erase state.

Especially, fast erase speed is meaningful for a floating channel device because string erase is needed instead of block erase.

V_{TH} windows increase with smaller nanowire size and higher P/E voltage as shown in Fig. 6 (a). Larger V_{TH} shift and faster program speed are resulted from enhanced electric fields of tunnel oxide and increased ratios of SiN volume to SiN circumference as nanowire size decreases as shown in Fig. 6 (b) [7]. In addition, electric fields of control oxide are reduced and back tunneling can be controlled. Fig. 7 compares V_{TH} shift with different tunnel oxide thicknesses. Thinner oxide shows higher V_{TH} shift at the same gate bias than thicker one.

Endurances with nanowire width and tunnel oxide thickness are shown in Fig. 8 and Fig. 9. V_{TH} window shrinkage after 10^4 cycles is less than 0.5 V. Endurance difference with nanowire width and tunnel oxide thicknesses are smaller than 0.2 V. Fig. 10 shows data retention times measured at room temperature. V_{TH} window of 4.0 V is guaranteed for 10 years retention time with nanowire of 7 nm width.

Fig. 11 shows SEM images and cross-sectional TEM images of single nanowires in SONOS NAND strings. Cylindrical nanowire of 7 nm width is obtained and nanowires are surrounded by ONO layers of 3.0/7.8/5.8 nm. Top and bottom gate length are 45 nm and ~50 nm, respectively. Fig. 12 shows that similar V_{TH} are obtained with 8 word lines and 4 bit lines. In Fig. 13, low read current is obtained due to large S/D resistance. Sufficient read current can be expected with optimized S/D doping condition. V_{TH} window of 5.2 V is achieved at V_G of 14 V and P/E time of 100us and 100ms in a NAND string as shown in Fig. 14. To confirm the width effect, P/E speeds with different nanowire sizes are compared in Fig. 15. Back tunneling is effectively suppressed and V_{TH} window increases due to reduced electric fields of control oxide with small nanowire as shown in Fig. 6 (b).

Conclusion

GAA MOSFET with single nanowire of 7 nm is fabricated and applied to NAND strings for the first time. Large V_{TH} shift and fast P/E speed are obtained. As nanowire size and tunnel oxide thickness decrease, V_{TH} window and program speed are improved. V_{TH} window of 4 V is achieved after 10^4 cycles and extrapolated retention time of 10 years at room temperature. NAND strings with GAA nanowire SONOS shows promising results for a device with narrow width of sub 10 nm.

References

[1] S. D. Suk et al., IEDM Tech., p.717, 2005. [2] K. H. Yeo et al., IEDM Tech., p. 539, 2006. [3] K. H. Cho et al., IEDM Tech., p. 543, 2006. [4] F.-L. Yang et al., Symp. on VLSI Tech., p.196, 2004. [5] N. Singh et al., IEDM Tech., p. 547, 2006. [6] C. Friederich et al., IEDM Tech., p. 963, 2006. [7] S. D. Suk et al., Symp. on VLSI Tech., p.142, 2007. [8] H.Lee et al., Symp. on VLSI Tech., p.144, 2007. [9] J. Fu et al., IEDM Tech., p. 79, 2007.

978-1-4244-1802-2/08/$25.00 ©2008 IEEE

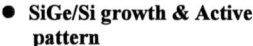

- SiGe/Si growth & Active pattern
- SiN trim & Trench etch
- STI process for isolation
- Damascene gate pattern
- Field oxide recess at gate region
- SiGe layer removal
- ONO & Gate deposition
- S/D mask removal

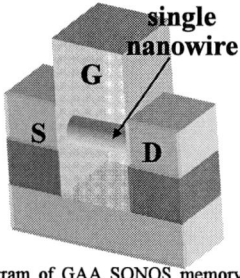

Fig. 1. Process flow and schematic diagram of GAA SONOS memory with single nanowire. The single NWFET has narrow source/drain (S/D) regions as well as the nanowire channel.

Fig. 2. Single nanowires of 10 nm to 25 nm width are formed after removing SiGe layer and surrounded by ONO layers of 3.0/6.8/6.7 nm. Height of S/D region is taller than that of channel region while widths are almost same.

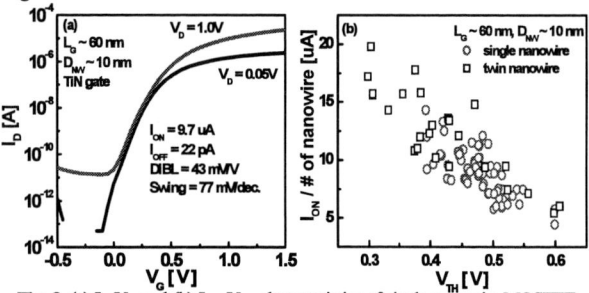

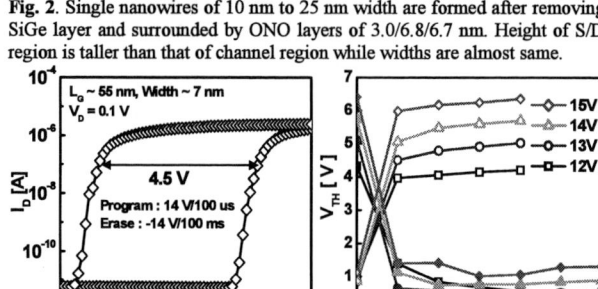

Fig. 3. (a) I_D-V_G and (b) I_{ON}-V_{TH} characteristics of single nanowire MOSFET. Single NWFET shows similar driving currents per a nanowire with twin NWFET due to well optimized processes.

Fig. 4. SONOS device with single nanowire shows large V_{TH} window of 4.5 V at V_G of 14V.

Fig. 5. P/E speeds using FN tunneling are very fast at all gate bias.

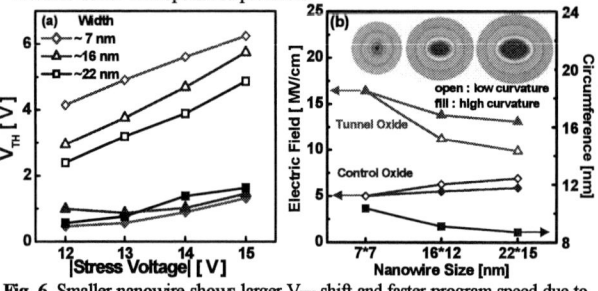

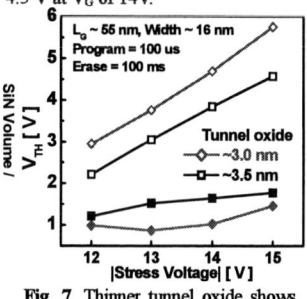

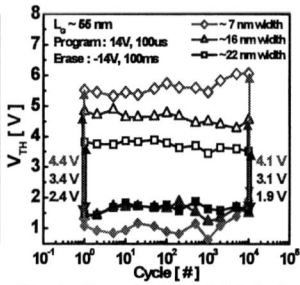

Fig. 6. Smaller nanowire shows larger V_{TH} shift and faster program speed due to larger ratios of SiN volume to circumference. In addition, electric field of tunnel oxide increases while that of control oxide decreases as nanowire size decreases.

Fig. 7. Thinner tunnel oxide shows higher and faster V_{TH} shift than thicker one.

Fig. 8. V_{TH} window of 4.1 V is maintained up to 10K P/E cycles with nanowire of 7 nm width.

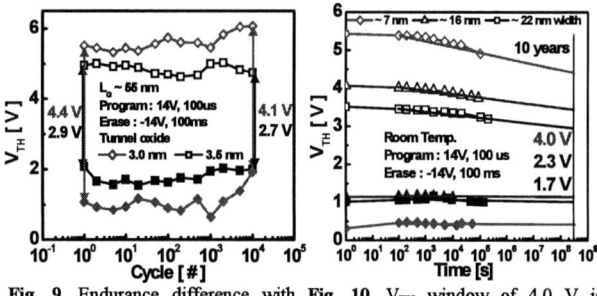

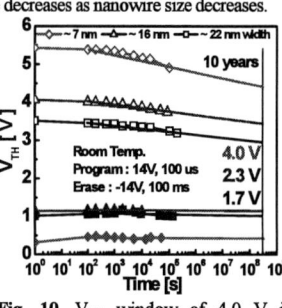

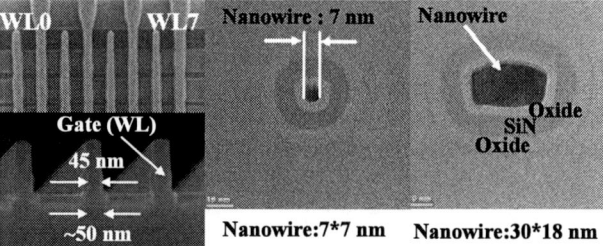

Fig. 9. Endurance difference with different T_{TO} is smaller than 0.1 V.

Fig. 10. V_{TH} window of 4.0 V is obtained with nanowire of 7 nm width.

Fig. 11. SEM and TEM images of single nanowire in a SONOS NAND string. Cylindrical nanowire of 7 nm width is obtained and nanowires are surrounded by ONO layers of 3.0/7.8/5.8 nm.

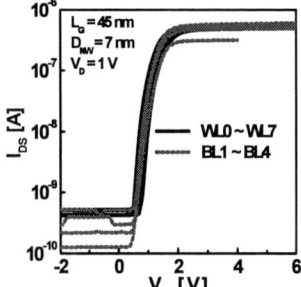

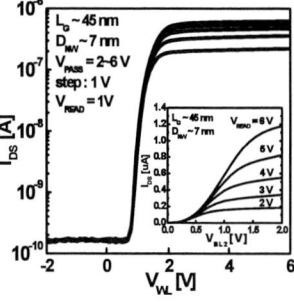

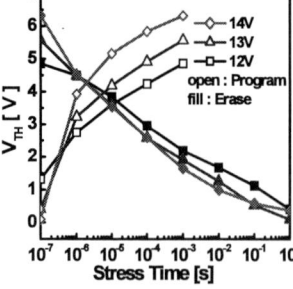

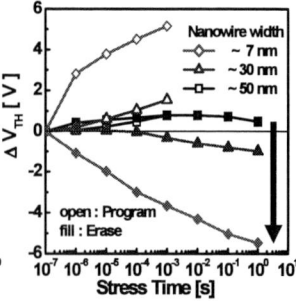

Fig. 12. Similar initial V_{TH} are obtained in 8 NAND strings of single nanowire SONOS memory.

Fig. 13. Low read current in 8 NAND string is obtained due to large S/D resistance.

Fig. 14. V_{TH} shift of 5.2 V is obtained at V_G of 14 V and P/E speed of 100 us and 100 ms in a NAND string.

Fig. 15. Back tunneling is effectively suppressed and V_{TH} shift increases with small nanowire.

108

A Novel Junction-Free BE-SONOS NAND Flash

Hang-Ting Lue, Erh-Kun. Lai, Y. H. Hsiao, S. P. Hong*, M. T. Wu*, F. H. Hsu*, N. Z. Lien*, S. Y. Wang*, L.W. Yang*, T. Yang*,
K.C. Chen*, K.Y. Hsieh, Rich Liu, and Chih-Yuan Lu
Emerging Central Lab, *Technology Development Center,
Macronix International Co., Ltd., 16 Li-Hsin Road, Hsinchu Science Park, Hsinchu, Taiwan   E-mail: htlue@mxic.com.tw

Abstract

We have successfully demonstrated a novel junction-free BE-SONOS NAND Flash. Junction-free devices greatly improve the short channel effect and thus promise scaling of NAND Flash below 20nm node. Instead of S/D junctions a very small space (< 30nm) is left between adjacent devices. Junction is formed only at the outer region of NAND array, while there is no junction inside the array. Fringe field from the gate inverts the Si under the narrow space allowing conduction without a diffusion junction. Successful n-channel, p-channel and TFT BE-SONOS NAND devices are demonstrated using this technique. Simulation results suggest that this novel junction-free technique is scalable beyond 20nm node. Moreover, the junction-free devices are unaffected by the thermal budget in the 3D TFT devices. This new device can be implemented in the current NAND Flash process without introducing new masks.

I. Introduction

Short channel effect (SCE) has increasingly become a scaling bottleneck for NAND Flash since it is difficult to scale the equivalent oxide thickness (EOT) below ~ 10nm. Shallow junctions help reduce the SCE but they are difficult to form, and become challenging in the future 3D integration [1,2] due to the large thermal budget.

In this work we propose a junction-free NAND Flash structure. In this novel device a simple narrow (<30nm) space replaces the conventional junction, and the fringe field from the poly gate induces inversion in the junctionless S/D area causing them to conduct. This is especially suitable for charge-trapping (CT) device since there is no complication from floating gate interference. Here we report our successful demonstration of junction-free BE-SONOS NAND Flash using n-channel, p-channel and TFT devices that are suitable for future 3-D layering [1, 2]. Simulation results are also presented and compared with experimental results.

II. Process

The typical device cross-sectional view of the junction-free BE-SONOS device is shown in Fig. 1(a). The device is fabricated in a 0.15 μm poly pitch. After patterning the hard mask of poly, an additional oxide liner is formed to reduce the poly space (originally >70 nm), followed by the poly etching to define the narrow poly space. Figure 1(b) shows the top view of the array. No abnormal poly short or line-end breaking is observed. The narrow space (S) can be accurately controlled by the spacer oxide thickness.

In Fig. 2, the narrow space (S) is then completely filled in by a second oxide spacer process. Next we carry out a tilt-angle implantation to form the junction outside the array. Due to the shadowing of the thick poly gate, the devices in the array do not have junction. This process is completely compatible with current NAND process, without introducing any new mask.

All the devices are evaluated using a 16-WL NAND array. Typical O1/N1/O2/N2/O3 are 13/20/25/60/60 Å, respectively.

III. Bulk-Si Device Characteristics

(a) Simulation results:

Figure 3 shows the simulated electron density of the junction-free BE-SONOS NAND. For simplicity, only results from a 4-WL NAND are shown here. In Fig. 3(a), when select gate (V_G) = 0V, and V_{PASS}= 6V, the select device is in accumulation and turned-off. On the other hand, when V_G=6 V (Fig. 3(b)), all the channels are connected together even as there is no junction. This is caused by the gate fringe field. Figure 3(c) shows the effect of various p-well doping. A lighter well doping indeed provides larger

electron density, enabling more current flow. Figure 3(d) shows the effect of space (S). When S is increased, the electron density is slightly decreased, leading to lower current.

(b) DC characteristics of n-channel NAND:

Figure 4 shows the measured initial IV curve of the n-channel devices. Junction-free devices have similar subthreshold behavior as the with-junction devices. However, the drain current is slightly decreased for the junction–free device (inset). Moreover, larger S shows slightly lower current. Figure 5 indicates that a heavier well doping increases the Vt, consistent with Fig. 3(c).

(c) Program/Erase characteristics of n-channel NAND:

Figure 6 compares the +FN ISPP programming and –FN erasing. The junction-free device shows similar behavior as the with-junction device. This is because +/- FN injection is governed by the intrinsic ONONO property, and is independent of the junction.

(d) P-channel NAND:

We have also successfully fabricated p-channel BE-SONOS NAND with the same ONONO stacks. Figure 7 shows that the junction-free device has larger Vt and much smaller current than the with-junction device. The reason is possibly that the with-junction device is not properly optimized and shows large Vt roll-off effect.

For the p-channel NAND, the program/erase polarities are reversed compared to the n-channel NAND [3]. Figure 8 shows that –FN ISPP program and +FN erase of the p-channel BE-SONOS NAND are well-behaved.

(e) Endurance comparison:

Figure 9 compares the cycling endurance of n-channel devices. Junction-free devices do not show any reliability degradation compared to the with-junction device.

IV. TFT Device Characteristics

Figure 10 shows the IV curve of TFT BE-SONOS devices. In order to study the impact of thermal budget, we applied a 850 °C 20-min post annealing to emulate the thermal budget during 3D integration. It is clear from Fig. 10 that the TFT device is not affected by the extra annealing. This is reasonable since there are no junctions. These important results confirm that junction-free devices are especially suitable for 3D integration

V. Simulation of the scaling capability

Figure 11 shows simulation of the scalability of junction-free devices. We find that the junction-free devices are well controlled with insignificant Vt-roll-off. This is attributed to the larger effective channel length in the junction-free device.

We also simulate the effect of programmed states. At very small F (F = 1/2 poly pitch), the Vt shift after programming is degraded. This is because the device channel length is very short such that fringing field degrades the gate control capability.

It should be mentioned that the junction-free concept is more suitable for charge-trapping devices. In floating-gate devices, the small space (S) will induce serious FG-FG interference.

VI. Summary

A novel junction-free concept is successfully demonstrated in n-channel, p-channel and TFT devices. This novel technique is fully compatible with the current NAND process, and further enables scaling below 20nm nodes and 3D integration.

References

[1] E. K. Lai et al, in IEDM *Tech. Dig.*, 2006, pp. 41-44.
[2] S. M. Jung, et al, IEDM *Tech. Dig.* 2006, pp.39-43
[3] H. T. Lue, et. al, VLSI Tech. Digest, 2007, pp. 140-141

978-1-4244-1802-2/08/$25.00 ©2008 IEEE

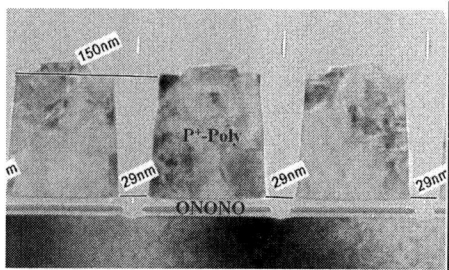

(a) Cross-sectional view

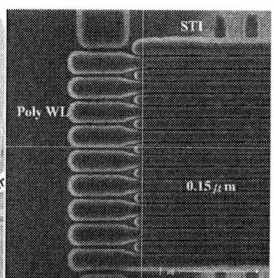

(b) Top View

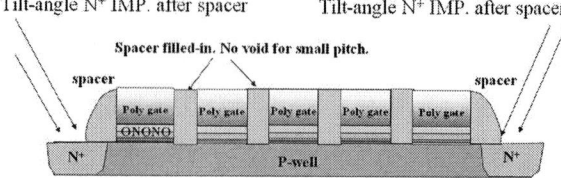

Tilt-angle N+ IMP. after spacer   Tilt-angle N+ IMP. after spacer

Fig. 1 (a) Cross-sectional view of narrow-space WL's for junction-free BE-SONOS NAND Flash. (b) Top-view of the poly WL and pad. Junction-free BE-SONOS NAND using N-channel, P-channel and TFT devices were fabricated.

Fig. 2 Schematic diagram for junction-free BE-SONOS NAND Flash process. First, narrow-space WLs are patterned. Then the spaces (S) between WLs are filled by conventional spacer formation. Next tilt-angle N+ implantation is carried out to form the junction outside the BL string. The narrow-space regions do not have junction because of the shadowing effect by the thick poly gates.

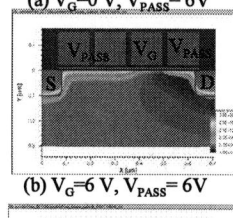

(a) $V_G=0$ V, $V_{PASS}=6$V

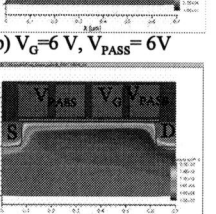

(b) $V_G=6$ V, $V_{PASS}=6$V

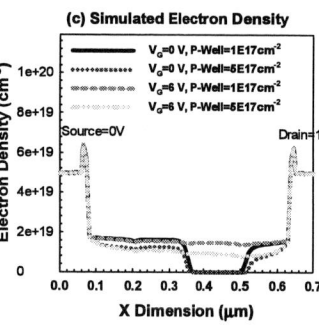

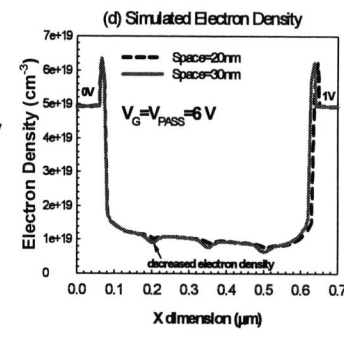

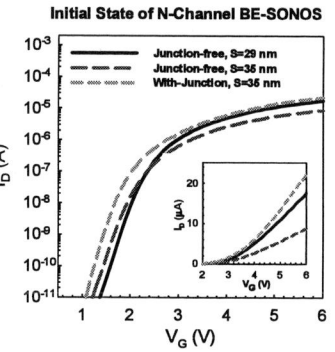

Fig. 3 Simulation of electron density of junction-free NAND device using Synopsis' Sentaurus tool. 4-WL NAND device with space (S) =20 nm are simulated. (a) V_G (select gate)=0 , V_{PASS}=6V. (b) V_G=6, V_{PASS}=6V. (c) Simulated electron density for various p-well doping density for S=20 nm. (d) Simulated electron density for S=20 and 30 nm. Larger S decreases the local electron density.

Fig. 4 Comparison of the experimental junction-free and with-junction n-channel BE-SONOS NAND devices. Inset shows the drain current in linear scale.

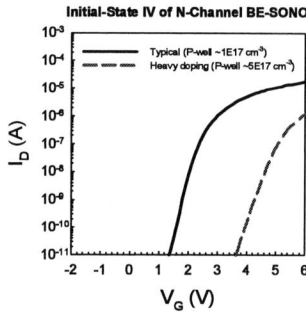

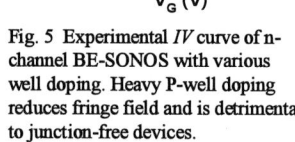

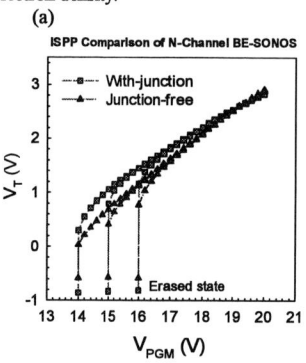

(a)

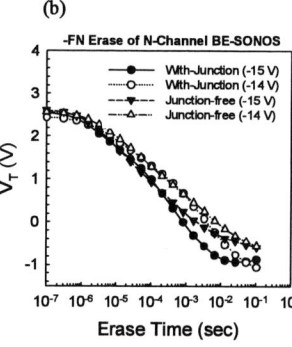

(b)

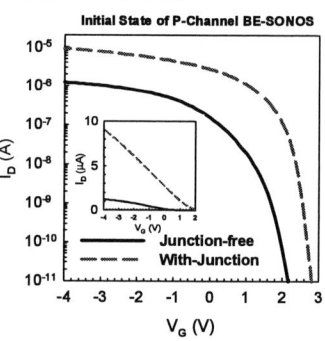

Fig. 5 Experimental *IV* curve of n-channel BE-SONOS with various well doping. Heavy P-well doping reduces fringe field and is detrimental to junction-free devices.

Fig. 6 Experimental (a) +FN ISPP program (b) –FN erase comparison of the junction-free and with-junction n-channel BE-SONOS NAND devices. Junction-free devices show similar behavior as the with-junction devices.

Fig. 7 Comparison of experimental junction-free and with-junction p-channel BE-SONOS NAND device. Inset shows the current in linear scale.

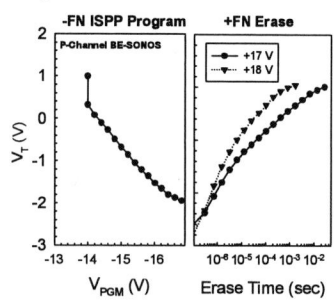

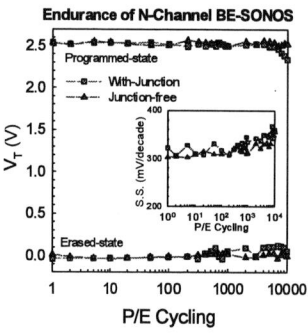

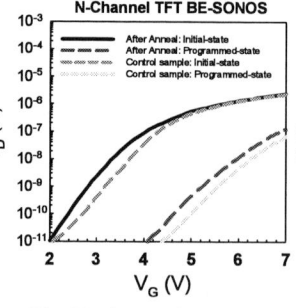

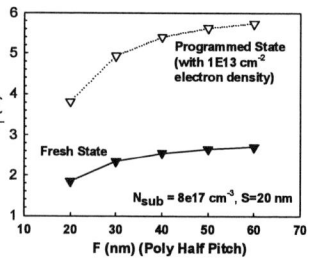

Fig. 8 Experimental -FN ISPP program and +FN erase for the junction-free P-Channel BE-SONOS NAND.

Fig. 9 Comparison of endurance of junction-free and with-junction N-Channel BE-SONOS NAND device. Inset shows the slight S.S. degradation after P/E cycling.

Fig. 10 The experimental *IV* curves of junction-free n-channel TFT BE-SONOS NAND. A post 850C 20-min anneal is applied for comparison.

Fig. 11 Simulation of the scaling capability of junction-free n-channel BE-SONOS NAND at various technology nodes. Junction-free device is viable down to at least 20 nm node.

110

Enhanced Endurance of Dual-bit SONOS NVM Cells using the GIDL Read Method

Alvaro Padilla*, Sunyeong Lee, David Carlton, and Tsu-Jae King Liu

Department of Electrical Engineering and Computer Sciences, University of California at Berkeley, Berkeley, CA 94720

*Phone: +1-510-643-2558, Fax: +1-510-643-2636, E-mail: apadilla@eecs.berkeley.edu

Abstract

Gate-induced drain leakage (GIDL) current is demonstrated to be more sensitive to charge stored locally within the gate-dielectric stack, as compared with the transistor threshold voltage (V_T). Thus the sensing of GIDL rather than V_T is advantageous for dual-bit SONOS NVM cell read operation, not only because it mitigates the complementary-bit disturb (CBD) issue and hence facilitates gate-length scaling, but also because it allows for reductions in stored charge and hence lower program/erase voltages for improved endurance.

Introduction

Dual-bit silicon-oxide-nitride-oxide-silicon (SONOS) non-volatile memory (NVM) cells recently have been developed to improve storage density. The SONOS cell design is advantageous as compared with the traditional floating-gate design because it avoids capacitive coupling interference between cells [1] and it allows for a thinner tunnel dielectric, hence thinner gate-stack equivalent oxide thickness (EOT), for more aggressive gate-length scaling; thus it is preferred for future high-density flash memory technologies [2]. For dual-bit storage per cell, however, the complementary-bit disturb (CBD) issue and increasing threshold voltage (V_T) variation with decreasing channel dimensions [3] present serious challenges for scaling to sub-50nm gate lengths. As a solution, we propose the use of a read method that is more sensitive to localized charge storage, which also allows for reduced program/erase voltage, hence better endurance. The advantages of this approach are demonstrated herein for silicon-on-insulator (SOI) FinFET SONOS NVM cells, but are also applicable to more conventional planar SONOS NVM cells.

Charge Detection (Read) Methods

To illustrate the two charge detection methods studied in this work, the I-V characteristics of an optimized SONOS FinFET NVM cell ($L_g=L_{eff}=100nm$, $T_{si}=40nm$) shown in **Fig. 1a**, obtained via 2-dimensional device simulations [4], are shown in **Fig. 1b**. The state of the bit next to the source (Bit 1) can be determined by the conventional method, which relies on a shift in V_T with stored charge. (Apply positive gate voltage ($V_{GS}\sim0.5V$) and moderate drain voltage ($V_{DS} = 1.5V$): low on-state current indicates that electrons are stored.) To determine the state of the bit next to the drain, a "reverse read" operation (in which the roles of the source and drain electrodes are interchanged) can be performed. Alternatively, the state of Bit 2 can be determined by sensing the cell's gate-induced drain leakage (GIDL) current, which increases with stored charge near to the drain [5,6]. (Apply negative gate voltage ($V_{GS}\sim-1.75V$) and moderate drain voltage ($V_{DS} = 1.5V$): high off-state current indicates that electrons are stored.) It should be noted that the sensing of band-to-band tunneling current has been proposed previously as an alternate charge detection method for bulk planar SONOS NVM cells [7] and for SOI FinFET NVM cells [5,6].

Note that V_T is affected by the state of the bit near to the drain because of short-channel effects (SCE) – even though this FET structure is properly designed to *suppress* these effects – so that the separation in V_T (ΔV_T) between programmed and erased Bit 1 states is reduced. In contrast, GIDL current is *not* significantly affected by the state of the bit near to the source. This shows that the GIDL read method is less susceptible to SCE, so that it can be used for aggressively scaled NVM cells.

Experimental

N-channel FinFET SONOS NVM cells were fabricated using a gate-first process flow similar to that described in [5]. The fabricated structures have an undoped silicon fin width (T_{si}) of ~50nm, and tunnel-oxide/nitride/control-oxide ('ONO') film thicknesses of 2.8nm/5.5nm/5nm.

Fig. 2 shows measured I_{DS}-V_{GS} characteristics for a SONOS FinFET NVM cell before and after Bit 2 was selectively programmed via hot electron injection (HEI) with various programming pulse durations (t_p) and relatively low programming voltages ($V_{GS}=6V$, $V_{DS}=6.5V$). As expected, an increase in off-state GIDL current, and no significant increase in V_T, is seen for forward read operation (**Fig. 2a**) with only Bit 2 programmed. For reverse read operation (**Fig. 2b**) low GIDL is seen, as expected, but no significant change in V_T is seen because the amount of stored charge is small. These measurements affirm that GIDL is much more sensitive to trapped charge than V_T [8].

Fig. 3a shows the measured change in reverse-mode V_T (defined at $I_{DS}=0.1uA$ for $V_{DS}=1.5V$) *vs.* Bit 2 HEI programming time, for various programming voltages. It can be seen that large programming voltages (e.g. $V_{GS}=8.0V$, $V_{DS}=8.5V$) are *required* to induce a significant shift (and thus a large separation) in the cell's V_T. This is *not* desirable, since the use of large programming voltages is detrimental to cell endurance. Additionally, the forward-mode V_T (indicative of the state of Bit 1) also increases slightly, highlighting another weakness of the conventional charge-detection method: V_T is sensitive not only to charge stored near to the source, but also to charge stored at the center of the channel; this makes it more difficult to achieve dual-bit operation with high stored charge density and/or very short L_g. **Fig. 3b** shows the measured change in GIDL current (defined at $V_{GS}=-1.5V$ and $V_{DS}=1.5V$) *vs.* Bit 2 HEI programming time, again for various programming voltages. The difference in GIDL current between the programmed and erased Bit 2 states, ΔI_{GIDL}, is large enough (~4 orders of magnitude) to allow proper identification of each state. It can be seen that the GIDL read method is compatible with lower programming voltages (which are adequate for inducing significant ΔI_{GIDL}) and is less susceptible to the CBD effect, as compared with the conventional V_T read method. Comparable erase voltages can be used for the two read methods, as seen from the measured erase characteristics shown in **Fig. 4**.

Measured cell endurance characteristics are shown in **Fig. 5**. Greater than 3 orders of magnitude operating window for the GIDL read method is maintained after 10^5 cycles, when moderate program/erase voltages are utilized. Note that GIDL shows less change than does V_T due to CBD over many program/erase cycles, indicating once again the reduced sensitivity of GIDL to charge stored *away* from the drain electrode. **Fig. 6** shows the measured retention characteristics for cells that have been cycled 10^5 times using moderate program/erase voltages. 10 years retention time can be achieved using the GIDL read method, but not using the V_T read method.

40nm-L_g Dual-Bit SONOS Cell Simulation

Fig. 7 shows simulated I_{DS}-V_{GS} characteristics of a FinFET SONOS cell with $L_g=40nm$, $T_{si}=20nm$. Even though the gate stack thickness has not been scaled down, ~3 orders of magnitude difference in GIDL current is maintained between programmed and erased Bit 2 states. This indicates that the GIDL read method can be used to detect charge stored in highly scaled dual-bit SONOS NVM cells.

Summary

The use of the more sensitive GIDL read method for dual-bit SONOS NVM cells is shown to mitigate the CBD effect and to provide for lower program/erase voltages, to enhance cell scalability and endurance.

978-1-4244-1802-2/08/$25.00 ©2008 IEEE

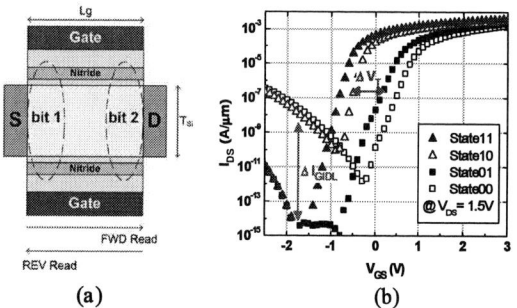

(a)

(b)

Fig.1: (a) 2D schematic cross-section of the dual-bit FinFET SONOS NVM cell. (b) Simulated I_{DS}-V_{GS} curves, showing that V_T (GIDL) can be used to distinguish the state of Bit 1 (Bit 2). In this figure, '0' ('1') refers to the programmed (erased) state. O/N/O film thickness is 3/6/5 nm, respectively, for simulation.

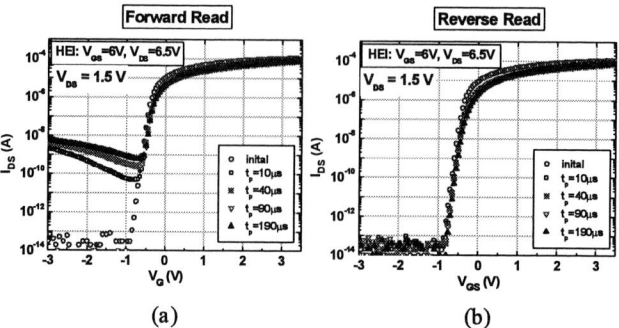

(a)   (b)

Fig.2: Measured I_{DS}-V_{GS} curves in (a) forward, and (b) reverse modes, of a FinFET SONOS NVM cell (L_g=350nm, T_{si}=50nm) before and after programming Bit 2 via HEI with various programming pulse durations (t_p).

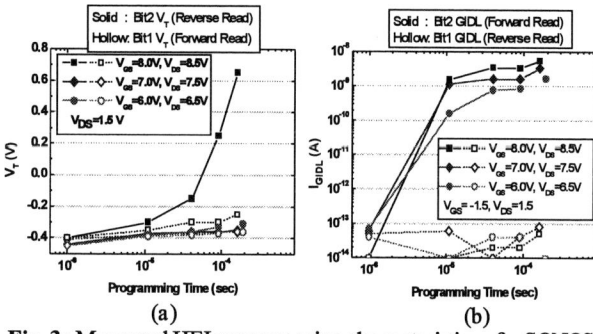

(a)   (b)

Fig.3: Measured HEI programming characteristics of a SONOS FinFET NVM cell, as measured by (a) ΔV_T, and (b) ΔI_{GIDL} read methods. L_g=350nm, W=0.1um, T_{si}=50nm.

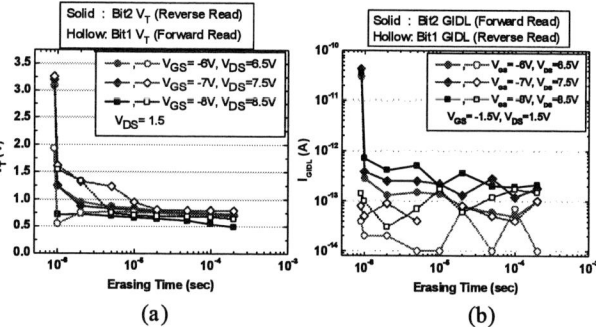

(a)   (b)

Fig.4: Measured BBHI erase characteristics (PGM: V_{GS}=7V, V_{DS}=7.5V, t_p=100ms), as measured by (a) ΔV_T, and (b) ΔI_{GIDL} read methods. L_g=330nm, W=0.1um, T_{si}=50nm.

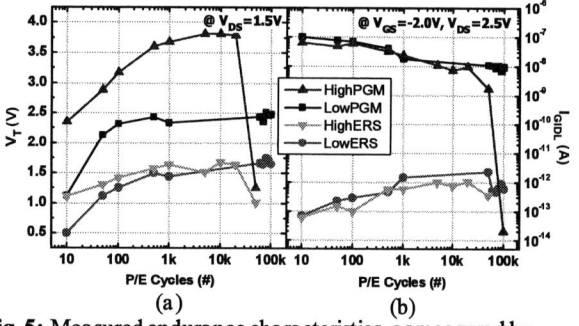

(a)   (b)

Fig.5: Measured endurance characteristics, as measured by (a) ΔV_T (Bit 1), and (b) ΔI_{GIDL} (Bit 2), read methods. L_g=330nm, W=0.1um, T_{si}=50nm. High PGM: V_{GS}=8V, V_{DS}=8.5V, t_P=50μs; High ERS: V_{GS}=-8V, V_{DS}=8.5V, t_E=50 μs; Low PGM: V_{GS}=7V, V_{DS}=7.5V, t_P=50μs; Low ERS: V_{GS}=-7V, V_{DS}=7.5V, t_E=50 μs.

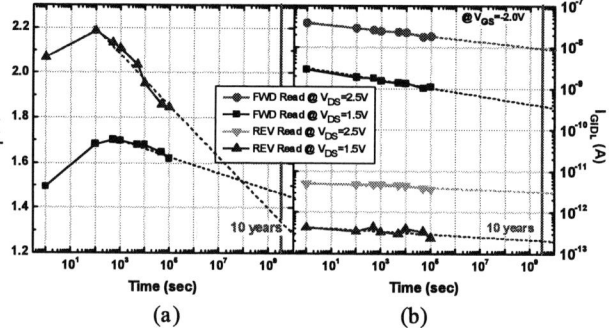

(a)   (b)

Fig.6: Measured charge retention characteristics (@ 85 ºC) of 100K cycled cell (PGM: V_{GS}=7V, V_{DS}=7.5V, t_p=50us), as measured by (a) ΔV_T, and (b) ΔI_{GIDL} read methods. L_g=330nm, W=0.1um, T_{si}=50nm.

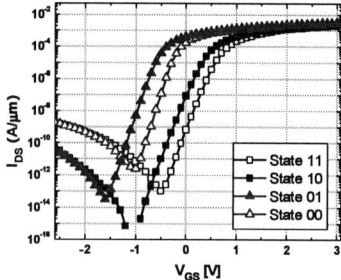

Fig.7: Simulated I_{DS}-V_{GS} characteristics (V_{DS}=1.5V) of a SONOS NVM cell with L_g=40nm, L_{eff}=60nm, T_{si}=20nm and O/N/O film thickness=3/6/5 nm, respectively.

References:

[1] J.-D. Lee *et al.*, *IEEE-EDL* **23**, 264 (2002)
[2] K. Kim, *IEDM Tech. Dig.*, Paper 13.5 (2005)
[3] M. Pelgrom *et al.*, *IEEE-J. Solid-St. Elect.* **38**, 450 (2003)
[4] Taurus User's Manual (Synopsys, Inc.)
[5] A. Padilla *et al.*, *VLSI-TSA Tech. Symp.*, pp 24 (2007)
[6] A. Padilla *et al.*, *IEEE-EDL* **28**, 502 (2007)
[7] C.C. Yeh *et al.*, *VLSI Tech. Symp.*, pp 116 (2005)
[8] E. Lusky *et al.*, *IEEE-TED* **51**, 444 (2004)

Planar Bulk+ Technology using TiN/Hf-based gate stack for Low Power Applications

G. Bidal[1,2], F. Boeuf[1] , S. Denorme[1], N. Loubet[1], C. Laviron[3], F. Leverd[1], S. Barnola[3], T. Salvetat[3], V. Cosnier[1], F. Martin[3], M. Grosjean[1] , P. Perreau[3], D. Chanemougame[1], S. Haendler[1], M. Marin[1], M. Rafik[1,2], D. Fleury[1,2], C. Leyris[1], L. Clement[1], M. Sellier[1], S. Monfray[1], J. Bougueon[1], M.P. Samson[1], J.D. Chapon[1], P. Gouraud[1], G.Ghibaudo[2] and T. Skotnicki[1].

frederic.boeuf@st.com ; tel : +33 (4) 38 92 36 88 ; fax : + 33 (4) 38 92 29 51, [1]STMicroelectronics, 850 rue Jean Monnet, 38920 Crolles Cedex, France; [2]IMEP, 3 parvis Louis Néel, BP 257, 38016 Grenoble Cedex 1, France; [3]CEA-LETI, 17 rue des Martyrs, 38000 Grenoble, France

Abstract

This work highlights the new Bulk+ technology using High-K dielectric, Single Metal Gate and Fully Depleted SON (Silicon On Nothing) channel for sub-45nm low cost applications. Thin Silicon channel (down to T_{si}=8nm) and thin BOX (T_{box}=15 to 25nm) are obtained using the SON process [1]. Transistor performance (W_{design}/L_{gate}=90nm/40nm) at V_{dd}=1.1V and I_{off}<2nA/µm is as high as 1298µA/µm for nMOS and 663µA/µm for pMOS. In addition, reliability, noise and 6T-SRAM bit cells down to 0.249µm² are characterized. Significant improvements with respect to conventional Bulk technology are demonstrated.

Introduction

While scaling CMOS devices below the 45nm node, several parasitic effects appear. In static regime, gate leakage when using SiON gate dielectric impedes scaling of gate capacitance. At the 45nm node, the first dual metal gate and high-K dielectric gate stack appeared for High-Performance applications [2]. Nevertheless, most of the low power products are still fabricated with the conventional Poly/SiON gate stack [3], mainly for low cost purpose, whereas a single metal - single high-K gate solution would be preferred. The second limitation appears in dynamic regime. As technology shrinks, circuit's load capacitances rely more and more on fixed parasitic capacitances that do not scale with the transistor width (e.g wires and gate to corner capacitances). This strongly impacts the delay of small width transistors, and impedes a proper width scaling. At the circuit level this leads to a big area penalty due to the large portion of those transistors for the sub-32nm technologies (Fig.1). Fig.2 shows the simulated delay of a NAND2x2 circuit as a function of nMOS transistor width (V_{th} variation and stress effects with W_{design} are not taken into account) and clearly shows a strong dependence in W_{design}. Addressing specifically the delay issue of those small width transistors is thus mandatory. We propose here the Bulk+ device structure using both single metal gate/high-K gate stack and featuring higher electrical width without penalty on the designed footprint.

New Bulk+ folded Device Strategy

We previously introduced the SON technology concept [1,5] aimed at low cost applications. The Bulk+ technology is using this SON device along with a single gate first TiN/High-K gate stack. The combination of a localized un-doped fully depleted channel fabricated on a bulk wafer with a TiN electrode, allows a proper adjustment of the threshold voltage together with a regular capacitance scaling. In order to increase the drive current of narrow devices without expense in designed footprint, we use the faceting properties of Silicon selective epitaxial growth (SEG) during the SON process. This allows maximizing the electrical width of the device and therefore increasing the drive current per unit of designed width W_{design}. TEM cross section in both L and W directions are shown in fig.3-4. Bulk+ device cumulates the advantage of planar UTB devices like SON or FDSOI with those a Tri-gate in terms of larger W per unity of silicon footprint.

Experimental

Fig.6 describes the process flow. After the STI formation, 8nm to 10nm Si and 25 to 15nm SiGe SEG are formed, defining the conduction channel and the sacrificial layer for BOX formation respectively. The thin conduction channel extends in the width direction thanks to the SiGe faceting. Fig.5 shows the relative increase in electrical W as a function of SiGe thickness for <110> channel devices. The width of the folded SON device can be

increased up to 30% for W_{design} below 0.15µm, depending on the SiGe thickness. This channel extension features a thinner thickness due to the smaller Si growth rate on (111) planes. Next Hf-based dielectric and 10nm TiN are deposited and patterned down to L_g=40nm. After SiN thin-spacer formation, junctions are recessed and SiGe is selectively removed using a hot HCl process [6]. BOX is formed using an ONO stack. After S/D re-growth, LDD are implanted in combination with deep halo used for Ground Plane formation below the BOX.

Transistor Electrical Results

Long channel (L>0.1µm) transistor threshold voltage is about 0.45V using un-doped channel, and can be reduced by using a lightly counter-doped channel (e.g. ΔVth= -150mV on pMOS, Fig. 7). Thanks to the thin channel and thin BOX (25nm), excellent electrostatic control is obtained on Bulk+ devices (Fig.8): DIBL < 80 mV/V for L=40nm with SS<80 mV/dec. I_{on}-I_{off} trade-off for L=40nm as a function of W_{design} (Fig.9), shows the increasing contribution of the conduction on lateral facets for width below 0.2µm. Best I_{ds}-V_{ds} and I_{ds}-V_{gs} characteristics of nMOS and pMOS (W_{design}=90nm) transistors are plotted on fig.10-11. At V_{dd}=1.1V and I_{off}<2nA/µm, drive current up to 1298µA/µm and 663µA/µm is obtained for nMOS and pMOS respectively.

SRAM Electrical Results

We fabricated 6T-SRAM bit cells from 0.374µm² down to 0.249µm² using Bulk+ devices with 15nm BOX (Fig.12). Fig.13 shows well-defined and symmetric butterfly curves for 0.299µm² bit-cell, down to V_{dd}=0.8V. At V_{dd}=1.1V, we obtained a SNM=180mV, WM = 320mV with I_{read}=23µA/cell and I_{stdb}= 172pA/cell (Fig. 14-15). The average SNM of 0.374µm² bit-cells is 246mV with σSNM= 62mV (Fig.16). Functional 0.249µm² bit cell were measured (Fig. 17).

Variability, Noise and Reliability Measurement

The measurement of σΔVt on transistor-pairs as a function of $1/\sqrt{(WL)}$ is shown on fig.18. For smallest devices σΔVt <60mV is measured. RTS noise characterization on SRAM devices is shown in fig.19. The comparison with a Bulk/Poly/SiON reference shows that using High-K and ONO Box does not degrade the RTS-noise characteristics. Finally, TDDB measurement on Hf-based/TiN gate stack shows a maximum power supply of 1.4V for 10-years lifetime (Fig 20).

Conclusion

Using the SON technology, we demonstrated the feasibility of a Bulk+ fully-depleted folded SON structure with TiN/High-K gate stack and specially designed to enhance the drive current of small width layout footprint, by extending the device physical width thanks to SEG. Competitive drive currents, SRAM functionality and reliability characterisation shows the potential of this architecture for 32nm and 22nm node.

Acknowlegment

This work was partially supported by the European Project IST-PullNano and MEDEA+ DECISIF.

References

[1] M. Jurczak et al., ., in Digest of Symp. VLSI 1999, p 29
[2] K. Mistry et al., in Tech. Digest of IEDM 2007, p 247
[3] E.Josse et al., in Tech. Digest of IEDM 2006, p 693
[4] F.Boeuf et al., in Digest of Symp. VLSI 2007, p 24
[5] S. Monfray et al., in Tech. Digest of IEDM 2004, p 635
[6] N. Loubet et al., in Proceedings of SSDM2007,p 716

978-1-4244-1802-2/08/$25.00 ©2008 IEEE

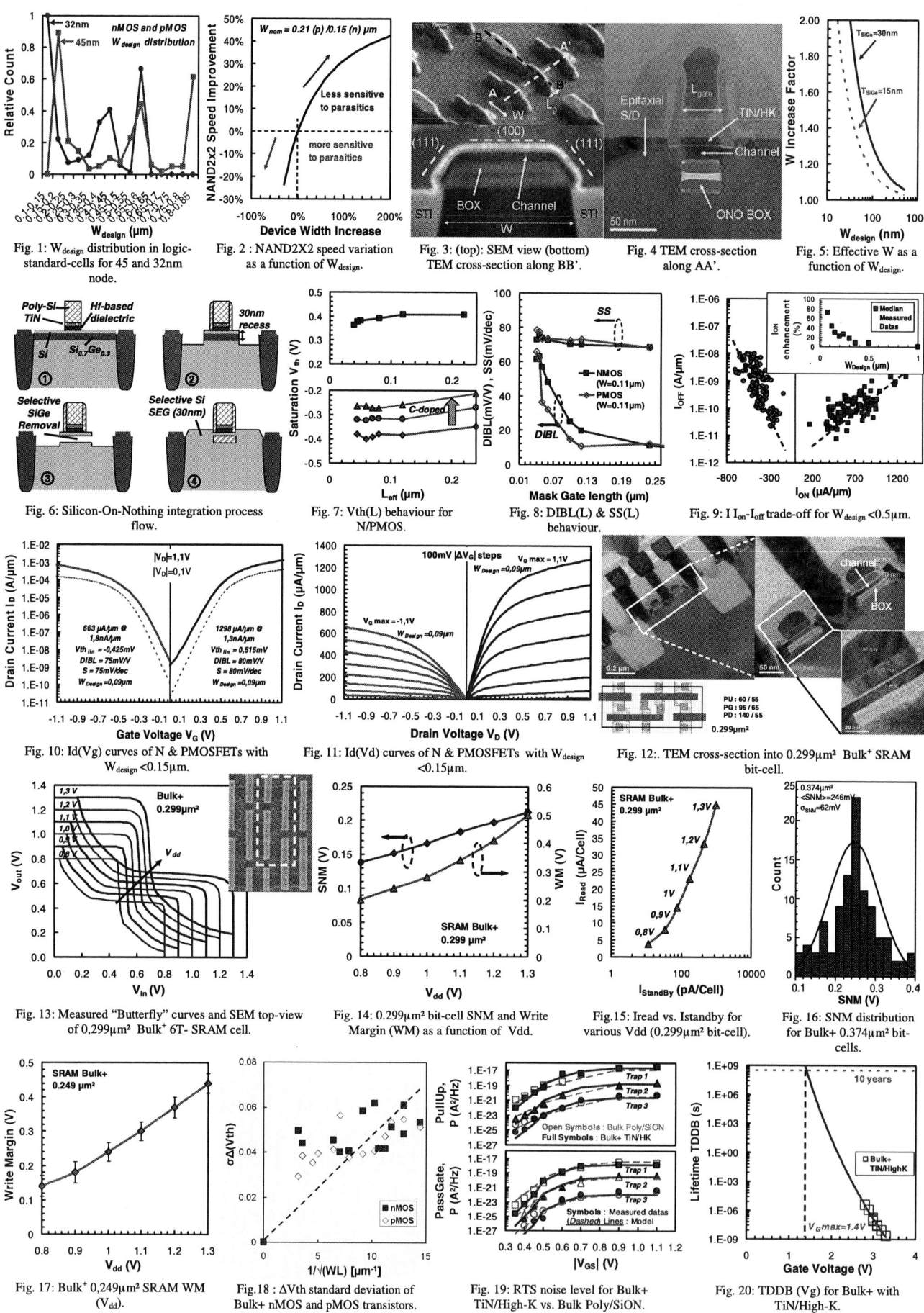

Fig. 1: W_design distribution in logic-standard-cells for 45 and 32nm node.

Fig. 2 : NAND2X2 speed variation as a function of W_design.

Fig. 3: (top): SEM view (bottom) TEM cross-section along BB'.

Fig. 4 TEM cross-section along AA'.

Fig. 5: Effective W as a function of W_design.

Fig. 6: Silicon-On-Nothing integration process flow.

Fig. 7: Vth(L) behaviour for N/PMOS.

Fig. 8: DIBL(L) & SS(L) behaviour.

Fig. 9: I I_on-I_off trade-off for W_design <0.5μm.

Fig. 10: Id(Vg) curves of N & PMOSFETs with W_design <0.15μm.

Fig. 11: Id(Vd) curves of N & PMOSFETs with W_design <0.15μm.

Fig. 12:. TEM cross-section into 0.299μm² Bulk+ SRAM bit-cell.

Fig. 13: Measured "Butterfly" curves and SEM top-view of 0,299μm² Bulk+ 6T- SRAM cell.

Fig. 14: 0.299μm² bit-cell SNM and Write Margin (WM) as a function of Vdd.

Fig.15: Iread vs. Istandby for various Vdd (0.299μm² bit-cell).

Fig. 16: SNM distribution for Bulk+ 0.374μm² bit-cells.

Fig. 17: Bulk+ 0,249μm² SRAM WM (V_dd).

Fig.18 : ΔVth standard deviation of Bulk+ nMOS and pMOS transistors.

Fig. 19: RTS noise level for Bulk+ TiN/High-K vs. Bulk Poly/SiON.

Fig. 20: TDDB (Vg) for Bulk+ with TiN/High-K.

High Performance Sub-35 nm Bulk CMOS with Hybrid Gate Structures of NMOS ; Dopant Confinement Layer (DCL) / PMOS ; Ni-FUSI by using Flash Lamp Anneal (FLA) in Ni-silicidation

H.Ohta, K.Kawamura*, H.Fukutome, M.Tajima*, K. Okabe*, K.Ikeda, K.Hosaka,
Y. Momiyama, S.Satoh and T.Sugii

Fujitsu Laboratories Ltd., *Fujitsu limited., 50 Fuchigami, Akiruno Tokyo 197-0833, Japan
Phone: +81-42-532-1253, FAX: +81-42-532-2513, E-mail:ohta.hiroyuk-03@jp.fujitsu.com

Abstract

We applied Flash Lamp Annealing (FLA) in Ni-silicidation to our developed Dopant Confinement Layer (DCL) structure for the first time. DCL technique is a novel Stress Memorization Technique (SMT). We successfully improved the short channel effect (SCE) with keeping a high drive current by FLA in Ni-silicidation. For pMOSFET, 2 layers Ni Fully-silicide (Ni-FUSI) was selectively formed on gates, and both effective work function (WF) control and thinner Teff are improved. On the other hand, unlike pMOS, Ni-FUSI process is not performed in nMOS. Both higher activation of halo and reduction of parasitic resistance in nMOSFET are improved by the combination of DCL structure and FLA in Ni-silicidation. Consequently, the higher drive currents of 1255 $\mu A/\mu m$ and 759 $\mu A/\mu m$ were obtained I_{off}=122 nA/μm and 112 nA/μm at $|V_{dd}|$=1.0 V for nMOSFET and pMOSFET, respectively.

Introduction

Applications of millisecond annealing (MSA), such as laser annealing and flash lamp annealing, to CMOSFETs have recently been reported [1,2]. MSA technique is powerful technique for a breakthrough of the trade-off line. MSA technique is mainly applied to S/D annealing and utilizes the higher carrier activation. However, there has been little work on applying it to silicide annealing. On the other hand, the SMT has been considered to be a promising stressor as a process-induced booster that has been used together with SIN stressed liner [3,4]. We have already reported the most effective solution, named as DCL structure to maximize its effect on device performances without adding new materials [5]. In this paper, we present the process feasibility and superior electrical characteristics of DCL CMOSFETs using FLA in Ni-silicidation.

Device Process Concept

Figure 1 shows the concept of our sub-35 nm CMOSFETs. The important features in this MOSFET are (a) aggressively scaled gate height of only pMOS with Ni-FUSI gate, (b) DCL / a-Si interface for stopping of nMOS Ni-silicidation, (c) DCL structure. The gate height of only pMOS transistor is aggressively scaled down for Ni-FUSI formation. Flash lamp annealing (FLA) is performed in the RTP-2 of Ni-salicide process to form a full silicide phase selectively on pMOS gates. The nitrided gate oxide is utilized in this device. The important feature is the use of FLA in Ni-silicidation added on CMOSFETs. Gate electrodes are fabricated by the lamination of DCL / a-Si layer. Strain in nMOS channel can be improved by optimizing ion implantation in DCL.

Results and discussion

Fig. 2 shows cross-sectional TEM images of nMOS and pMOS after full processing. It is clearly observed that 2 layers Ni-FUSI is formed in pMOS while the DCL in nMOS remains in the silicon phase. We speculate that the nickel silicide in the DCL of pMOS formed by FLA is an Si-rich phase from the contrast in the Dark field TEM image. Unlike pMOS, such a FUSI gate structure is not observed in nMOS due to the suppression of the Ni/Si reaction at DCL / a-Si interface. The V_{fb} shifts of pMOS are very small (20 mV) while C-V characteristics of nMOS are almost the same (Fig. 3). In this work, gate oxynitride of EOT=1.2 nm is used, with achieving the equivalent oxide thickness of inversion layer capacitance (T_{eff}) = 1.9 nm and 1.55 nm by FLA in Ni-silicidation for nMOS and pMOS, respectively. In general, a work function (WF) tuning range cannot be covered by Si band-gap [6,7]. Therefore, conventional Ni-FUSI gates cannot be applied to high performance devices which require low threshold voltages. Contrary to the conventional FUSI process, our technique can be applicable to high performance devices. This is because the Ni density is very low in the DCL and the threshold voltage is controlled with impurities in gate electrode. To achieve the Ni-FUSI, it is necessary to choose the proper condition of FLA energy and assist temperature as shown in Fig. 4, indicating that Ni-FUSI gates are formed in optimal region. Higher temperature or higher FLA energy causes the increase of the sheet resistance due to the silicide agglomeration while lower temperature or energy does not form the Ni-FUSI because of the sufficient reaction.

Since the temperature in the a-Si gate on insulator is probably higher than that of S/D active areas because of the large difference of thermal conductivity between silicon (1.5 W/cm/degree C) and SiO$_2$ (0.014 W/cm/degree C), Ni-FUSI is selectively formed on gate electrode, not in the S/D active areas.

High Performance Bulk CMOS Devices.

Fig. 5 shows saturation V_{th} roll-off characteristics of pMOSFETs. Our V_{th} is defined by constant current (@100 nA/μm). L_{min} significantly improves by using FLA in Ni-silicidation. This effect is due to the effective WF by Ni-FUSI formed using FLA in Ni-silicidation. Because the threshold voltage is controlled with impurities of high concentration in DCL, our Ni-FUSI can be applied to high performance devices. Fig. 6 shows I_{on}-I_{off} characteristics of pMOSFETs. The I_{on} of Ni-FUSI improved by 10 % compared to the control device (without FLA in Ni-silicidation). Additionally, good SCE immunity were achieved down to 30 nm gate length. We aggressively scaled the pMOS gate height selectively by using dry etch. As the result, pMOS gates selectively fully-silicided, leading to the dramatical improvement of T_{eff}. Fig. 7 shows V_{th} (@100 nA/μm) of Ni-FUSI using FLA in Ni-silicidation as a function of the active width. It is clearly observed that the dependence of the V_{th} on active width is very small in the gate lengths of 33,35 & 38 nm. This indicates that FLA resulted in homogeneous WF. Fig. 8 shows I_{on}-L_{min} plots of nMOSFETs with and without FLA in Ni-silicidation. Improvements of drive current (+3-5 %) and L_{min} (7 nm) were achieved in the FLA splits. Fig. 9 shows the junction capacitance (C_j) as a function of S/D voltage of both area and gate edge, respectively. FLA in Ni-silicidation affected C_j of both area and gate edge. From the fact that significant improvement of L_{min} was achieved by FLA, we suggests that halo deactivation was suppressed and extension impurities were further activated. Fig. 10 shows I_{on}-I_{off} characteristics of optimized CMOSFETs. As a result, 8.4 % of Ion improvement was achieved for our previous report [5]. Because the difference of L_{min}, it's not possible to compare this results to previous pMOS data. Well controlled sub-threshold slope of 93.6 mV/dec and 90.8 mV/dec were obtained for nMOS and pMOS, respectively (Fig. 11a). Very well controlled sub-threshold slope of pMOS is due to the thinner Teff=1.55 nm. Fig. 11b shows I_d-V_d characteristics of nMOS and pMOS. Highest Ion of 1255 $\mu A/\mu m$ for nMOS and 759 $\mu A/\mu m$ for pMOS at $|V_{dd}|$ =1.0 V were obtained. Table 1 shows the device performance summary in this work (FLA in Ni-silicidation). Table 1 shows the material of gate electrode, δVth, Ion improvement, Roll-off improvement, improvement mechanism for nMOS and pMOS, respectively.

Conclusions

We applied FLA in Ni-silicidation to the DCL CMOSFETs for the first time. Both drive current and L_{min} were improved due to the effective WF and thinner T_{eff} by Ni-FUSI formation for pMOS. Our Ni-FUSI can be applied to high performance devices because of small V_{fb} shifts. Applying FLA in Ni-silicidation for DCL structure, we achieved higher drive current of 1255/759 $\mu A/\mu m$ for sub-35 nm nMOS/pMOS at $|V_{dd}|$=1.0 V. Consequently, we believe that this method is a strong candidate to achieve high performance for the future technology node.

References

[1] S. Severi et al.: *IEDM Tech. Dig.*, pp. 859-862, 2006.
[2] T. Yamamoto et al.: *VLSI Symp. Tech.*, pp. 122-123, 2007.
[3] A. Wei et al.: *VLSI Tech.*, pp. 216-217, 2007.
[4] A. Eiho et al.: *VLSI Tech.*, pp. 218-219, 2007.
[5] H. Ohta et al.: *IEDM Tech. Dig.*, pp. 289-292, 2007.
[6] K. Takahashi et al.: *IEDM Tech. Dig.*, pp. 91-94, 2004.
[7] D. Aimé et al.: *IEDM Tech. Dig.*, pp. 87-90, 2004.

FLA in Ni-silicidation

Strain Engineering (Tensile/Compressive)
Aggressively scaled gate height for Ni-FUSI

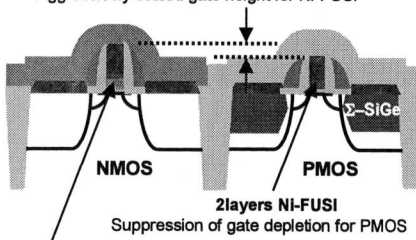

NMOS   **PMOS**
2layers Ni-FUSI
Suppression of gate depletion for PMOS

Dopant Confinement Layer (DCL) Technique
Strain Booster for NMOS

Fig.1 Device structure and concept of our Bulk CMOS technology.

Hybrid Gate Structures of NMOS ; DCL / PMOS ; Ni-FUSI
Aggressively scaled gate height

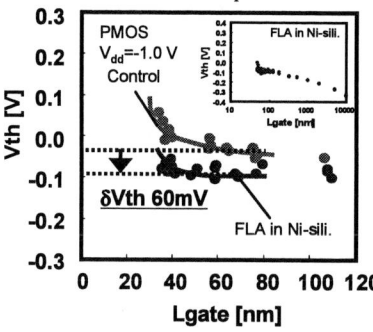

NMOS   **PMOS**
Selectively formed Ni-FUSI

DCL / a-Si interface for stopping of Ni-silicidation

Fig.2 TEM cross-sections of NMOS and PMOS, respectively. Ni-FUSI (NiSi / Si-rich NiSi) was selectively formed on PMOS. This Ni-FUSI using FLA in Ni-silicidation is Si-rich phase.

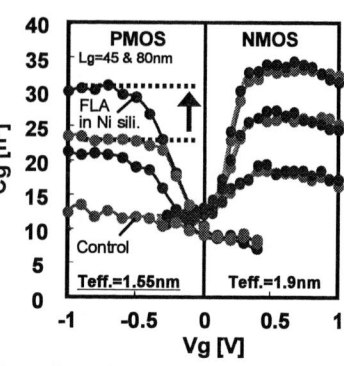

Fig.3 Effects of FLA in Ni-silicidation on gate capacitance. Ni-FUSI with slight Vfb shift (20mV) was selectively formed on PMOS.

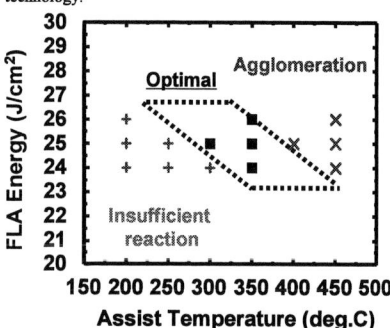

Fig.4 Optimal range of Ni-FUSI using FLA in Ni-silicidation. This graph shows optimal region of FLA energy and assist temperature for Ni-FUSI formation.

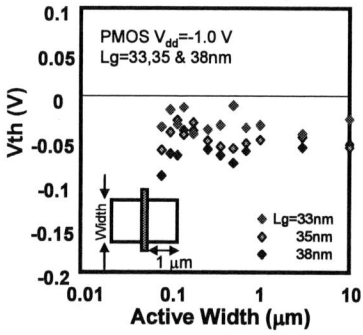

Fig.5 Saturation threshold voltage roll-off of pMOSFET's. Our Vth is defined by constant current (@100nA/μm). Lmin significantly improves by using FLA in Ni-silicidation. This effect is due to the WF by FUSI formed using FLA in Ni-silicidation.

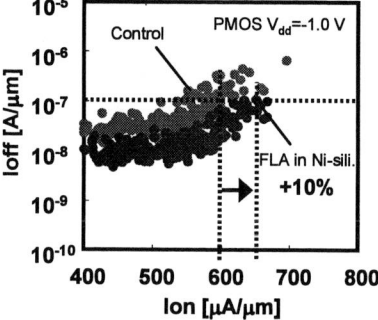

Fig.6 Drive current vs Off-leakage current characteristics of pMOSFETs at V_{dd}=1.0 V. Ion improves (+10%) by using FLA in Ni-silicidation.

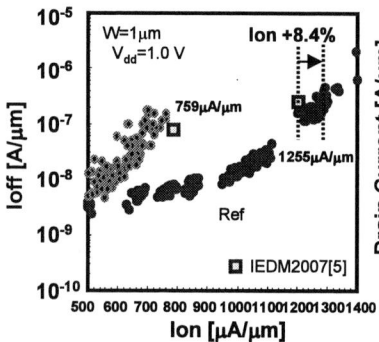

Fig.7 Active width dependency on Vth (@100nA/μm) of FUSI using FLA in Ni-silicidation. The active width dependency on Vth is very small in 3 kinds of Lg (33,35 & 38nm). It indicate that FLA resulted in homogeneous WF.

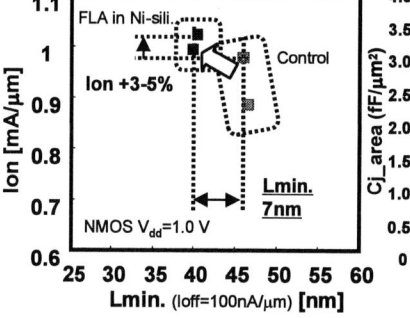

Fig.8 Ion-Lmin plots of nMOSFETs with and without FLA in Ni-silicidation. Lmin significantly improves by using FLA in Ni-silicidation. Improvements of Ion (+3-5 %) and Lmin (7 nm) were achieved.

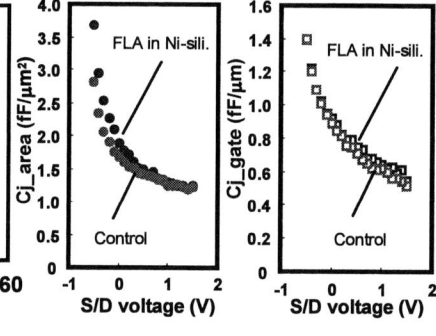

Fig.9 Junction capacitance (Cj) as a function of S/D voltage of both area and gate edge, respectively. . FLA in Ni-silicidation affected Cj of both area and gate edge. Halo deactivation was suppressed and extension impurities were further activated.

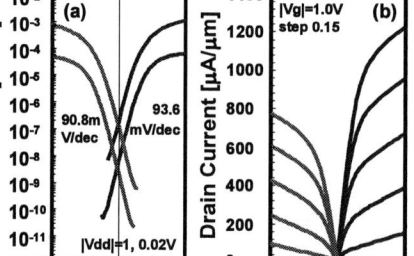

Fig.10 Optimized CMOS drive current vs off-leakage current characteristics at V_{dd}=1.0 V. The high drain current of 1255/759 μA/μm were obtained, respectively.

Fig.11 Id-Vg (a) and Id-Vd (b) characteristics of nMOS and pMOS with DCL structure using FLA in Ni-silicidation. The highest drain current of 1255/759 μA/μm were obtained, respectively.

Table 1 Device performance summary in this work (FLA in Ni-silicidation). Table 1 shows the materials of gate electrode, δVth, Ion improvement, Roll-off improvement, Improvement mechanism for nMOS and pMOS.

FLA in Ni-sili.	nMOS	pMOS
Gate electrode	Ni-silicide	Ni-FUSI
δVth	−	60mV
Ion	+3-5%	+10%
Mechanism (Ion)	Resistance	Teff.=1.55nm
Roll-off	7nm	Improvement
Mechanism (Lmin.)	Halo / extension	Work function

116

Cost-Effective Ni-Melt-FUSI Boosting 32-nm Node LSTP Transistors

H. Fukutome, *K. Kawamura, H. Ohta, K. Hosaka, T. Sakoda, Y. Morisaki and Y. Momiyama

*Fujitsu Laboratories Ltd., *Fujitsu Limited, 50 Fuchigami, Akiruno, Tokyo, 197-0833, Japan*
Phone: +81-42-532-1249, Fax: +81-42-532-2513, E-mail: fukutome@jp.fujitsu.com

Abstract

We demonstrated for the first time that novel Ni-FUSI process using FLA (Melt-FUSI) dramatically improved both electrical characteristics and cost-benefit performance of LSTP devices. Since the T_{inv} was aggressively scaled ($T_{inv} = 2.1$nm) with keeping SiON-gate leakage current and increasing hole mobility twice, we achieved the record I_{on} of 300 $\mu A/\mu m$ at the I_{off} of 20 pA/μm for the pMOS transistor with the L_g of 45 nm at V_d of -1.2 V.

Introduction

Various performance-booster techniques have been investigated for LSTP transistors for 45-nm node and beyond. Gate stack scaling is an effective method to improve their performance, however, it is also required to enhance effective carrier mobilities at low electric field because intermediate characteristic currents predominantly determine LSTP circuit performance [1]. From this realistic viewpoint, scaling procedure via Metal/SiON with good interface characteristics to Metal/High-K is suggested as the best scenario [2]. Fully-silicided (FUSI) gate is a promising technology to fabricate metal gates. Many FUSI transistors have already been reported [3-6], and most of them require additional processes to fabricate FUSI gate, *e.g.*, CMP, capping layer, elevated S/D, and so on. In contrast, such additional process steps should be as few as possible for cost-effective fabrication of LSTP devices. Moreover, it is also important to prevent increase of gate and junction leakage currents induced by FUSI formation (Fig. 1). In this study, we demonstrate for the first time that the Ni-FUSI/SiON gate stack fabricated by only one additional melting process (Melt-FUSI) improved electrical characteristics of LSTP transistors.

Ni-Melt-FUSI/SiON Gate Stack

Novel "Melt-FUSI" method dramatically improved electrical characteristics, in particular, in pMOS gate stack. Figure 2 shows a key part of the Melt-FUSI process. We found that the second anneal with flash lamp anneal (FLA) in Ni silicidation made the pMOS gates selectively fully-silicided. Such FUSI gates have homogeneous Si-rich phase from top to bottom (Fig. 3). It is speculated that non-equilibrium state at high temperature induced by the FLA enhanced Ni diffusion and resulted in homogeneous Ni-FUSI gate. It is found that the gate electrode was expanded upward by the Melt-FUSI method.

It is obvious that the Melt-FUSI process strongly affected gate capacitance of both n- and pMOS transistors (Fig. 4). The FLA process at low temperature increased the maximum gate capacitance (C_{gc}) by 7% without shift of threshold voltage (V_{th}) for both n- and pMOS transistors. It is considered that the FLA activated dopants in the gate bottom at non-equilibrium state and effectively suppressed gate depletion. In contrast, the FLA process at high temperature increased the maximum C_{gc} by 18% with a slight V_{th} shift for pMOS transistors in spite of Si-rich Ni-FUSI phase. This result suggests that the Melt-FUSI has less V_{fb} shift related with pinning states, which decrease effective carrier mobilities. We achieved the pMOS T_{inv} of 2.1 nm without serious increase of gate leakage current by the Melt-FUSI method (Fig. 5). Moreover, we confirmed TDDB reliability of the Melt-FUSI pMOS better than control under a constant electric field, which suggested no silicide spike induced by the Melt-FUSI method (Fig. 6). Consequently, the Melt-FUSI method is very attractive to cost-effectively improve the electrical characteristics of LSTP devices without degradation in gate leakage current and effective hole mobility.

Electrical Characteristics

First, effects of the Melt-FUSI process on electrical characteristics of the LSTP pMOS transistors were investigated. We achieved the record drive current of 300 $\mu A/\mu m$ at the I_{off} of 20 pA/μm for the LSTP pMOS transistor with the gate length (L_g) of 45 nm at V_d of -1.2 V. It was found that the Melt-FUSI process increased the drive current by 60% without degradation of the minimum L_g of the operating transistor (Fig. 7 and 8). Moreover, junction leakage currents were effectively reduced.

Next, we discuss effects of the Melt-FUSI process on electrical characteristics of low-V_{th} pMOS transistors with a slightly higher off-leakage current. The threshold voltage was simply shifted by optimizing channel concentration in the high-V_{th} pMOS transistor. As shown in Figs. 9 and 10, we achieved the drive current of 340 $\mu A/\mu m$ at the I_{off} of 100 pA/μm for the low-V_{th} pMOS transistor with the L_g of 45 nm at V_d of -1.2 V. Moreover, the Melt-FUSI gate significantly increased the linear current by 50% (Fig. 11). Such improvement in the pMOS transistor was considered to be mainly originated with two causes. The first is obviously the aggressively scaled T_{inv}. The second is considered to be drastic twice improvement of the hole mobility, in particular, for short-channel pMOS transistors with Melt-FUSI gates (Fig. 12).

In addition, the Melt-FUSI method never seriously degrade the electrical characteristics of the devices with narrow gate width. Although the Melt-FUSI gate slightly changed dependence of the V_{th} on gate width, it was revealed from Pelgrom plots that the V_{th} fluctuation of the FUSI LSTP pMOS transistor is smaller than that of the control device (Fig. 13). In contrast, even at V_d of -1.0 V, the drive current was 200 $\mu A/\mu m$ at the I_{off} of 20 pA/μm for the Melt-FUSI pMOS transistor (Fig. 14). Based on these results, the operation of the Melt-FUSI device with low power consumption could be also expected. Electrical characteristics of the Melt-FUSI pMOS transistors are summarized and compared with the published data in table 1.

Finally, electrical characteristics of the corresponding nMOS transistor are briefly discussed. Since the Melt-FUSI method also decreased the junction leakage current for nMOS transistors, we achieved the stable operation of LSTP nMOS transistor at the I_{off} of 20 pA/μm with keeping the minimum L_g. Consequently, it was confirmed that the Melt-FUSI process is very useful to boost LSTP device performance.

Conclusions

We demonstrated for the first time that the Melt-FUSI method dramatically improved both electrical characteristics ($T_{inv} = 2.1$ nm, $I_{on} = 300$ $\mu A/\mu m$ at the I_{off} of 20 pA/μm for the L_g of 45 nm, twice mobility for pMOS transistor) and cost-benefit performance (one-step modification) of LSTP devices. Since the T_{inv} was sufficiently scaled by the Melt-FUSI, the SiON gate film with the better interface characteristics could be used to easily improve LSTP device performance. Therefore, the Melt-FUSI method would be a promising technology in 32 nm node and beyond.

References

1) E. Yoshida, *et al.*, *IEDM Tech. Digs.*, **195** (2006).
2) K. Imai, *et al.*, *Ext. Abst. SSDM*, **252** (2007).
3) B. Tavel, *et al.*, *IEDM Tech. Digs.*, **825** (2001).
4) J. Kedzierski, *et al.*, *IEDM Tech. Digs.*, **247** (2002).
5) K. Hosaka, *et al.*, *Symp. VLSI Tech.*, **66** (2005).
6) T. Hoffmann, *et al.*, *IEDM Tech. Digs.*, **269** (2006).
7) G. Tsutsui, *et al.*, *Symp. VLSI Tech.*, **176** (2007).
8) H. Nakamura, *et al.*, *Symp. VLSI Tech.*, **198** (2006).
9) P. F. Hsu, *et al.*, *Symp. VLSI Tech.*, **14** (2006).
10) T. Hayashi, *et al.*, *IEDM Tech. Digs.*, **247** (2006).
11) H.-C. Wen, *et al.*, *Symp. VLSI Tech.*, **160** (2007).
12) S. K. Han, *et al.*, *IEDM Tech. Digs.*, **621** (2006).

978-1-4244-1802-2/08/$25.00 ©2008 IEEE

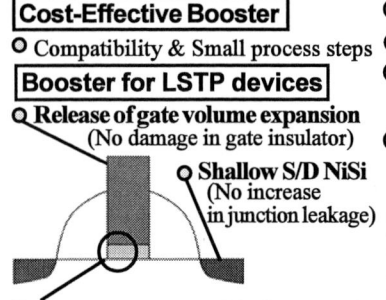

Cost-Effective Booster
- Compatibility & Small process steps

Booster for LSTP devices
- **Release of gate volume expansion** (No damage in gate insulator)
- **Shallow S/D NiSi** (No increase in junction leakage)
- **Low gate leakage with high mobility**

Fig. 1 Schematic of requirements on booster techniques (e.g., FUSI process) applied for low-cost LSTP devices.

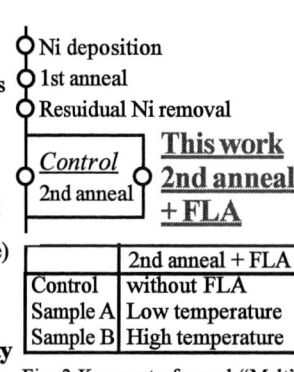

○ Ni deposition
○ 1st anneal
○ Resuidual Ni removal

Control 2nd anneal

This work 2nd anneal + FLA

	2nd anneal + FLA
Control	without FLA
Sample A	Low temperature
Sample B	High temperature

Fig. 2 Key part of novel "Melt" FUSI process for low-cost LSTP devices and list of samples.

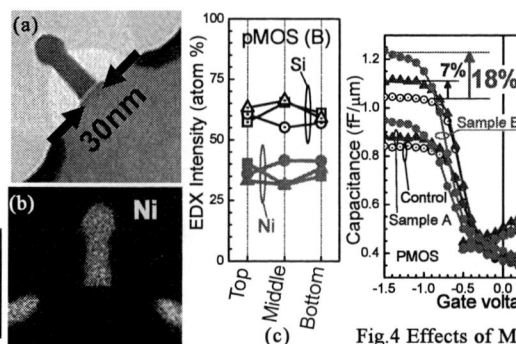

Fig. 3 Novel Ni-Melt-FUSI/SiON gate stack (a) pMOS TEM image. (b) Ni EDX image. (c) Ni/Si ratio from EDX.

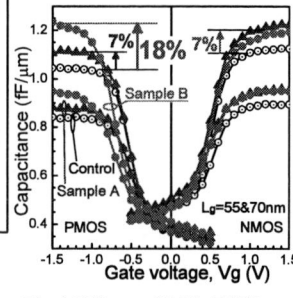

Fig.4 Effects of Melt-FUSI process on gate capacitance. PMOS in sample B was selectively fully-silicided with slight V_{th} shift.

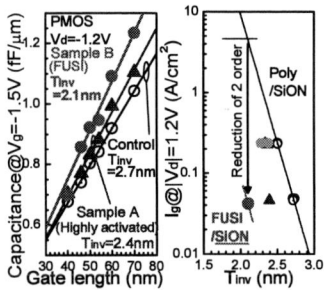

Fig. 5 Estimation of T_{inv} and gate leakage current as a function of T_{inv}. Melt-FUSI causes no degradation of I_g.

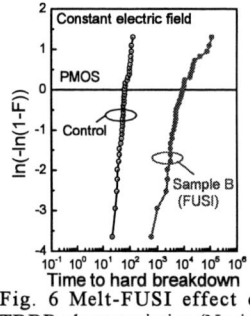

Fig. 6 Melt-FUSI effect on TDDB characteristics (No initial failure: No silicide spike).

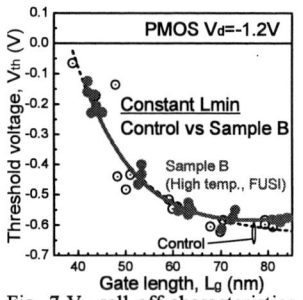

Fig. 7 V_{th} roll-off characteristics of pMOS LSTP transistors with L_{min} of 45 nm at V_d = -1.2 V.

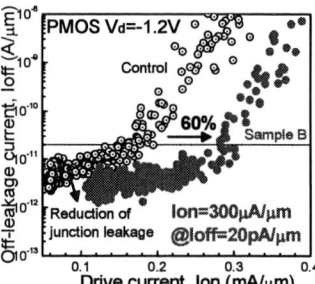

Fig. 8 Melt-FUSI effect on I_{on}-I_{off} characteristics of pMOS LSTP transistors at V_d = -1.2 V (increase by 60%).

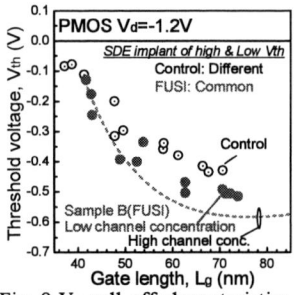

Fig. 9 V_{th} roll-off characteristics of pMOS transistors with low Vth at V_d = -1.2 V (L_{min} = 45 nm).

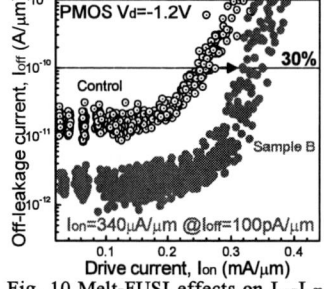

Fig. 10 Melt-FUSI effects on I_{on}-I_{off} characteristics of pMOS transistors with low-V_{th} at V_d = -1.2 V.

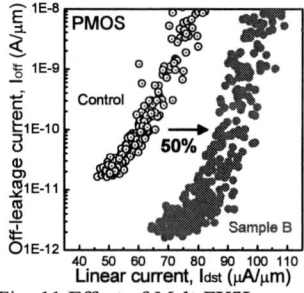

Fig. 11 Effect of Melt-FUSI process on linear current of pMOS transistors at V_d = -0.1 V (Increase by 50%).

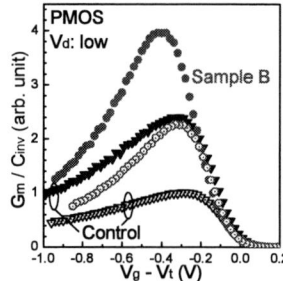

Fig. 12 Effect of Melt-FUSI process on G_m/C_{inv} of pMOS transistors. Filled: L_g=55nm. Open: L_g=80nm.

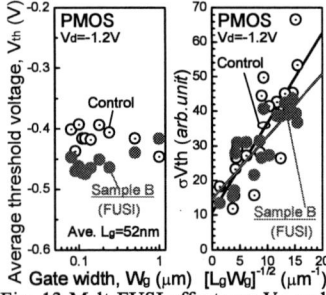

Fig. 13 Melt-FUSI effects on V_{th} and its fluctuation of pMOS LSTP transistors (corresponding Fig. 7).

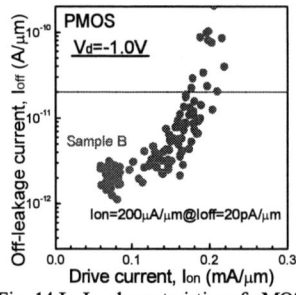

Fig. 14 I_{on}-I_{off} characteristics of pMOS transistors with Melt-FUSI gates at V_d = -1.0 V (good V_{dd} scaling).

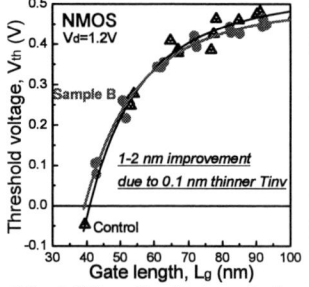

Fig. 15 V_{th} roll-off characteristics of nMOS LSTP transistors at V_d = 1.2V.

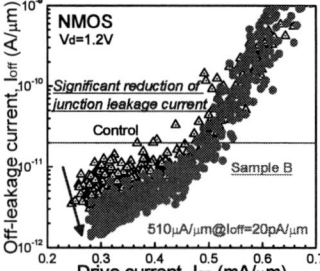

Fig. 15 Effect of Melt-FUSI process on I_{on}-I_{off} characteristics of nMOS LSTP transistors at V_d = 1.2 V.

Table 1. Comparison of low-cost pMOS LSTP transistor performance among this work and other published data.

	This work		ref.[6]	ref.[7]	ref.[8]	ref.[9]	ref.[10]	ref.[11]	ref.[12]
V_{dd}[V]	1.2	1.0	1.2	1.2	1.2	1.2	1.2	1.2	1.0
I_{on}(P)[μA/μm]	300/340	200	300	325	295/340	390	230	360	220
I_{off} [pA/μm]	20/100	20	20	100	20/200	200	20	20	20
L_g [nm]	45		68	40	50	55	60	60	65
Gate stack	Melt-FUSI/SiON		CMP FUSI/High-K	Poly/SiON+Hf	Poly/HfSiON	MIPS/High-K	MIPS/High-K	MIPS/High-K	Metal/High-K

Design and Demonstration of Very High-k (k~50) HfO$_2$ for Ultra-Scaled Si CMOS

S. Migita[1], Y. Watanabe[2], H. Ota[1], H. Ito[2], Y. Kamimuta[2], T. Nabatame[2], and A. Toriumi[1,3]

[1]MIRAI-AIST, AIST Tsukuba West 7, 16-1 Onogawa, Tsukuba 305-8569, Japan
Phone: +81-29-861-5943 Fax: +81-29-849-1529 E-mail: s-migita@aist.go.jp
[2]MIRAI-ASET, AIST, Japan, [3]The University of Tokyo, Japan

Abstract

We have demonstrated very high-k (k~50) HfO$_2$ films for ultra-scaled CMOS application. Higher symmetric crystalline structure enables us to achieve higher–k HfO$_2$. We present a feasible method to obtain sub-nm EOT with very high-k HfO$_2$ under actual process conditions, together with an underlying mechanism.

Introduction

1-nm EOT is within a reach of the current high-k CMOS technology [1]. The next challenge is how to achieve much higher-k gate dielectrics with a reasonable band offset with regard to Si. High polarizability materials are generally associated with a smaller band gap [2]. Although it has been reported that HfO$_2$-based ternary dielectrics show higher-k without any expense of the band offset against Si, the highest-k value so far obtained is ~30 at most [3-4].

This paper demonstrates k~50 using the pure HfO$_2$ thanks to the structural phase transformation induced by an appropriate design of gate stack formation process. **Fig. 1** shows an overall comparison of the HfO$_2$ k-values including the present result.

Concept and proof of very high-k dielectric films

Through the analysis of higher-k materials research so far, we have put emphasis on the following two points: (1) The high polarizability materials should be used as long as the band offset requirement is satisfied. (2) Higher symmetric phase should be used for reducing the molar volume.

To meet these requirements, our strategy is to change the structural symmetry of HfO$_2$ without any other material doping. We have already revealed that the post-deposition anneal (PDA) of HfO$_2$ stacked with gate electrode enhances the k-value [5]. The origin of the k-enhancement has been considered to be the stress effect from the gate electrode, but not fully understood.

HfO$_2$ was deposited by the rf-sputtering at room temperature and the related process flow is shown in **Fig. 2**. Physical thickness was cross-checked by XRR, ellipsometer, and TEM. The PDA with and without electrode remarkably changes the HfO$_2$ crystalline phase to cubic and monoclinic, respectively (**Fig. 3**). Electron diffraction data of a plane-view TEM image after removing the electrode also show that the dominant phase of HfO$_2$ prepared by PDA with electrode was cubic, and only a small amount of other phases were included. The k-value was estimated by the linear relationship between the physical thickness and EOT of MOS capacitors, where the IL thickness was confirmed to be constant by XTEM irrespective of HfO$_2$ thickness in a fixed condition (**Fig. 4**). **Fig. 5** shows that k~50 at highest is obtained in the PDA with TaN, while k~17 in the conventional PDA. **Fig. 6** shows the SIMS profiles of Ti and Ta before and after PDA with gate electrode, showing that there is no diffusion of Ti or Ta into HfO$_2$, It clearly indicates that the k-enhancement is not caused by the formation of TiO$_2$ or Ta$_2$O$_5$, but caused by the higher symmetrical cubic structure. A big concern is the band gap reduction associated with the phase change of HfO$_2$. However, **Fig. 7** shows that the cubic phase HfO$_2$ film keeps the bandgap of conventional HfO$_2$.

Feasibility of very high-k HfO$_2$ gate stacks

We have found a noticeable point that the cubic HfO$_2$ phase comes out only in the case of starting from the amorphous phase. Once the monoclinic phase nucleated at molecular level, no cubic transformation was available. In practice, a faster ramp rate and a shorter hold time in the annealing process are better for the k-enhancement, as shown in **Fig. 8**. While a higher temperature PDA with gate electrode degrades the k-value (**Fig. 9**) which is caused by the transformation to lower symmetric non-cubic phases, confirmed by XRD. This fact indicates that this very high-k can be achieved under non-equilibrium condition. The results of **Figs. 4** and **10** show the interface layer (IL) growth accompanied with the k-enhancement, when starting from the HF-last Si substrate. Thus, although the gate electrode-induced compressive stress may trigger the structural phase transformation [5], some oxygen atoms released from HfO$_2$ are likely to react with Si substrate to form SiO$_2$ IL and which assists the HfO$_2$ symmetric phase transformation thanks to the oxygen vacancy formation [3], as illustrated in **Fig. 11**. By preparing the oxygen deficient HfO$_2$ at the initial stage, it will be possible to optimize the IL growth while keeping the HfO$_2$ quality. **Fig. 12(a)** summarizes the schematic PDA recipe in the progress of HfO$_2$ phases, and **(b)** three key parameters required for attaining very high-k cubic HfO$_2$.

Finally, FUSI/TaN/HfO$_2$/p-Si MOSFETs with cubic- and monoclinic-HfO$_2$ phases were fabricated by PDA with and without electrode, and the effective electron mobility is compared in **Fig. 13** (1.1 nm EOT for both cases). The structural symmetry of HfO$_2$ has little influence on the mobility as we recently reported [6], and provides no additional disadvantage in high-k MOSFET characteristics.

Conclusion

We have successfully engineered very high-k HfO$_2$ films without any expense of the band offset tradeoff. The essential point of the dielectric engineering is the structural phase transformation to higher symmetric phase of HfO$_2$ in the molecular level without introducing any dopants. Although some challenges remain to be solved, this is a dramatic progress in achieving very high-k dielectric films for ultra-scaled Si-CMOS.

Acknowledgement

This work is supported by NEDO.

References

[1] K. Mistry *et al.*, IEDM Tech. Dig. (2007), p.247.
[2] P.W. Peacock and J. Robertson, *J. Appl. Phys.*, **92** (2002) 4712.
[3] K. Kita *et al.*, *Appl. Phys. Lett.*, **86** (2005) 102906.
[4] A. Toriumi and K. Kita, "Material Engineering of High-k Gate Dielectrics," John Wiley & Sons, (2007), p.297.
[5] Y. Watanabe *et al.*, *ECS Trans.* Vol. 11(4), 2007, p. 35.
[6] H. Ota *et al.*, IEDM Tech Dig. (2007), p. 65.

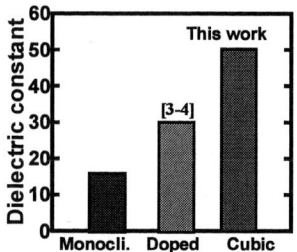

Fig. 1. Comparison of dielectric constants of HfO₂ films with monoclinic, doped [3-4], and cubic phases. In this work, a dramatic enhancement of dielectric constant is attained by the cubic HfO₂.

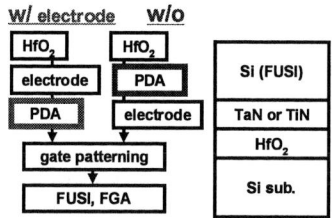

Fig. 2. Process flow and the gate stack structure examined. The PDA with electrode is the key process [5]. Film thicknesses are 2-20 nm (HfO₂), 5 nm (TaN), 10 nm (TiN) and 30 nm (Si), respectively.

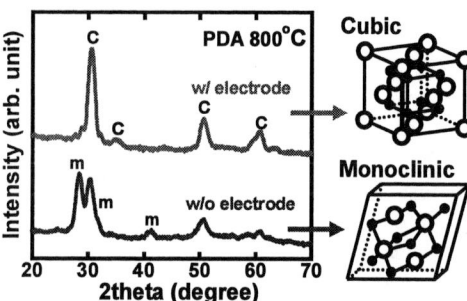

Fig. 3. XRD patterns of HfO₂ films formed by PDA at 800°C with and without electrode. The cubic HfO₂ is successfully formed by PDA only with electrode.

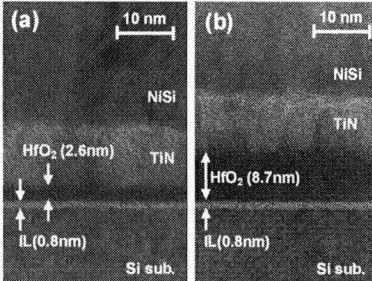

Fig. 4. Cross-sectional TEM images of FUSI/TiN/HfO₂/Si structures with HfO₂ thickness (a) 2.6 nm and (b) 8.7 nm, prepared by PDA at 800°C with electrode. IL with the same thickness (~0.8 nm) is formed irrespective of HfO₂ thickness.

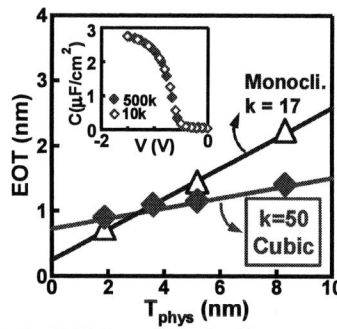

Fig. 5. EOT-T_{phys} plots of HfO₂ MOS capacitors of cubic and monoclinic HfO₂ films prepared by PDA at 800°C. Very high-k value ~50 is achieved by cubic HfO₂. The inset is the C-V characteristics of a cubic HfO₂ MOS capacitor (EOT=0.9 nm) showing very small frequency dispersion (freq.=500k and 10k Hz).

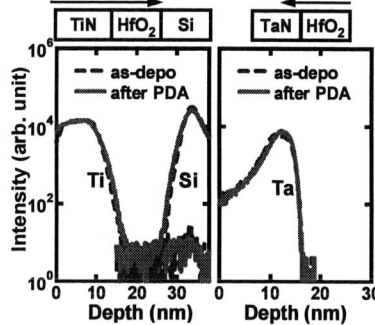

Fig. 6. Front-side SIMS of TiN/HfO₂/Si (left) and backside SIMS of TaN/HfO₂/Si (right), both show as-depo (dashed) and after (solid) PDA at 800°C with electrode. It is confirmed that metal is not diffused into HfO₂.

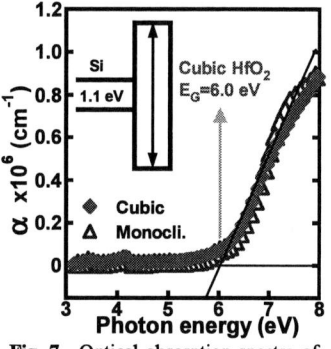

Fig. 7. Optical absorption spectra of cubic and monoclinic HfO₂ films (8.7 nm) measured by spectroscopic ellipsometry. The cubic HfO₂ has 6.0 eV optical bandgap.

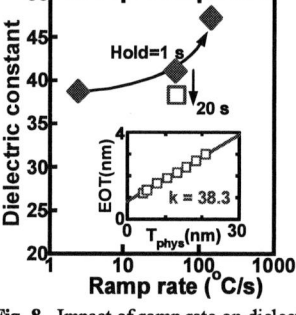

Fig. 8. Impact of ramp rate on dielectric constant in PDA at 800°C with electrode (hold time =1 s). Cooling rates were almost the same as respective ramp rates. A longer hold time (20s) degrades the k-value (open square), of which EOT-T_{phys} plot is shown in the inset.

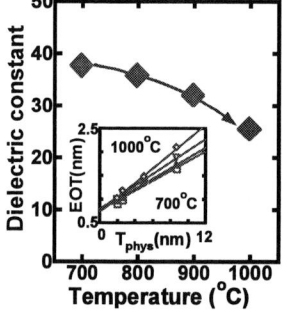

Fig. 9. Influence of PDA temperature (with electrode) on dielectric constant (hold time =20 s). The inset shows that IL thickness is kept unchanged.

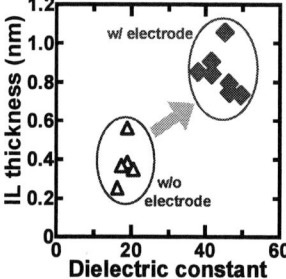

Fig. 10. Increase of IL thickness with k-enhancement. Results of several PDA (with electrode) conditions are plotted.

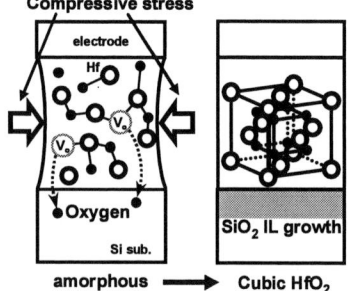

Fig. 11. Illustration of structural phase transformation by PDA with electrode, promoted by compressive stress and oxygen vacancy. Released oxygen reacts with Si and forms SiO₂ IL.

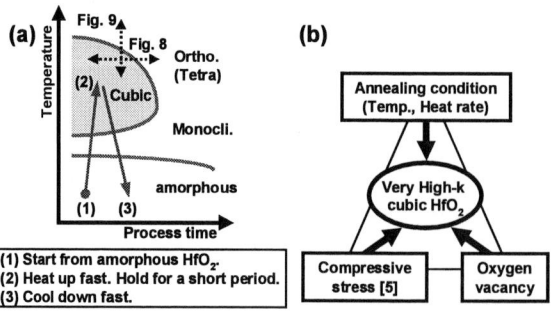

Fig. 12. (a) Progress of HfO₂ phases with temperature and process time. A schematic PDA recipe for cubic HfO₂ formation is shown. Dashed lines indicate experimental conditions of Figs. 8 and 9. (b) three key parameters for producing very high-k cubic HfO₂.

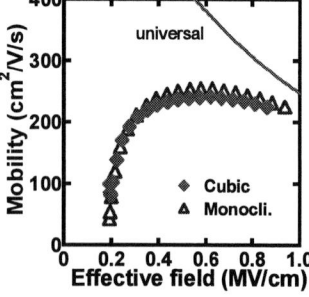

Fig. 13. Electron mobility compared between cubic and monoclinic HfO₂ gate dielectric MOSFETs (EOT=1.1 nm for both cases). Negligible difference suggests little influence of HfO₂ crystalline symmetry on mobility.

120

Analyses of 5σ V_{th} Fluctuation in 65nm-MOSFETs Using Takeuchi Plot

T.Tsunomura[1], A.Nishida[1], F.Yano[1], A.T.Putra[3], K.Takeuchi[1], S.Inaba[1], S.Kamohara[1],
K.Terada[2], T.Hiramoto[1,3], T.Mogami[1]

[1]MIRAI-Selete, [2]Hiroshima City University, [3]University of Tokyo

16-1 Onogawa, Tsukuba, Ibaraki, 305-8569, Japan (Phone: +81-29-849-1272, E-mail:tsunomura.takaaki@selete.co.jp)

Abstract

Using 1M DMA-TEG, the analyses of 5σ V_{th} fluctuation in 65nm-MOSFETs were carried out. Physical and electrical analyses confirmed that random dopant fluctuation is dominant though NMOSFET has larger fluctuation as compared with PMOSFET. To explain this phenomenon, a B clustering model is proposed. In the case of clustering with 5 to 6 B atoms in the channel, V_{th} fluctuation of NMOSFET can be explained.

(Keywords: V_{th} fluctuation, random dopant fluctuation)

Introduction

The minimum feature size of MOSFETs is now below 65 nm, and the variability in characteristics of MOSFETs has become a serious problem [1-4]. However, its mechanism has not been fully understood yet. In this paper, the causes of random V_{th} fluctuation are evaluated. The integrated physical analyses of the MOSFETs that show anomalous V_{th} in DMA (Device Matrix Array)-TEG with 1M MOSFETs, analyses of process condition modulation, and analysis of V_{BS} dependence on V_{th} fluctuation are executed. It is revealed that dominant cause is RDF (Random Dopant Fluctuation) and others causes have only small impact on V_{th} fluctuation.

V_{th} measurement by DMA-TEG

The measured V_{th} distribution of the 1M DAM-TEG is shown in Fig.1 [3]. It indicates high normality both in N- and PMOS-FETs at the range from -5σ to +5σ. Its random component is dominant as confirmed by 4th order polynomial fitting [5]. This feature justifies us to evaluate the causes of random fluctuation by the 1M DMA-TEG.

Cause of V_{th} fluctuation and analyzing method

To evaluate the cause of V_{th} fluctuation, B_{VT} is used. It is a new index of V_{th} fluctuation proposed in our previous work [1]. The plots shown in Fig. 2 are so-called Pelgrom plot [6]. Its slope A_{VT} is a conventional index of V_{th} fluctuation and has dependence on V_{th} and T_{INV}. The plots shown in Fig.3 are called "Takeuchi plot". B_{VT} is a slope of this plot. Feature of B_{VT} is that it does not depend on V_{th} and T_{INV} when RDF is a cause of V_{th} fluctuation. B_{VT} of PMOSFETs is close to that of the RDF model. However, B_{VT} of NMOSFETs is larger than that of the RDF model. These results indicate that PMOSFETs' V_{th} fluctuation is dominated by RDF and NMOSFETs' V_{th} fluctuation may be dominated by RDF and others causes. The candidates of others causes and the effect of each cause on B_{VT} are shown in Fig.4. Combining equations in Fig.4 with following analyses, the additional causes of NMOSFETs' V_{th} fluctuation are examined.

Physical parameters fluctuation

The I_d-V_g curves of N- and PMOSFETs having anomalous and typical V_{th} (Fig.1 (i), (ii), (iii)) in the 1M DMA-TEG are shown in Fig.5. These curves indicate that the 1M DMA-TEG properly performs. These transistors are analyzed by the integrated physical analyses including plain-view and cross-sectional TEM observations and nano-beam-diffraction analyses. The analyzed parameters are summarized in Table 1. Since the L_g and W_g difference between the MOSFETs with higher and lower V_{th} is so small, it is impossible to explain V_{th} difference. No difference in T_{OX} among examined MOSFETs is observed. The channel stress difference is small compared to the accuracy (130MPa) and there is no clear tendency between V_{th} and channel stress. Although the difference of the grain structure is observed, extreme phenomena like one large grain being placed on one MOSFET [7] does not occur. These results indicate that there is not enough difference among the physical parameters of MOSFETs with anomalous V_{th} and accordingly, it is difficult to explain the V_{th} fluctuation by the L_g, V_{FB} and T_{INV} fluctuation (Eq. (1), (3), (4) in Fig.4).

Impact of other parameters

Fig.6 shows the relationship between A_{VT}, B_{VT} and depth of gate depletion layer (T_D). B_{VT} does not depend on T_D as compared with A_{VT}. Therefore, the T_{INV} fluctuation, which is caused by the T_D fluctuation dose not affects on V_{th} fluctuation as expressed by Eq. (4) in Fig.4.

Fig.7 shows the relationship between A_{VT}, B_{VT} and nitrogen concentration in the gate oxide film. B_{VT} does not depend on nitrogen concentration as compared with A_{VT}. Therefore, the fixed charge (N_{int}) generated by the gate oxide nitridation process does not affect on V_{th} fluctuation as expressed by Eq. (5) in Fig.4.

Fig.8 shows the V_{BS} dependence of A_{VT} and B_{VT} [1]. As -V_{BS} increases, B_{VT} does not change as compared A_{VT}. Among causes of V_{th} fluctuation, only RDF has no V_{BS} dependence as shown in Fig.4. This result indicates that the causes other than RDF have small impact on V_{th} fluctuation.

Discussions

As examined by above mentioned analyses, possible causes of V_{th} fluctuation other than RDF are not crucial. To explain the large B_{VT} of NMOSFETs, a B clustering model is now proposed. In general, it is well known that B atoms with high concentration are clustered in Si [8]. Some B atoms of channel dopants gather with weak binding force and act as one B cluster. When n B atoms gather, B_{VT} is enhanced by a factor of $\sqrt{n}$ as shown in Fig.9. The relationship between the number of binding B atoms and B_{VT} is shown in Fig.10. In the case of clustering with 5 to 6 B atoms, B_{VT} of NMOSFETs can be explained by this model.

Conclusions

From cause analyses of V_{th} fluctuation, the importance of RDF is indicated. The others candidates of causes of V_{th} fluctuation have small impact on V_{th} fluctuation. A new mechanism to explain V_{th} fluctuation of NMOSFETs, a B clustering model, is proposed.

Acknowledgement

This work is supported by NEDO.

References

[1]K. Takeuchi et al., IEDM, p.467, 2007. [2]A. Aenov, IEEE TED, 45, p.2505, 1998. [3]T. Hiramoto et al., MIRAI Project Meeting, 2007. [4]K. J. Kuhn, IEDM, p.471, 2007. [5]S. Ohkawa et al., IEEE Proc. ICMTS., p.70, 2003. [6]M. J. M. Pelgrom et al., IEDM, p.915, 1998. [7]Y. Yoshida et al., IEDM, p.475, 1999. [8]S. B. Herner et al., JAP, p.6182, 1998.

978-1-4244-1802-2/08/$25.00 ©2008 IEEE

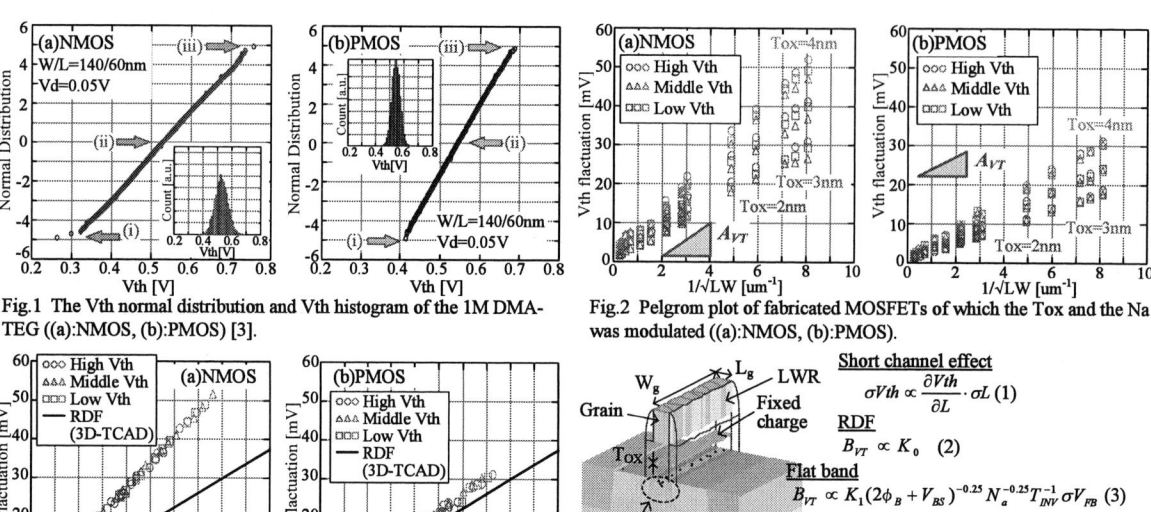

Fig.1 The Vth normal distribution and Vth histogram of the 1M DMA-TEG ((a):NMOS, (b):PMOS) [3].

Fig.2 Pelgrom plot of fabricated MOSFETs of which the Tox and the Na was modulated ((a):NMOS, (b):PMOS).

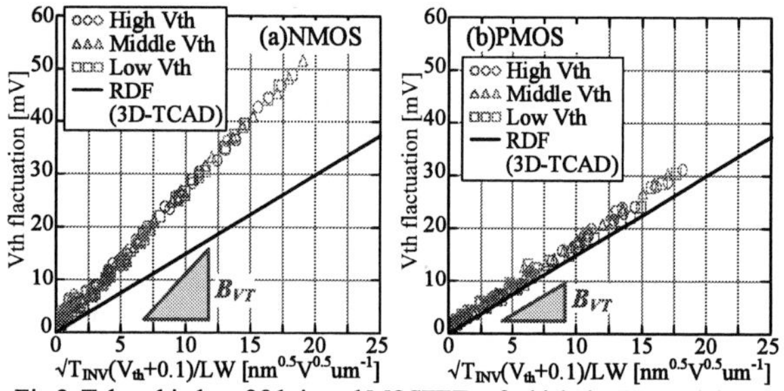

Fig.3 Takeuchi plot of fabricated MOSFETs of which the Tox and the Na was modulated ((a):NMOS, (b):PMOS).

Fig.4 The causes of characteristic fluctuation in MOSFET and equations expressing the effect of each cause on B_{VT}.

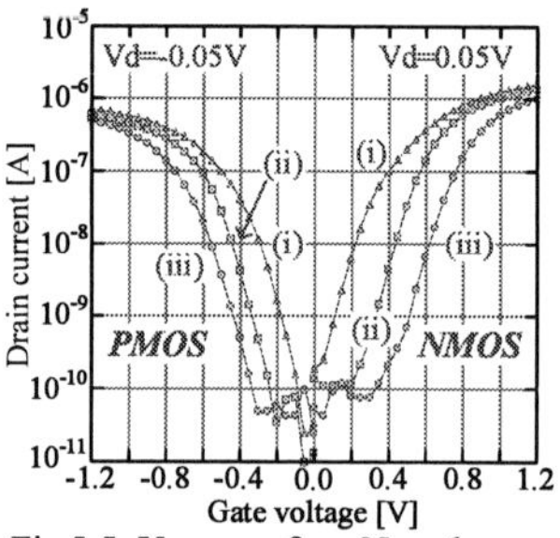

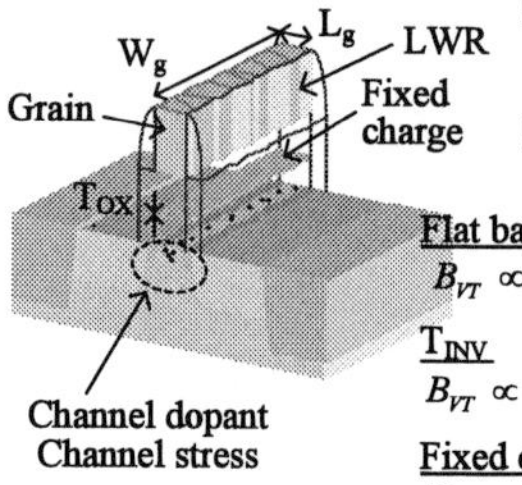

Short channel effect

$$\sigma Vth \propto \frac{\partial Vth}{\partial L} \cdot \sigma L \quad (1)$$

RDF

$$B_{VT} \propto K_0 \quad (2)$$

Flat band

$$B_{VT} \propto K_1 (2\phi_B + V_{BS})^{-0.25} N_a^{-0.25} T_{INV}^{-1} \sigma V_{FB} \quad (3)$$

T_{INV}

$$B_{VT} \propto K_2 (2\phi_B + V_{BS})^{0.25} N_a^{0.25} T_{INV}^{-1} \sigma T_{INV} \quad (4)$$

Fixed charge

$$B_{VT} \propto K_3 (2\phi_B + V_{BS})^{-0.25} N_a^{-0.25} N_{int}^{0.5} \quad (5)$$

Fig.5 I_d-V_g curves from N- and PMOSFETs with various V_{th} in the 1M DMA-TEG.

Table 1 Physical parameters of MOSFETs with various Vth (*:Strain indicates horizontal and vertical value)

MOS type	NMOS (W/L =140/60 nm)			PMOS (W/L =140/60 nm)		
Vth	0.335V (-5σ)	0.529V (median)	0.719V (+5σ)	0.425V (-5σ)	0.546V (median)	0.683V (+5σ)
Plan-view TEM image	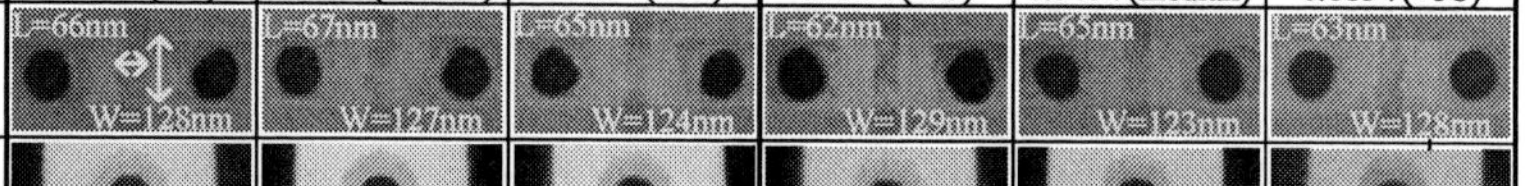 L=66nm W=128nm	L=67nm W=127nm	L=65nm W=124nm	L=62nm W=129nm	L=65nm W=123nm	L=63nm W=128nm
Cross-sectional TEM image	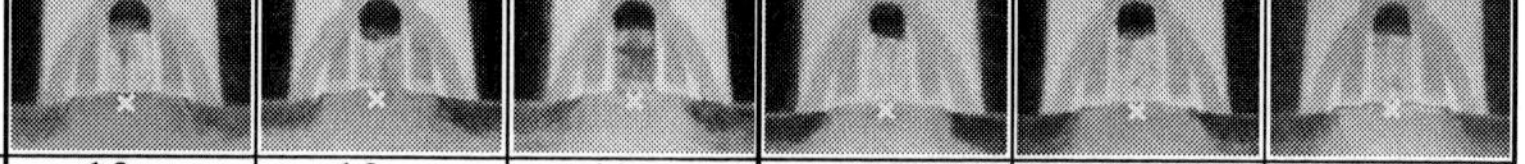					
Tox	1.9nm	1.9nm	1.9nm	1.9nm	1.9nm	1.9nm
Strain*	130/-130MPa	130/130MPa	130/130MPa	0/-260MPa	130/-130MPa	0/-260MPa
Gate poly-Si Grain						

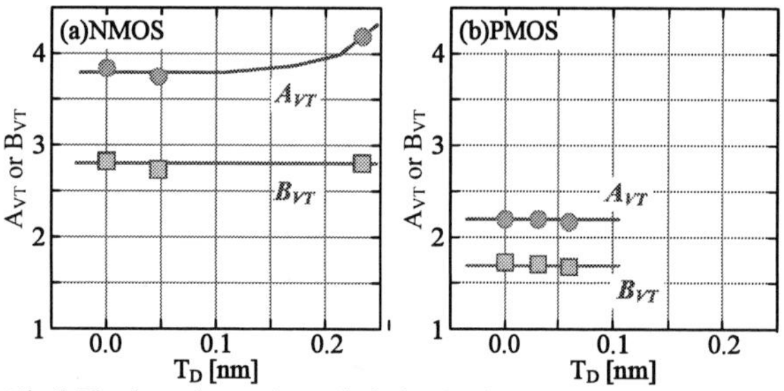

Fig.6 The dependence of gate depletion depth on A_{VT} and B_{VT} ((a):NMOS, (b):PMOS).

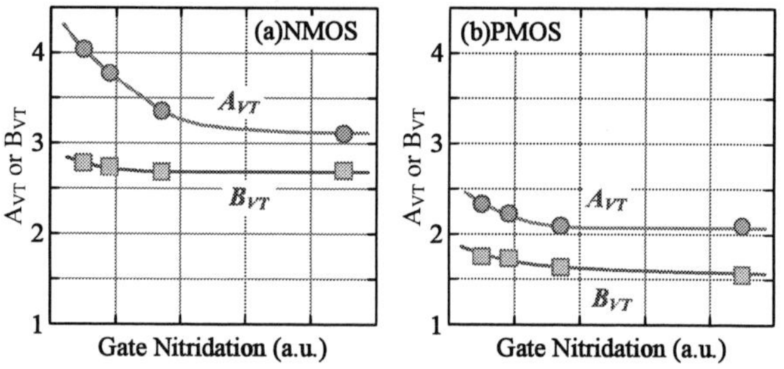

Fig.7 The dependence of nitridation concentration of gate oxide film on A_{VT} and B_{VT} ((a):NMOS, (b):PMOS).

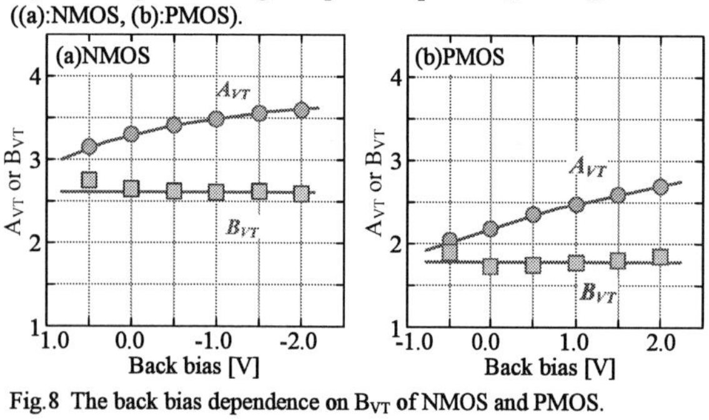

Fig.8 The back bias dependence on B_{VT} of NMOS and PMOS.

$$B_{VT} = \sqrt{\frac{q}{3\varepsilon_{OX}}}$$

$$q \rightarrow nq$$
$$Na \rightarrow \frac{Na}{n}$$

$$B_{VT} = \sqrt{\frac{nq}{3\varepsilon_{OX}}}$$

n: number of clustered B atoms

Fig.9 Explanation of B_{VT} enhancement by a B clustering model.

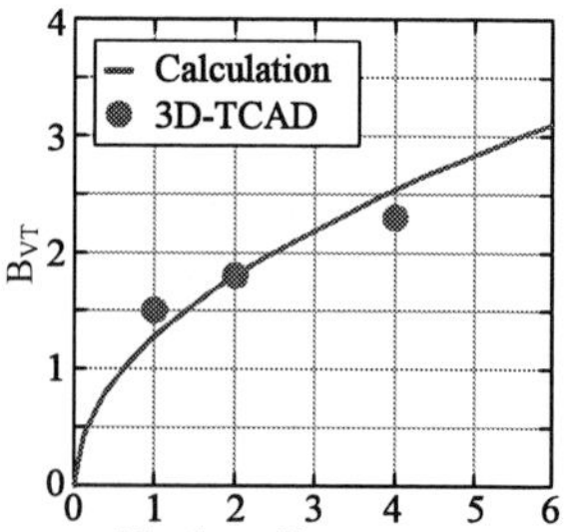

Fig.10 The analytical calculation and the 3D-TCAD simulation results of a B clustering model.

Reduction of Vth Variation by Work Function Optimization for 45-nm Node SRAM Cell

G. Tsutsui, K. Tsunoda, N. Kariya, Y. Akiyama, T. Abe, S. Maruyama, T. Fukase, M. Suzuki, Y. Yamagata, and K. Imai

Advanced Device Development Division, NEC Electronics Corporation

1120 Shimokuzawa, Sagamihara, Kanagawa 229-1198, JAPAN

Tel: +81-42-771-0707, Fax: +81-42-771-0952, e-mail: gen.tsutsui@necel.com

Abstract

Work function (WF) control is a key technology for the reduction of channel impurity concentration, which results in the decrease in intrinsic random dopant variation (IRDV). It is demonstrated that saturation V_{th} is affected not only by IRDV but also by S factor variation owing to fluctuation of DIBL. While channel impurity reduction by WF control decreases IRDV, DIBL is degraded in turn, and this enhances S factor variation. The optimal ΔV_{FB} by WF control is determined by two competing factors, S factor variation and IRDV. With this technique σV_{th} of 45-nm SRAM comparable to 65-nm, despite cell size reduction is obtained.

I. Introduction

Variation factors as represented in Fig. 1 are considered as crucial issues of VLSI technology that strongly affects both circuit performance and yield. In this work, we demonstrate an optimization technique of ΔV_{FB} by work function (WF) control from the viewpoint of random dopant fluctuation. Two variation factors are evaluated to determine the optimal ΔV_{FB}. One is an intrinsic random dopant variation, which appears both in linear and saturation V_{th}. The other is S factor variation which appears only in saturation V_{th}. S factor variation is induced by DIBL fluctuation due to a random fluctuation of channel dopant number or position. These two competing variation factors determine the optimal ΔV_{FB} that is around 100 mV. As a result, reduction of V_{th} variation and SRAM cell current variation is realized thanks to WF control by Hf-doped silicate.

II. V_{th} Variation Reduction by Hf-doped Sil. (measured)

A_{VT} [1] and B_{VT} [2] are used as variation indicator, which are represented as

$$\sigma V_{th} = A_{VT} \frac{1}{\sqrt{LW}} = B_{VT} \frac{\sqrt{T_{inv}\left(V_{th} + V_{FB} + 2\phi_B\right)}}{\sqrt{LW}}, \quad V_{FB} + 2\phi_B \sim 0.1V. \quad (1)$$

Fig. 2 illustrates measured pair transistors to evaluate V_{th} variation. Fig. 3 shows a normalized Pelgrom plot with Hf-doped silicate gate dielectric. Fig. 4 shows B_{VT} - channel dose characteristics. B_{VT} is independent of channel dose, because V_{th} change by channel dose and T_{inv} is normalized in B_{VT} as represented in eq. (1) [2].

II-i. Hafnium Content Fluctuation:

One concern of WF control by Hf-doped silicate is a fluctuation of Hf content in the silicate. Hf content fluctuation induces V_{FB} fluctuation leading to σV_{th} increase. Fig. 5(inset) shows A_{VT} - Hf content characteristics. It is clearly observed that A_{VT} does not change as an increase of Hf content. Note that T_{inv} change by Hf content change is less than 3 %, so an impact of T_{inv} change by Hf content change on A_{VT} is negligibly small. After all, it is concluded that V_{FB} variation induced by Hf content fluctuation is negligibly small in our Hf-doped silicate technology.

II-ii. Reduction of V_{th} Variation:

Fig. 6 shows B_{VT} - V_{th} characteristics. A reduction of B_{VT} is clearly observed with Hf-doped silicate gate dielectric. This is due to the reduction of channel impurity concentration thanks to V_{FB} control by Hf-doped silicate.

III. Evaluation of Optimal ΔV_{FB} (Simulation)

Measured data (Fig. 6) demonstrates that WF control by Hf-doped silicate effectively reduces channel impurity concentration resulting in a reduction of V_{th} variation. Here we argue how far channel impurity concentration can be reduced from the viewpoint of random dopant fluctuation. Fig. 7 shows σV_{th} - V_{th_median} characteristics. It is clearly observed in Fig. 7 that σV_{th} firstly decreases and then increases as channel dose increases. The decrease of σV_{th} as an increase of channel dose cannot be explained by intrinsic random dopant variation.

S Factor Variation: In order to further investigate the phenomenon observed, I_{ds} - V_{gs} and S factor characteristics are plotted (Figs. 7 and 8). It is observed in Fig. 8 that S factor variation becomes larger as channel dosage decreases. Note that S factor variation is induced by random fluctuation of dopant number or position (LER and systematic variation of gate length are not taken into account in the simulation) whose mechanism can be explained by ref [3]. This result indicates that random dopant fluctuation induces S factor variation owing to the fluctuation of DIBL. As a result, the two variation factors, S factor variation and intrinsic random dopant variation determine the optimal ΔV_{FB}. Fig. 9 shows σV_{th} - ΔV_{FB} characteristics. The minimum σV_{th} is obtained at ΔV_{FB} of around 100 mV, where DIBL is 125 mV/V. Further increase of ΔV_{FB} degrades σV_{th} owing to the DIBL increase (in other words, increase in S factor variation).

IV. 45-nm Node SRAM (measured)

σV_{th} of 45-nm node SRAM is evaluated by comparing σV_{th} with 65-nm node as shown in Fig. 10. Although the cell size reduction and additional halo dosage increases σV_{th}, σV_{th} of 45-nm node Drv. Tr. is comparable to that of 65-nm node. Fig. 11 compares SRAM cell current (Icell) variation between Hf-doped silicate gate dielectric and conventional SiON. Smaller Icell variation is observed with Hf-doped silicate. As a result, it is concluded that WF control by Hf-doped silicate effectively reduces V_{th} variation, which is promising for 45-nm node SRAM.

V. Conclusion

V_{th} variation factors as represented in Fig. 1 are evaluated. While channel impurity reduction by WF control decreases intrinsic random dopant variation, DIBL is degraded in turn, and this enhances S factor variation owing to the fluctuation of DIBL. DIBL fluctuation is induced by a random fluctuation of dopant number or position. It is demonstrated that WF control by Hf-doped silicate effectively reduces channel impurity concentration, which results in the reduction of V_{th} and SRAM cell current variation. This technique enables to achieve comparable σV_{th} of 45-nm node SRAM to 65-nm node.

978-1-4244-1802-2/08/$25.00 ©2008 IEEE

References

[1] M. J. M. Pelgrom et al., IEEE JSSC, vol. 25, p. 1433, 1989.

[2] K. Takeuchi et al., IEDM, 2007, p. 467.

[3] M. Miyamura et al., Symp. VLSI Tech., 2007, p. 22.

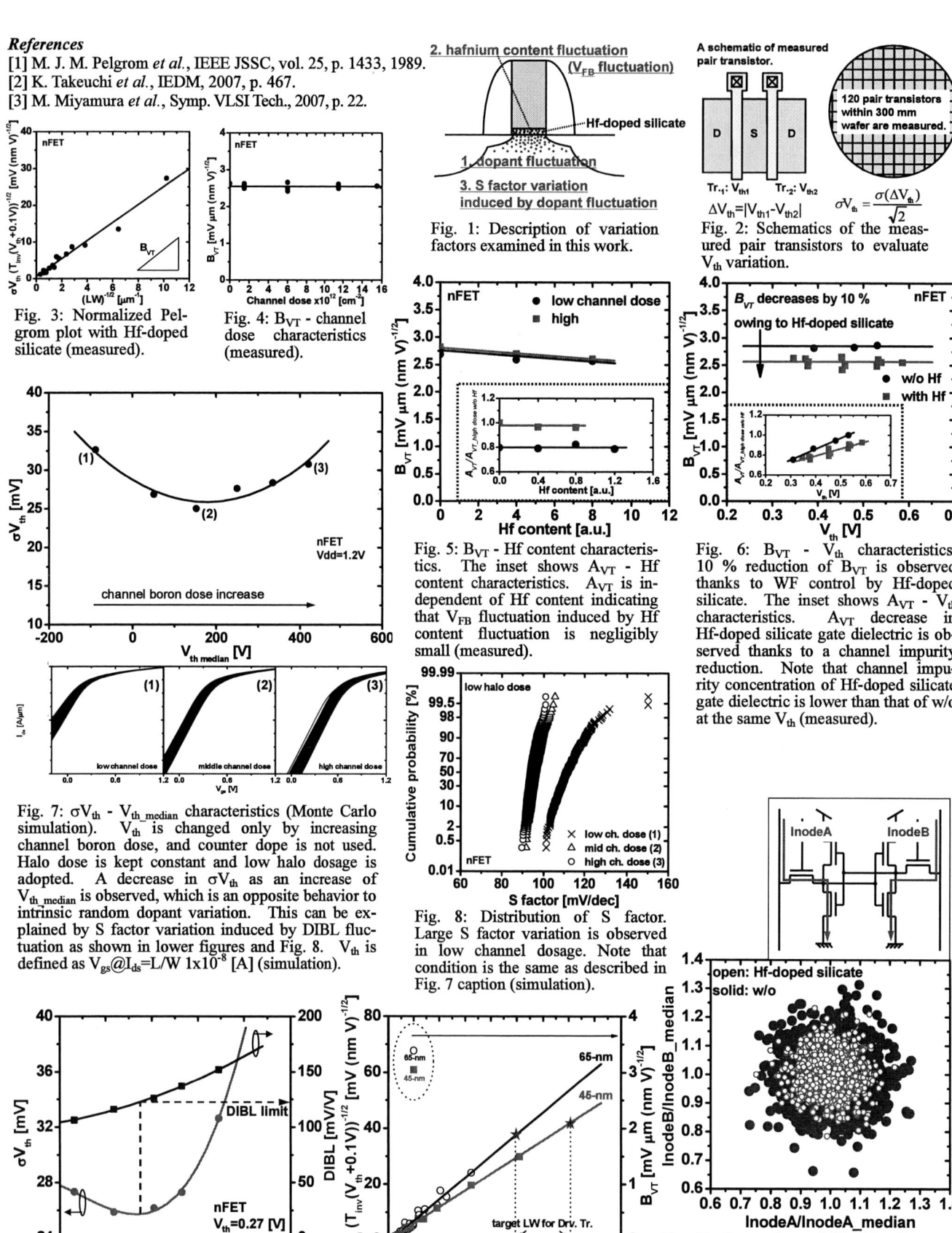

Fig. 3: Normalized Pelgrom plot with Hf-doped silicate (measured).

Fig. 4: B_{VT} - channel dose characteristics (measured).

Fig. 1: Description of variation factors examined in this work.

Fig. 2: Schematics of the measured pair transistors to evaluate V_{th} variation.

Fig. 7: σV_{th} - V_{th_median} characteristics (Monte Carlo simulation). V_{th} is changed only by increasing channel boron dose, and counter dope is not used. Halo dose is kept constant and low halo dosage is adopted. A decrease in σV_{th} as an increase of V_{th_median} is observed, which is an opposite behavior to intrinsic random dopant variation. This can be explained by S factor variation induced by DIBL fluctuation as shown in lower figures and Fig. 8. V_{th} is defined as $V_{gs}@I_{ds}=L/W\ 1\times10^{-8}$ [A] (simulation).

Fig. 5: B_{VT} - Hf content characteristics. The inset shows A_{VT} - Hf content characteristics. A_{VT} is independent of Hf content indicating that V_{FB} fluctuation induced by Hf content fluctuation is negligibly small (measured).

Fig. 6: B_{VT} - V_{th} characteristics. 10 % reduction of B_{VT} is observed thanks to WF control by Hf-doped silicate. The inset shows A_{VT} - V_{th} characteristics. A_{VT} decrease in Hf-doped silicate gate dielectric is observed thanks to a channel impurity reduction. Note that channel impurity concentration of Hf-doped silicate gate dielectric is lower than that of w/o at the same V_{th} (measured).

Fig. 8: Distribution of S factor. Large S factor variation is observed in low channel dosage. Note that condition is the same as described in Fig. 7 caption (simulation).

Fig. 9: σV_{th} - ΔV_{FB} characteristics. Minimum σV_{th} is obtained at ΔV_{FB} of around 100 mV where DIBL is 125 mV/V (simulation).

Fig. 10: Normalized σV_{th} comparison between 45-nm and 65-nm node. A comparable σV_{th} of 45-nm node to 65-nm is obtained, which is accomplished by introducing Hf-doped silicate (measured).

Fig. 11: Comparison of SRAM cell current (comparing the cell current of two nodes within SRAM cell) between Hf-doped silicate and conventional SiON gate dielectrics. Smaller Icell variation is observed with Hf-doped silicate thanks to a reduction of channel impurity concentration by WF control (measured).

45nm Low-Power CMOS SoC Technology with Aggressive Reduction of Random Variation for SRAM and Analog Transistors

S. Ekbote, K. Benaissa, B. Obradovic, S. Liu, H. Shichijo, F. Hou, T. Blythe, T. W. Houston, S. Martin, R. Taylor, A. Singh, H. Yang, G. Baldwin

Texas Instruments, MS 365, P.O. Box 655012, Dallas, TX 75265
Tel: 214-567-7730, Fax: 972-995-2770, e-mail: s-ekbote1@ti.com

Abstract

Mobile System-on-Chip (SoC) technologies require high-quality analog active and passive components along with low-power CMOS and dense SRAM. However, area scaling for both the SRAM bit cell and analog CMOS circuits is becoming increasingly difficult due to the impact of transistor random variation [1-5]. To avoid added cost, co-optimizing the process for low random variation along with high performance and low power is required. We report a 45nm low-power technology with significantly reduced random variation for high yielding $0.255\mu m^2$ SRAM arrays and analog transistors. Flexible RF and passive components for mobile SoC's are also described. These process techniques enable continued 50% area scaling at 45nm and beyond.

Introduction

This technology includes a complete suite of core, IO and analog transistors (Table 1), $0.255\mu m^2$ SRAM cell transistors and passive and active components for analog (Table 2). High-performance 1.8V I/O CMOS; high-V_T and low-V_T CMOS; fully isolated n-FET, and low-TCR poly resistors are available with additional reticles.

The core gate dielectric is 1.55nm EOT nitrided-oxide and the IO is 2.7nm EOT. For the logic n-FET performance attainment, emphasis was placed on optimizing inversion oxide thickness (T_{INV}); short-channel effect (SCE) control through halo implant and co-implant optimization; and channel mobility enhancement through stress memorization technique (SMT) and contact-etch-stop-layer (CESL). T_{INV} optimization and RSD improvement for the p-FET device was achieved through co-optimizing source-drain (S/D) implants and S/D RTA. In addition, abrupt halo implant optimization was employed to further improve the logic p-FET performance.

Local Variation Reduction

For acceptable yield, SRAM cell stability and write margins at the worst case global corners must be $>5\sigma$ of the random (local) variation. The global variation is aggressively controlled leaving the local variation as critical, as can be seen in the trend for local V_T variation (σV_T) as a percent of total variation in Fig. 1. Without significant changes to alter the trend, local variation would become as high as 92% of the total variation at 45nm node; too high to achieve the SRAM cell area goals. For this technology, both logic and SRAM transistors were used to investigate processes to reduce σV_T. The logic transistors have global variation minimized including proximity effects. Significant improvement in local variation was achieved by reducing the random dopant fluctuation (RDF).

For the logic n-FETs, 30% reduction in local σV_T (Fig. 2) is achieved using carbon with halo and MDD implants. Carbon is known to reduce diffusion of interstitial diffuser species such as boron through annihilation of interstitials. Carbon dose, energy and implant angle optimization along with halo dose and energy optimization is done to maintain the required V_T. Removal of poly doping prior to gate etch reduces local σV_T by ~10% as shown in Fig. 2. For logic p-FETs, at a given V_T, the most significant process variable for improving local σV_T is the halo implant angle. Reducing the angle by 15 degrees reduces local σV_T by as much as 40% as shown in Fig. 3. Further RDF reduction can also be achieved through halo species optimization as demonstrated for the PMOS load transistor shown in the column labeled halo species in Fig. 4.

It is also critical to ensure that there are no layout-specific sources of local variation in the SRAM. For example, the load transistor has nearly 50% reduction in local σV_T with the removal of pre-gate etch doping (Fig. 4), which was the source of dopant diffusion through the poly from driver to load transistor resulting in variations in T_{INV}. Since σV_T is directly proportional to T_{INV} [2], the load transistor σV_T is significantly reduced by eliminating this source of variation. The most significant local variation reduction processes were individually applied to the SRAM transistors and are compared in Fig. 4 along with the combination of all of the process improvements, resulting in a final process that meets the required total variation as demonstrated by $0.255\mu m^2$ SRAM array yield of >90% at production V_{MIN} and V_{MAX} screening conditions. An approximately 20% larger cell would have been required for equivalent yield without the process improvements for local σV_T reduction.

Analog Integration

For analog CMOS transistors with less stringent requirements for specific V_T targeting compared to the SRAM transistors, the RDF reduction processes identified above can be applied to a much greater degree to build a 1.8V analog transistor with substantial improvements in mismatch compared to a standard 1.8V IO transistor as shown in Fig. 5. The same processes that reduced RDF also enable improved flicker noise (Fig. 6). With a low V_T, good gm/gds, and high Ft, this analog transistor is integrated with no cost adders and is the workhorse for many RF/analog applications.

To support additional analog functions, a compensated-well pnp bipolar transistor has been optimized for improved mismatch and increased hfe compared to standard substrate pnp. For RF applications, scalable inductors are provided with flexible layout styles to allow optimization for specific design requirements. Multiple different types of optimization have been characterized and modeled. For example, the impact of dummy metal on inductor performance is shown in Fig. 7. A non-optimized metal fill (#2 in Fig. 7) will substantially deteriorate the inductor Q compared to a metal fill algorithm designed for improved RF performance (#1 in Fig. 7), which is close to the ideal case with no dummy metal. RF characterization of the 1.1V transistor, 1.8V analog transistor and 3.3V DENMOS transistors is shown in Fig. 8. The high Ft performance of the core transistor even at low currents allows RF design for low power applications. The data is shown for a case with minimum contacted pitch and dense interconnects typical of area-optimized SoC layouts. The RFCMOS layout is allowed to be flexible so that applications that may be less area restrictive can have minimized parasitics for even higher performance. The 3.3V DENMOS couples good RF performance with high voltage capability without cost adders for power amplifier applications. Fig. 9 shows an integrated 3.5G baseband and multimedia applications processor which uses the 45nm low power process described in this paper.

Conclusion

High yielding $0.255\mu m^2$ SRAM arrays on a low-power 45nm SoC technology are achieved through aggressive local variation reduction. The same techniques are applied to an even greater extent on a 1.8V IO analog transistor to obtain significant mismatch improvement without adding cost to the base flow. These results, along with good RF performance and flexible RF layout capability for transistors and inductors with no cost adders provides an excellent technology platform for a wide array single-chip SoC designs for mobile applications.

References

[1] M.J.M. Pelgrom, et. al, IEEE Journal of SSC, Vol. 24, No.5, Oct. 1989

[2] T. Mizuno, et. al, IEEE Trans. Electron Devices, Vol. 41, 1994

[3] M. Kanno, et. al, VLSI Technology Symposium, pp. 88, 2007

[4] R. Morimoto, et. al, VLSI Technology Symposium, pp. 28, 2007

[5] M. Miyamura, et. al, VLSI Technology Symposium, pp. 22, 2007

[6] G. Gammie, et. al, ISSCC, Feb. 2008

Transistors	Lg (nm)	Ion(N) (μA/μm)	Ion(P) (μA/μm)	Ioff(N) (nA/μm)	Ioff(P) (nA/μm)
1.1V Low-power	38	733	370	2.6	2.8
1.8V I/O Analog	120	840	648	10	1000
1.8V I/O	120	664	410	0.33	1.58
1.8V High-Perf. I/O (+2 reticles)	120	670	380	0.1	0.11
1.8V LVDEMOS	330	480	192	10	0.1
3.3V HVDEMOS	330	630	380	0.04	0.63
5V VHVDENMOS	1200	475	--	0.003	--
1.1V Low-power N,PMOS, 1/f noise (W=10μm)	-15.9, -16.7 A^2/Hz at Vd=0.55V, Id=0.299mA(N), Id=0.304mA(P), 500Hz				
1.1V Low-power N,PMOS, Gds	65, 63 μS/μm at Vg=Vtsat+0.25V, Vd=0.5V				
1.8V IO Analog N,PMOS, Gds	19.2, 12 μS/μm at Vg=Vtsat+0.25V, Vd=0.5V				
1.8V LVDE(N,P)MOS, Gds	6, 12 μS/μm at Vg=Vtsat+0.25V, Vd=0.5V				
3.3V HVDE(N,P)MOS, Gds	5.5, 5 μS/μm at Vg=Vtsat+0.25V, Vd=0.5V				

Table 1. Digital and analog transistor parameters.

RF/Analog Components	Description
M6 Inductor	Ldiff=1.0 nH, Qdiff=9.2 at 5GHz
1.8V Npoly/Nwell Varactor	Tuning range = 4, Q≥25 at 5GHz
MIM Capacitor	C=3.3 fF/μm^2
1.1V Npoly/Nwell Capacitor	C=16.9 fF/μm^2 at 1.1V
1.8V Npoly/Nwell Capacitor	C=10.5 fF/μm^2 at 1.8V
Nwell Resistor	Rs=585 Ω/sq
Silicide Resistor	Rs=16 Ω/sq
Bandgap transistor	Vertical PNP bipolar, hfe=10
N-Poly Resistor (+1 reticle)	Rs=233 Ω/sq, TCR=<20 ppm/°C

Table 2. Summary of key component parameters.

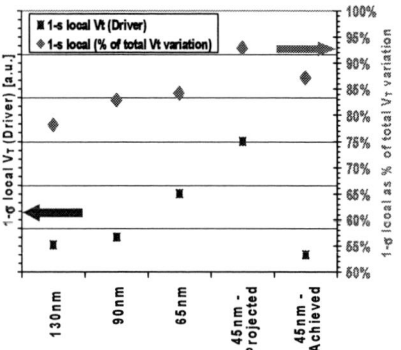

Figure 1. Trend of local V_T variation and as a percentage of total variation

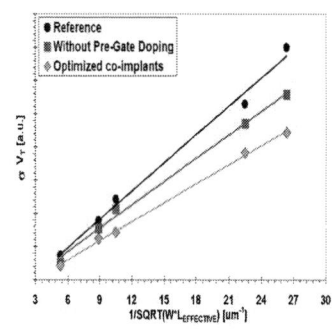

Figure 2. n-FET local V_T variation improvement processes.

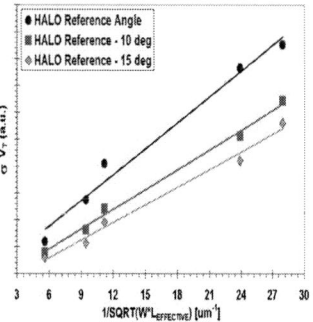

Figure 3. p-FET local V_T variation improvement processes.

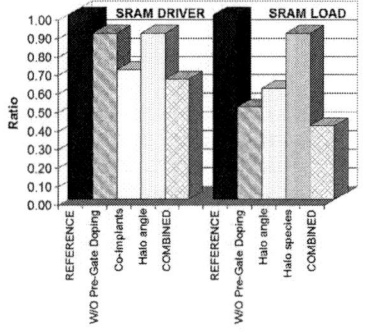

Figure 4. Ratio of SRAM local variation improvement processes to non-optimized starting process (REFERENCE).

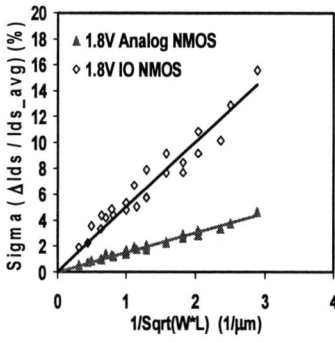

Figure 5. Analog transistor Ids mismatch compared to standard IO.

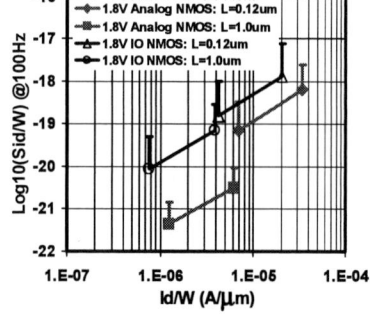

Figure 6. Analog transistor flicker noise compared to standard IO.

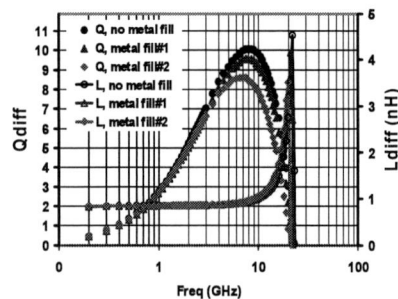

Figure 7. Inductor Q improvement with optimized dummy metal fill.

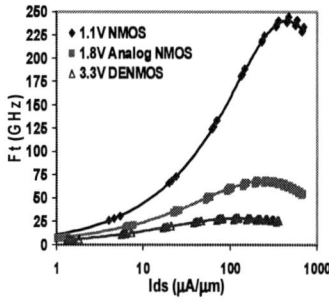

Figure 8. Ft v. Ids for the key RF transistors: 1.1V core NMOS, 1.8V Analog NMOS and 3.3V DENMOS.

Figure 9. 45nm integrated 3.5G baseband and multimedia applications processor [6]

Understanding and Prediction of EWF Modulation Induced by Various Dopants in the Gate Stack for a Gate-First Integration Scheme

X. P. Wang[1,2], H. Y. Yu[1,6], Y.-C. Yeo[2], M.-F. Li[2,3], S.-Z. Chang[4], H.-J. Cho[5], S. Kubicek[1], D. Wouters[1], G. Groeseneken[1], and S. Biesemans[1]

[1] IMEC, [4] TSMC, [5] Samsung assignee to IMEC, Kapeldreef 75, B-3001, Leuven, Belgium, Phone: +3216288781; Email: xinpeng@imec.be,
[2] SNDL, Dept. of ECE, NUS, Singapore, 119260, [3] Dept. of MicroE., Fudan Univ., Shanghai, China 201203, [6] School of EEE, NTU, Singapore.

Abstracts: For the first time, after considering the thermodynamic properties (evaluated by the molar Gibbs energy of oxide formation, $\Delta_{Oxide}G$) and the electronegativity (χ) for both the dopants (via ion implantation, thin capping layer or co-deposition) and host materials in the gate stack, a practical model to understand the effective work function (EWF) modulation induced by various dopants is proposed. It is found that the dopant oxide will determine the EWF if the $\Delta_{Oxide}G$ of dopant ($\Delta_{Ox-dop}G$) is more negative than that of host gate oxide ($\Delta_{Ox-host}G$). Or else, χ difference between dopants and host materials will play a more critical role for determining the EWF. This model can serve as a guideline for understanding the EWF modulation by various dopants and to select appropriate gate stack materials for the gate-first technology.

Introduction: Recently, $45nm$ logic technologies with metal gate (MG)/high-k stacks were implemented by either a gate-last or a gate-first approach [1, 2]. It is noted that a comprehensive study of V_{fb} (or V_{th}) shift in high-k CMOS fabricated by a gate-last process was reported [3]. Such an investigation for a gate-first process is also critical to implement MG/high-k in the future technology nodes. As illustrated in **Fig. 1**, several regions can affect the V_{th} of a MOS device. The top interface would be more important for EWF when the bottom interfacial layer (IL) is further scaled due to the strict EOT requirement [4]. Therefore, in this work, the interface between MG and high-k is thoroughly examined and a practical model involving $\Delta_{Oxide}G$ and χ is proposed to understand and predict the EWF modulation induced by various dopants for the gate-first process.

Experiments, Results and Discussion:

A. Description of the model: MG's EWF is determined by both vacuum WF and the underlying gate dielectric due to the existence of intrinsic or extrinsic interface states [5-7]. Extrinsic states are reported to arise from defects or the interfacial reactions due to a high temperature anneal [7]. Therefore, the thermal stability of interface is critical in the gate-first technology for determining the EWF. A thermodynamic parameter, $\Delta_{Oxide}G$, is employed in this model to predict reactions at a temperature of 1000°C between the dopant and host dielectric and compare the stability of the dopant oxide and host oxide in the gate stack, where $\Delta_{Oxide}G$ at 1000°C (1273K) is evaluated from Eq. (1) by assuming only solid-phase is involved during reactions, i.e. the items of standard molar enthalpy ($\Delta_{Oxide}H^0{}_{298}$) and standard molar entropy ($\Delta_{Oxide}S^0{}_{298}$) of oxide formation are independent of temperature (T).

$$\Delta_{Oxide}G_{1273} = \Delta_{Oxide}H^0{}_{298} - T*\Delta_{Oxide}S^0{}_{298} \qquad (1)$$

The model details are thus illustrated in **Table I**. Electronegativity (χ), which has been widely suggested to explain the EWF shift induced by different dopants [8,9], would play a dominant role for MG EWF modulation in the case of $\Delta_{Ox-dop}G_{1273} > \Delta_{Ox-host}G_{1273}$. Further, from both $\Delta_{Oxide}G_{1273}$ and χ, the tendency of EWF shift can be qualitatively classified into three categories after dopant incorporation, and these three cases shall be examined respectively in the following section. **Table II** summarizes χ and calculated $\Delta_{Oxide}G_{1273}$ [10] for the most widely-studied elements applied in the advanced gate stacks.

B. Validation of the model: Our proposed model is validated using the existing experimental data. N-FETs with Poly-Si/TaC (100nm/10nm) electrode were fabricated using a conventional transistor process flow [11] and different dopants (Al & La) was introduced by a thin capping layer (~1nm) on the top of host dielectric (HfSiON), as illustrated in **Fig. 2(a)**. The V_{fb}, extracted from transistor C-V curves (**Fig. 2(b)**), shifts to different directions after Al or La incorporation (in the case of a single cap layer) as compared to the case w/o cap layer. The consistent V_{th} extracted from I_d-V_g curves (**Fig. 2(c)**) indicates the EWF modulation of TaC electrode due to the different dopants is the main reason for the V_{th} (or V_{fb}) shift. **Fig. 3** shows schematic band diagrams for the afore-mentioned two cases (increase or decrease of EWF). Because both $\Delta_{Al2O3}G_{1273}$ and $\Delta_{La2O3}G_{1273}$ are more negative than $\Delta_{HfO2}G_{1273}$, i.e. Al_2O_3 and La_2O_3 have better thermal stability than HfO_2, the TaC/Al_2O_3 or TaC/La_2O_3 interface are

then more stable (or less reactive) than the $TaC/HfSiON$ interface. From this viewpoint, the intrinsic states at the top interface of the gate stack would be dominated by the TaC/Al_2O_3 or TaC/La_2O_3 interface rather than the $TaC/HfSiON$ interface. Due to the variation of intrinsic states arising from Al or La [5,6], the EWF of the gate stack would be thus effectively increased (for Al case) or decreased (for La case). Moreover, La (or Al) doped at the MG side also shows a similar impact on EWF modulation as doped at the dielectric side (**Fig. 4** & ref. 12-15). In **Fig. 5**, La-O bonds determining the final EWF of the gate stack at the top interface for the devices shown in **Fig. 4** are indicated by EDX study. Likewise, Al-O bonds, which are supposed to establish the final EWF based on our model, were also reported in ref. 13-15. What would happen in case of combined Al & La dopants into an existing MG/HfO system? Due to $\Delta_{Al2O3}G_{1273} > \Delta_{La2O3}G_{1273}$, our model predicts that LaO will be in charge of the final EWF. Indeed the results in **Fig. 2** (for the case of dual cap layers) and **Fig. 6** confirm the above-mentioned prediction. It is further noted that the intermixing or diffusion between dopant and host dielectrics during high temperature treatment could also contribute to the EWF modulation. In **Fig. 7**, X-ray reflectivity (XRR) study demonstrates the different intermixing behavior between La_2O_3 cap layer and HfSiO or HfO_2, leading to the different final EWF value [16] despite of the same MG and LaO cap layer. Equally the variation on V_{th} modulation due to LaO/AlO cap layers in the respective laser annealed $Ta_2C/HfSiO$ or $TaCNO/HfSiON$ stack can be explained by the various thermal budgets (**Fig. 8**) [17]. However, it should be emphasized that the trend of EWF modulation predicted by our model remains valid. This concludes the discussion on the *case-(1)* listed in the **Table I**. It should be noted that our model can be extended to other elements other than La & Al examined in this work. As an example, EWF modulation data for Er, Gd, Yb dopants in TaN on HfSiON dielectric [18] can also be explained using our model.

For *cases (2) & (3)* listed in **Table I**, i.e. when $\Delta_{Ox-dop}G_{1273} > \Delta_{Ox-host}G_{1273}$, the χ difference between the dopants and the host material atoms which would be partially replaced by the dopants at the MG or dielectric side would play an influential role to set the final EWF. This claim is supported by an example of N incorporation induced V_{th} modulation, as shown in **Fig. 9** as well as in ref. 9: N would partially replace O at the interface, and due to $\chi_N < \chi_O$ and hence electrons tend to transfer from the high-k side to the MG side through the interface, EWF would be decreased. Similarly, the EWF modulation induced by Si or N [9,19,20] can be interpreted by this model as well, based on χ in **Table II**.

C. Guideline based on the model: According to the proposed model, several methodologies are recommended to tune the EWF for a given gate stack with a refractory metal nitride electrode (MNx) and HfO_2 dielectric, as shown in **Table III**. It is noted that lower N concentration would be required to increase the EWF no matter where it is doped into. F incorporation in the dielectric side would be helpful to increase the EWF as well.

Conclusions: By combining the effect of $\Delta_{Oxide}G_{1273}$ and χ, a practical model is proposed for the first time to understand and predict the EWF modulation induced by various dopants for the gate-first process.

References: [1] K.Mistry et al., IEDM, p.247, 2007. [2] M. Chudzik et al., VLSI, p.194, 2007. [3] Y. Kamimuta et al., IEDM, p.341, 2007. [4] M. Takahashi et al., IEDM, p.523, 2007. [5] Y.-C. Yeo, EDL, p.342, 2002. [6] S. B. Samavedam et al., IEDM, p.307, 2003. [7] H. Y. Yu et al., EDL, p.337, 2004. [8] Y. Tsuchiya et al., IEDM, 27.1.1-4, 2005. [9] F. Y. Yen et al., EDL, p.201, 2007. [10] *David R. Lide, ed., CRC Handbook of Chemistry and Physics, http://www.hbcpnetbase.com,* 2007. [11] L. Ragnarsson et al., EDL, p.486, 2007. [12] X. P. Wang et al., TED, p.2871, 2007. [13] H. N. Alshareef et al., APL, v.88, 072108, 2006. [14] H.-C. Wen et al., VLSI, 9A-4, 2007. [15] R. Singanamalla et al., EDL, p.1089, 2007. [16] V. S. Chang et al., IEDM, p.535, 2007. [17] S. Kubicek et al., IEDM, p.49, 2007. [18] H. R. Harris et al., IEDM, p.633, 2006. [19] D. G. Park et al., IEDM, p.671, 2001. [20] P.-H. Tsai et al., APL, v.90, 132101, 2007. [21] P. Sivasubramani et al., VLSI, p. 68, 2007.

978-1-4244-1802-2/08/$25.00 ©2008 IEEE

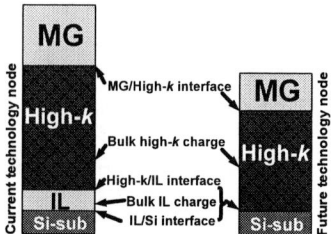

Fig. 1 Change of possible regions influencing V_{th} shift with continuous CMOS scaling down.

Table I Procedures to determine EWF shift after dopant incorporation, where $\Delta_{Ox\text{-}dop}G_{1273}$, $\Delta_{Ox\text{-}host}G_{1273}$, χ_{dop} and χ_{rep} represent the molar Gibbs energy of oxide formation for the dopant oxide and host dielectric at 1000°C and electronegativity of the dopant and replaced elements in the gate stack due to the incorporation of dopant, respectively.

I	Comparison of $\Delta_{Oxide}G$ at 1000°C	\multicolumn{2}{l}{$\Delta_{Ox\text{-}dop}G_{1273} < \Delta_{Ox\text{-}host}G_{1273}$}	\multicolumn{4}{l}{$\Delta_{Ox\text{-}dop}G_{1273} > \Delta_{Ox\text{-}host}G_{1273}$}
II	Doped region	High-k side	MG side	High-k side	MG side	High-k side	MG side
III	Comparison of χ	$\chi_{dop} >$ or $< \chi_{rep}$	$\chi_{dop} >$ or $< \chi_{rep}$	$\chi_{dop} < \chi_{rep}$	$\chi_{dop} > \chi_{rep}$	$\chi_{dop} > \chi_{rep}$	$\chi_{dop} < \chi_{rep}$
IV	Electron transfer direction through the interface and the tendency of EWF shift.	\multicolumn{2}{l}{(1) The oxide of dopant would be dominant at the top interface and the final MG's EWF would be determined by the intrinsic states between the MG and the oxide of dopant, no matter it is doped into the high-k side or into the MG side. The magnitude of EWF shift will depend on the concentration of dopant at the interface and the effect of χ on EWF shift in this case is not very critical.}		(2) Electron would transfer from the high-k side to the MG side due to the χ difference and the MG's EWF tends to decrease subsequently.		(3) Electron would transfer from the MG side to the high-k side due to the χ difference and the MG's EWF tends to increase subsequently.	

Table II Summary of χ and calculated $\Delta_{Oxide}G_{1273}$ for the most widely-studied elements applied in the advanced gate stacks based on database from [10].

Atomic Number	7	8	9	13	14	21	22	38	39	40	57	66	68	72	73
Symbol	N	O	F	Al	Si	Sc	Ti	Sr	Y	Zr	La	Dy	Er	Hf	Ta
Electronegativity, χ	3.04	3.44	3.98	1.61	1.9	1.36	1.54	0.95	1.22	1.33	1.1	1.22	1.24	1.3	1.5
Solid oxide formula				Al_2O_3	SiO_2	Sc_2O_3	TiO_2	SrO	Y_2O_3	ZrO_2	La_2O_3	Dy_2O_3	Er_2O_3	HfO_2	Ta_2O_5
Molar Gibbs energy, $\Delta_{oxide}G_{1273}$ (KJ/mol)				-1741	-964	-2007	-1008	-661	-2031	-1165	-1956	-2054	-2096	-1220	-2228

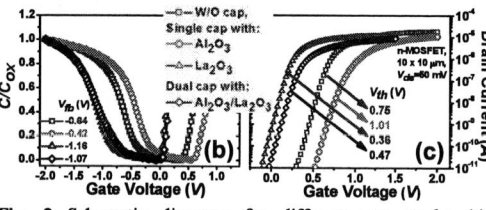

Fig. 2 Schematic diagrams for different gate stacks (a), comparison of typical C-V curves (b) and I_d-V_g characteristic (c) for Poly-Si/TaC/HfSiON gate stack with different cap layers.

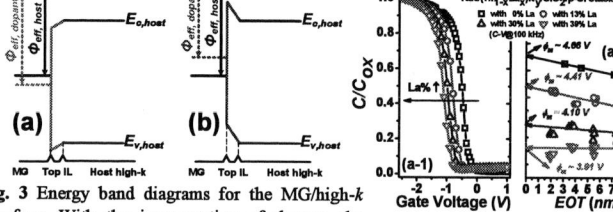

Fig. 3 Energy band diagrams for the MG/high-k interface. With the incorporation of dopant, the intrinsic states could vary due to the change of underlying dielectric layer so that EWF (Φ_{eff}) could shift following either (a) or (b).

Fig. 4 (a) C-V and (b) V_{fb}-EOT for $TaN/(Hf_{1-x}La_X)N_Y$ /SiO_2 stacks with different La%.

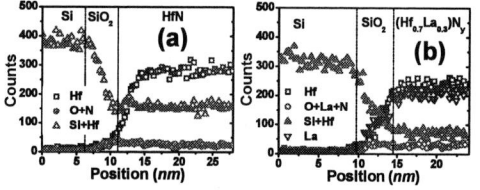

Fig. 5 EDX depth profile for (a) HfN/SiO_2 and (b) $(Hf_{0.7}La_{0.3})N_y/SiO_2$ stack. Intermixing of La and Hf with SiO_2 was found in (b), while no Hf diffusion into SiO_2 for (a).

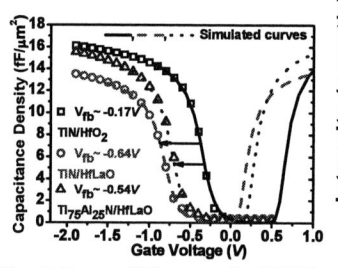

Fig. 6 Decent C-V curves for MOSCaps with TiN (or $Ti_{75}Al_{25}N$)/HfLaO & TiN/HfO_2 stacks, indicating LaO bond would be still dominant even though Al exists at the interface due to $\Delta_{La2O3}G_{1273} < \Delta_{Al2O3}G_{1273}$.

Fig. 7 XRR spectra for La_2O_3/HfSiO and La_2O_3/HfO_2 (inset) stacks. The change of the fringes after 1000°C anneal indicates intermixing.

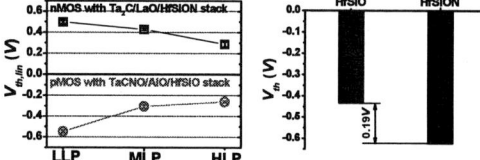

Fig. 8 Linear V_{th} for n- & p-FET as a function of thermal budget, where LLP, MLP & HLP means low, medium & high laser power.

Fig. 9 V_{th} of p-FETs with TaCNO/HfSiO and HfSiON stacks, V_{th} decreases with the incorporation of N.

Table III Possible dopants for modulating EWF towards Si BE for CMOS applications in a gate-first flow. (Note: the gate stack here is given as metal nitride (MNx)/HfO_2.

\multicolumn{2}{l}{n-MOSFET}		\multicolumn{2}{l}{p-MOSFET}	
If $\Delta_{Ox\text{-}dop}G_{1273} < \Delta_{HfO2}G_{1273}$	If $\Delta_{Ox\text{-}dop}G_{1273} > \Delta_{HfO2}G_{1273}$		If $\Delta_{Ox\text{-}dop}G_{1273} > \Delta_{HfO2}G_{1273}$	If $\Delta_{Ox\text{-}dop}G_{1273} < \Delta_{HfO2}G_{1273}$
(n-1) The final EWF would tend to shift to the value determined by the gate stack with MNx and with the pure dopant oxide as gate dielectrics. La, Sc, Er, Dy and/or their oxides have been demonstrated to modulate EWF towards Si E_v [12,21]. So, the other elements from IIIB group in the periodic table with lower $\Delta_{Oxide}G_{1273}$ may also have the similar impact on EWF modulation for MNx/HfO_2 stack.	(n-2) Those elements with χ lower than χ_M (or even χ of nitrogen) would be dopant candidates to decrease EWF.	MG side (MNx)	(p-1) Those elements with χ lower than χ_M would be dopant candidates to increase EWF, but $\Delta_{Ox\text{-}dop}G_{1273}$ should be larger than $\Delta_{HfO2}G_{1273}$ in this case.	(p-3) The final EWF would tend to shift to the value determined by the gate stack with MNx and with the pure dopant oxide as a gate dielectric. Currently, Al or its compounds (AlO or AlN) seems the only candidate to produce this effect.
	(n-3) Those elements with χ lower than χ_{Hf} or lower than χ_O if the dopant tend to replace oxygen rather than Hf (such as N) would be dopant candidates to decrease EWF. But it's noted that $\Delta_{Ox\text{-}dop}G_{1273}$ in this case should be larger than $\Delta_{HfO2}G_{1273}$.	High-k side HfO₂	(p-2) Those elements with χ higher than χ_{Hf} (such as Si) or higher than χ_O if the dopant tend to replace oxygen rather than Hf (such as F) are dopant candidates to increase EWF.	

Fig. 10 The EWF modulation for Si and N as dopants at the MG side or dielectric side [9,19,20].

Smallest V_{th} Variability Achieved by Intrinsic Silicon on Thin BOX (SOTB) CMOS with Single Metal Gate

Y. Morita, R. Tsuchiya, T. Ishigaki, N. Sugii, T. Iwamatsu*, T. Ipposhi*, H. Oda*, Y. Inoue*, K. Torii and S. Kimura

Central Research Laboratory, Hitachi, Ltd., 1-280, Higashi-koigakubo, Kokubunji-shi, Tokyo 185-8601, Japan

*Renesas Technology Corp., 4-1 Mizuhara, Itami-shi, Hyogo 664-0005, Japan

Phone: +81-42-323-1111, Fax: +81-42-327-7768, e-mail: yusuke.morita.yh@hitachi.com

Abstract

A "silicon on thin BOX" (SOTB) CMOS with a 50-nm single metal (FUSI) gate has been developed. By employing an intrinsic channel and a metal gate, this SOTB achieves the smallest V_{th} variability ever reported. The measured Pelgrom coefficients of the SOTB were 1.8 and 1.5 for NMOS and PMOS, respectively, even in the case of relatively thick EOT of 1.9 nm. Both multi-V_{th} control as well as suppression of short-channel effects were carried out simply by adjusting the impurity concentration beneath the BOX while keeping the channel almost intrinsic. Inverter delay and off-current were optimized by controlling gate-overlap length by means of a dual-layer offset spacer. It is shown that, within planar-type low-power CMOS devices, the SOTB is the most scalable because of its capability of multi-V_{th} and excellent matching characteristics.

Introduction

Variability of threshold voltage (V_{th}) has been widely recognized as an obstacle that hinders the evolution of CMOS performance beyond the 45-nm node. Dopant fluctuation, the poly-grain effect of the gate electrode, and line-edge roughness are thought to be the major causes of the V_{th} variability [1, 2]. An approach that uses an intrinsic channel and a metal gate is thus a possible solution to this variability problem. FinFETs or their affiliates are suitable structures for this approach; consequently, it was reported that FinFETs exhibit smaller V_{th} variability than bulk CMOS [3].

On the other hand, multiple V_{th} control, which is inevitable for SoC applications, is hard to achieve with these devices. To overcome this control issue, a tough process development aiming at multiple-workfunction-metal-gate electrodes is required. It was shown that a fully depleted silicon-on-insulator (FD-SOI) CMOS with ultrathin buried oxide (BOX) [4-7], called a "silicon on thin BOX" (SOTB), has many advantages, as summarized in Fig. 1. In this paper, we show the smallest variability ever reported, and we demonstrate that all the advantages hold true for the SOTB down to and beyond a 50-nm gate length. This structure is scalable down to 30 nm, which is thought to be an ultimate size for the planar-type low-power CMOS architecture, while keeping low V_{th} variability.

Device fabrication and characterization

Figure 2 shows cross-sectional TEM images of the SOTB with a NiSi FUSI gate. The thicknesses of the epitaxial silicon layer for the elevated S/D structure and NiSi were carefully controlled to avoid source/drain shorting. A dual offset-spacer process combined with a "simultaneous" FUSI process was used to control gate-overlap length precisely.

Multiple V_{th} control was done by the conventional well-implantation process through the SOI and BOX layers. This well layer also acts as a back gate. A SIMS depth profile after the well formation is shown in Fig. 3. Thanks to the BOX layer, a super-steep retrograde-channel profile was successfully obtained. (Note that $N_{SOI}<1\times10^{17}$ cm^{-3} even though $N_{back-gate}=1\times10^{18}$ cm^{-3}.) As previously predicted by simulation [8], this channel profile is essential to suppress V_{th} variability both due to statistical fluctuation of dopants and by process variation in, for example, SOI and BOX thicknesses and gate length.

Symmetrical I_d-V_g characteristics were obtained (Fig. 4) for a gate length L_g of 50 nm. Note that GIDL is well suppressed by the overlap control. I_d-V_d curves in Fig. 5 demonstrate that the SOTB is free from the self-heating effect thanks to the thin BOX. Figures 6 and 7 plot V_{th}, DIBL, and S against L_g for N- and PMOS at V_{ds} of 1.2 V, respectively. V_{th} values were adjusted to high- or low-V_{th} specifications for LSTP by selecting $N_{back-gate}$ of 1×10^{18} or 1×10^{16} cm^{-3}, respectively. The short-channel effect was suppressed down to 50 nm for both doping levels.

I_{on}-I_{off} characteristics for the dual V_{th} values are shown in Fig. 8. It is clear that I_{on}-I_{off} characteristics can be controlled by $N_{back-gate}$. As a result, I_{on} values of 500 (NMOS) and 300 (PMOS) µA/µm, which exceed a typical value for conventional 65-nm bulk CMOS for LSTP operation, at I_{off} of 20 pA/µm were obtained for high V_{th}. As shown in Fig. 9 adaptive back-bias (V_{bg}) control is demonstrated. Each point corresponds to a typical value for each chip on a wafer. Scattered points (representing the case without back-gate control) can be converged to the desired value by chip-to-chip back-gate control. Reduction in the worst I_{off} value of roughly 20% to 10% is demonstrated with the adaptive V_{bg} control of 0.1-V step.

Optimization of inverter delay τ_{pd} was done by controlling gate-overlap length L_{ov} with a dual offset spacer. Dependences of τ_{pd}, I_{on}, and C_{ov} as a function of L_{ov} are shown in Fig. 10. With decreasing L_{ov}, both I_{on} and C_{ov} decrease, but change in τ_{pd} (C_{ov}/I_{on}) is small. The best L_{ov} value is thus thought to be around 1 to 2 nm, taking $1/\tau_{pd}I_{off}$ as a figure of merit into account. The optimized τ_{pd} value is competitive with that of a bulk LSTP CMOS of the same size.

Variability suppression by SOTB

V_{th} distribution under the saturation condition (V_{ds}=1.2 V) was evaluated. A typical cumulative probability plot (Fig. 11) indicates that the distribution is random and short-channel effect is suppressed even down to 49 nm. The Pelgrom plot of our best sample is shown in Fig. 12. The slope of the plot, the Pelgrom coefficient A_{Vt}, is 1.8 for NMOS and 1.5 for PMOS. To our knowledge, these are the lowest levels ever reported [9]. When the EOT value (i.e., 1.9 nm) is reduced to the level to match the high-performance specification (EOT=1.0 nm), the A_{Vt} value should be smaller than 1.0. Note that the value for $\sigma(V_{th}^{F}-V_{th}^{R})$ measured by S/D exchange (i.e., the indicator of local variation caused by dopant fluctuation [10]) was also small because of the nearly intrinsic channel. The Pelgrom coefficients for various devices are summarized in Figs. 13 and 14. These figures clearly show that using neither a halo nor a poly-gate is essential to decrease variability (Fig. 13). Suppression of gate depletion with the metal gate also contributes to decreasing A_{Vt} (by about 20%). Impurities below the BOX have a small impact on the variability, but channel doping significantly increases it (Fig. 14). These results suggest robustness of the SOTB in terms of variability.

The scalability of the SOTB was assessed by ATLAS simulation. In this work, we fabricated the SOTB of 12-nm-thick, and 50-nm-L_g operation was confirmed. The minimum SOI thickness to avoid mobility degradation is thought to be 6 nm. Given that value, minimum L_g is about 30 nm (Fig. 15). We consider this length is the ultimate value for a low-power planar CMOS with small V_{th} variability. The SOTB can thus be considered the last runner in the era of "terminus ad quem" in device miniaturization.

Summary

It was shown that an SOTB achieves small V_{th} variability, multiple-V_{th} control, and comparable speed (i.e., inverter delay) to bulk CMOS at the same time. An ultrathin-body SOTB enables the planar CMOS architecture to survive until the last minute of device scaling.

Acknowledgments

The authors would like to thank Prof. Hiramoto of the University of Tokyo for his valuable comments, the staff of Renesas Technology Corp. for fabrication of the SOTB device, and T. Onai of Hitachi Ltd. for his encouragement. Special thanks are due to M. Odaka and K. Kasai of Hitachi for their useful comments. This work was partly supported by the Japanese Ministry of Education, Culture, Sports, Science and Technology.

References

[1] B. Hoeneisen et al., Solid-State Electron., 15, p. 819 (1972). [2] A. Asenov et al., T-ED 47, p. 805 (2000). [3] A. V-Y. Thean et al., IEDM 2006, p. 881. [4] R. Tsuchiya et al., IEDM 2004, p. 631. [5] T. Ishigaki et al., SSDM 2007, p. 886. [6] S. Monfray et al., IEDM 2007, p. 693. [7] R. Tsuchiya et al., IEDM 2007, p. 475. [8] T. Ohtou et al., EDL 28, p. 740 (2007). [9] K. J. Kuhn, IEDM 2007, p. 471. [10] T. Tanaka et al., VLSI 2000, p. 136.

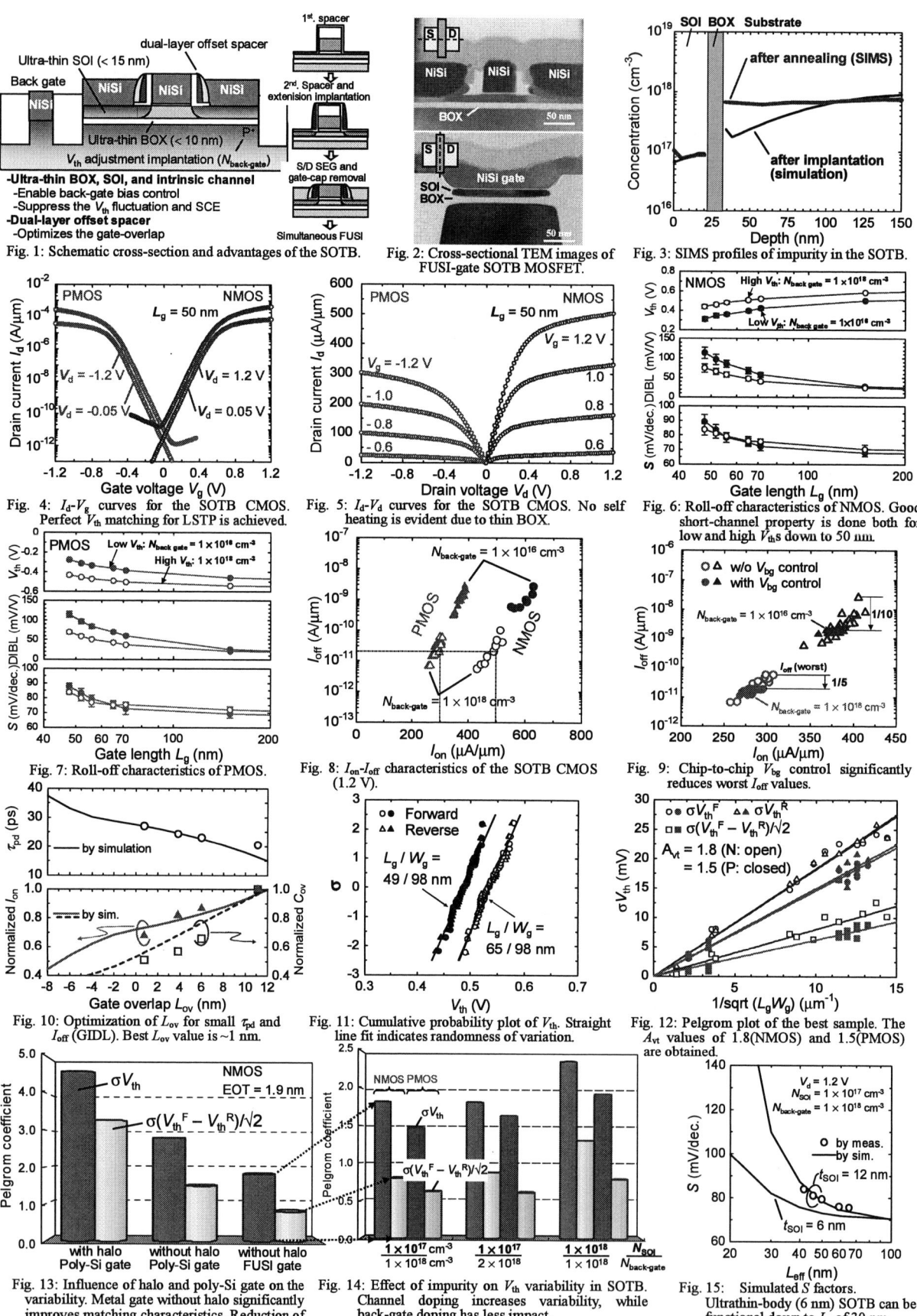

Fig. 1: Schematic cross-section and advantages of the SOTB.

Fig. 2: Cross-sectional TEM images of FUSI-gate SOTB MOSFET.

Fig. 3: SIMS profiles of impurity in the SOTB.

Fig. 4: I_d-V_g curves for the SOTB CMOS. Perfect V_{th} matching for LSTP is achieved.

Fig. 5: I_d-V_d curves for the SOTB CMOS. No self heating is evident due to thin BOX.

Fig. 6: Roll-off characteristics of NMOS. Good short-channel property is done both for low and high V_{th}s down to 50 nm.

Fig. 7: Roll-off characteristics of PMOS.

Fig. 8: I_{on}-I_{off} characteristics of the SOTB CMOS (1.2 V).

Fig. 9: Chip-to-chip V_{bg} control significantly reduces worst I_{off} values.

Fig. 10: Optimization of L_{ov} for small τ_{pd} and I_{off} (GIDL). Best L_{ov} value is ~1 nm.

Fig. 11: Cumulative probability plot of V_{th}. Straight line fit indicates randomness of variation.

Fig. 12: Pelgrom plot of the best sample. The A_{vt} values of 1.8(NMOS) and 1.5(PMOS) are obtained.

Fig. 13: Influence of halo and poly-Si gate on the variability. Metal gate without halo significantly improves matching characteristics. Reduction of local variation is evident ($V_{th}^F - V_{th}^R$).

Fig. 14: Effect of impurity on V_{th} variability in SOTB. Channel doping increases variability, while back-gate doping has less impact.

Fig. 15: Simulated S factors. Ultrathin-body (6 nm) SOTB can be functional down to L_g of 30 nm.

Selenium Co-implantation and Segregation as a New Contact Technology for Nanoscale SOI N-FETs Featuring NiSi:C formed on Silicon-Carbon (Si:C) Source/Drain Stressors

Hoong-Shing Wong, Fang-Yue Liu, Kah-Wee Ang*, Shao-Ming Koh, Alvin Tian-Yi Koh, Tsung-Yang Liow, Rinus Tek-Po Lee,
Andy Eu-Jin Lim, Wei-Wei Fang, Ming Zhu, Lap Chan, N. Balasubramaniam*, Ganesh Samudra, and Yee-Chia Yeo.

Silicon Nano Device Lab., Department of Electrical and Computer Engineering, National University of Singapore, Singapore 117576.
*Institute of Microelectronics, 11 Science Park Road, 117685 Singapore.

Phone: +65 6516-2298, Fax: +65 6779-1103, Email: yeo@ieee.org

Abstract

We report a novel contact technology comprising Selenium (Se) co-implantation and segregation to reduce Schottky barrier height Φ_{Bn} and contact resistance for n-FETs. Introducing Se at the silicide-semiconductor interface pins the Fermi level near the conduction band, and achieves a record low Φ_{Bn} of 0.1 eV on Si:C S/D stressors. Comparable sheet resistance and junction leakage are observed with and without Se segregation. When integrated in nanoscale SOI n-FETs with Ni-silicided Si:C S/D, the new Se-segregation contact technology achieves 36% reduction in total series resistance and 32% I_{ON} enhancement. Linear transconductance G_{MLin} also shows large enhancement in the sample with Se-segregated contacts.

1. Introduction

N-FET performance enhancement brought by strain engineering is constrained by high extrinsic series resistance R_{SD} of source/drain (S/D) regions [1][2]. Thus, there is industry-wide interest in reducing R_{SD}, especially the contact resistance R_{Con} component. A specific contact resistivity ρ_C of less than 2×10^{-8} $\Omega.cm^2$ [3] is required for 22 nm technology node and beyond. Reducing Φ_{Bn} offers significant advantages for achieving this objective (Fig. 1). However, there is very little work on interface Φ_{Bn} engineering.

This work demonstrates a new contact technology that employs Selenium (Se) co-implantation and segregation for reducing Schottky barrier height and contact resistance. The first integration of Se-segregated contact in SOI n-FETs with Si:C S/D is also reported. Record low Φ_{Bn} of 0.1 eV is obtained on Si:C S/D, enabling significant performance gains. Given that high interfacial n-type dopant concentration N_{Inf} may not be easily achievable on Si:C due to C-induced n-type dopant deactivation, the technology shown here for reducing Φ_{Bn} on Si:C would be of particular importance.

2. Φ_{Bn} Modulation using Selenium Segregation

The viability of Se segregation at the NiSi:C/n-Si$_{0.99}$C$_{0.01}$ interface is explored. A 40 nm thick Si$_{0.99}$C$_{0.01}$ layer was epitaxially grown on n-Si substrate, followed by 10 keV 3×10^{13} cm^{-2} As implant. The fabricated diodes were then implanted with different Se doses (0, 6×10^{13} cm^{-2}, 2×10^{14} cm^{-2}) at energy of 15 keV. All diodes underwent rapid thermal annealing (RTA) at 950°C, 30 s. 10 nm Ni was deposited for a single-step silicidation at 450°C, 60 s in N$_2$ ambient. Fig. 2 shows Φ_{Bn} is smaller for higher Se doses. A very low Φ_{Bn} of 0.1 eV was extracted for the sample with 2×10^{14} cm^{-2} Se dose using low temperature energy-activation measurement (Fig. 2). SIMS profile (Fig. 3) confirms Se segregation at the NiSi:C/n-Si$_{0.99}$C$_{0.01}$ interface. A higher Se dose contributes to a higher concentration of segregated Se and a lower Φ_{Bn}. A model for Φ_{Bn} modulation using Se segregation at the NiSi:C/n-Si$_{0.99}$C$_{0.01}$ interface is given in Fig. 4. Segregated Se pins the silicide Fermi level E_{Fm} close to E_C and modulates Φ_{Bn}. Fig. 5 shows the successful NiSi:C formation for samples with and without Se co-implant. The NiSi:C sheet resistance is independent of Se dose [Fig. 6(a)], suggesting minimum impact of Se segregation on metal-silicide formation. Fig. 6(b) illustrates the excellent thermal stability of

NiSi:C films formed with or without Se segregation. Comparable junction leakage is observed (Fig. 7) for diodes employing NiSi:C contacts with or without Se segregation, indicating that Se does not introduce defects that cause junction leakage problems.

3. Device Integration, Results and Discussion

We then integrate the novel Φ_{Bn} reduction technique in nanoscale SOI n-FET with NiSi:C Si$_{0.99}$C$_{0.01}$ S/D (Fig. 8). After Si S/D recess etch and selective epitaxy of 40 nm Si$_{0.99}$C$_{0.01}$ S/D, As implant (1×10^{15} cm^{-2} at 10 keV and 2×10^{14} cm^{-2} at 14 keV) as well as a Se co-implant (15 keV, 2×10^{14} cm^{-2}) were performed. S/D were activated. Ni silicidation was formed at 450°C, 60 s. Control SOI n-FETs with 40 nm Si$_{0.99}$C$_{0.01}$ S/D were implanted with As only [5], and did not receive Se implant. SOI silicon body thickness is 40 nm and gate width W_G is 0.5 μm.

Fig. 9 shows I_{DS}-V_{GS} characteristics of two closely matched <110> channel orientation SOI n-FETs with gate length L_G of 65 nm and comparable DIBL, subthreshold swing and off-state leakage. The n-FET with Se segregation shows 32% saturation drain current I_{ON} enhancement over the control at gate overdrive of 1.2V (Fig. 10). Devices with Se segregation demonstrated 36% lower R_{Total} over control devices [Fig. 11(a)]. Fig. 11(b) clearly shows the R_{Total} reduction was mainly due to reduction of R_{SD} or R_{Con}. This ultimately leads to I_{ON} enhancement. Fig. 12 shows that I_{ON} enhancement increases for decreasing L_G due to the increasing benefit of reduced R_{Con} at short L_G. A plot of I_{ON} as a function DIBL shows that the devices with Se segregation demonstrate 28% I_{ON} enhancement over the control devices at a DIBL of 0.2 V/V (Fig. 13). The peak linear transconductance G_{MLin} of devices with Se segregation shows an impressive 53% enhancement as compared to control devices (Fig. 14). Fig. 15 shows 41% effective drive current I_{Eff} enhancement for devices with Se segregation over the control. I_{Eff} has been used to accurately predict inverter delay for CMOS devices beyond 90 nm technology node [6]. Fig. 16 reveals that the I_{ON} enhancement for n-FETs with <100> and <110> channel orientations are roughly the same. Fig. 17 summarizes the key device performances achieved in nanoscale <110> channel orientation SOI n-FETs with Se segregation as compared to the control.

4. Conclusion

We demonstrated a novel CMOS-compatible Selenium co-implantation and segregation contact technology that achieves a record low Φ_{Bn} of 0.1 eV for the NiSi:C/Si$_{0.99}$C$_{0.01}$ interface. Sheet resistance and junction leakage were not degraded. Integration of this technology with SOI n-FETs with Si$_{0.99}$C$_{0.01}$ S/D was successfully demonstrated, leading to 36% R_{Total} reduction and I_{ON} enhancement of 32%. Se co-implantation and segregation is a promising technology for 22 nm technology node and beyond and is compatible with an advanced Si:C S/D stressor technology.

References

[1] H. N. Lin et al., IEEE EDL, vol.27, p. 659, 2006. [4] K. K. Ng et al., IEEE TED 37, p.1535, 1990.
[2] S. D. Kim et al., IEDM, p. 155, 2005. [5] K.-W. Ang et al., Symp VLSI, pp. 80, 2006.
[3] ITRS 2006, FEP 2006 Update. [6] M. H. Na et al., IEDM, p. 121, 2002.

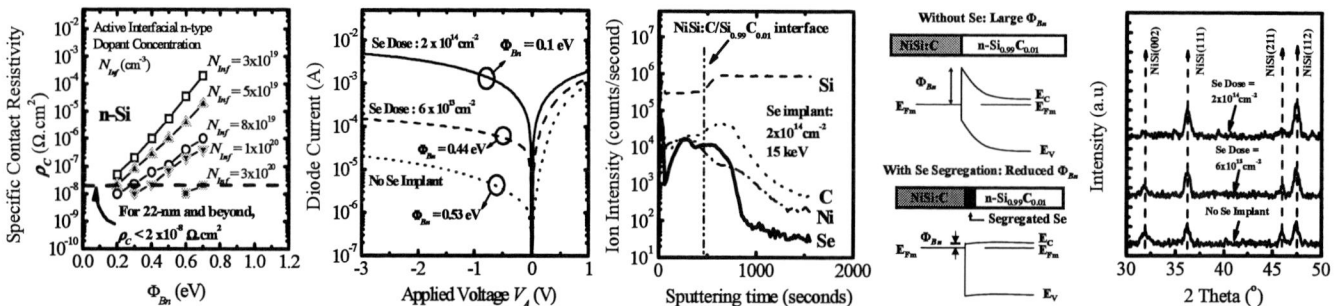

Fig. 1: To achieve a given ρ_C, theoretical prediction [4] requires a higher active interfacial n-type dopant concentration N_{Inf} for higher Schottky barrier height Φ_{Bn}.

Fig. 2: Higher Se dose leads to smaller Schottky barrier height Φ_{Bn}. Record low Φ_{Bn} of 0.1 eV for NiSi:C/n-Si$_{0.99}$C$_{0.01}$ interface was experimentally achieved.

Fig. 3: SIMS analysis showing the successful segregation of Se at the interface between NiSi:C and Si:C.

Fig. 4: Energy band diagram for illustrating the Φ_{Bn} reduction due to metal Fermi level pinning at NiSi:C/Si$_{0.99}$C$_{0.01}$ interface using Se segregation.

Fig. 5: XRD data confirm the formation of NiSi:C on n-Si$_{0.99}$C$_{0.01}$ for samples with and without Se co-implantation

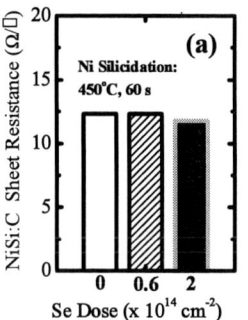
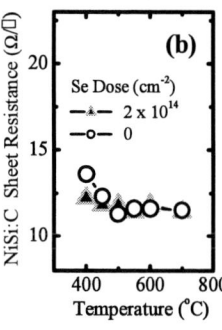
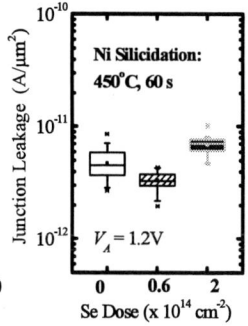

Fig. 6: (a) Measured sheet resistance is similar for NiSi:C with different implanted Se dose. (b) Measurements show comparable sheet resistance for NiSi:C with and without Se segregation for different Ni silicidation temperatures.

Fig. 7: Comparable junction leakage is observed for NiSi:C n+/p junction with and without Se.

Fig. 8: (a) Process flow for forming transistor with selenium-segregated contacts. A SOI n-FET with Se-segregated NiSi:C/Si$_{0.99}$C$_{0.01}$ S/D was formed. (b) Se$^+$ was co-implanted with As$^+$ dopants during S/D formation. (c) Se segregated contacts with low contact resistance were achieved after S/D anneal and Ni salicidation.

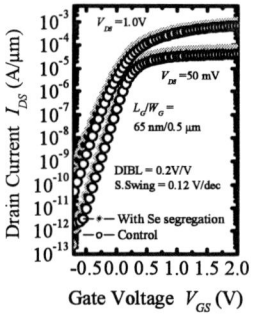
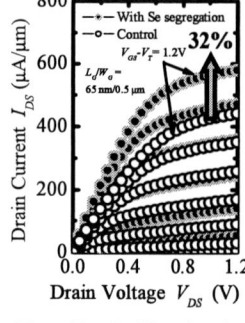
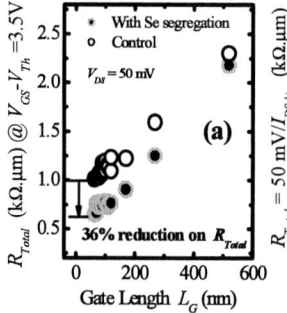
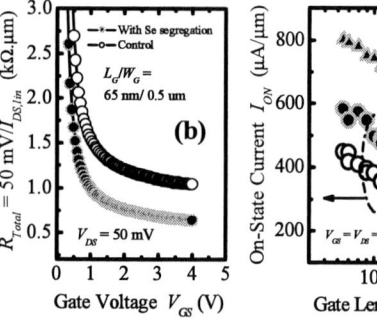

Fig. 9: I_{DS}-V_{GS} characteristics of two closely matched n-FETs with and without Se segregation (control).

Fig. 10: I_{DS}-V_{DS} showing 32% I_{ON} enhancement of n-FET with Se segregation over control at gate overdrive of 1.2V.

Fig. 11: (a) n-FETs with Se segregation show significant reduction in R_{Total} as compared to control n-FETs. (b) R_{Total} vs. V_{GS} plot for 65-nm gate length n-FET with and without Se segregation. R_{Total} in high V_{GS} regime mainly consists of external series resistance R_{SD}.

Fig. 12: I_{ON} enhancement due to Se segregation increases as L_G is reduced because of increasing dominance of R_{SD}.

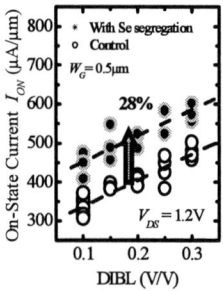
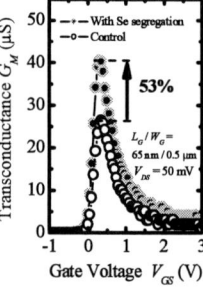
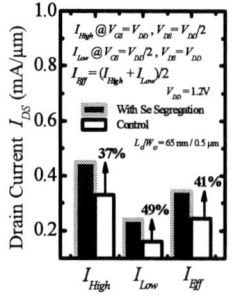
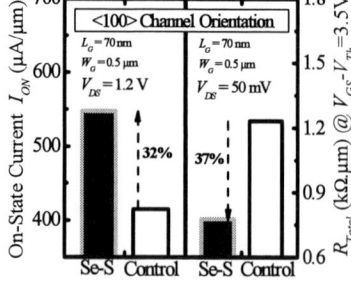
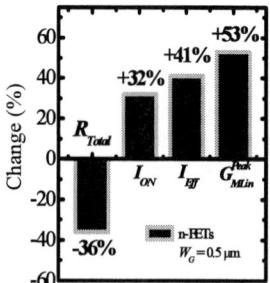

Fig. 13: Significant I_{ON} enhancement is observed in n-FETs with Se segregation at comparable DIBL.

Fig. 14: Linear region G_M, vs. gate voltage is shown for both n-FETs with and without Se segregation (control).

Fig. 15: I_{Eff} is the average of I_{high} and I_{low} [6]. The enhancement for each drain current component is shown.

Fig. 16: Enhanced I_{ON} for <100> channel n-FETs with Se segregation (Se-S) is demonstrated. This is also due to significant reduction in R_{SD} and hence R_{Con} in the <100> channel n-FETs.

Fig. 17: Summary of device performance improvements for n-FETs with Se segregation as compared to those without Se (control).

132

Mobility of strained and unstrained short channel FD-SOI MOSFETs: New insight by magnetoresistance

M. Cassé, F. Rochette, N. Bhouri, F. Andrieu, *D.K. Maude, †M. Mouis, G. Reimbold, F. Boulanger

CEA-Leti MINATEC, 17 rue des martyrs, 38054 Grenoble cedex 9, France (phone: (+33) 4 38 78 44 91, fax: (+33) 4 38 78 51 40, mikael.casse@cea.fr)
*CNRS-GHMFL, 25 rue des martyrs, BP 166 , 38042 Grenoble cedex 9, France
†IMEP CNRS/INPG/UJF, 3 parvis Louis Neel, BP 257, 38016 Grenoble cedex 1, France

Abstract

Electron mobility in short and long channel, strained or unstrained FD-SOI MOSFETs is deeply investigated in linear regime, by careful magnetoresistance measurements down to 40nm gate length, and down to 20K. This method differs from standard ones because i) it does not require any data on the short channel gate capacitance and gate length; ii) it is more accurate at low inversion charge; iii) the temperature dependence of the Coulomb Scattering limited mobility is higher. Additional mobility scattering has been thus confirmed for short channel undoped FDSOI, and unambiguously identified as Coulomb scattering (CS). A 50% mobility gain for strained Si MOSFETs is still observable even in this dominant CS regime.

Introduction

A strong reduction of the mobility in Si MOSFETs is almost universally observed as the gate length is scaled [1-3] (Fig.1). The reason of this degradation is not clearly understood although scattering by neutral defects is often advanced [1]. Standard methods like CV-split or Y-function [1,2,4] relies on the precise knowledge of the effective gate length L_{eff} or oxide capacitance C_{ox}. However the correct determination of these two crucial parameters is particularly difficult for short channel due to strong parasitic capacitances [4] (Fig.2) and possible non-scaling of oxide thickness for such short length. In contrast the extraction of mobility from magnetoresistance (MR) relies on the dependence of the channel resistance on a magnetic field B perpendicular to the channel plane and does not require any of these parameters (Tab.1) [6,7]. We used this method to bring new light on transport on short channel SOI transistors. Following the same way we have used MR to study the effect of strain on transport in short channel sSOI transistors.

Experimental devices and set-up

The channel mobility can be measured through the geometric MR effect in devices with an aspect ratio Width/Length>>5. μ_{MR} is simply extracted from the linear dependence of the drain current I_d on B^2 (Tab.1,Fig.3) at different temperatures and gate biases. The magnetic field produced by a superconducting magnet was swept from 0T up to 11T. We have used this method on strained and unstrained ultra short channel transistors made from thin film (10nm) Silicon On Insulator (SOI) substrate, and biaxially strained SOI (sSOI) substrate processed using a relaxed $Si_{0.8}Ge_{0.2}$ starting layer corresponding to roughly 1.4GPa. The gate stack was made of 3nm ALD HfO_2 on a thin SiO_x layer (~0.8nm), and covered by 10nm TiN metal gate (16Å EOT).

Experimental results and analysis

First we have compared the different mobility extraction procedures (Fig.4): μ_Y by Y-function method [9,2], μ_{eff} by split-CV [8], and μ_{MR} on long devices from 300K to 20K in linear regime (low V_D). MR mobility μ_{MR} and drift mobility μ_{eff} differ by the integration over energy of the scattering time [6]. At low temperature and/or high field μ_{MR} and μ_{eff} are identical as expected for T~0K and for a degenerate gas. Differences between μ_{MR} and μ_{eff} appear for energy dependent scattering mechanisms and T≠0K [6]. On Fig.4, we observed clear differences at low to medium field, i.e. for phonon scattering and Coulomb scattering (CS). To explain this, we have calculated μ_{MR} and μ_{eff} arising from Coulomb interactions due to remote charges [10,11] (Fig.5). The location or nature of charges is not important; the dependence with inversion carrier density N_{inv} and temperature results only from the coulombic nature of the interaction. We found that $\mu_{MR,CS}$ is higher than $\mu_{eff,CS}$, and differs by less than 10% at 20K, and by 180% at 300K, which explains the difference observed at low field and 300K on Fig.4. Furthermore this high temperature dependence of $\mu_{MR,CS}$ provides a very interesting way to identify CS, especially considering that CS is effective at rather low field where CV-split fails. μ_Y, reconstructed through μ_0 extraction (Tab.1), and μ_{eff} are equivalent providing CS limited mobility μ_{CS} is evaluated elsewhere

(at low temperature and low V_g-V_{th} for instance) and added by Mathiessen's rule to μ_Y (Fig.4).

Short channel SOI.– Figs.6,7 show raw μ_{MR} measured on nMOS for L down to 40nm, at 300K and 20K. A reduction of mobility was observed especially at low temperature for short nMOS. However, μ_{MR} was extracted through I_d measurement, which means that raw data included the effect of series resistance R_{SD}. Differential MR method [7] was used to correct R_{SD} influence by measuring two devices (labelled 1 and 2) with close gate lengths following

$$(R_2(B) - R_1(B))/(R_2(0) - R_1(0)) = 1 + \mu_{MR,corr.}^2 B^2$$

The procedure is similar to the $R_{tot}(L)$ method [1,12] and eliminates contributions which do not depend on gate length (SD and region near SD) leaving only channel length dependent contributions. In particular the effect of ballisticity sometimes introduced by a L-independent resistance R_{bal} is eliminated [3]. On Figs.6,7 the correction leads to R_{SD}~20Ω and μ_{RSD}~90cm²/Vs similarly to [7], and increases μ_{MR} at high V_g as expected. The discrepancy between μ_Y and μ_{MR} increases as L decreases (Figs.1,8) with a smaller mobility reduction for μ_{MR}. This difference is mainly due to the lack of accuracy in L_{eff} and C_{ox} determination for μ_0, and to the effect of CS on μ_0 extraction procedure. The extraction has to be performed at high V_g to avoid the proximity effects of the g_m maximum region [9]; experimentally μ_0 reaches μ_{MR} at high field only at very low T (Fig.9).

A slight positive temperature dependence of μ_{MR} is observed for V_g-V_{th}>0.5V, which corresponds to a much lower power law dependence of phonon contribution μ_{ph} than traditionally measured on long transistors (Figs.8,9). For lower gate biases, when T decreases μ_{MR} decreases following the temperature dependence predicted for μ_{CS} (+180% from 300K to 20K on Fig.5). μ_{MR} corrected from R_{SD} for short devices at 20K presents strong degradation compared to long MOS, and increases with N_{inv} as predicted for CS (Fig.5). At low temperature phonon scattering is negligible and μ_{MR} gives directly μ_{CS}, as in that case surface roughness limited mobility is much higher. The mobility of short nMOS is thus almost purely due to μ_{CS}. The measurements on short devices at 20K can be very well reproduced including CS equivalent to N_{it}~2.10¹³cm⁻² (Fig.7). A possible explanation is the increase effect of long-range Coulomb interactions for very short channels (proximity of SD) [10].

sSOI vs. SOI.– Short channel sSOI nMOS behaved like SOI ones (Fig.10). A strong mobility reduction was observed at 20K (as on Fig.7) for L<100nm. The R_{SD} corrected mobility still presents the characteristics of a CS dominant mechanism. A gain for strained Si was observed from 300K to 20K, and on the whole range of N_{inv} (Figs.11,12). The origin of the enhancement was identified from its temperature dependence which differs along with N_{inv}. At high carrier density although most of carriers are in Δ_2 valleys a gain as high as 35% was observed at 20K due to surface roughness and carrier distribution [13,14]. Above 20K the effect of phonon scattering enhances this gain [14]. At lower carrier density, where CS dominates, the gain drops as N_{inv} decreases and T increases (Fig.11). For short channel MOSFETs, despite the predominance of CS, a mobility gain is still observable for sSOI at room temperature as high as 50% at high V_g (Figs.12,13). This shows that strain has an effect even on μ_{CS} [15], and/or that less defects are induced in sSOI.

Conclusion

We have confirmed by original MR data that mobility degradation actually occurs for short channel SOI and sSOI nMOS. We have demonstrated without ambiguity that this degradation is due to stronger CS in nanoscaled transistors, which explains also the lower temperature dependence generally observed. This CS extends along the whole channel length. Strain nevertheless still enhances the mobility in this CS dominant regime, even if reduced compared to long channel nMOS (80% reduced to 50% at room temperature).

Acknowledgements: This work has been partly supported by the NAM Rhônes-Alpes Project, Sinano, MEDEA and 2T101 SILONIS.

References
[1] K. Rim *et al.*, IEDM (2002)
[2] A. Cros *et al.*, IEDM (2006)
[3] M. Zilli *et al.*, EDL 28, 1036 (2007)
[4] Toriumi *et al.*, IEDM (2006)
[5] L. Thevenod *et al.*, APL 90, 152111 (2007)
[6] R. Meziani *et al.*, JAP 96, 5761 (2004)
[7] W. Chaisantikulwat *et al.*, SSE 50, 637 (2006)
[8] K. Romanjek *et al.*, EDL 25, 583 (2004)
[9] G. Ghibaudo *et al.*, Elec. Lett. 24, 543 (1988)
[10] M. Fischetti *et al.*, JAP 89, 1205 & 1232 (2001)
[11] M. Cassé *et al.*, TED 53, 759 (2006)
[12] G. Niu *et al.*, TED 46, 1912 (1999)
[13] O. Bonno *et al.*, VLSI (2007)
[14] S. Takagi *et al.*, JAP 80, 1567 (1999)
[15] O. Weber *et al.*, VLSI (2007)

CV-split: $\mu_{eff} = \dfrac{I_d \cdot L^2}{V_d \cdot Q_{inv}}$

MR: $\dfrac{I_d(0)}{I_d(B)} = 1 + \mu_{MR}^2 B^2$

Y-function (Y=I_d/gm$^{0.5}$) with R_{SD} correction [9]:

$\mu_Y = \dfrac{\mu_0}{1 + \theta_{1,0}(V_g - V_t) + \theta_2(V_g - V_t)^2}$

with $\mu_0 = \beta L_{eff}/(WC_{ox})$

Inclusion of CS: $\mu_{tot}^{-1} = \mu_Y^{-1} + \mu_{CS}^{-1}$

Tab.1 Main equations of the *common* methods used to measure or extract the mobility. μ_{CS} can be numerically calculated and included in reconstructed μ_Y using Mathiessen's rule.

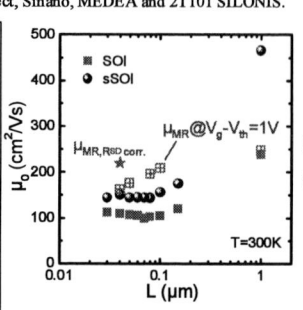

Fig.1 Evolution of μ_0 with L showing mobility reduction as L decreases for both SOI and sSOI. μ_{MR} at V_g-V_{th}=1V for SOI is also given for comparison.

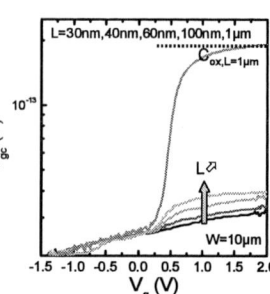

Fig.2 Gate to channel capacitance $C_{gc}(V_g)$ for L=1µm to L=30nm long nMOS showing the effect of parasitic capacitance on C_{ox} determination.

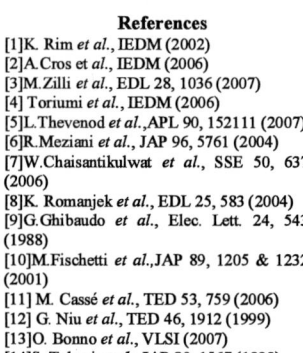

Fig.3 Change of resistance as a function of B^2. The slope gives directly μ_{MR}. The knowledge of L_{eff} or C_{ox} is not required.

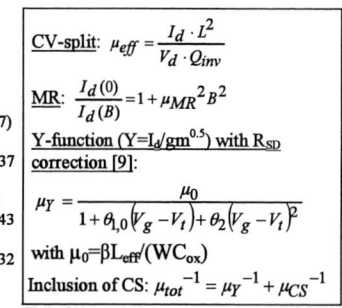

Fig.4 Comparison of μ_{MR}, μ_{eff} and μ_Y as a function of V_g-V_{th} at room and low temperature on long SOI nMOS (SiO$_2$/HfO$_2$/TiN).

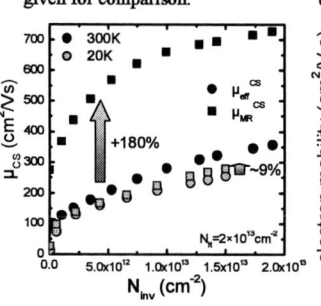

Fig.5 Contribution of CS due to remote charges calculated for μ_{eff} and μ_{MR}. 180% of mobility increase is observed at 300K for μ_{MR}, which reduces to less than 10% at 20K.

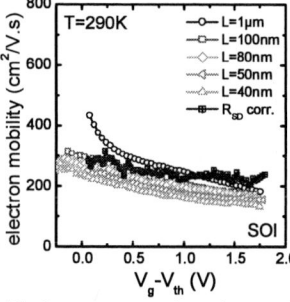

Fig.6 μ_{MR} measured on long and short nMOS at 290K. Raw data are presented for different L, as well as μ_{MR} corrected from series resistance (R_{SD} corr.) from L=40nm and L=50nm data.

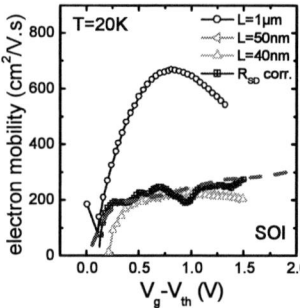

Fig.7 Same as Fig.5 but at T=20K. The dotted line is a fit with the CS model presented in Fig.5.

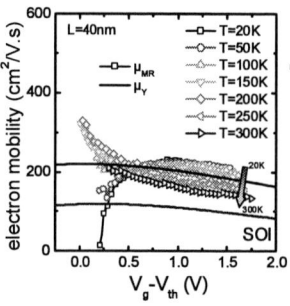

Fig.8 μ_{MR} for L=40nm for different temperatures from 20K up to 300K. The bold lines represent μ_Y calculated from μ_0, θ_1 and θ_2 at 20K and 300K (see Tab.1).

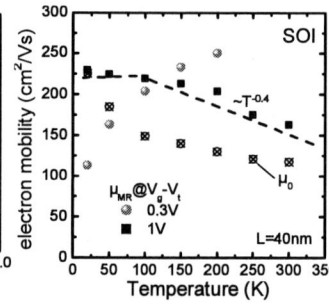

Fig.9 μ_{MR} and μ_0 measured as a function of temperature and at low and higher V_g-V_t for the 40nm long SOI nMOS.

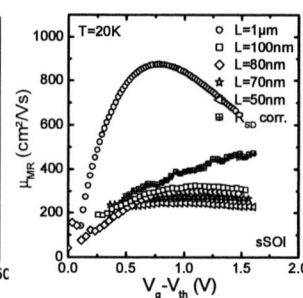

Fig.10 Raw data and R_{SD} corrected μ_{MR} at 20K for L=40nm up to 1µm long sSOI nMOS.

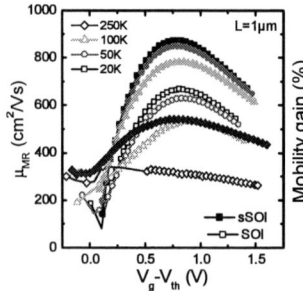

Fig.11 μ_{MR} measured on long SOI and sSOI nMOS for different temperatures showing the effect of strain on long transistors.

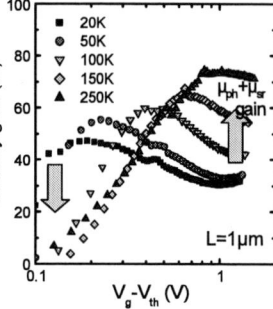

Fig.12 Temperature dependence of mobility enhancement $\Delta\mu_{MR}/\mu_{MR}$ between strained and unstrained long devices (from data of Fig.11).

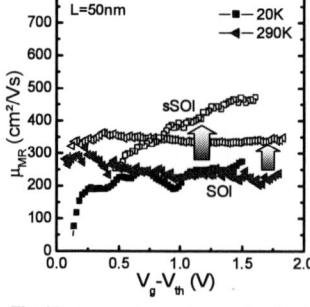

Fig.13 Comparison of μ_{MR} for SOI vs. sSOI on short nMOS corrected from R_{SD} from L=50nm and L=70nm measurements.

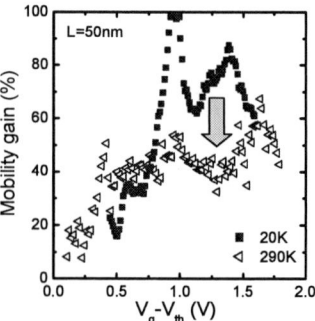

Fig.14 Temperature dependence of mobility enhancement $\Delta\mu_{MR}/\mu_{MR}$ for short devices (from data of Fig.13).

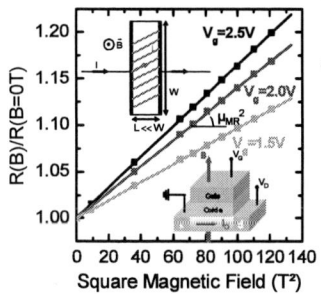

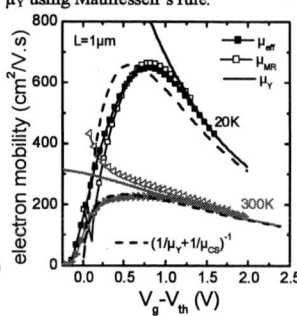

134

On Implementation of Embedded Phosphorus-doped SiC Stressors in SOI nMOSFETs

Zhibin Ren, G. Pei[§], J. Li, B. (F.) Yang[§], R. Takalkar, K. Chan, G. Xia, Z. Zhu, A. Madan, T. Pinto, T. Adam, J. Miller, A. Dube, L. Black[§], J. W. Weijtmans[§], B. Yang[§], E. Harley, A. Chakravarti, T. Kanarsky, R. Pal[§], I. Lauer[†], D.-G. Park[†] and D. Sadana[†]

IBM Semiconductor Research & Development Center, Hopewell Junction, NY 12533
[†]IBM T. J. Watson Research Center, Yorktown Heights, NY 10598, [§]AMD Inc., 2070 Rte. 52, Hopewell Junction, NY 12533
e-mail: zhibinr@us.ibm.com, phone: 845-892-1735

Abstract

We report a successful implementation of epitaxially grown Phosphorus-doped (P-doped) embedded SiC stressors into SOI nMOSFETs. We identify a process integration scheme that best preserves the SiC strain and minimizes parasitic resistance. At a substitutional C concentration (C_{sub}) of ~1.0%, high performance nFETs with SiC stressors demonstrate ~9% enhanced I_{eff} and ~15% improved I_{dlin} against the well calibrated control devices. It is found that the tensile liner technique provides further performance improvement for nFETs with SiC stressors, whereas the Stress Memory Technique (SMT) does not provide performance gain in a laser annealing process that is used to preserve SiC strain. The material quality of the SiC stressors strongly affects strain transfer.

Introduction

To continuously enhance nMOSFET drive current or to replace existing stress techniques that may no longer be effective in next-generation devices, the SiC stressor approach has been evaluated in recent years. SOI nFETs with undoped epitaxial SiC stressors were demonstrated by Ang et al. [1], and bulk nFETs with SiC stressors formed through carbon implants were shown by Liu et al. [2]. Strain preservation, material defect and resistance control were identified among the key issues to address. The most recent work by Grudowski et al. reported SOI nFETs with P-doped SiC stressors [3]; however, high parasitic resistance is still listed as the major factor that suppresses the potential benefit of the SiC stressors. After systematically assessing the drawbacks and advantages of different process schemes, we implemented a process that utilizes insitu P-doped epitaxial SiC stressors [3]. The process minimizes the integration changes from the conventional process flow, best preserves SiC strain, keeps the parasitic resistance under control, and provides a close stressor-to-channel proximity for enhanced strain transfer into the channels.

Processes and results

Two integration schemes were examined. Scheme 1 utilizes undoped epitaxial SiC stressors (grown in recessed source/drain cavities), followed by halo, extension and source/drain implants (Fig. 1, scheme 1). The final annealing can be a laser-annealing (LSA) + a rapid-thermal-anneal (RTA) + LSA, or a LSA-only. XRD analyses indicate that 1) implants degrade C_{sub}, particularly in the top region of the stressors next to the channel; 2) a LSA can not fully recover the C_{sub} loss; and 3) a high thermal budget RTA (>950C, or long time at lower T) decreases C_{sub}. The material study result is shown in Fig.2. Scheme 2 utilizes P-doped epitaxial SiC stressors. The stressors are grown after implants (gate, halo and extension), a high temperature RTA and recess etch (Fig. 1, scheme 2). The stressors only receive a LSA for dopant activation. In this scheme C_{sub} is fully preserved and the LSA enhances C_{sub} by a small amount [3].

This work uses scheme 2 process. The device structure and material study result is shown in Fig. 3. C_{sub} of ~1.0%, P of ~3-4.8x10^{20}cm^{-3}, and stressor-to-channel proximity of ~10-20nm are achieved. The performance assessment was done for two groups of devices built on the same chip with different gate length ranges (Fig. 3a, left and right). The control devices follow the same process, except that the P-SiC stressors are replaced with arsenic-implanted Si sources/drains. XRD analyses indicate that the SiC film is fully strained after metal 1 process. Parasitic resistance components are low and comparable between devices with SiC stressors and without SiC stressors. Low parasitic resistance and well preserved SiC strain has enabled a more accurate assessment of SiC stressor benefit.

To understand the channel strain benefit due to the SiC stressors, XRD and NBD (Nano-Beam-Diffraction) techniques were undertaken to correlate C_{sub} of the SiC stressors and the channel strain of devices (gate lengths of ~40-60nm). Table I summarizes the results. C_{sub} degradation measured on 200μmx200μm pads correlates to the channel strain (current flow direction <110> average strain in the top ~20nm channel layer) decrease measured in the nested devices on the same sample. A 1065C RTA fully relaxes SiC stressors, leading to no channel strain in the devices (row 1), while SiC stressors with well preserved strain (C_{sub}~1.6%) increase the channel <110> lattice constant by 0.54% (row 2). Presence of stacking faults in SiC stressors results in channel strain degradation (row 3 vs. row 4).

For devices with relatively long L_g (Fig. 3a, right), NBD measures <110> lattice constant increase of ~0.32% in the channels (L_g ~ 60nm, row 4 of Table I). Electrical results of the long L_g devices are shown in Fig. 4 through Fig. 10 (SiC effect only, no other stress techniques). Fig. 4 compares R_{on}-L_g for nFETs with and without SiC stressors. The two groups of devices have similar parasitic resistance (Y-axis intercept), whereas the channel mobility of the devices with SiC stressors is ~13% higher than the control devices. Fig. 5 shows that the mobility improvement transfers into ~15% linear drive current increase at L_g ~ 60nm. Fig. 6 exhibits that the trans-conductance enhancement increases to ~14% as the design gate length decreases from ~500nm to ~100nm, suggesting that the source/drain stressor effect becomes more evident as the gate length reduces. All these comparisons are done between the two groups of devices with approximately the same overlap capacitance, gate oxide, and short channel effects (SCEs), as shown in Fig. 7. I_{eff}-I_{off} is plotted in Fig. 8, demonstrating I_{eff} gain of ~9%. I_d-V_d, I_d-V_g groups and trans-conductance vs. V_g are plotted in Figs. 9 and 10, in which the devices show well behaved electric characteristics in both regions below and above V_t.

Discussion

Higher C_{sub} would further improve the SiC stressor strain benefit (row 2 of Table I), provided that resistance can be reduced. Closer stressor-to-channel proximity may also help. It is noted that P-SiC stressor-to-gate proximity of ~10nm does not degrade SCE in devices with L_g of ~40nm. Material quality is important, as well. We found that the short L_g devices (Fig.3b, left) suffer SiC material degradation after metal 1 process, ending up with less performance gain (Fig. 11). Stacking faults were seen in the SiC stressors (TEMs not shown) and NBD measures <110> lattice constant change of ~0.2% (row 3 of Table I), much lower than that measured in the long L_g devices discussed above (0.32%). No defects were seen in the TEMs of the long L_g devices. The root cause of SiC material degradation is under investigation. Fig. 11 also shows that the tensile liner technique provides further performance gain for nFETs with SiC stressors; however, SMT benefit disappears when a laser annealing is used to preserve SiC strain (the same stress liner responses are observed in the long L_g devices).

Summary

We demonstrated high performance SOI nFETs with epitaxially grown P-doped SiC source/drain stressors. For the first time, the fully preserved SiC strain and low parasitic resistance make possible a clean understanding of the SiC stressor effect. With a C_{sub} of ~1%, stressor-to-channel proximity of ~20nm, and embedded depth of ~50nm, mobility enhancement of ~13% was observed, transferring into ~9% I_{eff} gain over the well calibrated control devices at a gate length of ~60nm. The tensile liner technique provides further performance gain, but the SMT technique does not.

978-1-4244-1802-2/08/$25.00 ©2008 IEEE

Acknowledgements

This work was performed by Research Alliance Teams at various IBM Research and Development Facilities. We wish to thank ASM America for processing support and G. Shahidi and T.C. Chen for management support.

References:

[1] K.-W. Ang, *et al.*, IEDM 2005, p.503.
[2] Y. Liu, *et al.*, Symp. on VLSI Tech., 2007, p.44.
[3] P. Grudowski, *et al.*, SOI Conf. Proc., 2007, p.17.

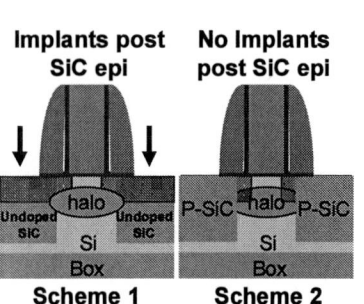

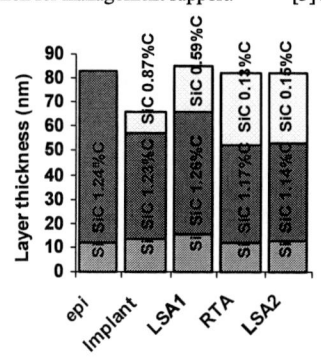

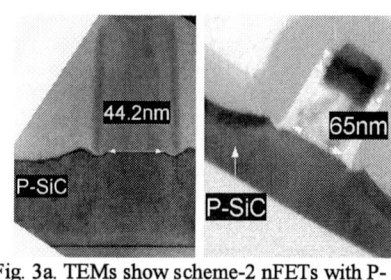

Fig. 3b. The post epi SIMS results of C and P (bottom left), and XRD results taken after epi and after metal 1.

Fig. 1. Process schemes: in scheme 1, implants are performed after the growth of undoped SiC stressors, followed by a LSA + RTA + LSA, or a LSA-only; in scheme 2, implants and a high T RTA are done before P-SiC growth, followed by a LSA for dopant activation.

Fig. 2. Scheme 1 undoped SiC stressor strain loss through processing. XRD measurement was done at different processing steps. Simulation of XRD data shows that C_{sub} decreases, particularly in the top region. A 3-layer model is used in the simulation.

Fig. 3a. TEMs show scheme-2 nFETs with P-SiC stressors (left, a short L_g device after epi, and right, a long L_g device after metal 1).

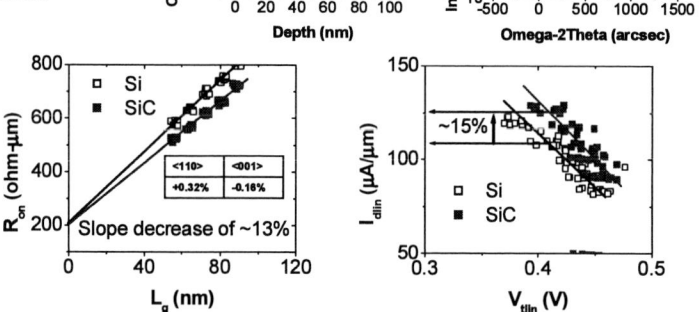

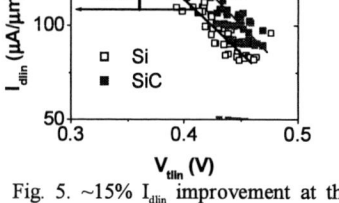

Experiments	C_{sub} (as deposited)	C_{sub} (post process)	Channel strain in devices
P-SiC + implants + Hi T RTA	~0.7%	~0% SiC relaxed	0.057% <110> -0.053% <001>
P-SiC + implants + Low T RTA	~1.6%	~1.6%	0.54% <110> -0.25% <001>
P-SiC + no implants + LSA	~0.9%	~1%	0.197% <110> -0.16% <001> (stacking faults seen in devices)
P-SiC + no implants + LSA	~0.9%	~1%	0.32% <110> -0.16% <001>

Table I. Correlation between C_{sub} obtained from fitting XRD data and channel strain obtained from NBD. Gate lengths are ~40-60nm. <110> is the current flow direction, <001> is the direction normal to the wafer surface.

Fig. 4. R_{on} vs. L_g. R_{on} is measured at a gate over-drive of 0.7V and V_{ds}=50mV. The insert shows NBD channel strain in SiC devices of L_g~60nm.

Fig. 5. ~15% I_{dlin} improvement at the same V_t. V_d=50mV, V_g=1.5V.

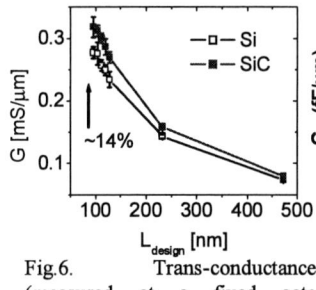

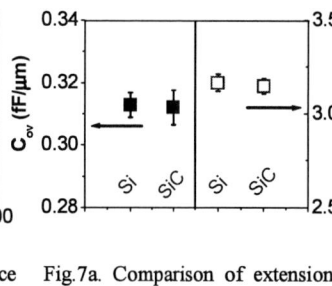

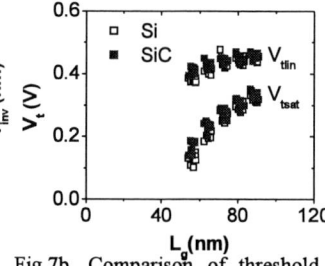

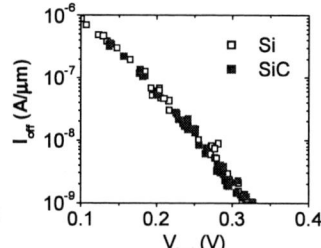

Fig.6. Trans-conductance (measured at a fixed gate overdrive) vs. design gate length.

Fig.7a. Comparison of extension-to-gate overlapping capacitance and gate inversion layer thickness.

Fig.7b. Comparison of threshold voltages vs. gate length.

Fig. 7c. Comparison of off state current vs. V_{tsat}. V_{dd}=1.5V.

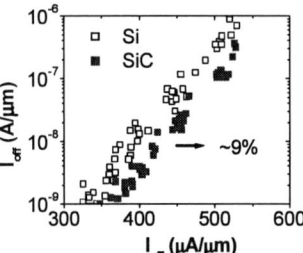

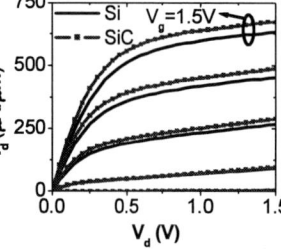

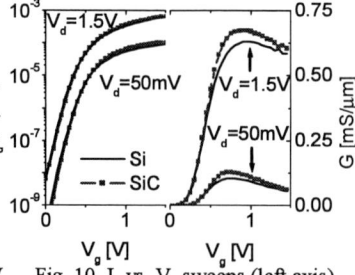

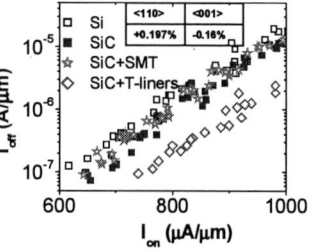

Fig. 8. ~9% I_{eff} enhancement is seen in devices with SiC stressors. V_{dd} is 1.5V.

Fig. 9. I_d vs. V_d sweeps, with a V_g step of 0.3V, up to 1.5V. L_g~60nm.

Fig. 10. I_d vs. V_g sweeps (left axis), and the corresponding conductance (right axis), with V_d of 50mV and 1.5V. L_g~60nm.

Fig. 11. I_{on} vs. I_{off}, V_{dd}=0.9V, shows effects of different stress techniques. The insert shows NBD channel strain in SiC devices (nominal L_g of ~40nm).

A Designer Friendly 45nm High Performance Technology with In-situ C-doped e-SiGe & Dual Stress Liner in SRAM

R. Khamankar, C. Bowen, H. Bu, D. Corum, I. Fujii, Y. Gu, B. Hornung, T. Kim, B. Kirkpatrick, K. Kirmse, A. Krishnan, C. Lin, L. Liu, T. Lowry, C. Montgomery, O. Olubuyide, S. Prins, D. Riley, S. Yu, J. Blatchford, C. Machala, C. O'Brien, G. Shinn, T. Grider

Texas Instruments, MS 365, 13121 TI Boulevard, Dallas, TX, USA

Tel: +1-972-995-4778, Fax: +1-972-995-6383, e-mail: r-khamankar1@ti.com

Abstract

A 45nm high performance technology with 11 level metallization is presented for SOC applications. High performance and density are maintained through new process optimizations that allow the use of less restrictive layouts by eliminating defect generation from strain enhancing processes. Additionally, technology modeling has been made simpler through optimization of key processes to minimize context dependences while simultaneously providing a competitive technology. High drive currents of 1150uA/um and 720uA/um are obtained for nMOS and pMOS, respectively at 1.0V and Ioff of 100nA/μm. The first yielding SRAMs incorporating both in-situ C-doped e-SiGe and dual stress liner (DSL) in the SRAM are demonstrated.

Introduction

Digital signal processors (DSP), SerDes and other SOCs require high performance and high logic density at low cost. To achieve high performance, strain engineering techniques including e-SiGe, DSL, stress memorization technique (SMT) and STI fill are routinely used in the industry [1-3]. However, the maximum performance gain that can be extracted from these techniques diminishes with increasing logic circuit density due to defect generation issues as well as context dependence of transistor performance. This paper describes the innovative use of key processes to generate a low cost but competitive high performance technology that eliminates defect and context dependent restrictions on designers.

Process details

Key technology components and targets are summarized in Table 1. Up to 11 levels of Cu metallization (ULK: M1-3 (k=2.6), OSG: M4-7 (k=2.9), FSG: M8-9 (k=3.6), TEOS for M10-11) are supported. High aspect ratio trench with SACVD fill provides excellent inter- and intra-well isolation. Plasma nitridation is used for a scaled 11.5A gate dielectric. Immersion lithography is utilized at crucial levels, including 35nm gates (Fig. 1). NiPt silicide is utilized for low Rcont and Rsheet.

Stressor Integration: C-doped e-SiGe, SMT, DSL

Aggressive gate-length scaling and multiple strain enhancers are utilized to achieve performance consistent with best-in-class industry trends. e-SiGe is utilized for substantial pMOS Idrive improvement. A vertical recess etch (VRE) is implemented for SiGe recess to minimize pitch dependence of Ion (Fig. 2). High-temp Source/Drain laser anneal is ideally needed for NMOS performance enhancement, but restrictions arise due to wafer warpage issues, when used in conjunction with e-SiGe [4]. For the first time, in-situ C-doped e-SiGe is utilized. This eliminates warpage issues (Fig. 3), and allows for a wide range of laser anneal conditions. Additionally, in-situ C-doping provides excellent boron diffusion control enabling substantial reduction in Gate-SiGe space (Fig. 1b), maximizing stress transfer.

High stress processes, such as SMT, have also been used to boost nMOS performance. The interaction of high stress, implant damage, high thermal budget and layout variations can lead to defects (Fig 4) that nucleate in the end-of-range (EOR) region of the implant due to extensive interstitials. This can lead to restrictions on the optimal use of each of the above factors. Defect propagation is eliminated through a novel nitrogen implant process that removes damage faster than dislocation growth; pinning dislocations. Nitrogen energy and dose are optimized to eliminate yield loss from stressor induced defects

(Fig 5). The elimination of defects has allowed the use of stressors and implants to their maximum potential while minimizing layout restrictions on designers.

The DSL process involves deposition and selective etching of 1.6GPa tensile and 2.8GPa compressive nitride contact etch stop layers. To achieve SRAM yield, the liner etch patterns, angles and selectivity were optimized to maintain adequate space of the etch boundary from N/P active areas in the tight bit-cell space and for poly contact etch process margin (Fig. 6).

Context (Layout) Dependences

Stress enhancers can result in an increase in device variability, especially as a function of the device environmental context [5]. The context dependent impact of performance enhancers is evaluated and optimized for this technology. Laser anneal and DSL layers are optimized to generate a strong reduction (up to 55%) in the Length of Diffusion (LOD) effect. Moreover, for pMOS transistors, the impact of varying gate space is also reduced through optimization of e-SiGe, laser anneal and compressive liner, yielding a 20% reduction in Idsat variation (Fig 7). For nMOS (Fig. 8), optimal laser annealing is critical in maintaining low variability. The reduction in context dependent variability has simplified technology modeling.

Transistor Performance, Reliability, and Yield

NMOS short channel effects are mitigated using a novel In+B+C co-implanted halo process. Amorphization caused by indium implant reduces boron channeling while enhancing the ability of carbon to retard B-diffusion via interstitial clustering. TCAD simulations were calibrated to capture these effects, producing quantitatively predictive simulations (Fig. 9), aiding transistor optimization.

Ion-Ioff characteristics for nMOS and pMOS transistors are shown in Fig. 10, with competitive Ion of 1150uA/um and 720uA/um respectively at 100nA/um Ioff. The 35nm gate length transistors exhibit well behaved sub-threshold characteristics (Fig. 11) and low DIBL for good low voltage operation. PMOS NBTI is not impacted by optimized DSL liner or e-SiGe stressors (Fig. 12).

e-SiGe and DSL process loops can easily inhibit SRAM yield. Hence, reports of SRAM yield with both e-SiGe and DSL present in the SRAM are rare. In this work, with the above discussed optimizations, good yield with good static noise margin is obtained (Fig. 13) with both C-doped e-SiGe and DSL present in the SRAM. Also key to this result is the utilization of a new FEOL Defect Density (DD) test vehicle designed to monitor key device and process components and evaluate early systematic process problems. This FEOL DD methodology, with short cycle time, enabled execution to the originally planned SRAM yield roadmap (Fig. 14).

Summary

A competitive 45nm SOC technology has been demonstrated. This technology enables simpler technology modeling and is designer friendly. C-doped e-SiGe and DSL are successfully integrated into yielding SRAMs with good Static Noise Margin.

References

[1] P. R. Chidambaram et al., VLSI Tech Digest, 2004, p. 48
[2] T. Ghani et al., IEDM Digest 2003 p. 978
[3] R. Khamankar et al, VLSI Tech Digest, 2004, p. 162
[4] T. Yamamoto et al., VLSI Tech Digest, 2007, p. 123
[5] V. Chan, et al., IEEE CICC, 2005, vol 23, pg. 667-674

Transistors	Lg (nm)	Ion(N) (uA/um)	Ion(P) (uA/um)	Ioff (nA/um)
1.0V SVT	35	1150	720	100
1.0V LVT	35	1350	850	1000
1.5V I/O	100	650	420	5

Additional Components	Description
NWell Resistor	Rs=585 Ohms/sq
Silicide Resistor	Rs=16 Ohms/sq
N-Poly Resistor	Rs=250 Ohms/sq
Npoly/Nwell Capacitor	C=11 fF/um2 @ 1.5V
Npoly/Nwell Varactor	Tuning range = 4, Q>25 at 5GHz
Inductor	L=0.91, Q=16.8 @ 5GHz
Bipolar Transistor	Vertical PNP, hfe = 10

Table 1. Key technology components and targets.

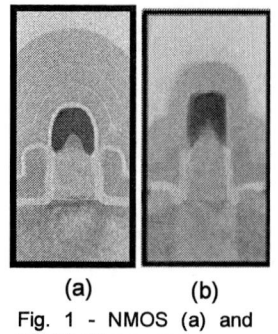

(a)   (b)

Fig. 1 - NMOS (a) and PMOS (b) x-sections.

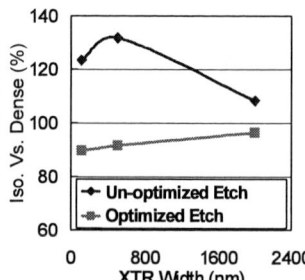

Fig. 2 - Reduction of Idsat pitch & LOD dependence with optimized SiGe etch.

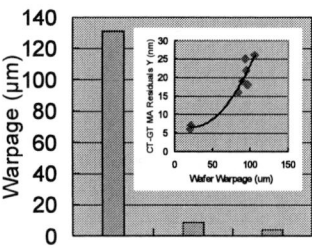

Fig. 3 - Impact of in-situ C-doping in eliminating warpage from Laser anneal. Overlay issues are also resolved (inset).

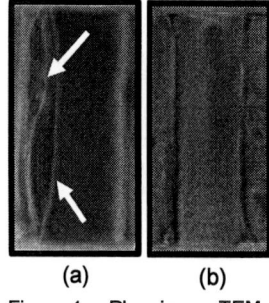

(a)   (b)

Fig. 4 Planview TEMs showing (a) Dislocation, which were then (b) eliminated through an optimized N2 implant at NSD

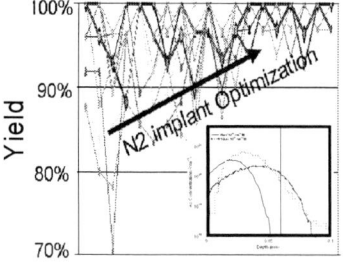

Fig. 5 Yield improvement through N2 implant optimization @ NSD (inset : N2 energy optimized for profile to be just below EOR region)

Fig. 6 - SEM image of dense SRAM post DSL formation (inset : SEM image post SiGe formation)

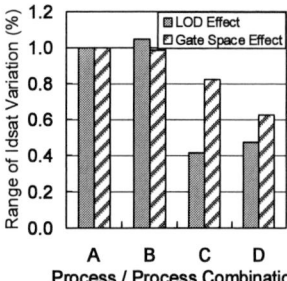

Fig. 7 - Reduction in context dependent variation for PMOS transistors

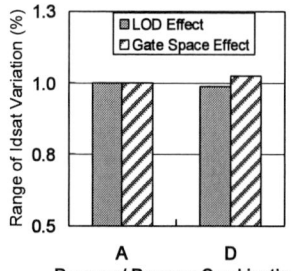

Fig. 8 - No impact on context dependence for NMOS due to context dependence optimization for PMOS.

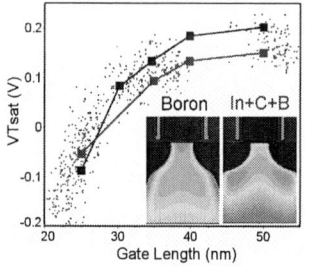

Fig. 9 - Simulated (lines) and measured (dots) NMOS VT rolloff for Boron (blue) and In+B+C (red) halos. Simulated doping profiles inset.

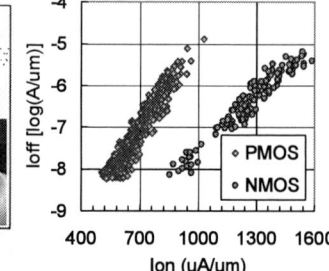

Fig. 10 – Universal Ion-Ioff characteristics for PMOS and NMOS transistors.

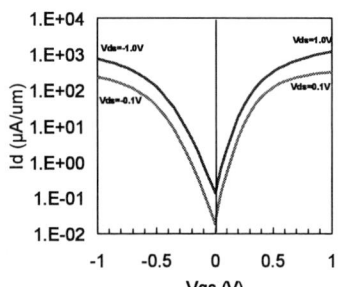

Fig. 11 - Id-Vg plots showing well behaved sub-threshold characteristics for 35nm gate-length transistors.

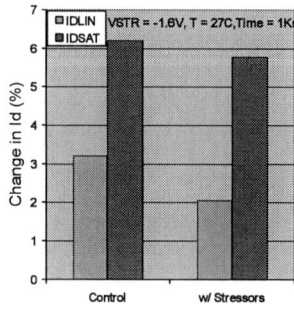

Fig. 12 - No degradation in pMOS NBTI is observed with stressor integration.

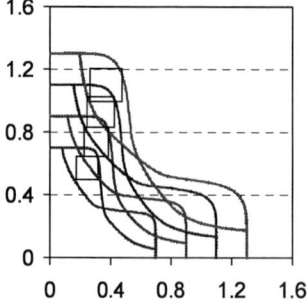

Fig. 13 - Butterfly curves for SRAM with both e–SiGe and DSL present.

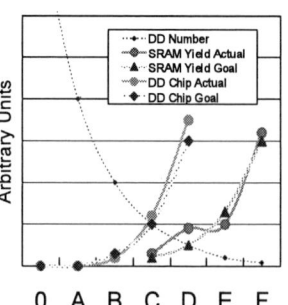

Fig. 14 - FEOL DD and SRAM yield vs. learning cycle. Use of DD methodology enabled SRAM yield ramp that matched original plan.

Higher hole mobility induced by twisted Direct Silicon Bonding (DSB)

M. Hamaguchi, H. Yin[1], K. L. Saenger[2], C. Y. Sung[2], R. Hasumi, R. Iijima, K. Ohuchi, Y. Takasu, J. A. Ott[2], H. Kang[1], M. Biscardi[1], J. Li[1],
A. G. Domenicucci[1], Z. Zhu[1], P. Ronsheim[1], R. Zhang[1], N. Rovedo[1], H. Utomo[1], K. Fogel[2], J.P. de Souza[2], D.K. Sadana[2],
M. Takayanagi, D. Park[1], G. Shahidi[1] and K. Ishimaru

Toshiba America Electronic Components Inc., Yorktown Heights, NY 10598, U.S.A.

[1]IBM Semiconductor Research and Development Center, Systems and Technology Group, Hopewell Junction, NY 12533, U.S.A.

[2]IBM Research Division, T.J. Watson Research Center, Yorktown Heights, NY 10598, U.S.A.

Tel: 914-945-2546, Fax: 914-945-3920, E-mail: masafumi.hamaguchi@taec.toshiba.com

Abstract

Twisted Direct Silicon Bonded (DSB) substrate demonstrates a higher hole mobility advantage over (110) bulk substrate for PFET. The mobility shows a (110) layer thickness dependence with the thinner DSB layer having a higher hole mobility. 25% on-current improvement is obtained for thin DSB PFETs at long channel ($L_g= 2$ μm), 10% higher at short channel ($L_g = 36$ nm) compared to (110) bulk PFETs. Moreover, we found that the thinner DSB shows better V_t roll-off characteristics. On the other hand, NFETs on DSB are as good as (100) bulk NFETs. Thin DSB substrate demonstrates 11% faster ring oscillator speed over thick DSB substrate and 30% faster over (100) bulk due to higher mobility and lower capacitance.

Introduction

DSB is a bulk CMOS hybrid orientation technology [1,2] that exploits the higher electron and hole mobility expected from silicon (100) and (110) surfaces, respectively. In a previous paper, we focused on the integration scheme and explored the scalability mainly in terms of defect control [3]. We revealed that thinning the DSB layer thickness is the key to reducing the size of the defective regions at the boundaries of the (100) and (110) Si regions in a scheme known as SPE-before-STI, which forms (100) Si by solid phase epitaxy (SPE) before STI formation.

In this paper, we use a 45deg twisted (100) wafer for the DSB base substrate instead of the normal (100) base wafer used in the previous work, an approach which crystallographic modeling suggests should provide an improvement in the morphology of the defective boundary regions. Moreover, we investigate DSB layer thickness dependence on electrical device characteristics in detail.

Experimental

The process flow of SPE-before-STI is illustrated in Fig. 1. PFET regions are covered with resist and NFET regions are amorphized beyond the depth of the bonding interface by Ge ion implantation. NFET amorphous regions are recrystallized to (100) by performing SPE. STI is used to replace the defective regions form at the boundary of (110) and (100) orientations during the SPE process. 45nm node CMOS process with conventional RTA and dual stress liner is resumed after STI formation. eSiGe is not implemented for PFETs. Figure 2(a) shows the schematic of DSB with 45deg twisted (100) base wafer. The crystal direction along the channel direction is different between (110) and (100) for this wafer, <110> for PFET but <100> for NFET (Fig. 2(b)). We prepared three DSB substrates with different (110) Si layer thickness (Fig. 3) $T_3>T_2>T_1$, and compared them to (110) bulk wafer.

Results and discussion

Figure 4 shows SEM images of line edge morphologies for normal (0deg) and 45deg twisted DSB wafers. The lateral dimension of the defective region is proportional to DSB thickness (T_{DSB}) of which 45deg twisted DSB wafers have narrower defective region in each direction [4]. Moreover, the benefit of using 45deg twisted DSB wafer is also the elimination of defective triangular regions for (110)/<100> shown in Fig.4 (b).

Figure 5 shows I_{on}-I_{off} plots of PFET's with $L_g= 2$ μm devices on (100) bulk wafer, (110) bulk wafer, 0deg normal DSB wafer and 45deg twisted DSB wafers with three different thickness. (110) bulk shows 2.5 times higher I_{on} comparing to that of (100) bulk, and normal DSB wafer shows the same I_{on} as that of (110) bulk. Moreover, it is found that 45deg twisted DSB wafers show I_{on} enhancement over (110) bulk wafer. This enhancement also depends on the (110) layer thickness. The thinner DSB has higher I_{on}, and the thinnest DSB(T_1) shows 25% I_{on} enhancement over (110) bulk wafer (Fig. 5). Transconductance (G_m) vs V_g-V_t plots also show the thickness dependency (Fig. 6), which is consistent with I_{on}-I_{off}. Since the equivalent gate oxide thickness is confirmed from the same gate leakage current for all wafers, this result suggests that I_{on} enhancement for DSB wafers is due to mobility improvement. Though the magnitude of the enhancement is not as large as long channel devices, 10% I_{on} enhancement is also obtained at short channel region with thin DSB (T_1) device (Fig. 7).

Nanobeam diffraction (NBD) measurements of lateral strain vs. depth under the gates of 2μm-gate-length devices showed no depth dependence for devices in (110) bulk, and a small (0.13%) strain differential between the top (relatively tensile) and bottom (relatively compressive) of the DSB layer for devices in T_3 twist-bonded substrates. This may suggest that the strain in channel region is induced by the twisted bonding interface and is more strained when the thickness of DSB layer decreases. Therefore, thin DSB wafer with the channel being closer to the bonding interface has higher mobility though the strain value is small to explain the results.

V_t roll-off characteristic and DIBL also improve significantly as we decrease T_{DSB} (Figs. 8 and 9). Lower gate overlap capacitance (C_{ov}) with decreasing T_{DSB} suggests longer effective channel length at the same physical gate length, resulting in better short channel control for thinner DSB wafer (Fig. 10). These results imply further L_g scaling without T_{inv} reduction. Not only C_{ov}, but also junction capacitance (C_j) has the thickness dependency, the thicker DSB wafer has higher C_j (Fig. 11). We consider that these dependencies come from the change of boron profile near the bonding interface. Figure 12 shows boron SIMS profile for (110) bulk wafer and DSB(T_3). Boron is piled up at the bonding interface for DSB(T_3). We speculate that 2D profile should change as shown in Fig.13. For thinner DSB(T_1), reduced boron concentration at Si surface induces C_{ov} reduction, while higher butting concentration with well impurities induces higher C_j due to the bonding interface near junction depth for thicker DSB(T_3).

Considering CMOSFET performance, NFET performance is also important. I_{on}-I_{off} of NFET on each DSB wafer is as good as (100) bulk NFET (Fig. 14). This means that conversion from Si (110) to (100) Si does not affect on NFET performance, regardless of T_{DSB}. I_d-V_g curve of NFET and PFET with L_g=36nm show good subthreshold characteristic (Fig. 15). In addition to I_{on} benefit, the lower C_j and C_{ov}, and even equivalent NFET performance make it possible for thin DSB wafer to demonstrate 11% faster ring oscillator speed over thick DSB wafer and 30% faster over (100) bulk (Fig. 16).

Conclusions

CMOSFET performance of twisted DSB substrate has been investigated. We found that thinner DSB wafer has higher hole mobility over (110) bulk. The thin DSB wafer shows better short channel characteristics and 30% faster ring oscillator speed over (100) bulk thanks to lower C_j and C_{ov} in addition to higher hole mobility.

Acknowledgement

This work was performed by the Research Alliance Teams at various IBM Research and Development Facilities.

References

[1] C.-Y. Sung et al., IEDM Tech. Dig. p.225 (2005), [2] H. Yin et al., IEDM Tech. Dig. p.75 (2006), [3] H. Yin et al., VLSI Tech Dig. P. 222 (2007), [4] K.L. Saenger et al., J. Appl. Physics, 101, 084912 (2007)

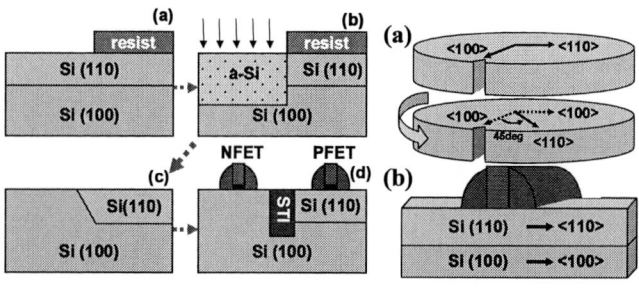

Fig.1. Schematics of the SPE-before-STI scheme: (a) PFET regions are covered with resist, (b) NFET regions are amorphized by Ge I/I, (c) NFET regions are recrystalized to (100). (d) CMOS processes are resumed after STI is formed.

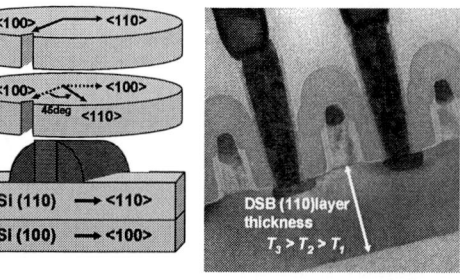

Fig.2. (a) Schematics of the DSB structure with 45deg twisted base wafer. (b) Diagram of PFETs cross section with 45deg twisted bese wafer.

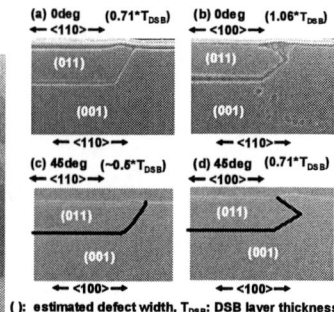

Fig.3. TEM image of DSB wafer. Compared three DSB substrates with different (110) Si layer thickness.

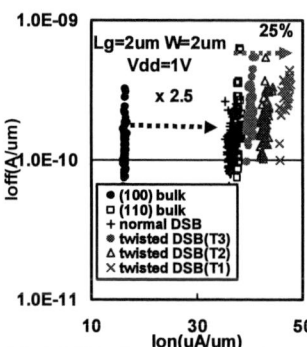

Fig.4. SEM images of line edge morphologies for normal (0deg) and 45deg twisted DSB wafers. The crystal orientations indicated are for the line edges; the line edges of (c) and (d) are narrower than they appear because the samples are cleaved at 45deg to the lines.

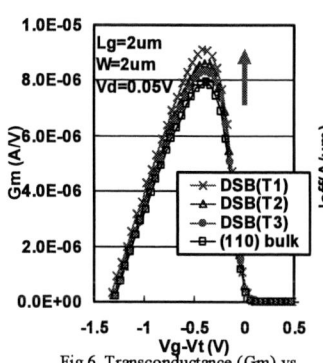

Fig.5. Ioff-Ion from PFETs with 2 μm gate length: (100) bulk vs (110) bulk, 0deg DSB, 45deg twisted DSB with three different thickness. On-current depends on DSB thickness.

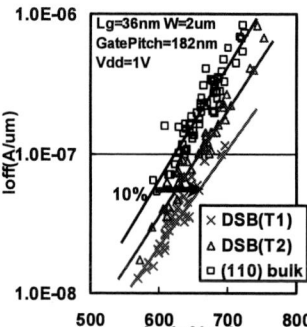

Fig.6. Transconductance (Gm) vs Vg-Vt from PFET with 2 μm gate length: (110) bulk vs DSB with three different thickness.

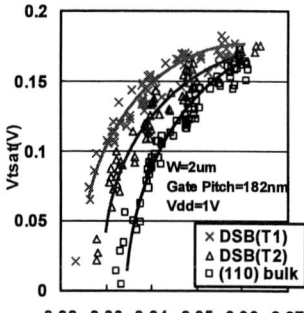

Fig.7. Ioff-Ion from PFETs with 36nm gate length: (110) bulk vs DSB with two different thickness.

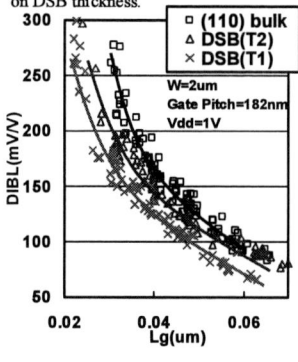

Fig.8. Vt vs Lpoly from PFETs: (110) bulk vs DSB with three different DSB thickness. The thinner DSB has the better Vt roll-off.

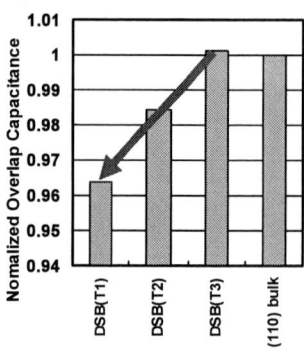

Fig.9. DIBL from PFETs: (110) bulk vs DSB wafers. The thinner DSB has better DIBL.

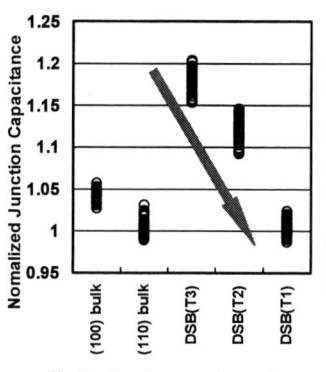

Fig.10. Overlap Capacitance depends on DSB thickness. This causes DSB thickness dependency on Vt roll-off.

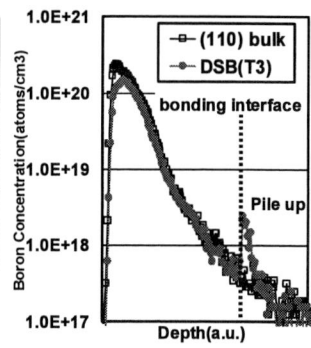

Fig.11. Junction capacitance from PFETs: (100) bulk, (110) bulk, DSB with three different thickness.

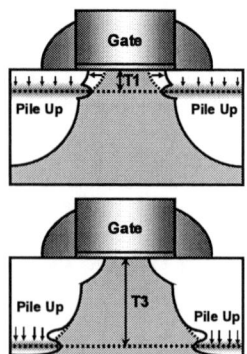

Fig.12. SIMS profiles from DSB(T₃) and (110) bulk. Boron piles up near the bonding interface.

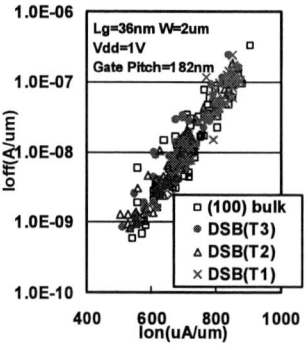

Fig.13. Schematics of 2D profile for DSB(T₁) and DSB(T₃). The change of 2D profile induced by Boron pile up at the bonding interface affects the electrical characteristics.

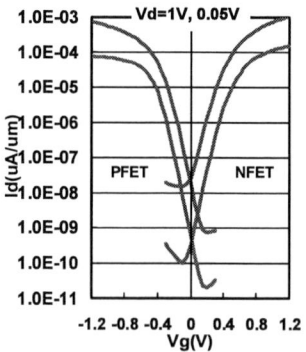

Fig.14. Ioff-Ion from NFETs with 36nm gate length: (100) bulk vs DSB wafers. Ion does not depend on original DSB thickness for NFET.

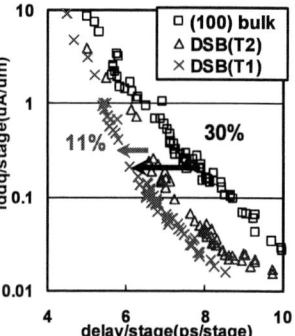

Fig.15. Id-Vg curve data for PFET and NFET with 36nm gate length on DSB(T₁).

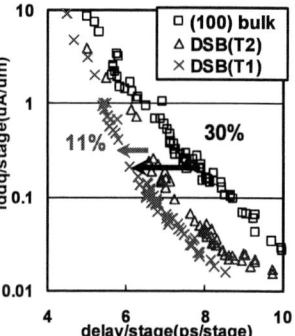

Fig.16. Thin DSB(T₁) achieves 11% faster ring oscillator speed than DSB(T₂).

140

32nm Device Architecture Optimization for Critical Path Speed Improvement

R.Gwoziecki [1), 2)], S. Kohler [1)] and F.Arnaud [1)]

[1)] ST Microelectronics, 850, rue J.Monnet, 38926 Crolles, France. [2)] CEA-LETI, Minatec, 17 rue des Martyrs, 38054 Grenoble
ph : +33 4 38 92 27 71, e-mail : franck.arnaud@st.com

Abstract

This study investigates key elements improving CMOS critical path speed. We proposed a full analysis of input signal slope impact on the switching current trajectories depending on Vt centering. Based on inverter output characteristics shape, we demonstrated that speed of low-Vt (LVT) path preferred higher drive current (I^{ON}) whereas high-Vt (HVT) cells speed is enhanced by lower Drain Induced Barrier Lowering (DIBL). Finally, we proposed a link with transistor architecture by optimizing halos and Light-Doping-Drain (LDD) design to improve logic gate as a function of Vt options.

Introduction

CMOS circuits performance need specific transistor architecture to enhance speed of electronics system. Many authors have already addressed the impact of device optimization on overall circuits performances by demonstrating that linear current is more important that saturation current , leading to definition of effective drive current or characteristic drive current [1],[2]. Impact of stacked transistors was also addressed recently [3] and device performance optimization was also studied [4] but only in frame of high performance devices. In this study, we demonstrate the clear impact of both input signal slope and transistor type on switching current trajectory. We determine an inverter chain model as a function of both DIBL and I^{ON}. Based on this electrical analysis, we change process flow scheme allowing specific halos and LDD architecture versus VT options.

Inverter switching current trajectories

Figure 1 represents inverters chain including variable Fan-In (FI) and Fan-Out (FO) ratio along the chain. We focused our analysis on transition stage between inverter 4 (INV4) and inverter 5 (INV5) reporting different FO/FI. In fig2 INV4 (FO/FI=4) reports a degraded signal slope versus INV5 (FO/FI=1). As reported by figure 3, FO/FI ratio modifies the switching characteristic of inverter by changing input and output signal slopes: higher FO/FI is, smoother is signal slope. In fig4a, it is shown that when INV4 is heavy loaded (FO/FI=8), the switching occurs mainly for Vgs=Vdd, whereas INV5 (fig4b) is switching for Vgs<<Vdd. Fig5 demonstrate also the impact of Vt flavor on switching, for values of Vt reported in table1. It is shown that the switching behavior depends mainly of FO/FI ratio and that Vt impact only the current level. Note that already previous proposed currents, I_L, I_H, I_{CH}, I^{ON} [1-4] to monitor the inverter chain speed are well representative (fig5a) of INV4 switching (heavy loaded gate) but not that of INV5. This implies that these characteristics currents are not sufficient to model whole critical path which mix different FO/FI ratios. In order to map more accurately switching current trajectory required in case of complex critical path, we propose to monitor not one but two parameters, I^{ON} and DIBL, to optimize device speed.

DIBL / I^{ON} trade-off for speed

Following equation describes a simple model used to work out gate delay (τ_p) along the inverter chain

$$\tau_p = \int_{Vcc}^{V_o-Fi} \frac{C(V_{Ds})}{\beta(V_{Ds}) \cdot (V_{Gs} - Vt(V_{Ds}))^\alpha} dV_{Ds} + \int_{Vcc-Fi}^{0.5V_o} \frac{C(V_{Ds})}{\beta(V_{Ds}) \cdot (V_{Gs} - Vt(V_{Ds}) - 0.5 V_{Ds}) V_{Ds}} dV_{Ds}$$

$\beta(V_{ds})$ and $Vt(V_{ds})$ are key parameters for current and DIBL, respectively. We built a complete design of experiment based

DIBL variation of +/-50mV and I^{ON} variation of +/-15%. Figures 6 show switching time abacus as a function of both I^{ON} and DIBL for multiple VT flavors. For LVT option we found a slope of 5.2mV/% to obtain an iso-delay. In this case, a drive current change impact significantly inverter speed. For HVT type iso-delay slope achieved only 3mV/% which means that switching time is mainly driven by DIBL. Those figures demonstrate clearly that key element to enhance critical path delay are not similar, depending on Vt centering. We examine a selection of process parameters to optimize DIBL/ION trade-off.

Device architecture optimization

To modulate both β (i.e. I^{ON}) and DIBL, we propose experimental variation of both halos and LDD dosage. As shown in fig7, low LDD combined with high halo mitigate I^{ON} and DIBL. Figure 8 quantifies the effect of halo/LDD couple inside I^{ON}/DIBL map based on experimental results. We found that halo condition allows a change in the compromise, whereas LDD dosage plays a role of DIBL improvement without drive current penalty. We add in fig9.a experimental points inside iso-delay area determine by the design of experiment. To optimize inverter gate delay in case of LVT transistor, we found that high LDD dosage is required because of low impact of DIBL on delay. Results are more complex in case of HVT (fig9b). Due to the fact that switching current is often is saturation mode but at low gate bias, HVT transistor speed is mainly driven by DIBL, and consequently low LDD dosage is required to enhance the speed. In order to optimize both LVT and HVT transistors type, we propose to change the process sequence from pre-gate Vt scheme to post-gate Vt scheme, as depicted by figure 10. Initial flow uses specific Vt value but with common LDD, whereas optimized process sequence propose separated LDD and halos. Based on this architecture and model analysis, new halos and LDD experimental points have been investigated (fig11) for 3 Vt options. We add lines of iso-delay extracted from the model in case of high and low LDD demonstrating the strong interest of low LDD for HVT device. Finally we examine a LDD/halos optimization to improve DIBL from loop1 to loop5 (see. fig. 13). We show gate delay improvement for HVT and SVT by using loop5 condition helping DIBL and consequently inverter speed. Same benefit is obtained for LVT using loop 2 conditions. Optimized process data are summarized by table 2.

Conclusion

We demonstrate the significant impact of both signal slope and Vt centering on the current switching trajectory of inverter chain. We verified that process condition between HVT and LVT options can not be identical to optimize inverter speed, because devices have not similar response inside DIBL/I^{ON} map. Finally we propose a table of LDD/halos based on post-gate scheme.

References

[1] M.H. Na, E.J. Nowak, W. Haensch, J.Cai "The effective drive current in CMOS inverters", IEDM Tech. Dig. 2002
[2] E. Yoshida, Y. Momiyama, M. Miyamoto, T. saiki, K.Kojima, S. Satoh, T.Sugii, "Performance boost using a new device design methodology based on characteristic current for low-power CMOS", IEDM Tech. Dig. 2006
[3] K. von Arnim, C. Pacha, K. Hofmann, T. Schulz, K. Schrüfer, J. Berthold, "An effective switching current methodology to predict the performance of complex digital circuits", IEDM Tech. Dig. 2007
[4] E. Morifuji, P. Kapur, A.K-A. Chao, Y. Nishi, "New constraint for Vth optimization for sub 32nm node CMOS gates scaling", IEDM Tech Dig. 2005

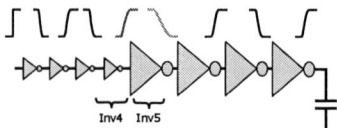

Wn=0.36μm / Wp=0.5μm / Lg=35nm
FI=fan-in=size INV4
FO=fan out=size INV5

Fig.1: schematic of inverter chain used for the study.

Flavors	LVT	SVT	HVT
Ion,n [μA/μm]	775	625	412
Ion,p [μA/μm]	385	315	205
Vt,lin [V]	0.31	0.40	0.49
Vt,sat [V]	0.15	0.24	0.33

Table 1: Vt centering for 32nm transistors (3 flavors).

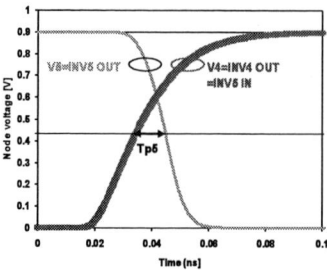

Fig.2: typical waveform with INV5 size equal to 4 time INV4 size (FO/FI=4).

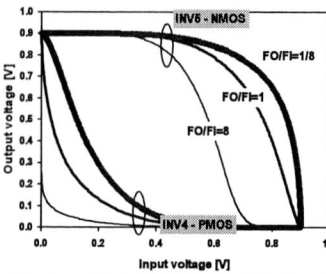

Fig.3: switching path for NMOS and PMOS device wrt FAN-IN and FAN-OUT ratio.

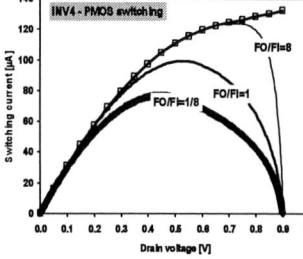

Fig. 4a: switching current of INV4 (PMOS) as function of FO/FI, for LVT device (see table1).

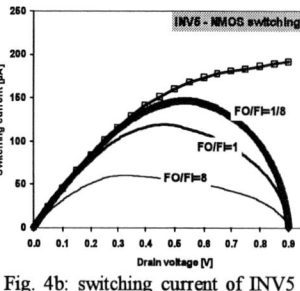

Fig. 4b: switching current of INV5 (NMOS) as function of FO/FI, for LVT device.

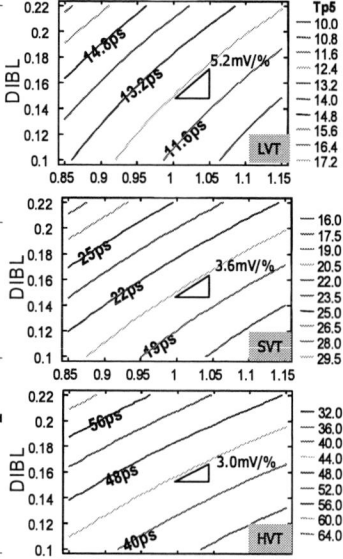

Fig.6: equivalent switching time as function of DIBL and Ion.

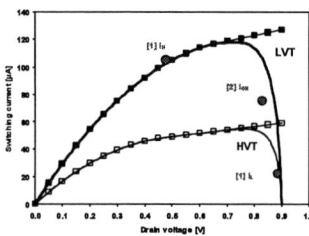

Fig.5a: switching current of INV4 (PMOS) as function of device type, for FO/FI=8.

Fig.5b: switching current of INV5 (NMOS) as function of device flavour, for FO/FI=8.

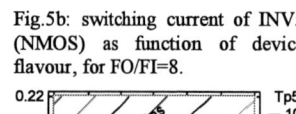

LDD dose : 6e14 to 1.3e15/cm²
HALO dose : 3.5e13 to 6e13/cm²

Fig.7: β (~Ion) and DIBL variation vs LDD and HALO doses.

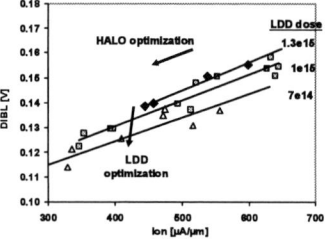

Fig.8: DIBL &Ion for LDD & HALO conditions of fig.7. LDD dose is an efficient way to play on DIBL.

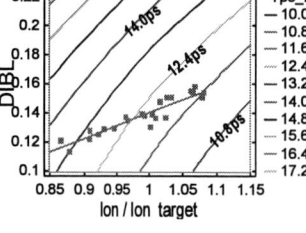

Fig.9a: impact of DIBL and Ion variations on switching time for LVT device.

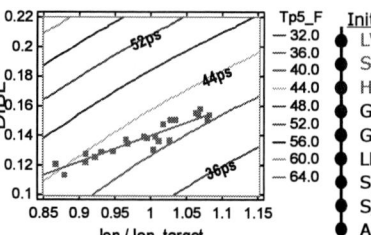

Fig.9b: impact of DIBL and Ion variations on switching time for HVT device.

Initial flow
- LVT channel
- SVT channel
- HVT channel
- Gate oxide
- Gate definition
- LDD/HALO impl.
- Spacers
- S/D implant
- Anneal

Optimized flow
- Channel
- Gate oxide
- Gate definition
- LVT LDD/HALO
- SVT LDD/HALO
- HVT LDD/HALO
- Spacers
- S/D implant
- Anneal

Fig.10: optimized versus initial flow. Dedicated HALO and LDD to optimize both HVT and LVT devices.

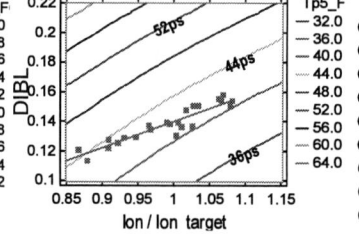

Fig.11: summary of experimental data to optimize DIBL and Ion – 3 groups are shown, corresponding to LVT, SVT and HVT devices.

Fig.12: iso-delay lines in Tp (calculated from model) are plotted together with DIBL & Ion variation for given process trials.

Fig.13: results for main loop dedicated to optimized flow – more than 30mV gain in DIBL are reported for Loop5 with respect to initial flow at same Ion.

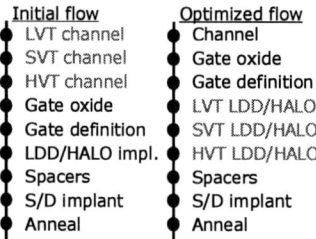

NMOS	LVT	SVT	HVT
HALO [/cm²]	2.8e13	3e13	4e13
LDD [/cm²]	1.1e15	8.2e14	7e14

PMOS	LVT	SVT	HVT
HALO [/cm²]	3e13	3.1e13	4.2e13
LDD [/cm²]	1e15	7e14	5e14

Table 2: optimized LDD and HALO conditions for each device flavor.

142

Strain Additivity in III-V Channels for CMOSFETs beyond 22nm Technology Node

[††]S. Suthram, [†]Y. Sun, [1]P. Majhi, [*]I. Ok, [*]H. Kim,, [2]H. R. Harris, N. Goel, [†]S. Parthasarathy, [†]A. Koehler, [†]T. Acosta, [†]T. Nishida, H.-H. Tseng, [**]W. Tsai, [*]J. Lee, [3]R. Jammy and [†]S. E. Thompson

SEMATECH Austin, Texas, USA; [††]Intern from University of Florida, [1]Intel Assignee, [2]AMD Assignee, [3]IBM Assignee, [*] University of Texas at Austin, [**]Intel Corp. Santa Clara, [†]University of Florida, Gainesville FL.

Abstract

For the first time strain additivity on III-V using prototypical (100) GaAs n- and p-MOSFETs is studied via wafer bending experiments and piezoresistance coefficients are extracted and compared with those for Si and Ge MOSFETs. Further understanding of these results is obtained by using multi-valley conduction band model for n-MOS and performing k.p simulations for p-MOS. For GaAs n-MOSFET, uniaxial tensile stress is shown to enhance performance only for small stresses biaxial tensile stress is shown to be more beneficial. Importantly uniaxial compressive stress is beneficial for GaAs pMOSFETs and the piezoresistance effect is much larger than that seen for Si MOSFETs along the <110> channel direction. This works shows that intrinsic mobility and stress induced mobility enhancement are key knobs for scaling of III-V CMOSFETs.

Introduction

High mobility, narrow band gap group IV and III-V materials are considered strong contenders to replace strained-Si channels for logic applications beyond the 22 nm node [1-3]. Recent work on III-V devices [2] suggests that high In content $In_xGa_{1-x}As$ is being investigated as a possible candidate for MOSFET applications. Even though these III-V based materials readily demonstrate high electron mobility, they are less promising for hole transport due to intrinsic lower hole mobility and disadvantages when compared with strained-Si [4]. Eventual integration of III-V based semiconductors as MOSFET channels requires both high performance n- and p-MOSFETs. This necessitates the use of performance boosters such as strain if promising especially for the p-MOS devices on III-V channels. The extreme complexity involved with the processing of InGaAs heterostructure MOSFETs [2] and difficulty in estimating the channel stress in such III-V devices makes the direct investigation of stress effects very challenging. The valence and conduction bands for GaAs, InAs and InGaAs being very similar [5-7] hints that the results of a similar investigation for bulk-GaAs channels which is relatively easier to form, can be extended qualitatively to InGaAs channels which are of technological importance. Hence in this work for the first time we present a comprehensive report on the effect of different kinds of stresses on bulk-GaAs ring-FETs and extract both n- and p-MOSFET piezoresistance co-efficients via wafer bending experiments. Further understanding about the trends and insight into the physics of stress effects is gained from k.p calculations.

Sample Preparation

The detailed device description and fabrication process of the GaAs devices used in this work is given in Ref [8]. Since the GaAs samples used are very small and brittle, they are first glued on to a Si substrate as shown in Fig 1(a). In order to ascertain the amount of stress transfer on to the GaAs sample using the flexure based wafer bending setup (Fig 1(b)) described in [9], a strain gauge is glued on a bare GaAs sample. The stress calibration plot from this measurement is shown in Fig 2. This stress calibration is used for subsequent device measurements.

Wafer Bending Experiments

It is known from past work [10-11] that III-V materials have two additional scattering mechanisms, namely polar optic phonon scattering and piezoelectric scattering due to their intrinsic polar nature. The temperature dependant mobility measurements for an n-type GaAs channel compared with that of Si [12] in Fig 3 does not show evidence of that. This discrepancy is thought to be due to the underestimation of mobility at low temperatures because room temperature split C-V measurements (which overestimate inversion charge density at low temperatures [13]) were used for mobility extraction. Fig 4 shows that GaAs n-MOSFETs show large enhancement for small uniaxial tensile stresses as compared with Si n-MOSFETs. The inset shows Id-Vg curves with and without a uniaxial stress applied. A similar measurement as a function of biaxial stress reveals that GaAs n-MOSFETs show enhancement for biaxial tensile stress, while biaxial compressive stress is detrimental to the performance as shown in Fig 5. This implies that when engineering a hetero-structure quantum well

InGaAs channel MOSFET, the layer below the InGaAs channel should have a larger lattice constant (and be relaxed) so as to induce the favorable biaxial tensile stress into the channel. Fig 6 shows wafer bending data for (100) p-MOSFETs. It is seen that GaAs p-MOSFETs show almost 2 times the enhancement with uniaxial compressive stresses as compared with Si p-MOSFETs. This is an encouraging result, for even though GaAs has poor hole mobility it is very sensitive to stress and hence process induced uniaxial stress can be used to achieve higher hole mobility in GaAs p-MOSFETs. Table 1 tabulates the piezoresistance co-efficients extracted from measurements in this work and compares it with previous published data on the same for Si and Ge [3, 14].

Discussion

Physical understanding of the origin of enhancement for GaAs n-MOS under stress is obtained by considering a multi-valley conduction band model, where the deformation potential values are taken from Ref [15]. The GaAs conduction band consists of three predominant valleys Γ, L and X, out of which most of the carriers populate the lowest mass Γ – valley. When we apply an uniaxial tensile stress the hydrostatic component of the strain increases the splitting between the Γ and L valleys, but the shear component of the strain splits the L – valleys into two groups, out of which the energy level of one group reduces and hence the effective band gap between the Γ and L valleys reduces as shown in the schematic diagram in Fig 7. Since the effective mass of the carriers in the L – valley is much larger than that of the Γ – valley, we see that under large uniaxial stresses there occurs a saturation point (Fig 8) and eventual degradation in electron mobility due to increased population in the L – valley as seen in Fig 4. In the biaxial tensile case no shear splitting of the L – valley occurs as shown in Fig 9. Hence we see increased enhancement with increasing biaxial tensile stress due to increased population of carriers in the low mass Γ – valley (Fig 10) and the opposite with biaxial compressive stress as seen in experimental results in Fig 5. To gain a similar understanding for the hole mobility case, we have to look at the strain effects on the constant energy contours of the top-most valence band, which is obtained by performing k.p simulations. Fig 11 shows that the band warping due to uniaxial compressive stress in GaAs is similar to that in Si from a qualitative perspective [4]. But calculation of actual mobility values in Fig 12 shows that the enhancement trend in GaAs is also similar to the Si case for stresses under 2 GPa. The reason for this discrepancy between simulation and experiment is not yet understood where experiment shows a much larger enhancement with uniaxial compressive stress.

Conclusions

We have investigated the stress effects on both n- and p-type III-V channels using prototypical GaAs MOSFETs and for the first time extracted piezoresistance coefficients for GaAs. It is shown that nMOS does not benefit from uniaxial stress but shows enhancement with biaxial tensile stress. Importantly, PMOS shows large improvement with uniaxial compressive stress as compared with a Si device though the physical origin for this is not yet understood. Hence strain enhanced III-V pMOS along with high mobility III-V nMOS show promise for being incorporated as MOSFET channels at the sub-22nm technology node. Even larger performance improvements with strain could probably be achieved by incorporating these III-V materials in non-planar finFET or tri-gate structures as is seen for Si [16].

References

[1] R. Chau, Proc. *IEEE Nanotechnology*, 2004 [2] M. Hudait, et. al., *IEEE IEDM*, 2007 [3] S. Suthram, et. al, *IEEE IEDM*, 2007 [4] S. E. Thompson, et. al., *IEEE TED*, May 2006 [5] S. T. Pantelides et. Al., *PRB* Apr 1975 [6] P. Lawaetz *PRB* Nov 1971 [7] J. D. Wiley et. al., *PRB* Jul 1970 [8] I. Ok, et. al., *IEEE IEDM*, 2006 [9] S. Suthram, et. al., *IEEE EDL*, Jan 2007 [10] D. Chattopadhyay *PRB* May 1986 [11] B. L. Gelmont *JAP* Jan 1995 [12] S. Parthasarathy et. al., submitted to *IEEE EDL* [13] A. Hairapetian, *IEEE TED* Aug 1989 [14] G. Sun et. al., *JAP* Oct 2007 [15] S. Adachi, "GaAs and Related Materials: Bulk Semiconducting and Superlattice Properties" 1994 [16] S. Suthram, et. al., *IEEE VLSI-TSA* 2008.

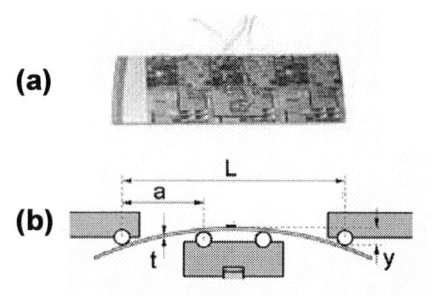

Fig 1. (a) Photograph of a bare GaAs sample glued to a Si substrate along with strain gauges for stress calibration. (b) Four-point wafer bending schematic.

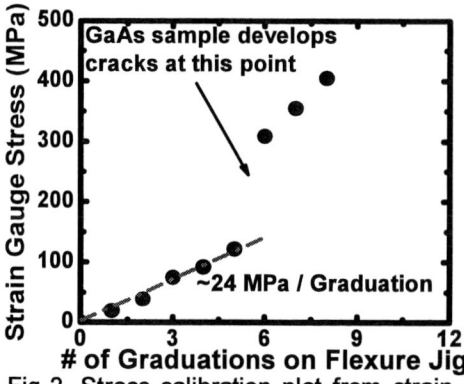

Fig 2. Stress calibration plot from strain gauge measurement.

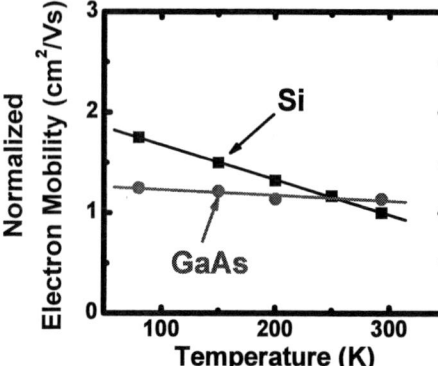

Fig 3. GaAs electron mobility temperature dependence at N_{inv} of $7 \times 10^{12}/cm^2$.

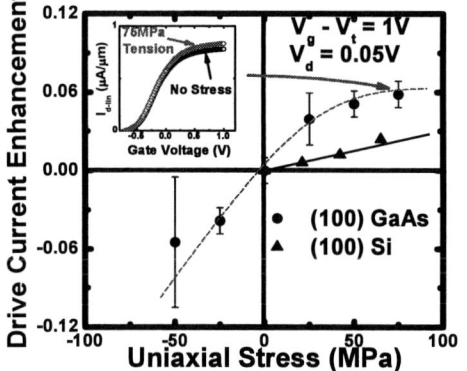

Fig 4. Drive current enhancement of (100) & (110) GaAs n-MOSFETs with uniaxial stress compared with (100) Si.

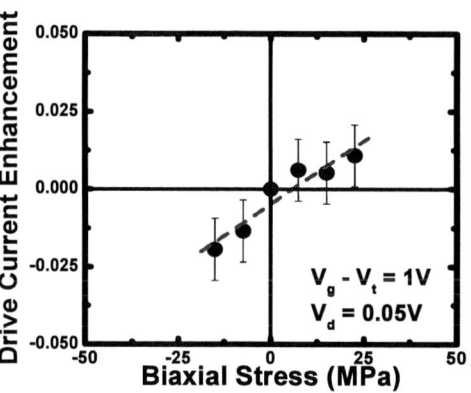

Fig 5. Drive current enhancement of (100) GaAs n-MOSFETs with biaxial stress.

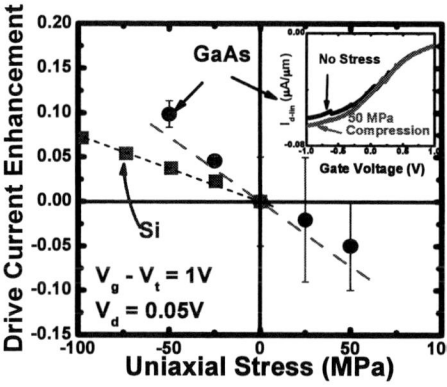

Fig 6. Drive current enhancement of (100) GaAs p-MOSFETs with uniaxial stress compared with (100) Si.

Channel Material	Longitudinal Piezoresistance Co-efficients Units: ($\times 10^{-11} Pa^{-1}$)			
	N-Type		P-type	
	Bulk	MOSFET	Bulk	MOSFET
Si	31.2	31.5	-77	-71
Ge	71.8	-	-48.1	-83.2
GaAs	-	~ 150	-	~ -150

Table 1. Longitudinal piezoresistance co-efficients of (100) surface n- and p-type MOSFETs for GaAs as compared with those for Si and Ge. GaAs shows much more stress sensitivity than both Si and Ge.

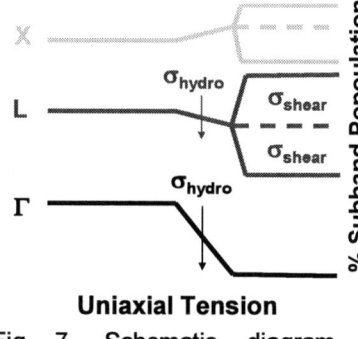

Fig 7. Schematic diagram showing valley shift under uniaxial tensile stress application.

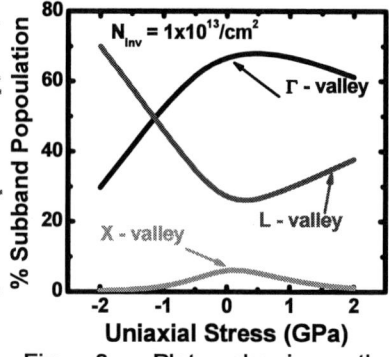

Fig 8. Plot showing the percentage population change in the (100) GaAs conduction band valleys with uniaxial stress.

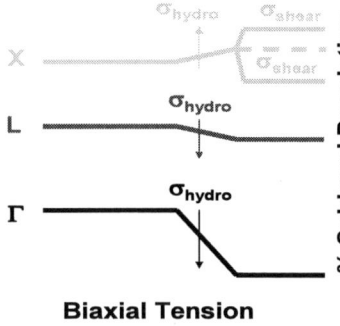

Fig 9. Schematic diagram showing valley shift under biaxial tensile stress application.

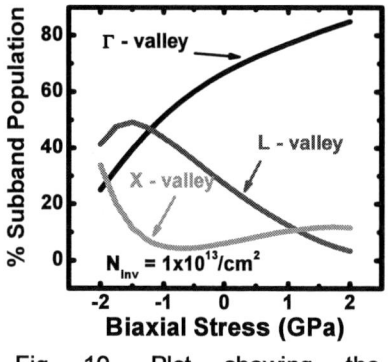

Fig 10. Plot showing the percentage population change in the (100) GaAs conduction band valleys with biaxial stress.

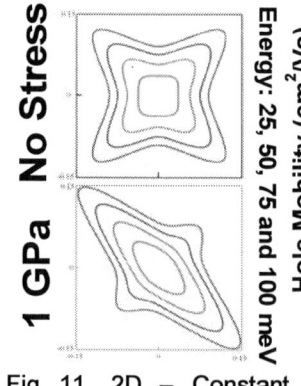

Fig 11. 2D – Constant Energy Contours of the topmost valence band in GaAs.

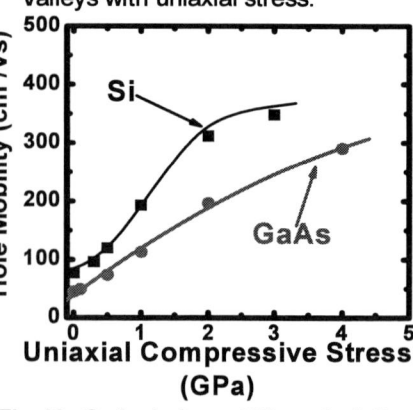

Fig 12. GaAs hole mobility calculation using k.p method as a function of uniaxial stress.

Laser-Annealed Junctions with Advanced CMOS Gate Stacks for 32nm Node: Perspectives on Device Performance and Manufacturability

C. Ortolland, T. Noda[1,3], T. Chiarella, S. Kubicek, C. Kerner, W. Vandervorst, A. Opdebeeck, C. Vrancken, N. Horiguchi, M. De Potter, M. Aoulaiche[2], E. Rosseel, S.B. Felch[4], P. Absil, R. Schreutelkamp[4], S. Biesemans and T. Hoffmann

IMEC, assignee at IMEC from [1]Matsushita, [2]IMEC and KU Leuven, Kapeldreef 75, B-3001, Leuven, Belgium
[3]Matsushita Electric Industrial Co., Ltd., 19 Nishikujo Kasuga-cho, Minami-ku, Kyoto, 601-8413, Japan
[4]Applied Materials, 974 E. Arques Avenue, Sunnyvale CA, USA
Ph: +32 16 28 76 20 - Fax: +32 16 28 17 06 - Email: claude.ortolland@imec.be

Abstract

In this paper, we report on the integration of laser-annealed junctions into a state-of-the-art high-κ/metal gate process flow. After implant optimization, we achieve excellent Lg scaling of 15/30nm over a Spike reference, for nMOS and pMOS respectively, without any performance loss. This enables to fabricate transistors with Lg_min meeting the 32nm node requirement. In addition, we highlight the implication of the metal gate integration flow ("Gate-First" vs. "Gate-Last") on the junctions design. Also, we demonstrate that a millisecond anneal only (MSA-only) process can fulfill even the stringent junction leakage requirement for low power applications. Finally, based on a combination of physical and electrical characterization, we show for the very first time that micro-uniformities specific to this diffusion-less process have a negligible electrical impact in nominal devices.

Introduction

Sub-45nm device technologies require very aggressive Ultra-Shallow Junctions (USJ). Rapid Thermal Anneal (spike – RTA) has been the main process to activate dopants up to now. Recently some papers propose to add a millisecond anneal, such as non-melt laser anneal, to obtain enhanced activation of dopants [1-2]. However, to achieve the 32nm requirements for sheet resistance Rs and junction depth Xj, ultra-low thermal budget processes (like laser-anneal) appear to be necessary (Figure 1) [3]. One caveat however is that the laser-only anneal requires pre-doped gates or metal gates. The goal of this study is to integrate this novel process with advanced gate stacks (FUSI or metal gates with high-k dielectrics) [4-5].

Junction Design with Laser-only

Integration of laser-only requires a complete redesign of the junction implant parameters as well as spacer. Figure 2 shows TCAD simulations for nMOS with different tilted Arsenic (As) implants for the Lightly Doped Drain (LDD) with laser-only annealing. Due to the nature of diffusion-less anneals, a tilted LDD implant is mandatory to obtain the correct gate overlap. To achieve the best compromise between Short Channel Effect (SCE) control and performance, we first performed the optimization with FUSI/SiON gate stack.

A. LDD

For nMOS both implant energy and tilt angle impact have been investigated. A physical gate length (Lg) reduction of 15nm at the same leakage level has been obtained with the help of laser-only anneal compared to spike anneal (Figure 3). For pMOS, Boron (B) is used for the LDD and show a large lateral straggling during low energy implantation and a significant transient enhanced diffusion (TED) during thermal annealing step. So, contrary to the nMOS with As, we can use a zero tilt implant and decrease Lg by 30nm without any performance degradation (Figure 3) upon going from spike to laser-only.

B. Co-implantation

It is well known that co-implantation is essential to reduce B diffusion and obtain correct SCE control for advanced technology nodes [6]. Germanium (Ge) is often used for pre-amorphizing implant (PAI) and Fluorine (F) co-implant can also have a PAI effect. PAI effect reduces the B channeling and F co-implant works effectively inside amorphous-Si layer [7]. Figure 4 demonstrates that a good compromise is to have F and Ge with low implant energy. However, both are needed to avoid a penalty in the minimum gate length. Note that B-only has not been plotted due to its deeper junctions and extremely high SCE degradation.

C. Laser power

The laser power is also an important parameter as we can see in Figure 5: pMOS performance is clearly improved as the power increases. Only the high power laser anneal enables similar performance as spike anneal.

Gate Dependency

We have achieved greatly improved threshold voltage roll-off with laser-only and FUSI advanced gate stack after junction optimization (Figure 6). Unfortunately the same implant condition does not produce a correct overlap with Metal Inserted Poly-Si Stack (MIPS), for nMOS (Figure 7). Figure 8 presents SEM/TEM cross-sections of different options for advanced gate stacks: FUSI [4], Replacement Gate (RPG) [8] & MIPS [5]. At the LDD and halo implant steps, only MIPS already has a metallic electrode and could have some footing at the bottom, due to the more complicated dry etch step. For "gate-last" approach (FUSI or RPG), the gate is still a conventional polysilicon at this step in the flow. Note that there is no issue for pMOS MIPS, thanks to the higher diffusivity of B. However, for As on nMOS and without a straight gate profile, this shadowing effect could be a problem. A re-optimization of process conditions results in the same performance as spike-only (Figure 9).

Junction Leakage Study

Co-implantation has become necessary for advanced technology nodes [6]. Germanium is one of the most commonly used species for this. The thickness of the amorphous silicon layer created by Ge implants is proportional to the implant energy (Figure 10). The density of end-of-range (EOR) defects, which are formed below the original amorphous/crystalline interface, is increased as a function of Ge implant energy [7&9]. Increasing the laser power also helps the defect evolution and reduces the defect density [10]. Those residual EOR defects could increase the diode leakage as a function of their location compared to the junction depletion region. When shallow Ge PAI is used, the residual EOR defects are dissolved easily [7&9], which enables to lower the leakage current to a level comparable to that of spike-only (Figure 11). The impact on gate edge junction leakage is shown on Figure 12: it is degraded as the Ge energy increases for laser-only. This leakage depends on the defect position (Figure 13) because laser-only cannot dissolve completely all of the defects, unlike spike-only. Nevertheless, we demonstrate that optimized Heavily Doped Drain (HDD) and Ge PAI conditions can achieve the junction leakage requirements for low power applications.

Reliability & Manufacturability

Figure 14 a) illustrates the laser scan process: the beam width is 11mm and the scan step size is 3.65mm. So, each part has nominally 3 exposures, but there could be some under-annealed or over-annealed regions which will create some stitching features (Figure 14 b). 10 by 10 μm^2 devices show some Vth sensitivity (about 1%) to this stitching with laser-only, contrary to spike-only (Figure 15). The total threshold voltage spread is dominated by the laser anneal for long devices, but its contribution is negligible for nominal devices (Figure 16). Finally, reliability performance comparable to spike anneal has been observed on advanced gate stacks using millisecond anneal (Figure 17) [12-13].

Conclusion

We have performed a comprehensive study of the integration of laser-only annealed junctions; compatibility with high-k / metal gate has been demonstrated and transistors with gate length required for the 32nm node have been successfully fabricated. Also the stringent junction leakage requirement for low power applications is met. Finally, no manufacturability issues have been encountered while integrating laser only annealing without absorbing layer.

Acknowledgement

The authors would like to thank CAPRES for performing micro-scale electrical measurements on laser-annealed blanket wafers.

References

[1] A. Pouydebasque *et al*, IEDM'05, p. 663
[2] T. Hoffmann *et al*, IWJT'07, p. 137
[3] T. Yamamoto *et al*, VLSI'06, p. 234
[4] A. Lauwers *et al*, IEDM'05, p. 646
[5] S. Kubicek *et al*, IEDM'07, p. 49
[6] B.J. Pawlak *et al*, Appl. Phys. Lett., **89**, 062110 (2006).
[7] T. Noda et al, IEDM'06, p. 377
[8] K. Mistry *et al*, IEDM'07, p. 247
[9] T. Noda *et al*, IEDM'07, p. 955
[10] S.B. Felch *et al*, MRS'06, Vol. 912, p.137
[11] D.H. Petersen *et al*, INSIGHT'07, p. 162
[12] P. Kalra *et al*, IEDM'07, p. 353
[13] S. Severi *et al*, IEDM'06, p. 859

978-1-4244-1802-2/08/$25.00 ©2008 IEEE

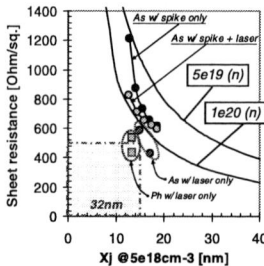

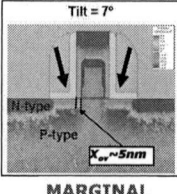

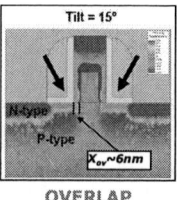

Figure 1: Sheet resistance vs. junction depth for n-type (Left) & p-type (Right) dopants with different anneal processes.

Figure 2: TCAD simulation for As implant (at 3KeV) with laser-only anneal and different tilt angles used for implantation.

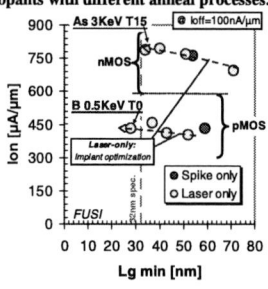

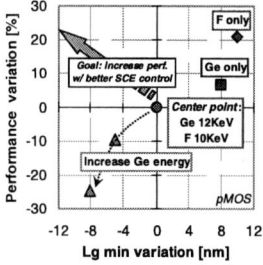

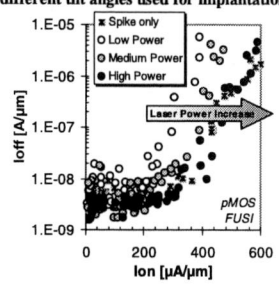

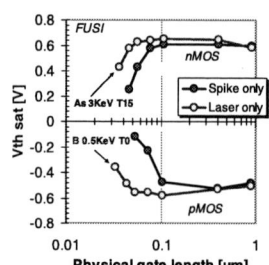

Figure 3: Ion vs. Lg min at fixed Ioff for nMOS and pMOS with different LDD implant and anneal conditions.

Figure 4: Performance variation vs. Lg min variation for pMOS with laser-only and different co-implant conditions.

Figure 5: pMOS Ion vs. Ioff with different anneal conditions (spike or laser-only).

Figure 6: Vth sat vs. gate length for nMOS and pMOS with spike or laser-only anneal and FUSI gates.

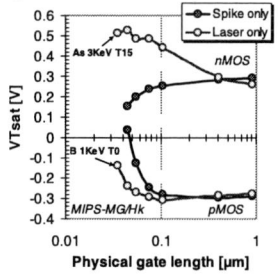

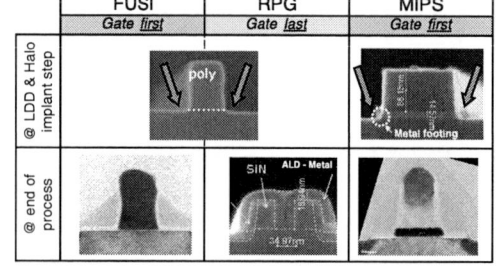

Figure 7: Vth sat vs. gate length for nMOS and pMOS with spike or laser-only anneal and Hk/MIPS-MG gate stack.

Figure 8: SEM or TEM cross-sections of different advanced gate stack process (FUSI, RPG & MIPS) at the LDD implant step and at the end of the process.

Figure 9: nMOS Ion vs. Ioff with different anneal conditions (spike or laser-only) and Hk/MIPS-MG gate stack.

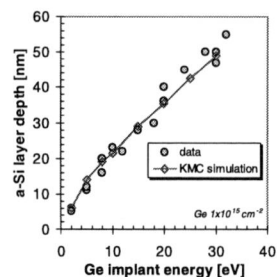

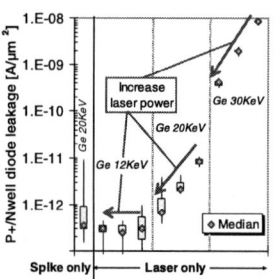

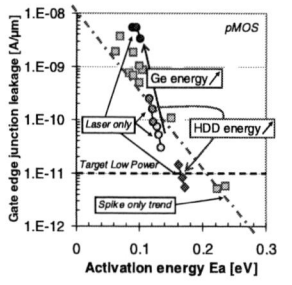

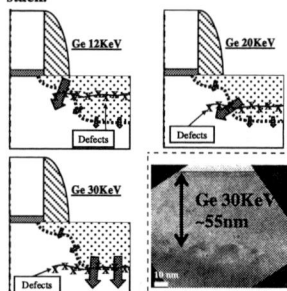

Figure 10: Amorphous silicon layer thickness vs. Ge implant energy [9].

Figure 11: P+/Nwell square diode leakage with different co-implantations and anneal conditions.

Figure 12: pMOS gate edge junction leakage vs. activation energy for spike or laser-only.

Figure 13: Schematic representation of the position of the defects. Bottom right: TEM after laser anneal.

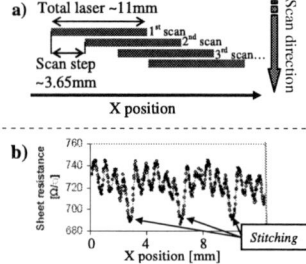

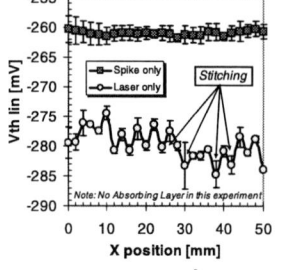

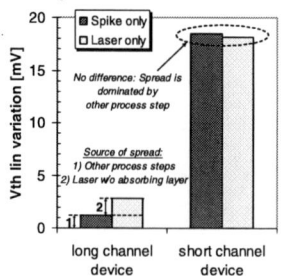

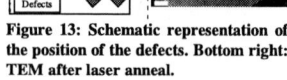

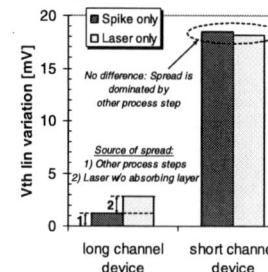

Figure 14: a) Schematic representations of laser scan. b) Sheet resistance measurement on silicon after laser-only annealing [11].

Figure 15: pMOS (10x10μm²) Vth lin vs. X position on wafer for spike or laser-only anneal.

Figure 16: Sources of Vth lin variation for long or short channel devices with spike or laser-only anneal.

Figure 17: Gate overdrive for 10 years NBTI lifetime extrapolated on Hk/MG gate stacks for different anneal conditions.

146

Advanced Junction Profile Design Scheme by Low-temperature Millisecond Annealing and Co-implant for High Performance CMOS

K. Ikeda, T. Miyashita, T. Kubo*, T. Yamamoto, T. Sukegawa*, K. Okabe*, H. Ohta, Y. S. Kim, H. Nagai*, M. Nishikawa*,
Y. Shimamune, A. Hatada*, Y. Hayami, K. Ohkoshi*, N. Tamura, K. Sukegawa, H. Kurata, S. Satoh, M. Kase* and T. Sugii
Fujitsu Laboratories Ltd. and *Fujitsu Limited, 50 Fuchigami, Akiruno Tokyo 197-0833, Japan.
Phone: +81-42-532-1249, FAX: +81-42-532-2513, E-mail: ikeda.keiji@jp.fujitsu.com

Abstract

We found that the relatively low temperature millisecond annealing at S/D activation for nFET is enhanced the co-implanted halo activation regardless of sequence of MSA and spike-RTA. Tilt-and-twist extension implantation technique with millisecond extension annealing for pFET was also performed to reduce the parasitic resistance. By combining these technique, an aggressively scaled high-performance bulk CMOS transistors with world competitive nFET and pFET drive currents of 1282/835 $\mu A/\mu m$ at $100nA/\mu m$ off-current at Vd = 1 V and Lg = 34 nm respectively, were developed with a conventional poly/SiON gate stack. The developed CMOS transistors not only have high-performance but also manufacturing friendly and cost-effective compared with metal/high-k stack devices.

Introduction

Millisecond annealing (MSA) such as laser-spike annealing (LSA) and flash lamp annealing (FLA) is indispensable technology on the current high performance CMOS and many papers have been reported[1-4]. However, these papers were intended to examine extension, deep S/D activation and reduction of gate depletion in a high temperature range and the effect on halo activation was not investigated so much. Consequently, we have investigated the activation characteristics of halo impurities by MSA, especially in a low temperature range, and found that the low temperature MSA at S/D activation is more effective to assist the activation of halo impurities with co-implant and this phenomena is common to different sequences of MSA and spike RTA. This low-temperature annealing is also effective to pFET with eSiGe S/D to avoid the strain relaxation. Moreover, we have been proposed that extension annealing method by laser spike annealing (LSA) with co-implant for pFET and it can drastically reduce extension sheet resistance with low sensitivity to LSA temperature[4]. By using this less-diffusivity activation technology, as-implanted extension profile design is of increasing importance to improve the device performance with overlap length control and parasitic resistance reduction. Thus, we have performed a tilt-and-twist extension ion-implantation with a fluorine co-implant to achieve a shallow and highly concentrated extension profile with an offset spacer which was optimized for nFET. At optimized extension implantation condition, 7 % Ion improvement was achieved with this technique.

Device fabrication

Figure 1 shows the cross-sectional TEM micrographs of our 45nm-node high performance CMOS devices. As a performance booster, we applied dual stress liner and Σ-shaped eSiGe S/D. Figure 2 shows the process flow used in this work. By using low-temperature SiGe growth technique[5], eSiGe growth was performed after extension/halo implantation and extension annealing. As S/D activation for nFET, we applied the 2-type flows, that is Pre-S/D MSA (MSA prior to spike-RTA) and Post-S/D MSA (MSA after spike-RTA).

Results and Discussion

Impact of low temperature S/D-MSA on nFET halo

Figure 3 shows Lmin-Ion characteristics (where Ion is at Ioff=100nA/μm and Lg is defined as gate length at Ioff=100nA/μm) of nFET with and without high temperature Pre-S/D MSA. In this temperature range, Lmin was degraded in the Lmin-Ion trade-off line. However, at low-temperature MSA Vth-rolloff characteristics were improved(Fig.4). Figure 5 shows the MSA temperature dependence of inversion oxide thickness (Teff) and overlap capacitance (Cov) for nFET in Pre-S/D MSA flow. To achieve the performance gain by Teff reduction, it is essential to have a high temperature MSA. However, Cov is more sensitive to MSA temperature compared with Teff and its control is necessary to maintain the Vth roll-off characteristics. Figure 6 shows the relation between junction capacitance of gate edge (Cjg) and Cov. By decreasing the MSA temperature, Cov was decreased while maintaining Cjg. This means that the DIBL was improved by lower temperature MSA condition.

To reveal this change of Cov, we investigated the MSA temperature dependence of halo implant (with and without co-implant) sheet resistance by Pre-S/D MSA and Post-S/D MSA as shown in Fig. 7. In both flows, it is found that the low temperature MSA is more effective in activating the halo impurities with co-implant. Figure 10 shows the Lmin-Ion characteristics of nFET fabricated by Pre-S/D MSA. By lowering the MSA temperature, Lmin-Ion characteristics split from the trade-off characteristics caused by assist of halo activation with MSA. Figure 11 (a) shows the Ion-Ioff performance of nFET fabricated with optimized MSA condition. Ion=1282$\mu A/\mu m$ at Ioff=100nA was achieved with a cost-effective conventional poly-Si/SiON gate stack. Even though the combination of Post-S/D MSA and co-implant, halo activation can improved in a low temperature range (Fig. 11(b)). In case of Post-S/D MSA, reduction of extension sheet resistance was small compared with Pre-S/D MSA(Fig.8). The Teff decrease caused by activation of gate dopant is attributed to the main cause of performance improvement(Fig.9). And it was also confirmed that the MSA at S/D activation for nFET was not exacerbate pFET (Fig.12) due to the sufficiently low temperature annealing to avoid the eSiGe strain relaxation.

Sophisticated extension profile design for pFET

Figure 13 shows the SIMS profiles of boron and co-implanted fluorine profiles of pFET extension with comparisons between LSA and FLA. It was found that the high activation level with reduced TED of the boron profile can be achieved by the both MSA method with same ion-implantation (I.I.) conditions. This means that as-implanted extension profile is still important to performance improvement of pFET in this less-diffusivity condition. Thus, we applied a tilt-and-twist extension I.I. technique as shown in Fig.14. With this configuration, overlapped shallow (A) and a highly concentrated (B) (twice as high as (A)) extension profile are achieved and high overlap controllability and reduction of parasitic resistance can be achieved. A 10 % reduction of sheet resistance (Fig. 15) and 36 Ω μm reduction of Rtotal for pFET (Fig. 16) were observed compared with the control at same dosage for area (A), respectively. Thanks to this parasitic resistance reduction, Ion was improved by 7 % with same Cov (Fig. 17). Figure 18 shows the Lmin-Ion characteristics of pFET fabricated by extension MSA with co-implant. Lmin-Ion characteristics are also improved by this technique. Figure 19 shows the Ion-Ioff performance of pFET fabricated by optimized tilt-and-twist implant with fluorine co-implanted MSA. Ion=835$\mu A/\mu m$ at Ioff=100nA was achieved.

Well controlled sub-threshold slopes of 94mV/dec and 97mV/dec were obtained for nFET and pFET, respectively (Fig. 20). Fig. 21 shows Id-Vd characteristics of 34nm gate nFET and pFET. The highest Ion of 1282 $\mu A/\mu m$ for nFET and 852 $\mu A/\mu m$ for pFET at Vd =1V were obtained. Finally, the device performance was compared with recently published conventional and metal/high-k gate stack devices which fabricated on <110>/(100) bulk substrate with gate first process (Fig.22). The results of this work are the best among conventional gate stack devices and competitive with metal/high-k gate stack.

Conclusions

We present high-performance bulk CMOS transistors with nFET and pFET drive currents of 1282/835 $\mu A/\mu m$ at $100nA/\mu m$ off-current at Vd=1V and Lg=34nm, respectively. A low temperature MSA with co-implant for nFET to enhance the halo activation and tilt-and-twist extension implantation technique with MSA and co-implant for pFET improved the device performance.

References

[1] A. Shima et al. : *VLSI Tech.*, pp. 174-175. 2004. [2] Z. Luo et al. : *VSLI Tech.*, pp. 16-17, 2007. [3] T. Sanuki et al. : *IEDM Tech. Dig.*, pp. 281-284, 2007. [4] T. Yamamoto et al. : *IEDM Tech. Dig.*, pp. 143-146, 2007. [5] Y. Shimamune et al. : *VLSI Tech.*, pp. 116-117, 2007. [6] H.T. Huang et al. : *IEDM Tech. Dig.*, pp. 285-288, 2007. [7] K. Cheng et al. : *IEDM Tech Dig.*, pp.243-246, 2007. [8] T. Miyashita et al.: *IEDM Tech. Dig.*, pp. 251-254, 2007.

978-1-4244-1802-2/08/$25.00 ©2008 IEEE

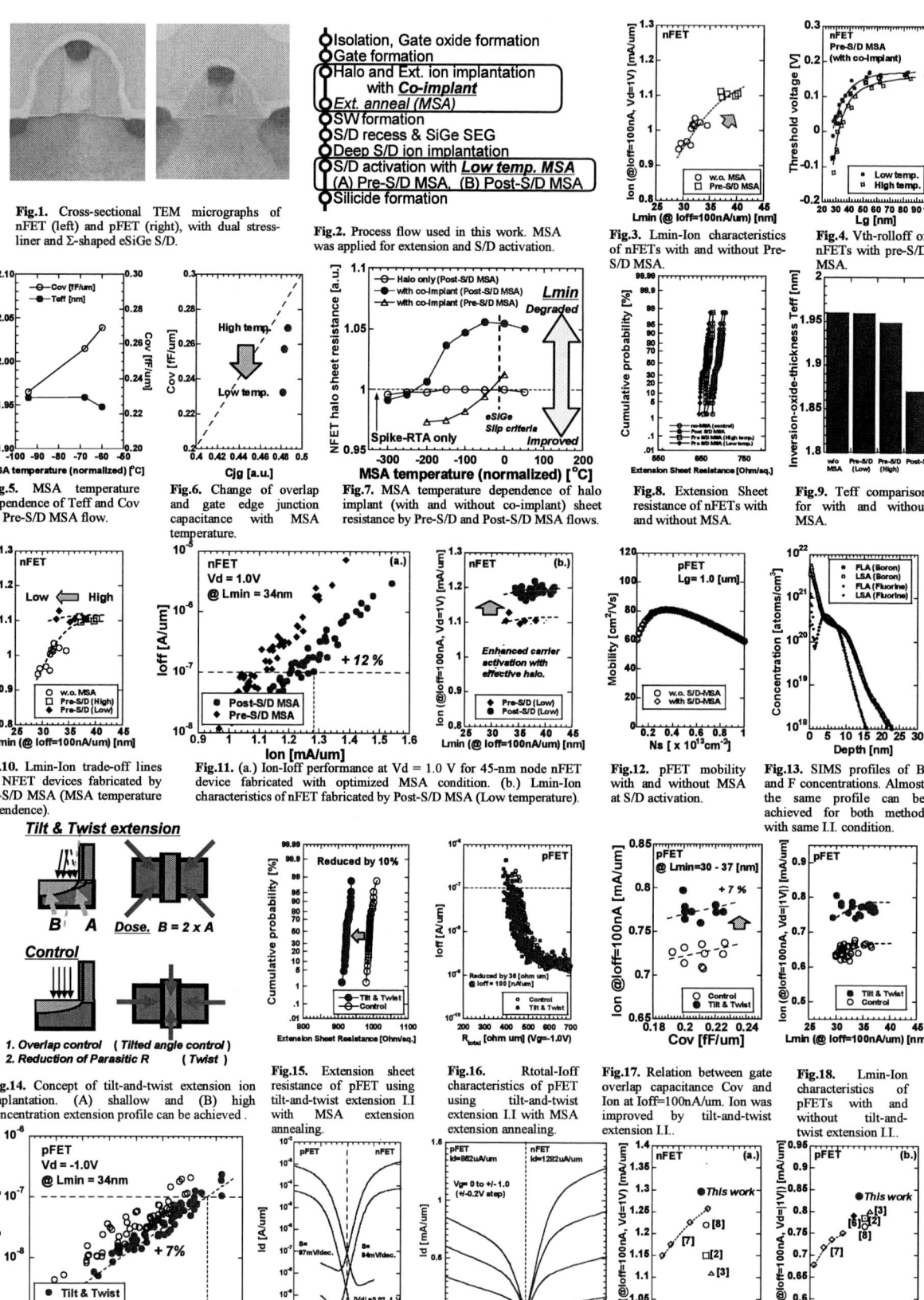

Fig.1. Cross-sectional TEM micrographs of nFET (left) and pFET (right), with dual stress-liner and Σ-shaped eSiGe S/D.

Fig.2. Process flow used in this work. MSA was applied for extension and S/D activation.

Fig.3. Lmin-Ion characteristics of nFETs with and without Pre-S/D MSA.

Fig.4. Vth-rolloff of nFETs with pre-S/D MSA.

Fig.5. MSA temperature dependence of Teff and Cov by Pre-S/D MSA flow.

Fig.6. Change of overlap and gate edge junction capacitance with MSA temperature.

Fig.7. MSA temperature dependence of halo implant (with and without co-implant) sheet resistance by Pre-S/D and Post-S/D MSA flows.

Fig.8. Extension Sheet resistance of nFETs with and without MSA.

Fig.9. Teff comparison for with and without MSA.

Fig.10. Lmin-Ion trade-off lines for NFET devices fabricated by Pre-S/D MSA (MSA temperature dependence).

Fig.11. (a.) Ion-Ioff performance at Vd = 1.0 V for 45-nm node nFET device fabricated with optimized MSA condition. (b.) Lmin-Ion characteristics of nFET fabricated by Post-S/D MSA (Low temperature).

Fig.12. pFET mobility with and without MSA at S/D activation.

Fig.13. SIMS profiles of B and F concentrations. Almost the same profile can be achieved for both method with same I.I. condition.

Fig.14. Concept of tilt-and-twist extension ion implantation. (A) shallow and (B) high concentration extension profile can be achieved .

Fig.15. Extension sheet resistance of pFET using tilt-and-twist extension I.I with MSA extension annealing.

Fig.16. Rtotal-Ioff characteristics of pFET using tilt-and-twist extension I.I with MSA extension annealing.

Fig.17. Relation between gate overlap capacitance Cov and Ion at Ioff=100nA/um. Ion was improved by tilt-and-twist extension I.I..

Fig.18. Lmin-Ion characteristics of pFETs with and without tilt-and-twist extension I.I.

Fig.19. Ion-Ioff performance at Vd=|1V| for 45-nm node pFET device fabricated with optimized MSA and tilt-and-twist extension I.I.

Fig.20. Id-Vg characteristics of 34-nm gate length devices.

Fig.21. Id-Vd characteristics of 34-nm gate length devices.

Fig.22. State of the art performance of (a.) nFETs and (b.) pFETs demonstrated by this work.

148

Low V_t Gate-First Al/TaN/[Ir$_3$Si-HfSi$_{2-x}$]/HfLaON CMOS Using Simple Laser Annealing/Reflection

C. C. Liao[a], Albert Chin[a,b], N. C. Su[c], M.-F. Li[d,e], and S. J. Wang[c]

[a]E. E. Dept., Nat'l Chiao-Tung Univ., Hsinchu, Taiwan, ROC; [b]Nano-Electronics Consortium of Taiwan achin@cc.nctu.edu.tw

[c]Inst. of Microelectronics, Dept. of Electronics Eng., National Cheng Kung Univ., Tainan, Taiwan ROC

[d]SNDL, ECE Dept., National University of Singapore, Singapore; [e]Microelectronics Dept., Fudan University, Shanghai, China

Abstract

We report low V_t Al/TaN/[Ir$_3$Si-HfSi$_{2-x}$]/HfLaON CMOS using simple laser annealing/reflection with self-aligned and gate-first process compatible with current VLSI. At 1.05 nm EOT, good ϕ_{m-eff} of 5.04 and 4.24 eV, low V_t of -0.16 and 0.13 V, high mobility of 85 and 209 cm^2/Vs, and small 85°C BTI ≤40 mV (10 MV/cm, 1 hr) are measured for p- and n-MOS.

Introduction

The toughest challenge for metal-gate/high-κ CMOS is to lower the undesired high V_t [1]-[7]. This is especially hard for p-MOS, since only Ir and Pt in the Periodic Table have the needed high effective work-function (ϕ_{m-eff}) gate >5.2 eV [7]. Previously we showed the possible mechanism for high V_t related to the interface reaction and inter-diffusion of HfO$_2$ and Si-channel during high temperature RTA [6], originated from close bond enthalpy of HfO$_2$ (802 kJ/mol) and SiO$_2$ (800 kJ/mol). These reactions then form non-stoichiometric oxides (HfO$_{2-x}$ and SiO$_x$) with dangling bonds and charged oxygen vacancy to cause V_{FB} roll-off. Since these reactions follow basic chemistry of Arrhenius temperature dependence, the low temperature processing will be the solution. This was confirmed by the low $|V_t|$ <0.1 V in HfLaO CMOS using <900°C solid-phase diffusion formed ultra-shallow junction (USJ) [6]. The vital USJ can also be formed by conventional ion-implantation with ultra-fast laser annealing, but the challenge is to lower V_{FB} roll-off by high temperature under gate dielectric. In this paper, we report the using simpler ion implantation and laser annealing/reflection to achieve low V_t in high-κ CMOS. At 1.05 nm EOT, the self-aligned and gate-first p-and n-MOS showed proper ϕ_{m-eff} of 5.04 and 4.24 eV, low V_t of -0.16 and 0.13 V, high mobility of 85 and 209 cm^2/Vs and good 85°C BTI reliability. This was achieved using laser annealing on ion-implanted source-drain and laser-reflection by Al-covered gate electrode. Here Al reflects as high as 91% of the KrF excimer laser power: this lowers the temperature under the gate and decreases the high-κ/Si interface reaction exponentially. The laser-annealed/reflected low V_t CMOS provides a simpler and lower cost process to Intel's gate dielectric first, poly-Si removal and filling gate electrode last process [1]. These device data compare well with other reports in Table 1 [2]-[7], with low V_t, small EOT, self-aligned and gate-first process compatible with VLSI line.

Experimental Procedure

Metal-gate/HfLaON CMOSFETs were made by depositing HfLaO using PVD, O$_2$ DPA, surface nitridation to form HfLaON [7], depositing amorphous-Si with various thickness (0~5 nm), Ir or Hf by PVD, 300 nm TaN, and 100 nm Al deposition. After gate patterning, self-aligned BF$_2^+$ or As$^+$ was implanted at 10 keV and 5×10^{15} cm^{-2} dose, followed by scanned KrF laser anneal (248 nm, 25 ns pulse). The top Al was selectively wet-etched with another 750~850°C RTA. Then source-drain metal contacts were added to form CMOS.

Results and Discussion

Fig. 1 shows the sheet resistance (R_s) for 10 keV BF$_2^+$ or As$^+$ implanted Si after different laser annealing condition. For both BF$_2^+$ and As$^+$ implantation, the R_s decreases rapidly

with increasing laser fluence (energy/area) to 0.36 J/cm^2 and fast levels off. This value is close to previous reported 0.32 J/cm^2 for 308 nm XeCl excimer laser anneal [8] that is due to the melt of very thin Si (<50 nm) and re-crystallization. This is useful for next generation USJ, but the high laser energy is also absorbed by TaN-covered gate to cause unwanted V_{FB} roll-off shown in the C-V and V_{FB}-EOT plots of Figs. 2~3.

To address this issue, we added a thin Al reflection layer on top of TaN. Fig. 4 shows the optical reflectivity (R) vs. light wavelength. The R increases with Al layer thickness and reaches high R of 87% and 91% at 30 and 100 nm, even at short 248 nm KrF laser. Using top Al laser-reflective gate, proper ϕ_{m-eff} of 5.04 and 4.24 eV are obtained with much improved V_{FB} roll-off compared with conventional top TaN gate (Figs. 2-3). Owing to the low 660°C melting temperature of Al, the laser energy should still be kept <0.55 J/cm^2 due to the distorted device pattern in Fig. 5. The low laser energy is also crucial to lower the junction edge leakage current that is poorer at higher laser fluence [8]. An EOT of 1.05 nm is obtained from quantum-mechanical C-V calculation in Fig. 2 with low leakage current of 6.7×10^{-4} and 5.4×10^{-4} A/cm^2 at ± 1V in Fig. 6. The ϕ_{m-eff} are the best reported data for CMOS at ~1.0 nm EOT; suggesting the low thermal budget under the gate is vitally important for metal-gate/high-κ CMOS. This is consistent with our previous very low V_t CMOS using low temperature solid-phase diffused USJ [6] and Intel's device with high-κ first and gate-electrode last process [1].

Using laser annealing on source-drain and laser reflection at gate in Fig. 7, good junction edge leakage comparable with 1000°C RTA is obtained shown in Fig. 8. The I_d-V_d, I_d-V_g and μ_{eff}-E of CMOS are shown in Figs. 9-11. Besides good transistor characteristics, low V_t of -0.16 and 0.13 V and high mobility of 85 and 209 cm^2/Vs are measured. The gate reliability is shown in the BTI data of Fig. 12, where small $|\Delta V_t|$ ≤40 mV occurs for CMOS stressed at 10 MV/cm and 85°C for 1 hr. Table 1 compares various metal-gate/high-κ CMOS data [2]-[7]. The merits of self-aligned and gate-first Al/TaN/[Ir$_3$Si-HfSi$_{2-x}$]/HfLaON CMOS with laser annealed shallow junction are proper ϕ_{m-eff} of 5.04 and 4.24 eV, low V_t of -0.16 and 0.13 V, high mobility of 85 and 209 cm^2/Vs, and small BTI ≤40 mV (85°C, 10 MV/cm & 1 hr). The V_t values are also lower than the reported 0.3~0.4 V and -0.35~-0.45 V $V_{t,lin}$ data using high-κ first and metal-gate last process [1]. These results are comparable with or better than the best reported data for self-aligned and gate-first metal-gate/high-κ CMOS, with small 1.05 nm EOT and using simpler process.

Conclusions

Using simple and low cost process, good 5.04 and 4.24 eV $\phi_{m,eff}$, small leakage and low V_t are obtained at 1.05 nm EOT.

References

1. K. Mistry et al, IEDM Tech. Dig., 2007, pp. 247-250.
2. V. S. Chang et al, IEDM Tech. Dig., 2007, pp. 535-538.
3. T. Hoffmann et al, IEDM Tech. Dig., 2006, pp. 269-272.
4. M. Kadoshima et al, IEDM Tech. Dig., 2007, pp. 531-534.
5. H. Takahashi et al, IEDM Tech. Dig., 2004, pp. 91-94.
6. C. F. Cheng et al, IEDM Tech. Dig., 2007, pp.333-336.
7. C. H. Wu et al, IEDM Tech. Dig., 2006, pp. 617-620.
8. B. Yu et al, IEDM Tech. Dig., 1999, pp. 509-512.

978-1-4244-1802-2/08/$25.00 ©2008 IEEE

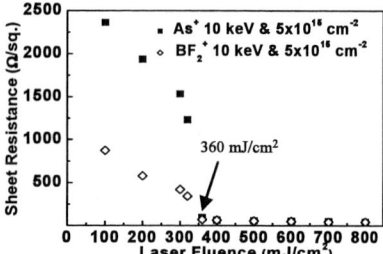

Fig. 1. Measured R_s of As$^+$ and BF$_2^+$ implantations after scanned KrF laser annealing. Sharp decrease of R_s is obtained at 360 mJ/cm^2 fluence (energy/area).

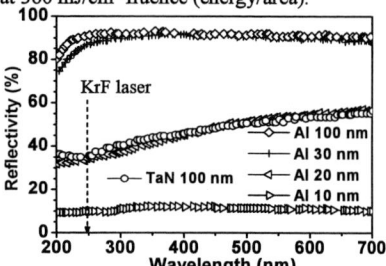

Fig. 2. C-V characteristics of n- and p-MOS capacitors after laser annealing with and without top Al layer. The V_{FB} roll-off is found without using top Al on gate.

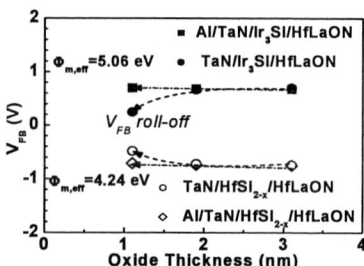

Fig. 3. V_{FB}-EOT plot of laser-annealed n- and p-MOS capacitors with and without top Al layer. Much improved V_{FB} roll-off is reached using simple Al coverage on gate.

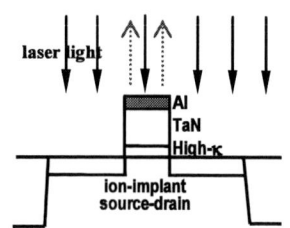

Fig. 4. Reflectivity (R) vs. light wavelength. High R of 91% are obtained for 100 nm thick Al but only 35% for TaN gate.

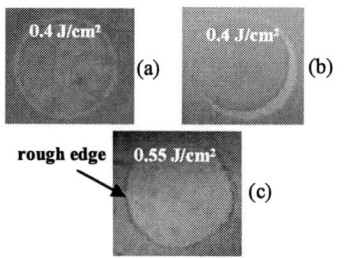

Fig 5. Device photos (a) with and (b) without top Al after 0.4 J/cm^2 laser anneal; (c) with Al but at higher 0.55 J/cm^2 fluence.

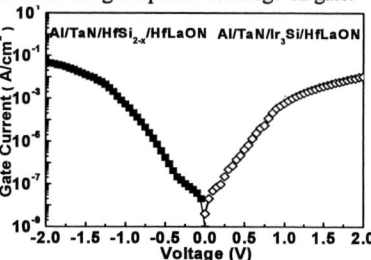

Fig. 6. J-V of Al/TaN/Ir$_3$Si/HfLaON and Al /TaN/HfSi$_{2-x}$/HfLaON p- & n-MOS devices after laser annealing and reflection at gate.

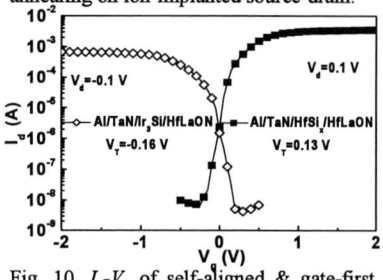

Fig. 7. Schematic diagram to show the laser reflection on Al-covered gate, during laser annealing on ion-implanted source-drain.

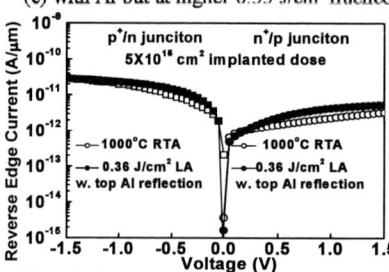

Fig. 8 Junction edge leakage current of laser annealing at 0.36 J/cm^2 and 1000°C RTA.

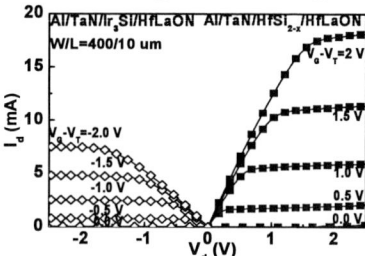

Fig. 9 I_d-V_d of self-aligned & gate-first Al/TaN/[Ir$_3$Si-HfSi$_{2-x}$]/HfLaON p- and n-MOSFETs after laser annealing/reflection.

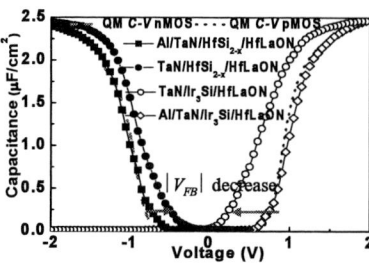

Fig. 10. I_d-V_g of self-aligned & gate-first Al/TaN/[Ir$_3$Si-HfSi$_{2-x}$]/HfLaON p- and n-MOSFETs after laser annealing on source-drain and laser reflection at gate.

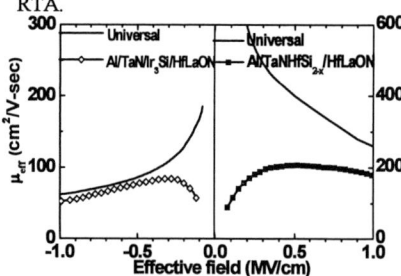

Fig. 11. Hole and electron mobility of self-aligned and gate-first p- and n-MOSFETs after laser annealing on source-drain and laser reflection at gate.

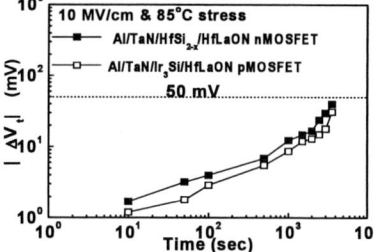

Fig. 12. The ΔV_t shift for laser-annealed Al/TaN/[Ir$_3$Si-HfSi$_{2-x}$]/HfLaON p- and n-MOSFETs stressed at 85°C and 10 MV/cm for 1 hour.

High-κ	Metal-Gate, p/n	EOT (nm)	$\phi_{m\text{-}eff}$ (eV), p/n	V_t (V), p/n	Process	Mobility (cm^2/Vs), p/n
This work HfLaON	**Al/TaN-covered Ir$_3$Si / HfSi$_{2-x}$**	**1.05**	**5.04 / 4.24**	**-0.16 / 0.13**	**Laser Annealing/ Laser Reflection**	**85 / 209**
Dy$_2$O$_3$/HfO$_2$ [2]	TaC$_x$N$_y$/TaC$_x$	1.4	4.9 / 4.2	-0.36 / 0.23	1050°C RTA	~80 / -
HfSiON [3]	Ni$_{31}$Si$_{12}$ / NiSi	1.5	~4.8 / ~4.5	-0.4 / 0.5	Low Temp. FUSI	~70 / ~240
HfSi(Al)ON [4]	TiAlN / TaSiN	1.0	4.8 / 4.44	~-0.5 / ~-0.5	1000°C RTA	~50 / ~220
HfSiON [5]	Ni$_3$Si / NiSi$_2$	1.7	4.8 / 4.4	-0.69 / 0.47	Low Temp. FUSI	65 / 230
HfLaO [6]	Ir / Hf	1.2	5.3 / 4.1	+0.05 / 0.03	<900°C SPD	90 / 243
HfLaON [7]	Ir$_3$Si / TaN	1.6	5.08 / 4.28	-0.1 / 0.18	1000°C RTA	84 / 217

Table 1. Comparison of device integrity data for various metal-gate/high-κ n- and p-MOSFETs.

Successful Enhancement of Metal Segregation at NiSi/Si Junction through Pre-amorphization Technique

Yoshifumi Nishi, Yoshinori Tsuchiya, Atsuhiro Kinoshita, Akira Hokazono[*] and Junji Koga

Advanced LSI Technology Laboratory, Corporate R&D Center, Toshiba Corporation
[*]Center for Semiconductor Research and Development, Semiconductor Company, Toshiba Corporation
8 Shinsugita-cho, Isogo-ku, Yokohama 235-8522, Japan Tel +81-45-776-5961 Fax +81-45-776-4113

Abstract

A new technique to enhance the metal segregation at NiSi/Si interface for reducing contact resistance in source/drain electrodes is proposed. It is demonstrated that metal segregation at the junction of pre-amorphized NiSi/Si using ion-implantation leads to reduction of Schottky barrier height by >0.2eV. This modulation width is far beyond the previous metal segregation technique [1] and allows 90% reduction of contact resistance in source/drain junctions.

Introduction

Reducing contact resistances at NiSi/Si junction in source/drain (S/D) electrodes is one of the difficult challenges for further scaling in CMOS. New silicide materials have been investigated to lower the contact resistance in nMOS [2-7] and pMOS [8,9]. Most of these technologies, however, achieve low Schottky barrier height (Φ_B) at the sacrifice of interface morphology, process complexity and cost. We proposed a novel metal segregation technique to lower Φ_B at NiSi/Si junction without degrading the interfacial morphology, which is simply incorporated into the present LSI process. Φ_B modulation width ($\Delta\Phi_B$) of 0.1eV from that of control NiSi has been demonstrated using this technique [1]. In this paper, we propose a new technique to improve metal segregation technique to achieve $\Delta\Phi_B > 0.1$eV and demonstrate short channel FET performance.

Strategy for Further Φ_B Modulation

When heat treatment is performed at a metal/NiSi/Si junction, the metal on the surface diffuses through the grain boundaries of NiSi and segregates at NiSi/Si interface. The segregated metal modulates Φ_B of the junction and $\Delta\Phi_B$ is linear to its concentration at the interface [1]. Hence $\Delta\Phi_B$ reaches its saturation after long time annealing, as shown Fig. 1, due to the solubility limit of the metal at the interface. However, it is considered that the efficiency of segregation can be enhanced and hence the saturation can be raised when the grain size of NiSi film is small as illustrated in Fig. 2, because the segregation occurs through grain boundary diffusion. In fact, when yttrium (Y) is employed as segregation metal for large- and small-grain NiSi/Si diodes (shown in Fig. 3), the modulation width of the reverse current in the I-V characteristics is larger in the small-grain case as shown in Fig. 4, suggesting that $\Delta\Phi_B$ is larger when the grain is smaller. For the enhancement of metal segregation, we propose a new technique in which ion-implantation for pre-amorphization is performed to make NiSi grains smaller.

Experimental

We performed two methods as shown in Fig. 5, NiSi pre-amorphization ion-implantation (NiSi PAI) process and Si pre-amorphization ion-implantation (Si PAI) process. After silicidation, Y film of 5nm thickness is deposited on the NiSi and annealed at 450°C for 60min in N_2 ambient for Y segregation in both processes. We selected Ge and Xe as implantation ions from group IV and noble gas group that are neutral in Si substrate. Fig. 6 compares $\Delta\Phi_B$ of NiSi and Si PAI processes for Ge and Xe. It can be seen that $\Delta\Phi_B$ is the largest in Ge-implanted NiSi PAI case. The junction of Xe-implanted NiSi PAI is leaky, indicating that NiSi/Si junction is damaged by heavy Xe ions. Therefore, we investigate NiSi PAI process using Ge implantation intensively. Simulated Ge ion profiles implanted into NiSi film of 20nm thickness on Si substrate are shown in Fig. 7. It can be seen that at the implantation energy of 10keV, most of the implanted ions are stopped inside the NiSi, indicating that the implanted ion does not affect the NiSi/Si junction. When the implantation energy is raised to 20keV, Ge ions penetrate into Si substrate and can influence the junction. We performed Ge implantation into NiSi 20nm/p-Si diodes at 10, 15 and 20keV. The doses are selected such that the Ge concentration at

the NiSi interface is around 10^{20} cm^{-3} (1, 2 and 5 x10^{15}cm^{-2}). We also performed Si implantation at 10keV and 1x10^{15} cm^{-2}.

Results and Discussions

Fig. 8 shows the I-V characteristics for the Y-segregated NiSi/p-Si diodes of Ge-implanted NiSi PAI processes at several implantation doses. The reverse current is more suppressed for the larger dose, indicating that $\Delta\Phi_B$ is more enhanced with the increase of Ge implantation dose as shown in Fig. 9 (Note that the reverse current in Fig. 8 reflects Φ_B for holes). Similarly, $\Delta\Phi_B$ becomes larger as Ge implantation energy is higher as shown in Fig. 10. It can be seen that $\Delta\Phi_B$ from Φ_B of control NiSi exceeds 0.2eV at 20keV. Reverse current enhancement, the ratio of the reverse current to that of non-PAI case is compared in Fig. 11 for the Si implanted case with the Ge cases (Fig. 10). The reverse current enhancement for Si implanted case is similar to that of the Ge 15keV case, which is consistent with the simulation results in Fig. 7 showing that the Si profile in NiSi is similar to that for Ge at 15keV. This fact implies that the enhancement of $\Delta\Phi_B$ arises from the mechanical effect of the ion-implantation, and the material properties of the respective ions are not influential.

TEM images of Y-segregated NiSi/Si junctions for Ge implanted NiSi PAI processes are shown in Fig. 12. The interfacial morphologies at NiSi/Si are not degraded and similar to that of control NiSi/Si interface. It was also confirmed that the resistivity of NiSi film is not degraded in this technique (not shown).

In order to investigate the origin of $\Delta\Phi_B$ enhancement, Y concentration at NiSi/Si interface is measured with SIMS analysis. Fig. 13 shows $\Delta\Phi_B$ as a function of Y concentration. Y concentrations in this work are larger than those of previous results [1], corresponding to $\Delta\Phi_B$ enhancement due to NiSi PAI. $\Delta\Phi_B$ of Ge 20keV case is larger than linear extrapolation of the previous work, implying that additional effect of PAI to Y segregation contributes to $\Delta\Phi_B$ enhancement.

Finally, 50nm gate length FET operations are demonstrated in Fig. 14. Although the channel profile is not optimized and the device suffers from short channel effect, it is suggested that I_d-V_d characteristics of Y-segregated NiSi PAI nMOSFETs are improved for larger $\Delta\Phi_B$.

Conclusion

It was demonstrated that metal segregation at NiSi/Si can be enhanced by NiSi PAI process. Increasing either the dose or energy for PAI can enhance $\Delta\Phi_B$. In the case of Y-segregated NiSi/Si junction with Ge implantation at 20keV and 1x10^{15}cm^{-2}, $\Delta\Phi_B$ from control NiSi exceeds 0.2eV, which corresponds to 90% reduction of contact resistance in junctions with practical dopant concentration (~10^{20}cm^{-3}). Therefore, it is concluded that metal segregation technique with pre-amorphized NiSi is promising for source/drain junction in 32-nm generation and beyond.

Acknowledgement

The authors thank to Shigeru Kawanaka, Nobutoshi Aoki, Takeshi Sonehara, Kazuya Ouchi, Haruko Akutsu, Kayo Nomura, Kyoichi Suguro, Yoshiaki Toyoshima, and Akira Nishiyama, Toshiba Corporation for supporting this study.

References

[1] Y. Nishi, et al., IEDM2007 p.135. [2] S. Zhu, EDL **25**, 565 (2004). [3] R.T.P. Lee, EDL **28**, 164 (2007). [4] R.T.P. Lee, IEDM2006 p.851. [5] R.T.P. Lee, 2007 Symp.VLSI Tech., 108. [6] W.-J. Lee, JAP**101**, 103710 (2007). [7] R.T.P. Lee, IEDM2007 p.685. [8] S.Y. Zhu, et al., EDL **25**, 268 (2004). [9] L.E. Calvet, et al., JAP **91**, 757 (2002).

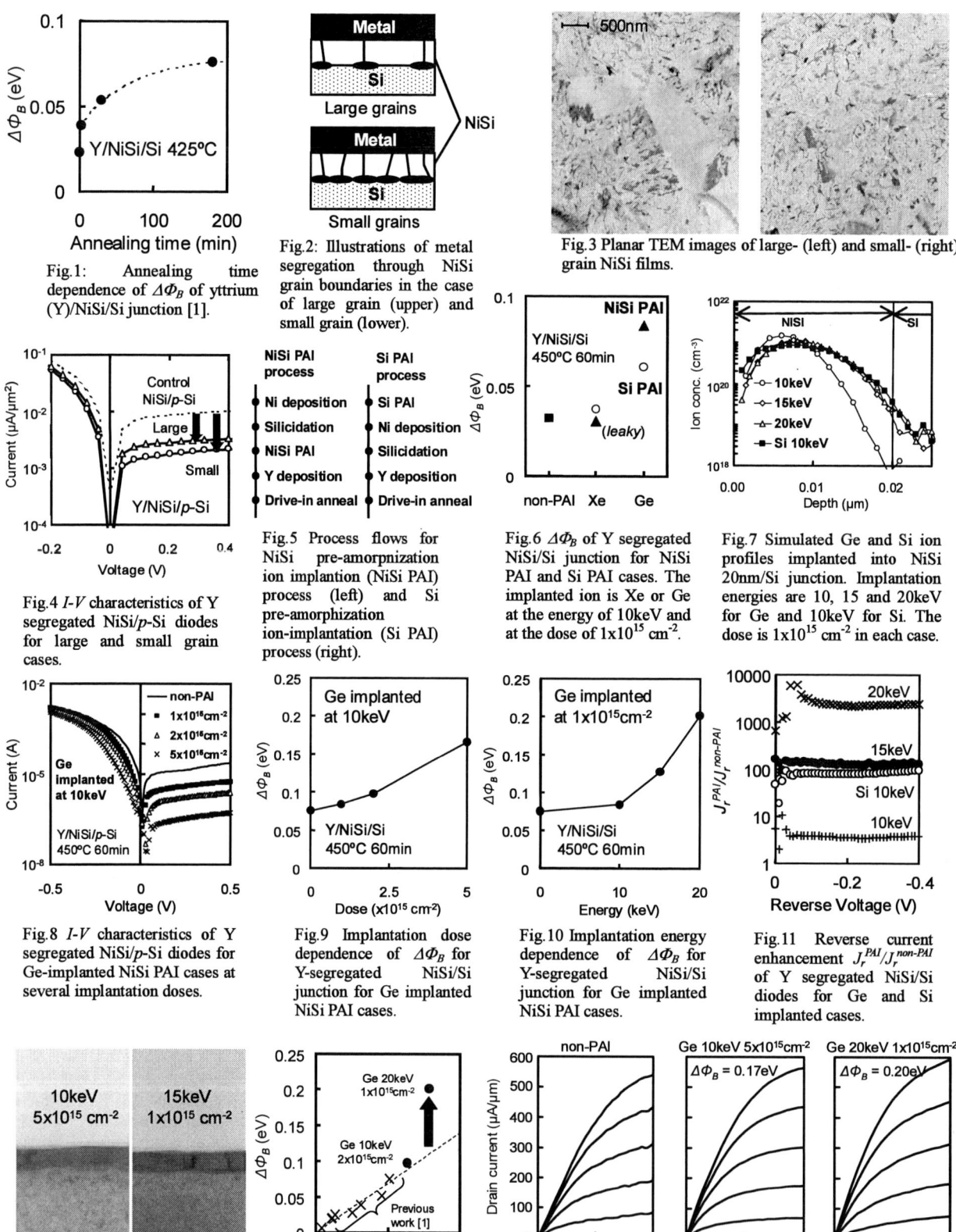

Fig.1: Annealing time dependence of $\Delta\Phi_B$ of yttrium (Y)/NiSi/Si junction [1].

Fig.2: Illustrations of metal segregation through NiSi grain boundaries in the case of large grain (upper) and small grain (lower).

Fig.3 Planar TEM images of large- (left) and small- (right) grain NiSi films.

Fig.4 I-V characteristics of Y segregated NiSi/p-Si diodes for large and small grain cases.

Fig.5 Process flows for NiSi pre-amorpnization ion implantion (NiSi PAI) process (left) and Si pre-amorphization ion-implantation (Si PAI) process (right).

Fig.6 $\Delta\Phi_B$ of Y segregated NiSi/Si junction for NiSi PAI and Si PAI cases. The implanted ion is Xe or Ge at the energy of 10keV and at the dose of 1×10^{15} cm^{-2}.

Fig.7 Simulated Ge and Si ion profiles implanted into NiSi 20nm/Si junction. Implantation energies are 10, 15 and 20keV for Ge and 10keV for Si. The dose is 1×10^{15} cm^{-2} in each case.

Fig.8 I-V characteristics of Y segregated NiSi/p-Si diodes for Ge-implanted NiSi PAI cases at several implantation doses.

Fig.9 Implantation dose dependence of $\Delta\Phi_B$ for Y-segregated NiSi/Si junction for Ge implanted NiSi PAI cases.

Fig.10 Implantation energy dependence of $\Delta\Phi_B$ for Y-segregated NiSi/Si junction for Ge implanted NiSi PAI cases.

Fig.11 Reverse current enhancement $J_r^{PAI}/J_r^{non\text{-}PAI}$ of Y segregated NiSi/Si diodes for Ge and Si implanted cases.

Fig.12 Cross sectional TEM images of Y segregated NiSi /Si junctions for Ge implanted NiSi PAI cases.

Fig.13 $\Delta\Phi_B$ as a function of Y concentration segregated at NiSi/Si interface.

Fig.14 I_d-V_d characteristics of Y-segregated NiSi PAI nMOSFETs with L_g =50nm for Ge implanted cases, where V_g-V_{th}= 0 ~ 1V.

New Global Shutter CMOS Imager with 2 Transistors per Pixel

Masaki Funaki, Takeshi Shimizu, Syuji Orihara, Hiroyuki Kawanaka,

Makoto Kurihara, Hidetoshi Sato, Noboru Katsumata, Munetoshi Oikawa,

Jun Higuchi, Kensei Oe, Raijiro Kuga, Kuniko Maki, and Toshihiko Nishibata

Micro-technology Center, Victor Company of Japan, Limited (JVC)

3-12 Moriya-cho Kanagawa-ku Yokohama 221-8528, Japan

funaki-masaki@jvc-victor.jp, shimizu-takeshi@jvc-victor.jp, orihara-shuuji@jvc-victor.jp

Abstract

We present a new global shutter CMOS imager with 2 transistors per pixel. The first transistor is a ring gate transistor for accumulating holes that modulate threshold voltage. The second one is a transfer gate transistor that transfers holes from a PD to the ring gate transistor at the same time in all pixels. Simple structure allows us to realize 5.4um pixel pitch, kTC noise free, and global shutter sensor using 0.35um technology.

Introduction

Pixel pitches of small CMOS imagers are below 2um [1]. But these imagers do not have a global shutter but have a rolling shutter. Because a global shutter imager based on a conventional structure requires 5 transistors per pixel whose size is 6um pitch in 0.18um technology [2].

In the meantime threshold voltage modulation devices such as CMD are developed [3, 4]. These imagers also have a rolling shutter, but their structure is very simple. They only have a ring gate (RG) transistor which can hold carriers and output a signal as a modulated threshold voltage and reset carriers.

In this work, we present a new imaging pixel structure which has only 2 transistors which are the RG and a transfer gate (TG) transistor. The TG transfers carriers from a photo diode (PD) to the RG simultaneously in all pixels (global shutter), then the RG hold carriers until readout. This structure is kTC noise free when the carriers are transferred and reset completely. An electronic shutter of the PD was made by over flow drain.

Pixel structure and operation flow

Fig.1 shows the layout of the pixel manufactured with 0.35um 2P3M technology. Pixel pitch is 5.4um x 5.4um. Shapes of the RG and the buried PD are hexagon. The PD opening rate is 26.7%. The RG is made by the first poly-Si, and the TG is made by the second poly-Si.

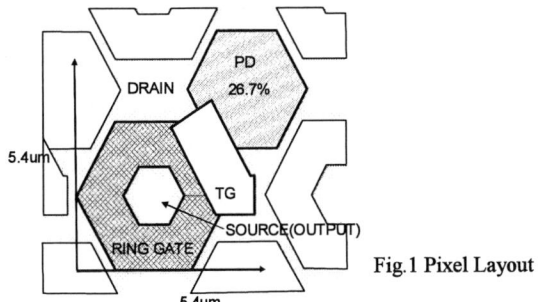

Fig.1 Pixel Layout

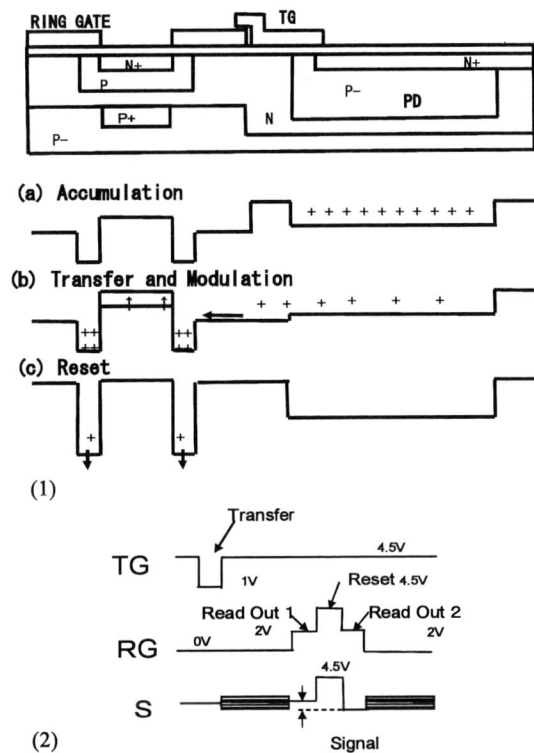

Fig.2 Pixel Operation (1) Potential figures (2) Timing chart

Fig.2 (1) shows a potential profile of a pixel operation. (a)When the TG is off, the PD accumulates holes. (b) When the TG is on, holes are transferred to the RG in all pixels at the same time (global shutter). Holes modulate threshold voltage of the RG. (c) After readout, voltage of the RG and source are raised high level and holes under the RG are discharged to a substrate (reset). If the holes are transferred completely from the PD to the substrate, the pixel has no kTC noise ideally.

Fig.2 (2) shows a timing chart of the operation. After transferring the holes, read out 1 is done with a source follower circuit when the RG voltage level is set at the middle level. Then high voltage of the RG and the source reset the holes. After reset, read out 2 is done without the holes. Voltage difference of read out 1 and 2 is a signal.

For the electronic shutter, the pixel has mechanism of over flow drain. When the drain voltage is set at high level (8V), the holes at the PD discharges to the substrate.

978-1-4244-1802-2/08/$25.00 ©2008 IEEE

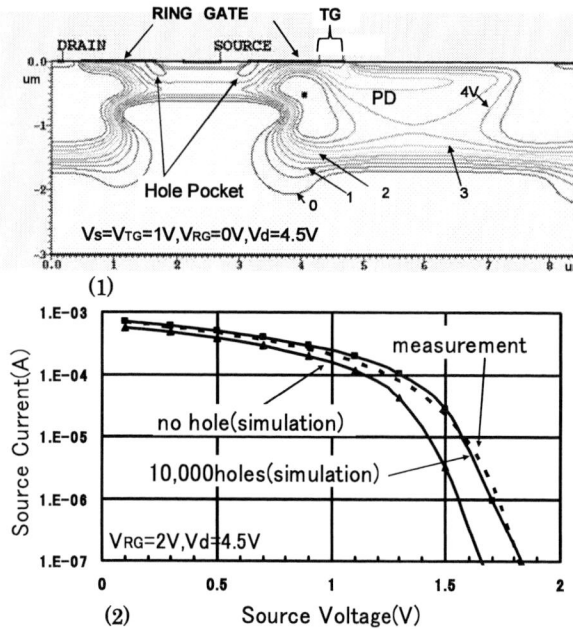

(1)

(2)

Fig.3 Simulation results (1) Potential profile at carrier transferring, (2) Source voltage vs. source current characteristic of the RG transistor at 2V for the ring gate and at 4.5V for the drain.

Simulation and measurement results

Fig.3 shows 2D simulation results. The potential profile (1) shows good transferring characteristic. The holes in the PD are transferred to hole-pocket in where potential is low around source diffusion layer. Simulated results for source voltage vs. source current of the RG transistor is well agreed with the measured results at the thermal equilibrium state (2). Simulation results also describe a modulation become 200 mV at 10,000 holes accumulation.

New sensor and the image of global shutter

Fig.4 shows top view of new sensor which has 920 x 690 pixels, CDS circuit, vertical and horizontal shift register, and column amplifier. A pattern of color filter is Bayer. The sensor outputs 30 frames per second.

Figure 5 shows comparison of images taken by the rolling shutter mode and the global shutter mode. Shutter speed was 1/300 second for the global shutter mode and 1/30 for the other one. The stable image with the global shutter clearly shows an advantage of the global shutter.

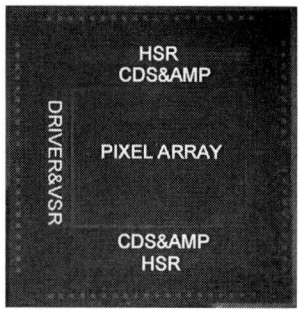

Fig.4 Top view of the new sensor

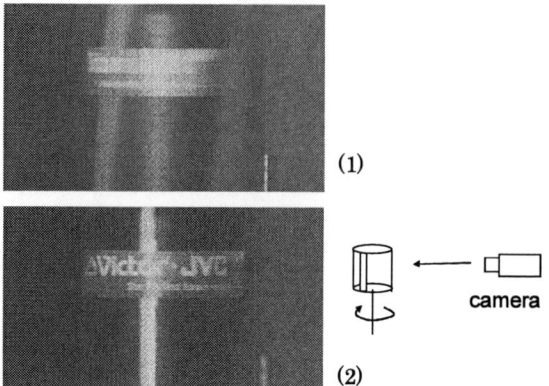

(1)

(2)

Fig.5 Images of new sensor. (1) Rolling shutter mode with ordinary 1/30 sec. accumulation. (2) Global shutter mode with 1/300 sec. electronic shutter.

No smear was observed for the new imager. The excess charges were absorbed to substrate and gave no influence to other pixels in the same column. This is a remarkable advantage among other imagers with a global shutter such as CCD.

Table 1 summarizes the characteristics of the new chip.

Table 1 New chip characteristics

Technology	0.35um, 2P3M
Pixel Size	5.4um x 5.4um
Tr./Pixel	2
Fill Factor	26.7%
Suply Voltage	4.5V/8V@OF Drain
Power consumption	75mW
Resolation	920×690
Sensitivity	10,000holes/lx*s (2850K light source with IR filter)
Electronic shutter	Overflow Drain
PD Suaturation charge	12,000 holes
Hole-pocket suturation charge	10,000 holes
Color pattern	Bayer
Smear	NO

Conclusion

We presented the new global shutter CMOS imager with 2 transistors per pixel. Our new sensor, which has 5.4um pixel pitch and manufactured by 0.35um technology, enables the global shutter operation and realizes the electronic shutter using the over flow drain. The simplicity of the new pixel structure shows a possibility to realize much smaller pixel size less than 2um.

References

[1] C. R. Moon, et al., "Dedicated Process Architecture and the Characteristics of 1.4um Pixel CMOS Image Sensor with 8M Density", 2007 Symposium on VLSI Technology Digest of Technology Paper, pp62-63
[2] N.Bock, et al., "A Wide-VGA CMOS Image Sensor with Global Shutter and Extended Dynamic Range", Proc. of IEEE Workshop on CCDs and AIS, Karuizawa, pp222-225, 2005.
[3] M.Ogata et al., "A small Pixel CMD Image Sensor", IEEE Transaction on Electron Device, vol.38, No.5, May, 1991.
[4] T.Miida et al., "1.5M Pixel Imager with Localized Hole-Modulation Method", ISSCC Digest of Technical Papers, vol.55, Feb., 2002.

35-nm Gate-Length and Ultra Low-voltage (0.45 V) Operation Bulk Thyristor-SRAM/DRAM (BT-RAM) Cell with Triple Selective Epitaxy Layers (TELs)

T. Sugizaki, M. Nakamura, M. Yanagita, M. Shinohara, T. Ikuta, T. Ohchi, K. Kugimiya, S. Kanda, K. Yagami, and T. Oda

Sony Corporation, 4-14-1 Asahi-cho, Atsugi-shi, Kanagawa, 243-0014, Japan
Phone: +81-46-230-5662 Fax: +81-46-230-5572 E-mail: Taro.Sugizaki@jp.sony.com

Abstract

We have successfully developed an alternative SRAM cell using a Bulk Thyristor-RAM (BT-RAM), which has a 35-nm gate-length with Triple selective Epitaxy Layers (TELs) for the anode, the n-base, and the cathode. The TEL BT-RAM reads and writes at an ultra low voltage of 0.45 V at 900 ps and reads and writes at a high speed of 100 ps at 0.9 V. It also has excellent scalability, a high I_{on}/I_{off} ratio, and good thermal stability even at 125°C. The TEL BT-RAM is therefore a promising alternative SRAM cell for the 35-nm gate length generation and beyond.

Introduction

Recently, 6T-SRAMs have encountered many problems with CMOS scaling such as decreased static noise margins, increased stand-by power consumption, and limits imposed by decreased cell areas [1]. Thyristor-based SRAMs are one of the most attractive candidates to overcome such problems. Thus far, thyristor-based RAMs using SOI wafers [2, 3] and BT-RAM [3-6] using bulk wafers have been reported. These devices have demonstrated excellent performances such as high-speed and low-voltage read/write, high I_{on}/I_{off} ratios, low standby current, and a small cell size. In addition, the BT-RAMs use bulk Si-wafers, which results in lower cost and compatibility with many applications. However, these devices had a minimum physical-gate-length ($\equiv L_{phy}$) of approximately 150 nm, and the scalability below it was estimated by simulation only. In this paper, we have successfully demonstrated BT-RAM with 35-nm L_{phy} for the first time by using EB lithography. The operation performance of this BT-RAM has surpassed the previously reported ones, which suggests the promising scalability in the future generations.

Device structure and Operations

A schematic of the structures of the TEL BT-RAM is shown in Fig. 1. It consists of a select-MOSFET and a MOS Effect Thyristor (MET) per bit. The MET has four diffusion layers: an anode, an n-base, a p-base with a MOS-gate on it, and a cathode, all of which are easily formed by conventional logic processes with four additional photo-masks. The anode, the n-base, and the cathode are formed using a selectively epitaxial CVD process.

There is a BT-RAM equivalent circuit shown in Fig. 2. A MET is a combination of npn and pnp bipolar transistors. Typical I-V characteristics of METs are plotted in Fig. 3 (a). The MET has a negative differential resistance region (NDR) and bi-stable states at gate-off while it does not have such a region at gate-on. The switching voltage is denoted by V_{FB}, and I_{hold} is the minimum turn-on current or stand-by current for the on-state. Data are kept as "1" at currents above I_{hold}. The basic operation sequence for the BT-RAM is outlined in Fig. 3 (b). Pulses for read/write operations are applied to 1) only MOS-gates at Read "1"/"0", 2) MOS and MET-gates at Write "1", and 3) MOS-gates and anodes at Write "0".

Experimental results and discussions

A cross-sectional TEM image of a fabricated 35-nm L_{phy} TEL BT-RAM cell is shown in Fig. 4. To optimize both doping profiles and I-V characteristics for the BT-RAM, we carried out a TCAD simulation (Fig. 5). The inset corresponds to I-V characteristics. The NDR regions were clearly observed at both RT and 85°C. The measured I-V characteristics of a MET with a 35-nm L_{phy} are plotted in Fig. 6. A clear NDR region can be recognized with an I_{hold} of 1E-12 A and a V_{FB} of 4.1 V even at this short L_{phy}, which is sufficient to obtain a large margin between the on and off states.

These results suggest that both the current gain (β) of npn-Tr. and V_{th} of MOSFET in MET were successfully optimized. Notice that one of the most challenging control to realize BT-RAM with very short L_{phy} (such as 35 nm) is to optimize both bipolar and MOS characteristics in MET, which is a hard problem that we do not have in shortening L_{phy} in conventional CMOS.

The V_{FB} roll-off characteristics of MET are plotted in Fig. 7. A slight roll-off is observed, but the V_{FB} is still high enough (4.1 V) even at the 35-nm L_{phy}. The simulation results suggest that the V_{FB} at the 35-nm L_{phy} is also high enough (2.8 V) at 85°C. The transient characteristics of the TEL BT-RAM are shown in Fig. 8. Large I_{on}/I_{off} ratios of more than 10^8 were obtained. The retention characteristics of the 35-nm L_{phy} TEL BT-RAM were measured for on and off states (Fig. 9). Both states were stably maintained for at least 1000 s even at 125°C, which showed good thermal stability. This suggests that the TEL BT-RAM is suitable for use in not only SRAM but also in DRAM cells.

We then evaluated the sub-ns response of BT-RAMs by using fabricated microwave guide patterns [4]. The MET's turn-on characteristics are plotted in Fig. 10. The turn-on time decreased as the MET-gate voltage (V_{gate}) increased. The turn-on time for the 35-nm L_{phy} reached 100 ps (the measurement limit in our system) at a V_{gate} of 0.65 V, which is approximately twice faster than that for the 200-nm L_{phy}. The MET's turn-off characteristics are plotted in Fig. 11. The turn-off time decreases as the anode voltage (V_{anode}) increases. The turn-off time for the 35-nm L_{phy} reaches 100 ps at a V_{anode} of -0.25 V, which is slightly slower than that for 200-nm, but is almost the same. We turned off the BT-RAMs very quickly by controlling the n-well bias to shunt excess carriers to the substrate [4]. We demonstrated the full memory operations of the 35-nm L_{phy} BT-RAM. The results carried out under an ultra low voltage below 0.45 V for write "1"/"0" and read "1"/"0" operations are shown in Fig. 12. A clear peak for read "1" appeared, but there was no peak for read "0". These results show secure RAM operations.

The minimum read/write speeds as a function of the maximum operation voltage are shown in Fig. 13. The operation lower limit of 35-nm L_{phy} BT-RAM is 0.45 V, which is much lower than that of the 200-nm L_{phy} BT-RAM (0.6 V). There was no deterioration in the output waveform even after 10-million operation cycles (Fig. 14). The main reason for the improvement in performance of the 35-nm L_{phy} BT-RAM is attributed to the improved β of npn-Tr. with a shorter p-base compared with that of the 200-nm L_{phy} BT-RAM.

Conclusion

We successfully developed a TEL BT-RAM with a 35-nm L_{phy} by optimizing the characteristics of both npn-Tr. and MOSFET in MET. The TEL BT-RAM had excellent performances in superior scalability, V_{FB} roll-off characteristics, and thermal stability. As a result, the TEL BT-RAM operates at a low voltage of 0.45 V or a high speed of 100 ps, and is a promising candidate as an alternative SRAM in the near future.

References

[1] A. Bhavnagarwala et al., IEDM Tech. Dig., p.675, 2005
[2] F. Nemati et al., IEDM Tech. Dig., p.273, 2004
[3] T. Sugizaki et al., Proc. of ESSDERC, p.323, 2007
[4] T. Sugizaki et al., IEDM Tech. Dig., p.157, 2006
[5] T. Sugizaki et al., VLSI Tech. Symp., p.170, 2007
[6] T. Sugizaki et al., IEDM Tech. Dig., p.933, 2007

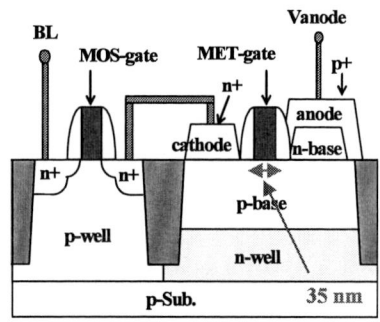

Fig. 1 Schematic of the structure of a BT-RAM cell with a TEL.

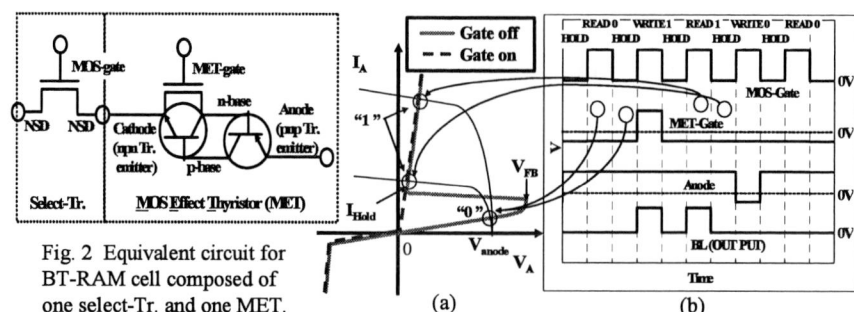

Fig. 2 Equivalent circuit for BT-RAM cell composed of one select-Tr. and one MET.

Fig. 3 Basic operation of BT-RAM: (a) I-V characteristic of MET (bold line) and select-Tr. I-V curves. (b) Timing chart for operation sequence

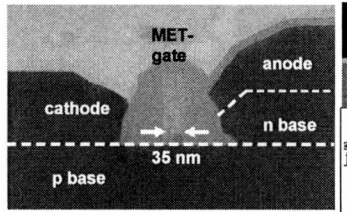

Fig. 4 TEM image of TEL BT-RAM. Gate length was 35-nm.

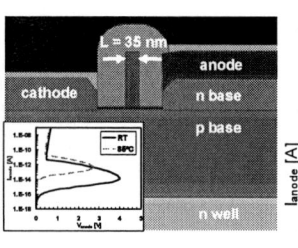

Fig. 5 Simulation results for doping profile of TEL with gate length of 35 nm. Inset is corresponding I-V curves.

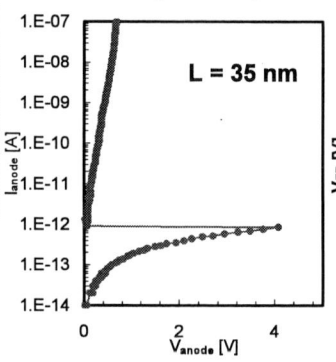

Fig. 6 Measured I-V characteristics of MET with gate length of 35 nm. Note the clearly NDR region.

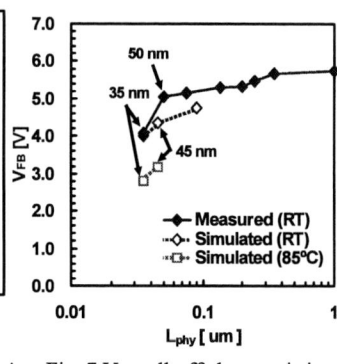

Fig. 7 V_{FB} roll-off characteristics of MET for TEL. V_{FB} at 35-nm gate length is sufficiently high even at 85°C

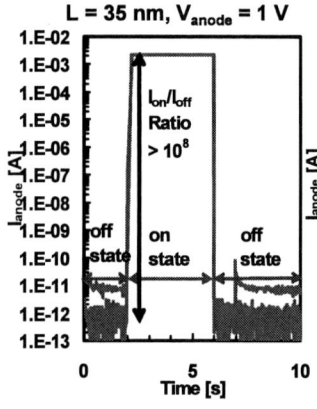

Fig. 8 Transient characteristics of a TEL.

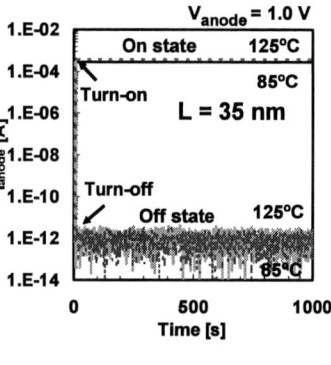

Fig. 9 Retention characteristics of TEL. It can maintain "0"/"1" states sufficiently long even at 125°C.

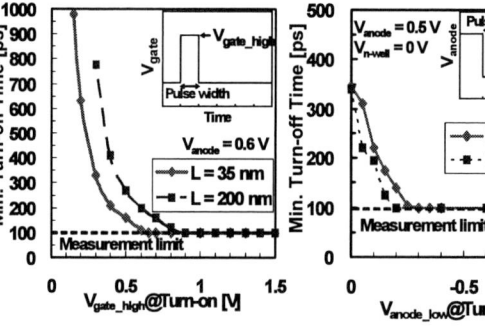

Fig. 10 Minimum turn-on times as function of MET-gate voltage (V_{gate_high}). Turn-on time for a 35-nm gate length TEL reached 100 ps at 0.65 V.

Fig.11 Minimum turn-off times as function of anode voltage (V_{anode_low}). Turn-off time for a 35-nm gate length TEL reached 100 ps at -0.25 V.

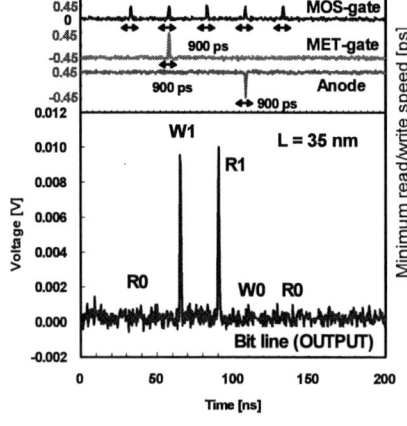

Fig. 12 Full operation of TEL BT-RAM. low voltage (below 0.45V) operation. All write "1"/"0" (W1/W0), and read "0"/"1" (R0/R1) operations were successful.

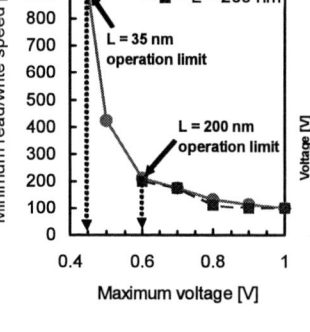

Fig. 13 Minimum read/write operation as a function of maximum voltage. The TEL BT-RAM with 35-nm gate length performs at 0.45 V, which is much lower than that of 200-nm gate length TEL BT-RAM.

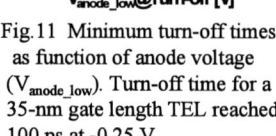

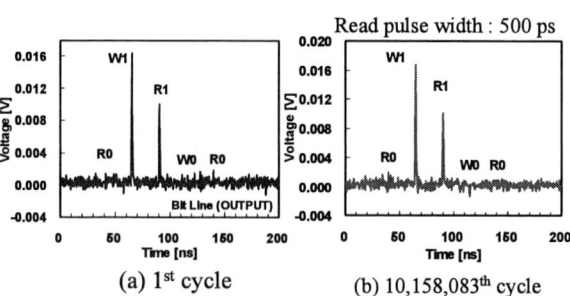

Fig. 14 Endurance of 35-nm gate length TEL BT-RAM: output waveforms of the full read/write operation at (a) 1st cycle, and (b) over the 10M[th] cycle. There is no degradation in the waveform after 10M operation cycles.

Band Offset FinFET-Based URAM (Unified-RAM) Built on SiC for Multi-Functioning NVM and Capacitorless 1T-DRAM

Jin-Woo Han[1], Seong-Wan Ryu[1], Sungho Kim[1], Chung-Jin Kim[1], Jae-Hyuk Ahn[1], Sung-Jin Choi[1], Kyu Jin Choi[2], Byung Jin Cho[1]
Jin Soo Kim[3], Kwang Hee Kim[3], Gi Sung Lee[3], Jae Sub Oh[3], Myong Ho Song[3], Yun Chang Park[3], Jeoung Woo Kim[3], and Yang-Kyu Choi[1]

[1]EECS, KAIST, Daejeon, Korea, [2]Jusung Engineering, Gwangju, Korea, [3]National Nanofab Center, Daejeon, Korea
E-mail: ykchoi@ee.kaist.ac.kr, Phone: +82-42-869-3477, Fax: +82-42-869-8565

Abstract

A FinFET-based unified-RAM (URAM) using the band offset of Si/SiC is demonstrated for the fusion of a non-volatile memory (NVM) and capacitorless 1T-DRAM operation. An oxide/nitride/oxide (O/N/O) gate dielectric and a floating body caused by the band offset are combined in a bulk FinFET to allow two memory operations in a single transistor. The device is fabricated on an epitaxially grown Si/SiC substrate and its process is fully compatible with a conventional bulk FinFET SONOS. Highly reliable NVM and high speed 1T-DRAM operation are confirmed in a single URAM cell.

Introduction

In the digital convergence era, the development of multi-functional or fusion memory holds particular attraction [1-3]. An example of fusion memory, NVM and 1T-DRAM can be operated in a single memory cell, and these operations are identified by the bias conditions of V_g and V_d as shown in **Fig. 1**. We recently proposed a FinFET SONOS URAM fabricated on a SOI substrate [3]. However, the SOI substrate is prone to heat dissipation, which degrades the sensing window for 1T-DRAM [4]. In the present work, we propose a band offset URAM architecture built on Si/SiC for effective management of heat and reduced fabrication cost.

A buried n-well layer that serves as a barrier to hold holes has been employed in a bulk substrate for 1T-DRAM whereas a buried oxide has been used in SOI [5-6]. However, with the deep implantation process for the buried n-well, it is difficult to accurately define the abrupt doping profile. Thus, epitaxially grown Si/SiC to hold holes by the band offset is proposed. A FinFET-based URAM is also fabricated on a buried n-well as a control group.

Device Fabrication

The process sequence is summarized in **Fig. 2**. First, a Si/Si$_{0.99}$C$_{0.01}$ layer is epitaxially grown for the band offset, and buried n-implantation is carried out in the case of the control group. The subsequent steps correspond with those of the bulk FinFET SONOS process flow [7]. The fabricated device dimensions and parameters are summarized in **Table 1**. SEM/TEM images are shown in **Fig. 3**. Si/SiC lattices are perfectly matched as shown in the fast Fourier transformed images. The SIMS profile for the buried n-well device is shown in **Fig. 4**.

Results and Discussion

NVM characteristics - **Fig. 5** shows the speed response of P/E carried out by a FN-tunneling mechanism. Both devices show similar threshold voltage (V_T) shift. **Fig. 6** shows the reliability characteristics. 10 year retention behaviors are observed along with a more than 3V V_T window; however, the Si/SiC device shows better data retention time. In addition, excellent endurances of more than 10^7 P/E cycles are observed without V_T window degradation for both devices.

1T-DRAM characteristics – A floating body effect in bulk devices is originated from the buried n-well or the valence band offset of Si/SiC. **Fig. 7** shows the 1T-DRAM operation mechanism. For programming, the holes generated by impact ionization are stored in the Si body on band offset. For erasing, accumulated holes are eliminated by the forward biased drain. As evidence of holes accumulation, a kink appears in the I_D-V_D curves, as shown in **Fig. 8**. A simulated contour of the hole concentration in **Fig. 9** clearly shows that both the buried n-well and the band offset of Si/SiC can store holes. **Fig. 10** shows the P/E characteristics for the 1T-DRAM operation.

Fig. 11 shows the P/E speed response. The Si/SiC device exhibits a wider sensing window than the buried n-well device. At $\tau_{PGM} = \tau_{ERS} = 20$nsec, the sensing window in Si/SiC ($\Delta I_S = 10\mu A$) allows a retention time of 50msec whereas the buried n-well ($\Delta I_S = 6\mu A$) retains data during 10msec. If the retention time is increased, faster operation can be possible because of the saved refresh time. One important advantage in bulk devices is a tunable substrate voltage. By modulation of the substrate voltage, the barrier height to hold holes can be adjusted. As shown in **Fig. 12**, positive V_{SUB} ($0.1V < V_{SUB} < 0.3V$) raises the barrier height resulting in increment of the retention time. In contrast, $V_{SUB} > 0.4V$ reduces the retention time due to a forwardly biased substrate-drain junction. Although doubled retention time was observed in the buried n-well device by optimal V_{SUB}, Si/SiC still shows superior performance to the buried n-well device. Furthermore, these characteristics the Si/SiC device can be improved by increment of the content of C in SiC, as this will provide an enlarged valence band offset [8]. Impact ionization for 1T-DRAM programming can adversely stimulate charges to be trapped in the O/N/O layer, and consequently an undesired soft program can be possible in NVM. In order to examine interference between these two operations, I_D-V_G characteristics are compared before/after 1T-DRAM operation. As shown in **Fig. 13**, interference by the impact ionization is found to be negligible.

Conclusions

A band offset FinFET-based unified-RAM (URAM) on Si/SiC substrate is demonstrated. URAM performs NVM and 1T-DRAM functions in a single transistor with benefits of low cost and high heat dissipation. NVM uses an O/N/O layer for charge trapping, and 1T-DRAM utilizes a floating body effect for capacitorless DRAM operation in the bulk substrate.

References

[1] C. W. Oh *et al.*, *VLSI*, p.58, 2006. [2] C. W. Oh *et al.*, *VLSI*, p.168, 2007. [3] J.-W. Han *et al.*, *IEDM*, p.929, 2007. [4] P. C. Fazan, *et al.*, *SPIE*, p.489, 2002. [5] R. Ranica *et al.*, *VLSI*, p.38, 2005. [6] R. Ranica *et al.*, *VLSI*, p.128, 2004. [7] J. R. Hwang. *et al.*, *IEDM*, p.154, 2005. [8] H. J. Osten *et al.*, *JAP*, p. 2716, 1998.

978-1-4244-1802-2/08/$25.00 ©2008 IEEE

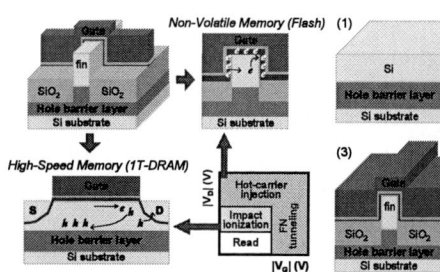

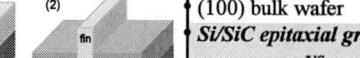

(100) bulk wafer
Si/SiC epitaxial growth
vs.
buried n-well implantation
channel implantation
fin patterning
STI formation
O/N/O and poly-Si formation
gate patterning
S/D formation

Parameter	Si/SiC	Buried n-well
V_T (V)	0.11	0.18
DIBL (mV/V)	110	115
SS (mV/dec)	93	95
L_g (nm)	230	
W_{fin} (nm)	50	
H_{fin} (nm)	50	
$T_{O/N/O}$ (nm)	4/6/4	

Fig. 1 : Schematics of URAM operation.

Fig. 2 : Process flow for bulk FinFET SONOS. An epitaxially grown Si/SiC layer or a buried n-well is used for the hole barrier.

Table 1 : Fabricated device dimensions and parameters of the measured device.

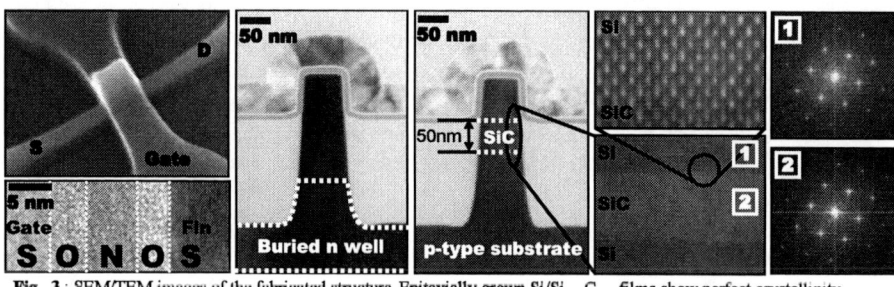

Fig. 3 : SEM/TEM images of the fabricated structure. Epitaxially grown $Si/Si_{0.99}C_{0.01}$ films show perfect crystallinity.

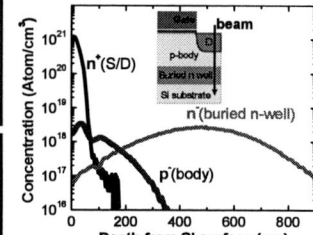

Fig. 4 : SIMS profile obtained on S/D to substrate region in buried n-well device.

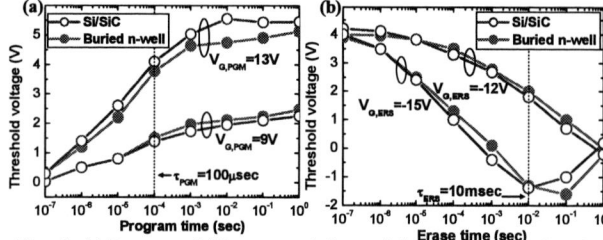

Fig. 5 : (a) Program and (b) erase speed characteristics for NVM operation. A FN tunneling mechanism is used for both program and erase.

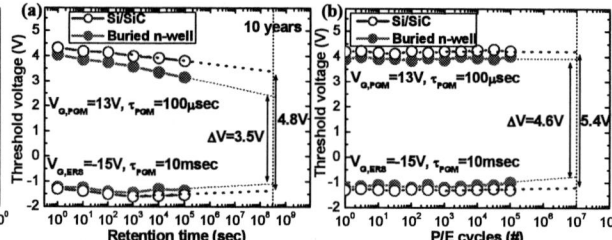

Fig. 6 : (a) Retention and (b) endurance characteristics for NVM operation. The Si/SiC device shows better retention characteristics. Both devices exhibit excellent endurance.

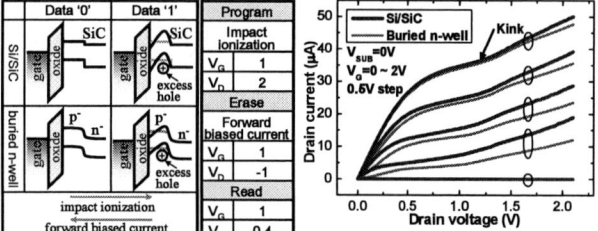

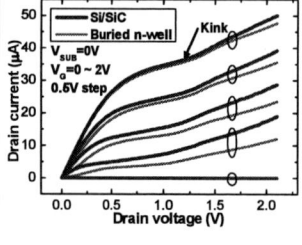

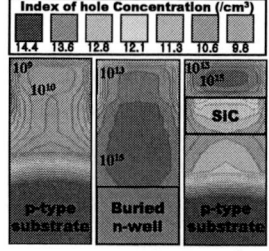

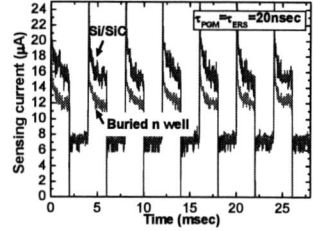

Fig. 7 : Schematic diagrams for P/E of 1T-DRAM and bias conditions used in this work.

Fig. 8 : I_D-V_D characteristics. Kink ensures a floating body effect.

Fig. 9 : Hole concentration at impact ionization condition in Fig. 7.

Fig. 10 : Program and erase characteristics for 1T-DRAM operation. Si/SiC shows a wider sensing window than buried n-well.

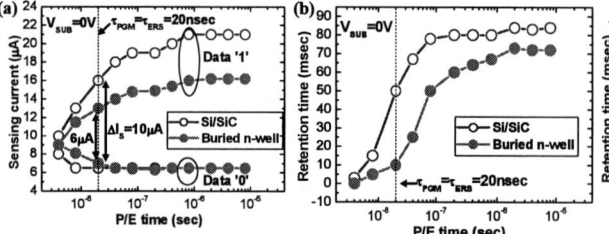

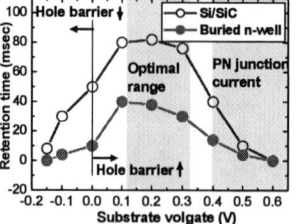

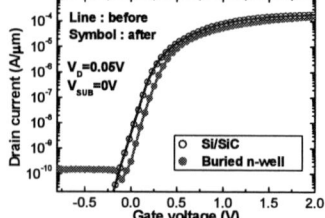

Fig. 11 : P/E speed for 1T-DRAM. (a) Data '1' state degradation is more severe than data '0' state. (b) Retention characteristics for 1T-DRAM. Si/SiC shows wider sensing window and longer data retention.

Fig. 12 : Retention time as a function of V_{SUB}. $0.1 < V_{SUB} < 0.3$ is optimal condition.

Fig. 13 : I_D-V_G curves before (lines) and after (symbols) 1T-DRAM programming.

Integrated Wafer-Scale Growth and Transfer of Directional Carbon Nanotubes and Misaligned-Carbon-Nanotube-Immune Logic Structures

Nishant Patil, Albert Lin, Edward R. Myers, H.-S. Philip Wong, and Subhasish Mitra

Stanford University {nppatil, mrlin, edmyers, hspwong, subh}@stanford.edu

Abstract

We successfully demonstrate essential components and their integration for large-scale Carbon Nanotube Field Effect Transistor (CNFET) technology: 1. First demonstration of full-wafer-scale growth of directional carbon nanotubes (CNTs) on 4" single-crystal quartz wafers. 2. First demonstration of full-wafer-scale CNT transfer from 4" quartz wafers to 4" silicon wafers for integration on silicon. 3. Integration of full-wafer-scale growth and transfer, together with metallic-CNT removal, for the first demonstration of misaligned-CNT-immune digital logic structures on a full-wafer-scale. Such logic structures guarantee correct logic functionality in the presence of a large number of misaligned and mis-positioned CNTs.

Introduction

Carbon Nanotube Field Effect Transistors (CNFETs) are promising extensions to silicon CMOS, and can provide 13X improvement in Energy-Delay product over 32 nm CMOS [1]. Despite major progress, e.g., ring oscillator using a single CNT [2], significant research is needed for design and integration of VLSI CNFET circuits. Essential components for such large-scale integration are: 1. Full-wafer-scale directional CNT growth (CNTs refer to Single-Walled CNTs): Full-wafer-scale CNT growth enables fabrication of CNFET circuits in mass scale using conventional lithography. CNTs grown on ST-cut quartz are significantly better aligned compared to silicon [3] or r-plane sapphire [4] for comparable CNT densities. Unfortunately, existing CNT growth techniques on quartz (e.g., [5]) do not allow such full-wafer-scale CNT growth; 2. Full-wafer-scale CNT Transfer: Large-scale silicon integration requires full-wafer-scale transfer of directional CNTs from quartz to silicon substrates; and, 3. VLSI-compatible Imperfection-Immune CNFET Circuits: Perfect alignment and positioning of all CNTs cannot be guaranteed at VLSI scale. Misaligned- and mis-positioned-CNT-immune logic structures, designed using principles described in [6], are required. However, such logic structures have not been experimentally demonstrated. CNFETs require semiconducting-CNTs. Metallic-CNTs create source-drain shorts and require removal using burning [7] or selective etching [8]. This paper successfully demonstrates the above essential components and their integration for large-scale CNFET circuits. We provide data on a large number of devices (cross-chip, cross-wafer) to illustrate the variability of process/device parameters relevant for VLSI integration consideration.

Wafer-Scale Directional CNT Growth on Quartz

High temperature (850°C) is required for growing directional CNTs [5]. CNT growth on quartz has only been demonstrated for wafer pieces because the transformation from alpha to beta quartz at ~573°C results in wafer fracture during ramp-up [9]. The key to achieving full-wafer-scale directional CNT growth is to control the temperature ramp rate near the phase transformation temperature (550°C – 620°C) to < 1°C / min (Fig. 1a). Unpatterned Ferritin was used as a catalyst for CNT growth. Prior to growth, the quartz wafer was annealed for 8 hrs in oxygen at 900°C using controlled ramp rate similar to Fig. 1a. Figs. 1b-e demonstrate that our new technique enables full-wafer-scale CNT growth on 4" quartz wafers. Fig. 1e also demonstrates that CNTs grew across the entire wafer with >94% of the measured devices being functional. The reproducibility of this technique is confirmed by multiple full-wafer-scale growth runs for various experiments presented in this paper. As shown in Fig. 2a, CNT alignment can be significantly improved by patterning the catalyst (similar to [10]).

Wafer-Scale CNT Transfer and CNFET Fabrication

Next, we demonstrate a new technique for full-wafer-scale CNT transfer from quartz to silicon for large-scale silicon integration. Previous CNT transfer techniques, e.g., [11, 12], did not demonstrate full-wafer scale transfer. Figs. 3a-f show the sequence of CNT transfer steps using thermal release tape. This low-temperature (120°C) technique preserves CNT directionality (Fig. 3e). Fig. 3g demonstrates that CNTs were successfully transferred across the entire wafer with >92% of the measured CNFETs being functional.

Metallic-CNT removal is essential for CNFET circuits. We applied electrical burning [7] to burn metallic-CNTs in CNFETs after full-wafer-scale transfer. Electrical burning is performed by applying high V_{ds} while turning off semiconducting CNTs using the gate. Metallic CNTs pass high current and are burnt resulting in improved I_{on}/I_{off} ratios of $10^4 – 10^5$ (Figs. 4a, b). CNFETs retain well-behaved I-V characteristics after electrical burning (Fig. 4c).

Misaligned- and Mis-positioned-CNT-Immune Logic Structures

It is nearly impossible to guarantee perfect alignment and positioning of all CNTs at VLSI scale. This can result in incorrect logic behaviors of fabricated logic structures (Fig. 5). In [6], we developed design principles for CNFET circuits that guarantee correct logic functions in the presence of a large number of misaligned and mis-positioned CNTs. For example, for a NAND pull-up, during layout design, we identify a lithographically-defined region from where CNTs are etched out (Fig. 6). Any structure consisting exclusively of series-connected CNFETs, e.g., NOR pull-up or NAND pull-down, is inherently immune to misaligned and mis-placed CNTs. The special layout design technique can be applied to any logic function, and is compatible with VLSI, i.e., it does not require special customization on individual die basis.

We present first experimental demonstration of misaligned-CNT -immune logic structures on both quartz after full-wafer-scale CNT growth (Fig. 7) and silicon after transfer of CNTs from quartz to silicon (Fig. 8). These logic structures correspond to pull-ups of NAND, NOR, AND-OR-INVERT and OR-AND-INVERT functions. For example, consider the NAND pull-up on quartz (Fig. 7a) and silicon (Figs. 8a-c). When both gates are off, the drive current (I_{drive}) is minimum. When both gates are on, I_{drive} is maximum, and when only one of the gates is on, I_{drive} is approximately half-way between minimum and maximum. Minimum I_{drive} is non-zero due to metallic-CNTs (metallic-CNTs also cause leakage in the NOR pull-up of Fig. 7b). To overcome this, we removed metallic-CNTs using electrical burning after fabrication of misaligned-CNT-immune logic structures. Fig. 9 demonstrates correct behavior of a misaligned-CNT-immune NAND pull-up on silicon after electrical burning of metallic-CNTs. In this case, when both the gates are off, the I_{drive} is very small. The I_{on}/I_{off} ratio is $> 10^3$, and well-behaved I-V characteristics are obtained.

Conclusion

Full-wafer-scale CNT growth on quartz, full-wafer-scale CNT transfer from quartz to silicon, their integration with metallic-CNT removal, and demonstration of misaligned-CNT-immune logic structures pave the way for VLSI CNFET technologies. It may be possible to integrate such an approach with silicon CMOS. For performance benefits of CNFETs over CMOS, the following open issues must be resolved: 1. High CNT density: One potential technique is to perform wafer-scale CNT transfer (Fig. 3) multiple times from multiple quartz wafers to the same target silicon wafer; 2. Scalable metallic-CNT removal [8]; and, 3. Metallic-CNT-tolerant circuits [13].

Acknowledgment

We thank FCRP GSRC, FENA, C2S2, NSF and Stanford Graduate Fellowship for support. We thank D. Akinwande, A. Badmaev, Prof. H. Dai, Dr. J. McVittie, Prof. Y. Nishi, K. Ryu, J. Zhang and Prof. C. Zhou for fruitful collaborations.

References

[1] J. Deng, et al., Proc. ISSCC, 70-71 (2007).
[2] Z. Chen, et al., Science, Vol. 311, 1735 (2006).
[3] A. Reina, et al., J. Phys. Chem. C 111, 7292 – 7297 (2007).
[4] S. Han, et al., J. of Am. Chem. Soc. 127, 5294 – 5295 (2005).

978-1-4244-1802-2/08/$25.00 ©2008 IEEE

[5] S.J. Kang, et al., Nature Nanotechnology, Vol. 2, 230-236 (2007).
[6] N. Patil, et al., Proc. Design Automation Conf., 958 - 961 (2007).
[7] P. G. Collins, et al., Science, Vol. 292, 706 – 709 (2001).
[8] G. Zhang, et al., Science, Vol. 314, pp. 974 – 977 (2006).
[9] S. Byers, Thesis, Case Western Reserve University (1974).
[10] C. Kocabas, J. of Am. Chem. Soc. 128, 4540 – 4541 (2006).
[11] S.J. Kang, et al., Nano Letters 7(11), 3343-3348 (2007).
[12] X. Liu, et al., Nano Letters, 6, 34-39 (2006).
[13] J. Zhang, et al., in press, Design Automation and Test in Europe (2008).

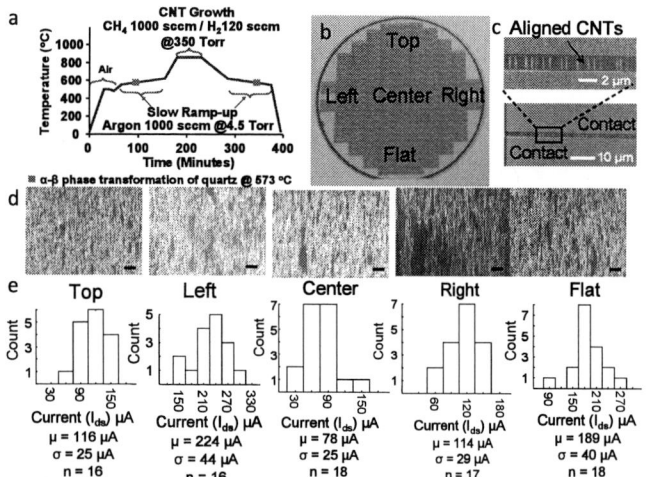

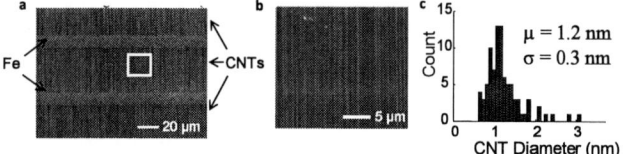

Figure 1. (a) Time course of full-wafer-scale CNT growth on 4" quartz wafers. (b) 4" quartz wafer after CNT growth and contact definition [Ti(1 nm)/Au(50 nm)]. (c) Aligned CNTs bridging two contacts. (d) SEM images of CNTs in five regions of the wafer. CNT density is ~2 CNTs/μm. Scale bars are 20 μm. (e) Current distributions at 1V bias for *n* functional devices (out of 18 on each die) (W=50 μm, L=1 μm) in 5 arbitrary dies (out of 133) over 5 regions of the wafer.

Figure 2. (a) CNT growth on quartz using 0.2 nm Fe catalyst in lithographically defined regions. (b) Zoomed inset from Fig. 2a. (c) CNT diameter distribution.

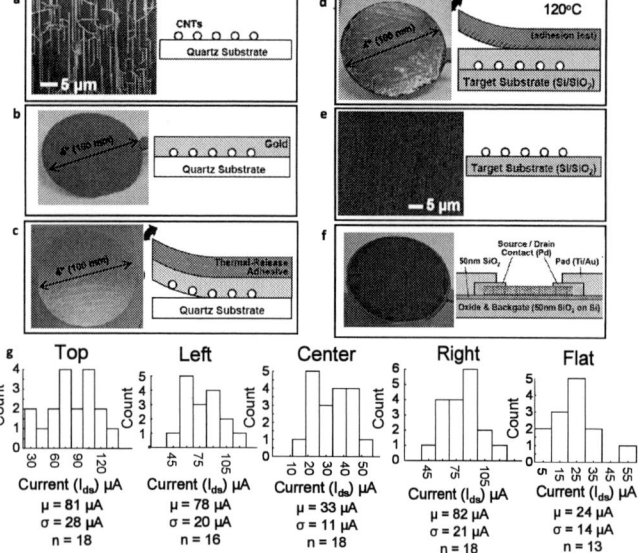

Figure 3. CNT transfer technique using Thermal Release Tape. (a) SEM of CNTs on quartz. (b) 100 nm of Au evaporated on 4" quartz wafer after CNT growth. (c) Thermal release tape is applied to the Au film and the tape/Au bilayer is peeled off. (d) SiO₂/Si Wafer with transferred Au after tape release at 120°C. (e) SEM images of SWNTs transferred from quartz to 50 nm SiO₂ on Si after gold etching (KI/I₂). (f) Si wafer after substrate-gated CNFET fabrication. (g) Current distributions (V_{ds} = 1V V_{gs} = -5V) for *n* (out of 18 in each die) functional CNFETs (W=50 μm, L=1 μm, t_{SiO2}=50 nm, Ti(5 nm)/Au(50 nm) contacts) in 5 arbitrary dies (of 133) in 5 regions of the wafer.

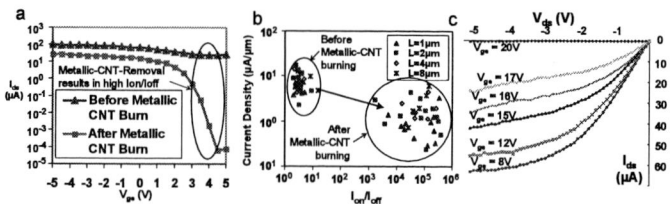

Figure 4. (a) CNFET current-voltage (I_{ds} vs. V_{gs}) before and after metallic-CNT burning (W=10 μm, L=1 μm, t_{SiO2}=50 nm). (b) Electrical burning of metallic CNTs improves I_{on}/I_{off} ratio to 10^4-10^5. (c) I_{ds} vs. V_{ds} after burning for another CNFET.

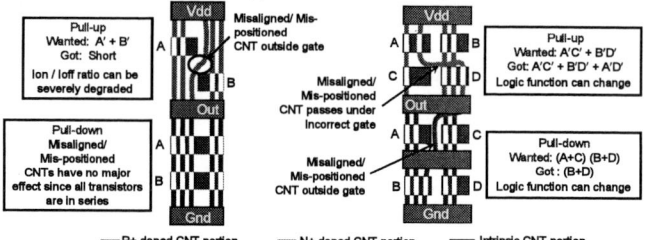

Figure 5. Incorrect Logic Functionality in the presence of misaligned CNTs.

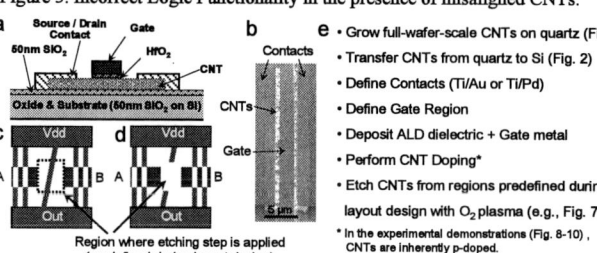

Figure 6. Misaligned-CNT-immune circuits using top-gated CNFETs. (a) Cross-section. (b) CNFET SEM. (c) Misaligned-CNT-immune NAND pull-up. (d) Misaligned-CNT-immune NAND pull-up after etching. (e) Process steps.

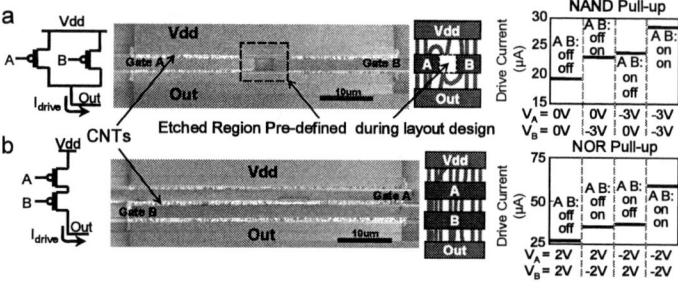

Figure 7. Misaligned-CNT-immune logic structures on quartz after full-wafer-scale CNT growth without metallic-CNT burning. NAND pull-up (a) and NOR pull-up (b) with SEM images and drive currents for all on/off input combinations. Contacts and Gate Metal: Ti(1 nm)/Au(50 nm), Gate dielectric: 10 nm HfO₂.

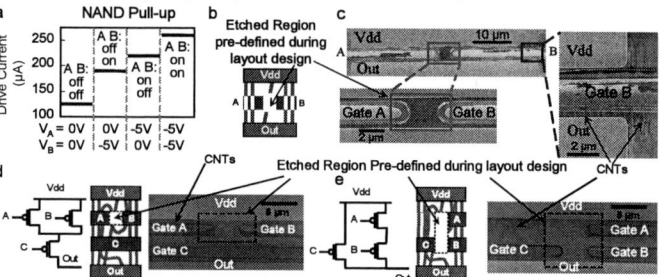

Figure 8. Misaligned-CNT-immune logic structures on silicon after CNT transfer and without metallic-CNT burning. (a)-(c) NAND pull-up. (d) AND-OR-INVERT pull-up and (e) OR-AND-INVERT pull-up logic structures and SEM images.

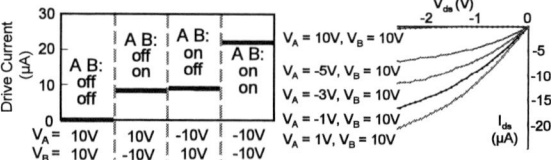

Figure 9. Misaligned-CNT-immune NAND pull-up on silicon after CNT transfer and metallic-CNT burning. Leakage currents are extremely small. Contacts and Gate Metal: Ti (1 nm)/Pd (50 nm), Gate dielectric: 50 nm HfO₂ (thick oxide needed to prevent oxide breakdown during metallic-CNT burning).

Performance enhancement schemes featuring lattice mismatched S/D stressors concurrently realized on CMOS platform: *e*-SiGeSn S/D for pFETs by Sn$^+$ implant and SiC S/D for nFETs by C$^+$ implant

Grace Huiqi Wang, Eng-Huat Toh, Xincai Wang[*], Debbie Hwee Leng Seng[**], Sudhinrajan Tripathy[**],
Thomas Osipowicz, Tau Kuei Chan, Ganesh Samudra, and Yee-Chia Yeo.

Silicon Nano Device Laboratory (SNDL), Dept. of Electrical and Computer Engineering, National University of Singapore (NUS), Singapore 117576.
*Singapore Institute of Manufacturing Technology, 71 Nanyang Drive, S638075. ** Institute of Materials Research and Engineering, 3, Research Link, S117602. Phone: +65 6516-2298, Fax: +65 6779-1103, Email: yeo@ieee.org

ABSTRACT

We report, for the first time, a simple and cost effective co-integration of strained p and n-FETs using Tin (Sn) and mono-carbon (C) implant in Source/Drain (S/D) of p- and n-FETs, respectively, to induce beneficial strain. For the first time, a single laser anneal step was employed to substitutionally incorporate the Sn and C atoms simultaneously into lattice sites. 7 at.% substitutional Sn concentration (the equivalent of adding 35% Ge to SiGe S/D stressors) was achieved in the Si$_{0.7}$Ge$_{0.3}$S/D of Si channel p-FET. A significant enhancement of up to 150% in hole mobility and 71% in drive current for a 50nm gate length device was observed. Mono C implanted S/D n-FETs show 19% current drive increase. With the simultaneous integration of Ni based FUSI gate, we provide a highly useful extension to future S/D technology for further $I_{D,sat}$ and mobility improvement.

INTRODUCTION

SiGe [1] and SiC [2] source and drain (S/D) stressor technologies has been investigated for introducing compressive and tensile strain in p- and n- channel transistors, respectively. Formation of S/D stressors with extremely high Ge or C concentration using epitaxy is challenging. For Si-channel FETs, there has been no investigation of S/D stressor formation by implant and laser anneal. For elements such as Sn and C with a large difference in atomic size compared to Si, the required concentration to achieve high strain levels can be easily introduced by implant. In addition, laser anneal can achieve solid solubility higher than equilibrium limits. Therefore, exploration of laser anneal of Sn- and C- implanted S/D for formation of stressors would be important.

In this paper, we report the first demonstration of a novel Si channel transistor with Si$_{0.63}$Ge$_{0.3}$Sn$_{0.07}$ S/D formed by Sn implant and laser anneal for achieving very significant strain effects. In terms of lattice constant, Si$_{0.63}$Ge$_{0.3}$Sn$_{0.07}$ is equivalent to SiGe with 65% Ge, thus providing an extremely large lattice mismatch with respect to the Si channel. In addition, the strained Si channel p-FETs with SiGeSn S/D were realized on (110) and (100) substrates for the first time. Dependence of performance on surface and channel orientations were investigated. The integration of metal silicidation process and Ni based FUSI gates for SiGeSn S/D devices is another first. This work also establishes Sn's compatibility with conventional CMOS salicidation process. Complementary to the strain benefits and drive current enhancements gained from the integration of Sn$^+$ for p-FETs, C$^+$ can be introduced on n-FET easily without adding additional mask, recess etch and Si:C epitaxy steps. This work also reports the first demonstration of utilizing laser anneal to incorporate carbon in S/D for enhanced electron mobility. The key contributions of this work are highlighted in Fig. 1.

DEVICE FABRICATION

The key features of the 50nm gate length CMOS structure is shown in Fig. 1. For p-FETs, devices were fabricated on (110) surface oriented SOI. For n-FETs, devices were fabricated on (100) surface oriented SOI. 20Å of gate oxide was formed by rapid thermal oxidation. After the gate stack and spacer formation, S/D recess etch of 15nm followed by selective epitaxial growth of Si$_{0.7}$Ge$_{0.3}$ was performed in p-FET. Then B/Sn were implanted in S/D of p-FETs and As/C in n-FETs. On both p-FET and n-FET transistors, pulsed laser annealing was then employed to activate the S/D dopants and substitutionally incorporate Sn in p-FETs and C in n-FETs respectively. The wafers then went through S/D Ni silicide formation at 400^{0}C, 30sec. Fig. 2 shows a TEM micrograph of a strained p-FET structure featuring SiGeSn S/D stressors and NiSiGeSn. CMP planarization was then adopted to expose the polysilicon and silicidation was carried out. High resolution TEM image featuring transistors with excellent gate dielectric interface with (110) substrate orientations after FUSI is shown in Fig. 3.

RESULTS AND DISCUSSION

A. Materials Analysis of (Si$_{0.7}$Ge$_{0.3}$)$_{1-x}$Sn$_x$ S/D stressors

Sn's substitutional concentration in Si$_{0.7}$Ge$_{0.3}$S/D is analyzed by RBS. At an optimal laser fluence of 400mJ cm^{-2}, a record high 7 at.% substitutional Sn concentration is achieved [Fig. 4]. This is the equivalent to adding 40% Ge in S/D. Inset further shows that SiGeSn is

epitaxial on the Si$_{0.7}$Ge$_{0.3}$ S/D region. This is indicated by the dip in the intensity for Sn. Comparable peak B concentration at the silicide-S/D interface suggests that drive current enhancement is enhancement is characteristically uniform. This further indicates that Sn is fully incorporated in the nickel silicide. Fig. 6 illustrates improved silicide sheet resistance in SiGeSn S/D over SiGe S/D. This enables lower contact resistance to be achieved in SiGeSn S/D p-FET. Fig. 7 compares the junction leakage characteristics of the SiGeSn S/D devices, when laser annealed at various fluences, and is benchmarked against devices without Sn implant. Hence, the optimized pulsed laser anneal conditions for this work are 2 irradiation pluses at 400 mJ/cm2.

B. P-FET with Si channel and (Si$_{0.7}$Ge$_{0.3}$)$_{1-x}$Sn$_x$ S/D stressors

Fig. 8 compares the $I_{D,sat}$ enhancement of the SiGeSn S/D devices on (110) and (100) surface orientations, at a gate overdrive of 1.0V. At an off-state leakage of 100nA/μm, the strained p-FET with SiGeSn S/D, on (100) surface orientation, exhibits 35% drive current enhancement over the control device with SiGe S/D. Additional 37% $I_{D,sat}$ enhancement is observed when (110) surface orientation is exploited on the strained p-FET with SiGeSn S/D. This $I_{D,sat}$ enhancement is further confirmed by the $I_{D,sat}$ –DIBL plot[Fig. 9]. 35% and 71% $I_{D,sat}$ enhancement is observed for strained SiGeSn S/D p-FETs fabricated on (100) and (110) respectively. To better relate to the strain enhancements observed with the introduction of Sn and (110) surface oriented wafers, hole mobility was extracted using μ$_{hole}$ = 1/[W.Q$_{inv}$.(dR$_{Total}$/dL$_G$)]. At low vertical effective field regime, strained SiGeSn S/D p-FET on (100) achieves 68% hole mobility enhancement [Fig. 10]. Even higher mobility gain of 150% is observed when (110) surface oriented p-FET was integrated with SiGeSn S/D. This is consistent with the observed transconductance $G_{m,lin}$ gain of 140% in the strained p-FET over control p-FET [Fig. 11]. Proximity of SiGeSn S/D stressor further impact strained devices performance gain over control. Fig. 12 suggests that at L_G = 50nm, maximum performance gain could be derived from the integration of SiGeSn S/D. No obvious degradation in DIBL and subthreshold swings are observed [Fig. 13] due to the introduction of Sn in SiGe S/D or (110) surface orientations. Fig. 14 compares the dependence of channel orientations in (110) surface oriented devices on performance gain. At a wide range of $I_{D,sat}$, (110) advantage is maintained because of its higher mobility than (100).

C. N-FET with Si channel and SiC S/D stressors

A complementary strained n-FET with SiC S/D stressors is fabricated by C$^+$ implant and laser annealing to substitutionally incorporate the carbon atoms. A carbon dose of 5×10^{15} cm^{-2} was implanted in Si S/D. Fig. 15 plots the $I_{D,sat}$-I$_{off}$ characteristics of a 50nm gate length SiC S/D n-FET. At I$_{off}$ 100nA/μm, 19% $I_{D,sat}$ improvement was obtained over its unstrained counterpart.

Fig. 16 and 17 summarizes the I_D-V$_D$ and I_D-V$_G$ characteristics of p-FETs and n-FETs. By integrating both SiGeSn S/D for p-FETs and SiC S/D for n-FETs simultaneously, and exploiting the best surface orientations, $I_{D,sat}$ enhancement is increased to 70% and 18% respectively. V$_t$ roll off comparison [Fig. 18] further verify that the strained and control devices have comparable short channel effects. A summary of the various strain enhancement schemes achieved with the integration of SiGeSn S/D for p-FET and SiC S/D for n-FET is illustrated in Fig. 19.

CONCLUSION

Successful integration of SiGeSn S/D for p-FET on (110) surface oriented SOI and its complementary realization of SiC S/D for n-FET on (100), by ion implantation of Sn and C respectively, is reported. 7 at.% substitutional Sn concentration (the equivalence of adding 35% Ge to SiGe S/D stressors) was achieved in the Si$_{0.7}$Ge$_{0.3}$S/D of Si channel p-FET. Record high Sn substitutionality in the SiGe S/D results in 150% hole mobility improvement. Substantial I_{dsat} improvement is observed for both p-FETs and n-FETs. Such structures could be potentially promising for realizing very high performance levels.

REFERENCES

[1] T. Ghani et. al., *IEDM Tech Dig.*, pp.978, 2003. [2] Y.-C. Yeo, Semicond. Sci. Tech. 22, p. S177, 2007. [3] G. H. Wang *et.al.*, *IEDM Tech Dig.*, pp.131, 2007.

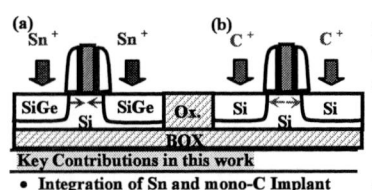

- Spacer Formation
- Recess Etch & SEG of $Si_{0.7}Ge_{0.3}$ ⎤ pFET
- Heavy Source/Drain Implant + Strain Engineered Implantation
  - (a) B, Sn ($[Sn]_{sub}$: 7%) ⎤ pFET
  - (b) As, C ($[C]_{sub}$: 1.5%) ⎤ nFET
- Excimer Laser Anneal at 400mJ cm^{-2}
- Nickel Silicidation at 400°C,30s
- Oxide Deposition and CMP to expose Gate
- FUSI Gate Formation

Key Contributions in this work
- Integration of Sn and mono-C Implant And L.A. process for Si channel FETs
- Development of metal silicidation process for SiGeSn S/D
- Realized (110) & (100) strained pFETs with SiGeSn S/D

Fig. 1. Key process steps featuring the first integration of strained p- and n-channel FETs concurrently by a simple and cost effective implant and anneal technique. Photoresist can be used to block the C$^+$ and Sn$^+$ implant in p-FET and n-FET.

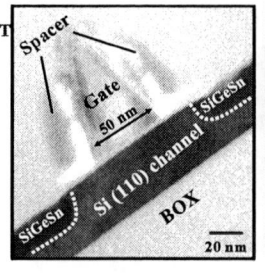

Fig. 2. TEM image featuring Ni silicided SiGeSn S/D p-FET prior to FUSI gate formation.

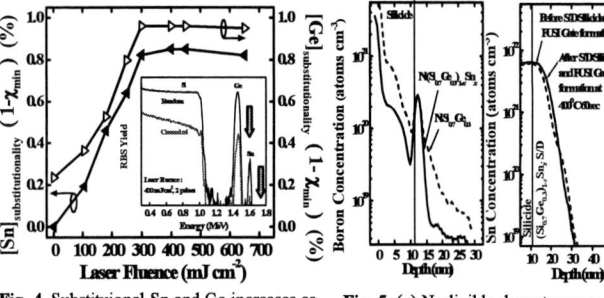

Fig. 3. (a)CMP to expose poly-gate for FUSI. (b) High resolution TEM image of FUSI and (110) channel.

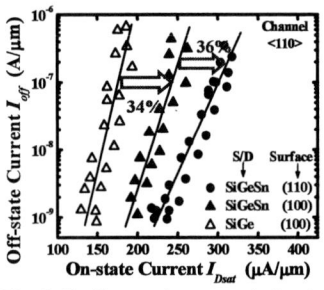

Fig. 4. Substituional Sn and Ge increases as higher laser fluence is used and Ge content in the SiGe S/D increases.

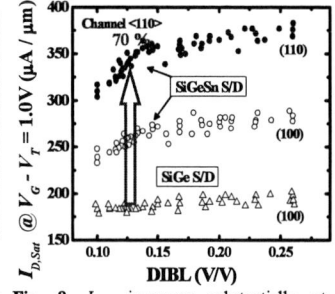

Fig. 5. (a) Negligible dopant consumption in silicide (b) Sn is distributed uniformly and is fully incorporated in the silicide

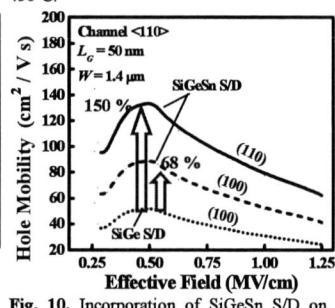

Fig. 6. Lower Rs of Ni silicided contact achieved in the presence of Sn. Superior morphological stability of NiSiGeSn was achieved at annealing temperatures below 450°C.

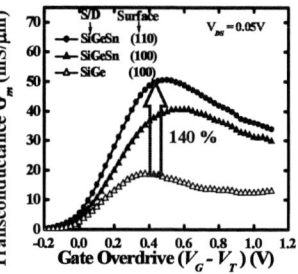

Fig. 7. No adverse impact on junction leakage when SiGeSn is incorporated in the S/D with Si channel. There is a slight increase in junction leakage but it is within the acceptable range

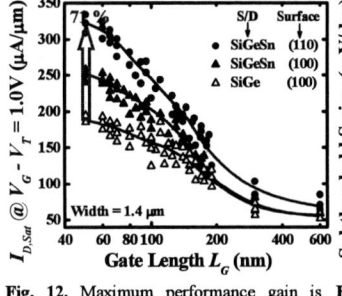

Fig. 8. Significant enhancement in I_{Dsat} is achieved with SiGeSn S/D and (110), attributed to enhanced μ_{hole}.

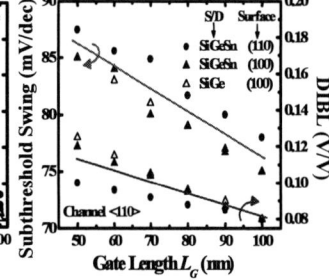

Fig. 9. I_{Dsat} improves substantially at comparable DIBL for SiGeSn S/D on (110) surface orientation.

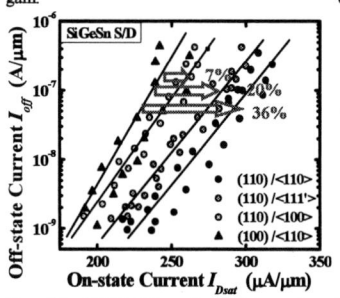

Fig. 10. Incorporation of SiGeSn S/D on (110) surface shows significant mobility gain.

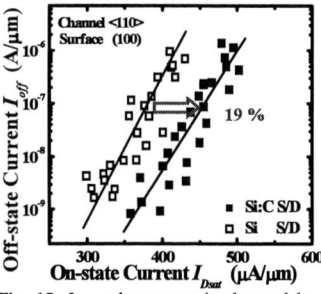

Fig. 11. Significant increase of 140% in G_m of 58% is observed for SiGeSn S/D pFET on (110) and (100) surface.

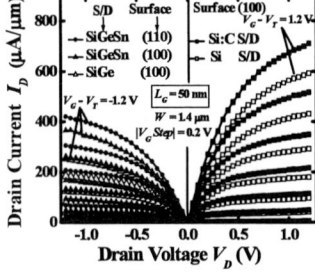

Fig. 12. Maximum performance gain is achieved at shorter L_G, consistent with stressors brought closer to channel

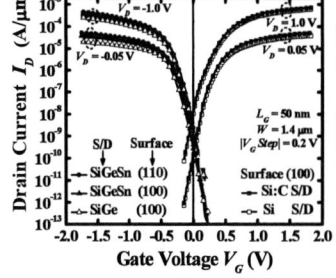

Fig. 13. Comparable subthreshold swing and DIBL achieved in all devices at each L_G.

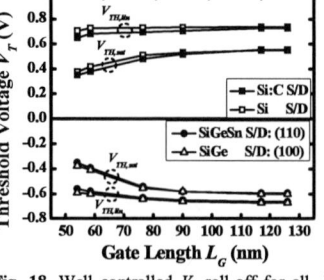

Fig. 14. (110)/<110> gives highest I_{Dsat} enhancement. Insensitive stress effect on (110)/<100>. results in smaller I_{Dsat} enhancement

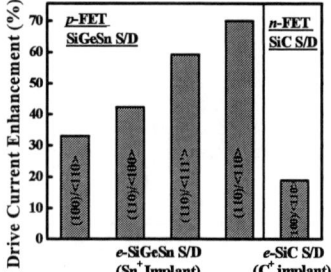

Fig. 15. I_{Dsat} enhancement is observed by employing Si:C S/D formed by mono carbon implant at 7KeV, using a carbon dose of 5× 10^{15} cm^{-2}.

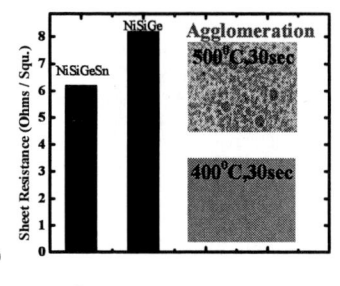

Fig. 16. Simple Implant and anneal steps allow both strained n- and p-channel FETs to be integrated easily.

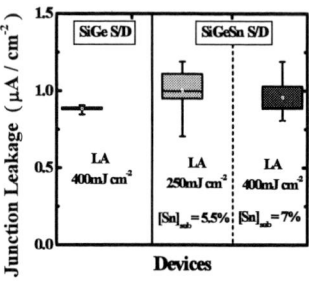

Fig.17. Subthreshold I_D-V_G of n-FET and p-FET using C$^+$ and Sn$^+$ implant. Well controlled subthreshold slope were obtained

Fig. 18. Well controlled V_T roll-off for all devices. SiGeSn and SiC S/D integration does not degrade short channel effects.

Fig. 19. Various strain enhancement techniques for realizing the n- and p-channel FETs.

162

Author Index

A

A, K. H. 79
Abe, T. 123
Absil, P. 49, 101, 145
Absil, P.P. 35
Acosta, T. 143
Adachi, K. 19
Adachi, T. 53
Adam, T. 135
Adelmann, C. 11, 35, 49, 101
Ahn, J.-H. 157
Ahn, W. S. 79
Aikawa, H. 69
Akheyar, A. 35, 101
Akiyama, K. 33, 61
Akiyama, Y. 123
Aminaka, T. 39
Ando, T. 97
Andrieu, F. 13, 133
Ang, K.-W. 131
Ang, K.-W. 21
Aoulaiche, M. 35, 49, 101, 145
Aoyama, T. 39, 51, 53, 85, 103
Aquilino, M. 67
Arita, K. 81
Arnaud, F. 141
Aussenac, F. 13
Auth, C. 99
Avci, U. E. 71
Awano, M. 19
Azuma, A. 103

B

Babich, K. 9
Bae, H. 107
Baiocco, C. 67
Baker, K. 105
Balasubramaniam, N. 131
Balasubramanian, N. 21, 29
Baldwin, G. 125
Ban, I. 71
Barnola, S. 113
Barral, V. 13
Baumann, F. H. 73
Benaissa, K. 125
Bender, H. 11
Bernard, E. 13
Bersuker, G. 45, 63
Bhouri, N. 133
Bidal, G. 113
Biesemans, S. 11, 35, 49, 127, 145
Biscardi, M. 139
Black, L. 135
Blatchford, J. 137

B (continued)

Blythe, T. 125
Boeuf, F. 113
Bougu, J. 113
Boulanger, F. 55, 133
Bowen, C. 137
Breitwisch, M. 73
Brus, S. 11, 35, 101
Bryant, A. 9
Bu, H. 137
Burr, G. W. 73
Byun, S. 93

C

Campbell, J. P. 57
Campidelli, Y. 13
Cappellani, A. 99
Carlton, D. 111
Cassé, M. 55, 133
Chakravarti, A. 135
Chan, D. S. H. 27
Chan, K. 135
Chan, L. 131
Chan, T. K. 161
Chanemougame, D. 113
Chang, J. 9
Chang, K.-M. 105
Chang, Myoung-Sik 65
Chang, P. L. D. 71
Chang, S. Z. 35, 49, 101, 127
Chang, V.S. 35, 101
Cheek, R. 73
Chen, C.-F. 73
Chen, H. W. 83
Chen, J. 25
Chen, K. C. 89, 109
Chen, X. 67
Cheung, K. P. 57
Chi, D.-Z. 23
Chiarella, T. 35, 49, 101, 145
Chikyo, T. 39
Chikyow, T. 53
Chin, A. 149
Chindalore, G. 105
Cho, B. J. 45, 63, 157
Cho, Gyu-Seong 65
Cho, H.-J. 35, 49, 101, 127
Cho, J. 93
Cho, K. H. 107
Cho, Moon-Ju 35
Choe, J.-D. 93
Choi, B.-I. 91
Choi, Bong-Ho 65
Choi, D. Y. 79
Choi, J. 91
Choi, K. J. 63, 157

Author Index

Choi, S.-J. 157
Choi, Y.-K. 157
Chow, S.-Y. 23
Chowdhury, M. 67
Chudzik, M. 67
Chun, J.-S. 99
Chung, Sung-Woong 65
Chung, U.-I. 95
Clement, L. 113
Collaert, N. 11
Conard, T. 11
Coolbaugh, D. 67
Coronel, P. 13
Corum, D. 137
Cosnier, V. 55, 113
Crupi, G. 41

D

Dalis, A. 99
Dasaka, R. 73
Davis, A. 99
De Meyer, K. 11, 35, 101
De Potter, M. 145
de Souza, J. P. 139
DeJaeger, B. 41
Delabie, A. 35, 49, 101
Delaye, V. 13
Deleonibus, S. 13
Demand, M. 11, 101
Denorme, S. 113
Dokumaci, O. 9
Domenicucci, A. G. 139
Dube, A. 135

E

Eguchi, K. 103
Eimori, T. 53
Ekbote, S. 125
Ellis, N. 105
Eneman, G. 41
Ercken, M. 35
Ernst, T. 13

F

Fang, W.-W. 23, 131
Fanton, A. 55
Favia, P. 49
Felch, S. B. 145
Ferain, I. 11, 41
Flaitz, P. L. 73
Fleury, D. 113
Fogel, K. 139
Fujii, I. 137
Fujii, K. 81

Fukase, T. 123
Fukushima, T. 103
Fukutome, H. 115, 117
Fuller, N. 9
Funaki, M. 153
Furutake, N. 81

G

Gao, B. 77
Garros, X. 55
Gasquet, H. 105
Gautier, P. 13
Ge, C. H. 83
Ghani, T. 99
Gilmer, D. C. 63
Glass, G. 99
Glassman, T. 99
Goel, N. 63, 143
Goto, M. 103
Graham, W. 9
Grider, T. 137
Groeseneken, G. 127
Grosjean, M. 113
Gu, Y. 137
Guillaumot, B. 13
Guillorn, M. 9
Gwoziecki, R. 141

H

Ha, D. 75
Haendler, S. 113
Haensch, W. 9
Hamaguchi, M. 139
Han, J. 67, 95
Han, J.-W. 93, 157
Han, R. Q. 77
Hane, M. 47
Harakawa, H. 103
Haran, B. 9
Harley, E. 135
Harper, M. 99
Harris, H. R. 45, 63, 143
Hartmann, J. M. 13
Hasumi, R. 139
Hatada, A. 147
Hattendorf, M. 99
Hayami, Y. 147
Hayashi, M. 39
Hayashi, Y. 81
Heh, D. 63
Hentges, P. 99
Herrick, M. 105
Heyns, M. 41
Higuchi, J. 153
Hiramoto, T. 25, 121

Author Index

Hirano, A. 33
Hobbs, C. 67
Hoffmann, T. Y. 35, 41, 49, 145
Hoffmann, T. 101
Hokazono, A. 87, 151
Hong, A. J. 95
Hong, C. M. 105
Hong, S. P. 89, 109
Hong, Sung-Joo 65
Hong, Y. K. 79
Hooker, J. 101
Hooker, J.C. 35
Horiguchi, N. 145
Hornung, B. 137
Hosaka, K. 115, 117
Hou, F. 125
Houston, T. W. 125
Hsiao, Y. H. 89, 109
Hsieh, K. Y. 89, 109
Hsu, F. H. 89, 109
Hsu, T.-H. 89
Huang, J. 63
Huang, Y. F. 89
Hung, P.Y. 63

I

Ichihara, R. 103
Iijima, R. 139
Ikeda, K. 115, 147
Ikeda, M. 33, 61
Ikezawa, T. 47
Ikuta, T. 155
Imai, K. 123
Inaba, S. 15, 87, 103, 121
Inoue, Y. 129
Inumiya, S. 103
Ipposhi, T. 129
Ishida, M. 5
Ishida, T. 103
Ishigaki, T. 129
Ito, F. 81
Ito, H. 119
Itokawa, H. 87
Iwai, M. 69
Iwamatsu, T. 129
Iwamoto, K. 33

J

J. Chen H. 67
Jaeger, D. 67
Jaloviar, S. 99
Jammy, R. 45, 63, 143
Jeon, Y. J. 75
Jeong, G. T. 75
Jeong, H. S. 75, 79

Jha, R. 67
Jiang, Y. 27
Jin, Y.-G. 93
Jin, Z. 67
Joseph, E. 73
Joshi, S. 99
Jung, D. J. 79
Jung, J. Y. 79
Jung, W. W. 79
Jung, Y.-J. 93
Jurczak, M. 11, 101

K

Kaczer, B. 11, 49
Kadoshima, M. 39, 53
Kalpat, S. 67
Kamimuta, Y. 33, 119
Kamiyama, S. 53
Kamohara, S. 121
Kanarsky, T. 135
Kanda, S. 155
Kaneko, A. 15
Kang, C. 91
Kang, C.-Y. 45
Kang, D.-H. 75
Kang, H. 139
Kang, J. F. 77
Kang, J. Y. 79
Kang, L. 67
Kang, S. K. 79
Kang, S.-T. 105
Kang, Y. M. 79
Kang, Y. 93
Kariya, N. 123
Kase, M. 147
Kato, S. 85
Katsumata, N. 153
Kauerau, T. 11, 35, 101
Kawada, M. 47
Kawahara, J. 81
Kawamura, K. 115, 117
Kawanaka, H. 153
Kawanaka, S. 87, 103
Kawano, T. 5
Kawasaki, H. 9
Kawase, Y. 19
Kelkar, P. 35
Kelly, D. Q. 63
Kencke, D. L. 71
Kerner, C. 35, 101, 145
Khamankar, R. 137
Khater, M. 9
Kies, R. 13
Kim, C.-J. 157
Kim, D.-W. 31, 107

Author Index

Kim, Dae-Young.................................65
Kim, H. H..79
Kim, H. S..79
Kim, H..91, 143
Kim, Hyung-Hwan..............................65
Kim, J. H....................................75, 79
Kim, J. S.......................................157
Kim, J. W.......................................157
Kim, J...95
Kim, Jin-Woong................................65
Kim, Junki......................................65
Kim, K. H.......................................157
Kim, M.-S.......................................107
Kim, N...67
Kim, S. Y..79
Kim, S.......................................93, 157
Kim, S.-H..31
Kim, T..137
Kim, Tae-Kyun..................................65
Kim, W...93
Kim, Wan-Soo...................................65
Kim, Y. S.......................................147
Kim, Young-Sik.................................65
Kimura, S......................................129
Kinoshita, A................................43, 151
Kinoshita, T.....................................15
Kirkpatrick, B.................................137
Kirmse, K.......................................137
Kirsch, P. D.....................................63
Kirsch, P..45
Kirshnan, S......................................67
Kitajima, M......................................39
Klaus, D..9
Klaus, J...99
Ko, C. H...83
Ko, H. K...79
Kobayashi, M.....................................43
Kobayashi, S.....................................59
Koehler, A......................................143
Koga, J...151
Koh, A. T.-Y................................23, 131
Koh, S.-M.......................................131
Kohler, S.......................................141
Koli, D...9
Kong, J. H.......................................75
Koo, J.-M..93
Kosemura, D......................................97
Kothandaraman, C.................................67
Koyama, M.......................................103
Krishnan, A.....................................137
Kuan, T. M.......................................83
Kubicek, S...................35, 49, 101, 127, 145
Kubo, T...147
Kuga, R...153
Kugimiya, K.....................................155
Kuhn, K..99

Kurata, H.......................................147
Kurihara, M.....................................153
Kurosawa, E......................................39
Kusunoki, N......................................87
Kwon, B..93
Kwong, D. L......................................27
Kyoh, S..69

L

Lage, C..67
Lai, D. M.-Y.....................................21
Lai, E.-K.......................................109
Lai, S. C..89
Lam, C...73
Lamorey, M.......................................73
Lauer, I..135
Lauwers, A.......................................35
Laviron, C......................................113
Lavoie, C...9
Lavric, D..99
Lee, B. H.....................................45, 63
Lee, C.-H.....................................91, 93
Lee, E. S..79
Lee, G. S.......................................157
Lee, Hae-Jung....................................65
Lee, Hyunjin.....................................65
Lee, J. J..93
Lee, J..143
Lee, J.-H..75
Lee, J.-M.......................................107
Lee, M. K..75
Lee, M.-H..73
Lee, R. T. P...............................23, 29, 131
Lee, S. Y..79
Lee, S..111
Lee, S.-H.....................................45, 63
Lee, T...93
Lee, W. C..83
Lee, W.-S...................................31, 91, 107
Lee, Y...67
Leroux, C..55
Leverd, F.......................................113
Leyris, C.......................................113
Li, J.......................................67, 135, 139
Li, M..31, 107
Li, M.-F....................................127, 149
Li, W..67
Liao, C. C......................................149
Liao, C. W.......................................89
Lien, N. Z...................................89, 109
Lim, A. E.-J................................23, 131
Lim, P.-C..21
Lin, A..159
Lin, C..137
Lin, Y.-S..95

Author Index

Liow, T. Y. 23, 27, 29, 131
Liu, F. 21
Liu, F.-Y. 131
Liu, L. F. 77
Liu, L. 137
Liu, R. 89, 109
Liu, S. 125
Liu, T.-J. K. 111
Liu, X. Y. 77
Lo, G. Q. 27
Lo, G.-Q. 23
Lofaro, M. 9
Loh, W.-Y. 45
Loubet, N. 113
Lowry, T. 137
Lu, C.-Y. 89, 109
Lu, M. 99
Lue, H.-T. 89, 109
Lung, H.-L. 73
Lysaght, P. 63

M

Ma, S. 95
Machala, C. 137
Madan, A. 135
Maffini-Alvaro, V. 13
Majhi, P. 45, 63, 143
Maki, K. 153
Malloch, W. 105
Manabe, K. 37
Mariappan, H. 99
Marin, M. 113
Martin, F. 55, 113
Martin, S. 125
Maruyama, S. 123
Massey, G. 67
Masuzaki, K. 37
Matsuki, T. 39
Matsuoka, F. 69
Maude, D. K. 133
Mayuzumi, S. 97
Mercha, A. 101
Meuris, M. 41
Migita, S. 119
Miller, J. 135
Mise, N. 39, 53
Mistry, K. 99
Mitard, J. 41
Mitra, S. 159
Mitsuhashi, R. 35, 101
Miyaji, K. 25
Miyano, K. 19
Miyashita, K. 19, 103
Miyashita, T. 147
Miyazaki, S. 39, 53

Mizubayashi, W. 33, 61
Mizuno, T. 17
Mizushima, I. 87
Mo, R. 67
Mogami, T. 121
Momiyama, Y. 115, 117
Monfray, S. 113
Montgomery, C. 137
Morifuji, E. 69
Morisaki, Y. 117
Morita, Y. 129
Moriyama, Y. 17
Morooka, T. 53
Motoyama, K. 81
Mouis, M. 133
Moumen, N. 67
Mueller, M. 49
Myers, E. R. 159

N

Nabatame, T. 33, 61, 119
Nagai, H. 147
Nagase, H. 81
Nagashima, N. 97
Nagatomo, K. 103
Nair, D. 67
Nakagawa, T. 37
Nakajima, K. 103
Nakamura, H. 81
Nakamura, M. 155
Nakayama, T. 19, 69
Nakazawa, E. 81
Nara, Y. 39, 51, 53, 85
Narayanan, V. 67
Neumeyer, D. 9
Newbury, J. 9
Nishi, Y. 43, 151
Nishibata, T. 153
Nishida, A. 121
Nishida, T. 143
Nishikawa, M. 147
Noda, T. 145
Nomachi, A. 103
Norris, B. 99
Nyns, L. 101

O

O'Brien, C. 137
O'Sullivan, B. J. 101
Oates, A. 57
Obradovic, B. 125
Oda, H. 129
Oda, T. 155
Oe, K. 153
Ogawa, M. 95

Author Index

Oguma, H. 103
Ogura, A. 97
Ogura, T. 37
Oh, H. 91
Oh, J. H. 75
Oh, J. S. 157
Oh, J. 63
Oh, J.-W. 45
Oh, S.-A. 21
Ohchi, T. 155
Ohji, Y. 39, 51, 53, 85
Ohkoshi, K. 147
Ohno, T. 97
Ohta, H. 115, 117, 147
Ohta, M. 69
Ohuchi, K. 139
Oikawa, M. 153
Ok, I. 143
Okabe, K. 115
Okabe, K. 147
Okada, N. 81
Okano, K. 15
Okuno, Y. 101
Okuno, Y. 35
Olubuyide, O. 137
Onizawa, T. 85
Onoda, H. 19, 103
Onodera, T. 81
Opdebeeck, A. 145
Orihara, S. 153
Ortolland, C. 101, 145
Osipowicz, T. 161
Ota, H. 33, 61, 119
Ott, J. A. 139
Ott, J. 9

P

Padilla, A. 111
Pal, R. 135
Pandey, S. 67
Pantisano, L. 41
Paraschiv, V. 35, 101
Park, B. 107
Park, C. 63
Park, C.-S. 45, 63
Park, D. 31, 107
Park, D.-G. 135
Park, J. H. 75
Park, J. 93
Park, S. 93
Park, Sung-Kye 65
Park, Sung-Wook 65
Park, W. I. 75
Park, Y. C. 157
Park, Y. 91, 93

Parthasarathy, S. 143
Parvais, B. 101
Patil, N. 159
Pei, G. 135
Peng, W. C. 89
Perreau, P. 113
Pinto, Mark R. 1
Pinto, T. 135
Poiroux, T. 13
Price, J. 63
Prins, S. 137
Putra, A. T. 121
Pyzyna, A. 9

R

Rafik, M. 113
Ragnarsson, L.-Å. 35, 49, 101
Rahhal-Orabi, N. 99
Rajendran, B. 73
Ranade, P. 99
Reddy, C. 67
Reimbold, G. 55, 133
Ren, Z. 135
Renault, O. 11, 55
Riley, D. 137
Rinkenberger, G. 105
Rochette, F. 133
Rohr, E. 35, 49, 101
Ronsheim, P. 139
Rooyackers, R. 11
Rosseel, E. 145
Roussel, P. J. 11
Rovedo, N. 139
Ryu, S.-W. 157

S

S, J. 99
Sadana, D. K. 139
Sadana, D. 135
Saenger, K. L. 139
Saito, S. 81
Saitoh, M. 15, 37, 59
Sakata, A. 69
Sakoda, T. 117
Salvetat, T. 113
Samavedam, S. 67
Samudra, G. S. 21, 23, 29
Samudra, G. 131, 161
Sanuki, T. 69
Saraswat, K. 43
Saraya, T. 25
Sasaki, T. 103
Sassman, B. 45
Sato, H. 39, 51, 53, 153
Satoh, S. 115, 147

Author Index

Sawada, K. 5
Sawada, T. 69
Scharm, T. 35, 49, 101
Schreutelkamp, R. 145
Schrott, A. 73
Sebai, F. 35
Sekine, M. 81
Sellier, M. 113
Seng, D. H. L. 161
Seo, J. 107
Seutter, S. M. 83
Severi, S. 41
Sheikh, A. 67
Shen, J. 105
Sherony, M. 67
Shichijo, H. 125
Shih, Y.-H. 73
Shimamune, Y. 147
Shimizu, K. 25
Shimizu, T. 153
Shinn, G. 137
Shinohara, M. 155
Shiraishi, K. 39
Shiraishi, K. 53
Sikorski, E. 9
Simoen, E. 11
Singanamalla, R. 101
Singh, A. 125
Singh, N. 27
Sivasubramani, P. 63
Skotnicki, T. 13
Son, N. J. 11
Song, E. B. 95
Song, Han-Sang 65
Song, M. H. 157
Song, Y. J. 75
Souifi, A. 13
Stein, K. 67
Su, N. C. 149
Suehle, J. S. 57
Sugii, N. 129
Sugii, T. 115, 147
Sugita, Y. 51
Sugiyama, N. 17
Sugizaki, T. 155
Suhail, M. 105
Suk, S. D. 31, 107
Sukegawa, K. 147
Sukegawa, T. 147
Sun, B. 77
Sun, Y. 143
Sunamura, H. 37
Sung, C. Y. 139
Sung, S.-K. 107
Suthram, S. 143
Suzuki, M. 123

Syzdek, R. 105

T

Tagami, M. 81
Tajima, M. 115
Takagi, S. 17, 47
Takalkar, R. 135
Takao, H. 5
Takasu, Y. 139
Takayana, M. 139
Takeda, H. 47
Takei, M. 97
Takeuchi, K. 121
Takeuchi, T. 81
Tamura, N. 147
Tan, B. L. H. 21, 29
Tan, K.-M. 23, 29
Tan, L. H. 27
Tan, Y. N. 63
Tateshita, Y. 97
Tatsumi, T. 37
Tatsumura, K. 103
Taylor, R. 125
Teh, Y. W. 67
Tekleab, D. 67
Terada, K. 121
Tezuka, T. 17
Thean, A. 67
Thompson, S. E. 143
To, B. 9
Toh, E.-H. 161
Torii, K. 129
Toriumi, A. 33, 61, 119
Tornello, J. 9
Toyoshima, Y. 15, 87, 103
Tripathy, S. 21, 161
Trojman, L. 41
Tsai, G. 83
Tsai, W. 143
Tseng, H.-H. 45, 63, 143
Tsuchiya, R. 129
Tsuchiya, Y. 151
Tsukamoto, M. 97
Tsunoda, K. 123
Tsunomura, T. 121
Tsutsui, G. 123

U

Uchida, K. 15, 59
Uedono, A. 53
Ueki, M. 81
Um, C. Y. 75
Umezawa, N. 53
Utomo, H. 139

Author Index

V

Van Elshocht, S. 11, 49, 101
Vandervorst, W. 145
Veloso, A. 11, 101
Vizioz, C. 13
Vos, R. 11, 35, 101
Vrancken, C. 35, 145, 101
Vulliet, N. 13

W

Wakabayashi, H. 97
Wallner, J. 67
Wang, G. H. 161
Wang, J. 97
Wang, K. L. 95
Wang, S. J. 149
Wang, S. Y. 89, 109
Wang, T. J. 83
Wang, W. 33, 61
Wang, X. P. 49, 127
Wang, X. 9, 21, 67, 161
Watanabe, H. 37
Watanabe, Y. 119
Wei, C. 79
Weijtmans, J. W. 135
Witters, L. 11, 101
Witters, T. 35, 49, 101
Wong, H. S. 21, 23, 43, 159, 131
Wouters, D. 127
Wu, C. H. 83
Wu, K. H. 83
Wu, M. T. 89, 109

X

Xia, G. 135
Xu, N. 77

Y

Yagami, K. 155
Yagishita, A. 9
Yako, K. 81
Yamabe, K. 53
Yamada, K. 39, 53
Yamagata, Y. 123
Yamakawa, S. 97
Yamamoto, H. 81
Yamamoto, T. 47, 147
Yanagita, M. 155
Yang, B. F. 135
Yang, B. 9, 135
Yang, H. S. 67
Yang, H. 125
Yang, L. W. 89, 109

Yang, T. 89, 109
Yano, F. 121
Yater, J. 105
Ye, C. N. 83
Yeo, K. H. 107, 31
Yeo, Y.-C. 21, 23, 29, 127, 131, 161
Yeoh, Y. Y. 31, 107
Yeung, F. 75
Yin, H. 139
Yong, A. M. 23
Yoo, I. 93
Yoon, T.-E. 93
Yoshimizu, Y. 103
Yoshimura, H. 19, 69
Young, C. 45, 63
Yu, B. 77
Yu, H. Y. 49, 127
Yu, H. 101
Yu, J. 75
Yu, R. 9
Yu, S. 137

Z

Zaleski, M. 67
Zhang, D. 67
Zhang, K. 83
Zhang, R. 139
Zhang, Y. 9
Zhu, M. 21, 29, 131
Zhu, Y. 73
Zhu, Z. 135, 139
Zhuang, D. 67

CURRAN ASSOCIATES INC.
proceedings
.com

9781424418022